SECOND EDITION

PHYSICAL

DONALD S. CLARK, Ph.D.

Professor of Physical Metallurgy
W. M. Keck Laboratory of Engineering Materials
California Institute of Technology

AND

WILBUR R. VARNEY, M.S.

Consulting Metallurgical Engineer

METALLURGY
FOR ENGINEERS

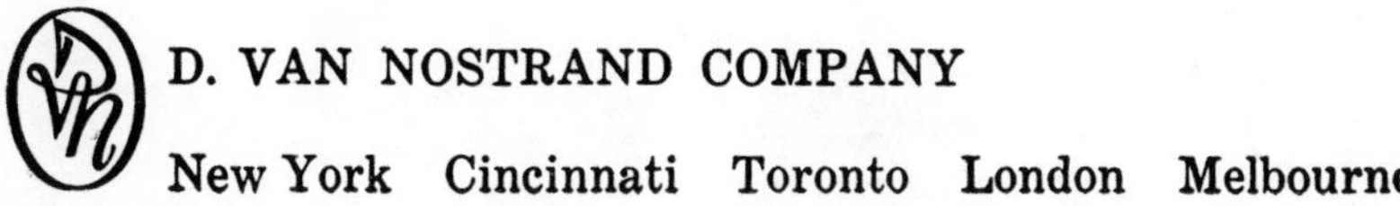
D. VAN NOSTRAND COMPANY

New York Cincinnati Toronto London Melbourne

D. VAN NOSTRAND COMPANY REGIONAL OFFICES: *New York* *Cincinnati* *Millbrae*

D. VAN NOSTRAND COMPANY INTERNATIONAL OFFICES: *London* *Toronto* *Melbourne*

ISBN: 0-442-01570-4

Published by D. VAN NOSTRAND COMPANY
450 West 33rd Street, New York, N.Y. 10001

Published simultaneously in Canada by
VAN NOSTRAND REINHOLD LTD.

10 9

PREFACE

This book is directed to the basic training of students of engineering in the field of physical metallurgy. One of the most serious problems confronting the engineer is the selection, treatment, and use of metals and alloys. These problems cannot be left entirely to the specialist known as the metallurgist, but must be dealt with to an appreciable extent by the engineer. No engineering project is without its problem of a material to do the job. To cope with this situation the engineer must have a basic understanding of the science and art of metallurgy as it is related to his interest. The authors have intended that this book should meet this need.

The engineering student should not be expected to learn practices in vogue, except as an illustration of the application of principles. It is on this basis that this book has been written. Every attempt has been made to present basic concepts insofar as they are within the scope of a course for young engineers.

A familiarity with metals and alloys that will be best adapted to certain applications comes through experience and continued contact with the field. This cannot be learned from a book, but the basic principles which help in understanding these things can be clarified.

Many phases of this subject are controversial and, hence, for the young engineer, are not pertinent to his practice. These matters have been presented when necessary for completeness of the discussion in the light of current majority opinion. The material can be covered in one semester or one quarter and should be correlated with courses in design.

While the purpose of this book in its second edition remains the same as in the first edition, there is a tendency to shift the emphasis to an even more fundamental point of view. An even greater emphasis has been given to the fundamentals by adding three new chapters, namely, *The Structure of Matter, Physical Properties,* and *Mechanical Properties.* Although these chapters are not complete in their coverage of the subjects, they give the engineering student an adequate foundation for the study of physical metallurgy. Some instructors may wish to omit these three chapters from their course. Such an omission will not cause any difficulty in presenting or understanding the remainder of the text.

A new chapter has been added which is concerned with some aspects of the metallurgy of nuclear engineering. References are provided from which more details can be secured on this subject.

All chapters of the book have been reworked, and in many instances the sequence of presentation has been changed. The tabular material has been checked in an attempt to bring it up to date. These data are included for the purpose of illustration and the solution of problems, and they are not intended to be complete or to take the place of reference books.

The authors have drawn heavily on many sources for material, and it is their hope that the approach to the subject and the presentation will prove to be of value to the young engineer. The authors express their appreciation to the many publishers and companies who have so graciously given permission to use their material. Credit is given wherever material has been derived from such sources.

Pasadena, California
September, 1961

DONALD S. CLARK
WILBUR R. VARNEY

CONTENTS

I

METALLURGY IN ENGINEERING

1.1 The Art and Science of Metallurgy. The advancement of civilization has been largely brought about by man's ability to adapt the elements to his service. Primitive man of the Stone Age undoubtedly learned to use certain metals before 3500 B.C. The ancient metalsmith was able to work native copper, gold, and silver into ornamental trinkets and to use meteoric iron-nickel alloys for making weapons.

Toward the end of the Stone Age, man discovered the art of smelting. It seems likely that the first casting was accidentally produced in the ashes of his campfire. Charcoal served as a reducing agent in the primitive smelting process, and the first crude bronzes were probably the result of accidental roasting of mixtures of copper and tin ores. About 2500 B.C., with the start of the Bronze Age, it is believed that the art of extracting relatively pure tin had advanced to a point where intentional additions to copper were possible. Thus bronze has been identified as the first alloy actually cast by man. Brass was introduced about 500 B.C. by the smelting of copper and zinc ores, but the widespread use of this alloy did not occur until developments of the eighteenth century made possible the availability of metallic zinc.

The first smelting of iron ore is believed to have taken place about 1500 B.C. and heralded the actual start of the Iron Age. Man-made iron was chiefly used during this period for coinage, cooking utensils, and implements of war. Metallurgists of the early Iron Age undoubtedly discovered the cementation process for steelmaking and the art of quenching steel for use in weapons.

Metallurgical progress was relatively slow until about A.D. 1300 when the Catalan forge was developed in Spain as the forerunner of the modern hearth furnace. For the first time in history, it was possible to produce a large quantity of iron in one heat. The first continuous shaft furnace to incorporate the basic principles of the modern iron blast furnace was developed in Germany about A.D. 1323. The high-carbon product of this furnace, with its lower melting point, became known as "cast iron" and greatly extended the use of iron castings.

Man's natural curiosity led to the birth of modern science early in the sixteenth century. The first published work to record the overall progress in the field of applied metallurgy and ore reduction was the *Pirotechnia* by Vannoccio Biringuccio, which appeared in 1540, followed by *De Re Metallica* by Georgius Agricola in 1556. It is beyond the scope of this text to review the history of science over the past 450 years; however, certain highlights of the past are important for orientation purposes.

The transition of metallurgy from an art to a science was relatively slow, compared to the technological progress brought about by the growth of civilization. Improvements in mining and ore processing, adapted from military engineering methods, exposed ore sources at locations and depths previously inaccessible to man. The introduction of the steam pump to remove mine water in 1704 led to the improved steam engine of James Watt in 1780 and the beginning of the industrial revolution or the Power Age.

The Power Age created a demand for larger tonnages of metal and, in turn, provided power for higher production. The first version of the modern rolling mill was invented by Cort in 1783. Since cementation and crucible steels were too expensive for large-scale use, wrought iron was the predominant structural metal until the invention of the Bessemer converter in 1855. The development of the Siemens-Martin, open-hearth furnace in the period 1861-1864 provided an additional source of tonnage steel, which, with Bessemer steel, met the demands of an expanding railroad industry. The subsequent development of the electric-arc furnace and the availability of cheap electric power made possible the high-quality carbon and alloy steels required by the automotive and machine-tool industries after the turn of the twentieth century.

Scientific man attempted to expand his knowledge by laboratory experimentation to explain countless phenomena in the art of metallurgy which had been regarded with ignorance and superstition down through the ages. Faraday, Lavoisier, and others in the eighteenth and nineteenth centuries formulated the quantitative conception of chemistry, thus providing a basis for later metallurgical work. The electric battery, developed by Volta in 1800, made possible the first separation of the light metals by Davy in 1806 and was the start of the electrometallurgical industry. The metallurgical microscope proposed by Sorby in 1864 extended man's vision in the study of metals and remains today the most valuable instrument at the disposal of the metallurgist for research and production control. The periodic table of the elements drafted by Mendelyeev in 1869 and the phase rule stated by

Gibbs in 1876 continue to guide man in his search for new alloy systems. Man's scientific vision was extended by the discovery of X-rays by Roentgen in 1895 and radium by Curie in 1898, which later provided a means of nondestructive examination of metals for internal defects. The development of X-ray diffraction techniques focused attention on the internal structure of metallic crystals.

The engineer is called upon to utilize metals and alloys under a wide variety of conditions. To make the best possible selection of an alloy for an application, he must be familiar with the factors that control the properties of metals and alloys and how these properties can be varied by certain treatments. For a long time, little attention was devoted to the relation of the structure of metals and alloys to their properties. One of the first treatises on the metallurgy of steel which had a bearing on engineering applications was written by Henry Marion Howe in 1890. In the early 1890's, Albert Sauveur was actively engaged in trying to introduce metallurgical control into industry. At first, he was quite unsuccessful, but as a result of his persistence, metallurgy was gradually recognized as an aid to industry. Through the many years since then, the value of metallurgical knowledge in industrial operations has been proved.

Sauveur, with others, proposed theories to explain the change of properties that resulted from the heat treatment of steel. Of course, many of these theories were incorrect, and, when the developments in physics were applied, better theories were developed and tested experimentally. The work of Bain, Davenport, Mehl, and many others has contributed to a better understanding of the behavior of metals and alloys and has instilled the scientific approach in this field that for centuries had been purely an art. At no time has there been such a close relationship between physics and metallurgy as there now exists; it is clear that this relationship will be even closer in the future developments of the Atomic and Space Age.

In the application of solid state physics to a better understanding of metallurgical engineering, a basic knowledge of fundamentals must be recognized as more important to the engineer than the specific properties of individual metals and alloys which are in a continual state of change. Although the engineer must make specific recommendations as to the application of materials and treatments based on experience, an appreciation of fundamentals makes experience more meaningful.

1.2 Divisions of Metallurgy. Metallurgy has been defined as the art and science that deals with the preparation and application of metals and alloys. The layman thinks of metallurgy as dealing with the reduction of metals from their ores. The field of metallurgy is actually broader than this conception and may be conveniently divided into three branches: (1) chemical, (2) physical, and (3) mechanical metallurgy.

Chemical metallurgy involves the reduction of metals from their ores and the refining and alloying of such metals. This branch is sometimes referred to as process metallurgy and includes such extractive processes as hydrometallurgy, pyrometallurgy, and electrometallurgy, together with methods of ore preparation and concentration, including cyaniding, calcination, flotation, etc. Since the engineer is chiefly concerned with the application of metals, this phase of metallurgy will not be discussed in this text.

Physical metallurgy deals with the nature, structure, and physical properties of metals and alloys, together with the mechanism of varying such properties. The subject of physical metallurgy includes metallography, mechanical testing, and heat treatment. Metallography is the study of the structure of metals and alloys with the aid of the metallurgical microscope, X-ray diffraction equipment, etc., and provides a means of correlating structure with physical and mechanical properties. No attempt will be made in this text to develop the techniques of metallography. The results obtained by these methods will be studied and interpreted in terms of engineering application.

Mechanical metallurgy covers the working and shaping of metals and alloys by casting, forging, rolling, drawing, extruding, etc.

These three branches of metallurgy are interrelated. Variations of operations during refining processes may be reflected in variations of properties and structure of the alloys produced. The operations that are applied to an alloy in the solid state have a marked influence on the properties. Consequently, a fundamental knowledge of chemistry, physics, thermodynamics, and mechanics is necessary for a proper understanding of metallurgy.

The engineer is not expected to function as an expert in all phases of metallurgy, but he must have a thorough knowledge of those phases of physical metallurgy, involving the selection, treatment, and use of metals and alloys, which are closely related to his work.

1.3 Metallurgy in Industry. The place of metallurgy in an industrial organization depends largely on whether the company is a prime producer of metals and alloys or a manufacturer of products requiring metals in their construction. Metallurgy occupies a position of paramount importance in organizations of the first type, since metallurgical operations are of major interest. Metallurgy serves in an advisory capacity to industries of the second type where engineering design and production methods govern the success of manufacturing enterprises. In view of the interrelation between the metallurgical and manufacturing industries, the engineer should be familiar with the organization of each. Line charts illustrating the relationship of metallurgy in representative organizations of both types are shown in Figs. 1.1 and 1.2. In comparing these charts the engineer should recognize that organizations within different industries, and even in different com-

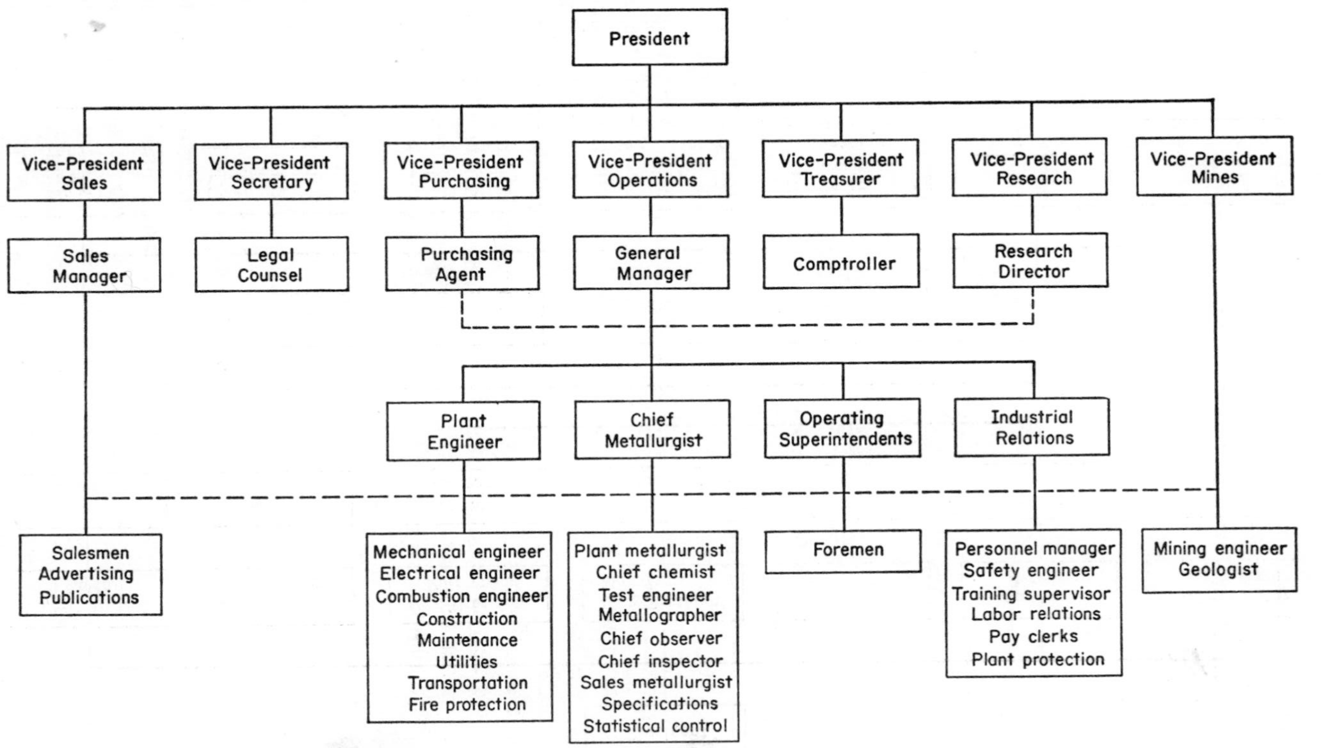

FIG. 1.1 Typical metallurgical plant organization.

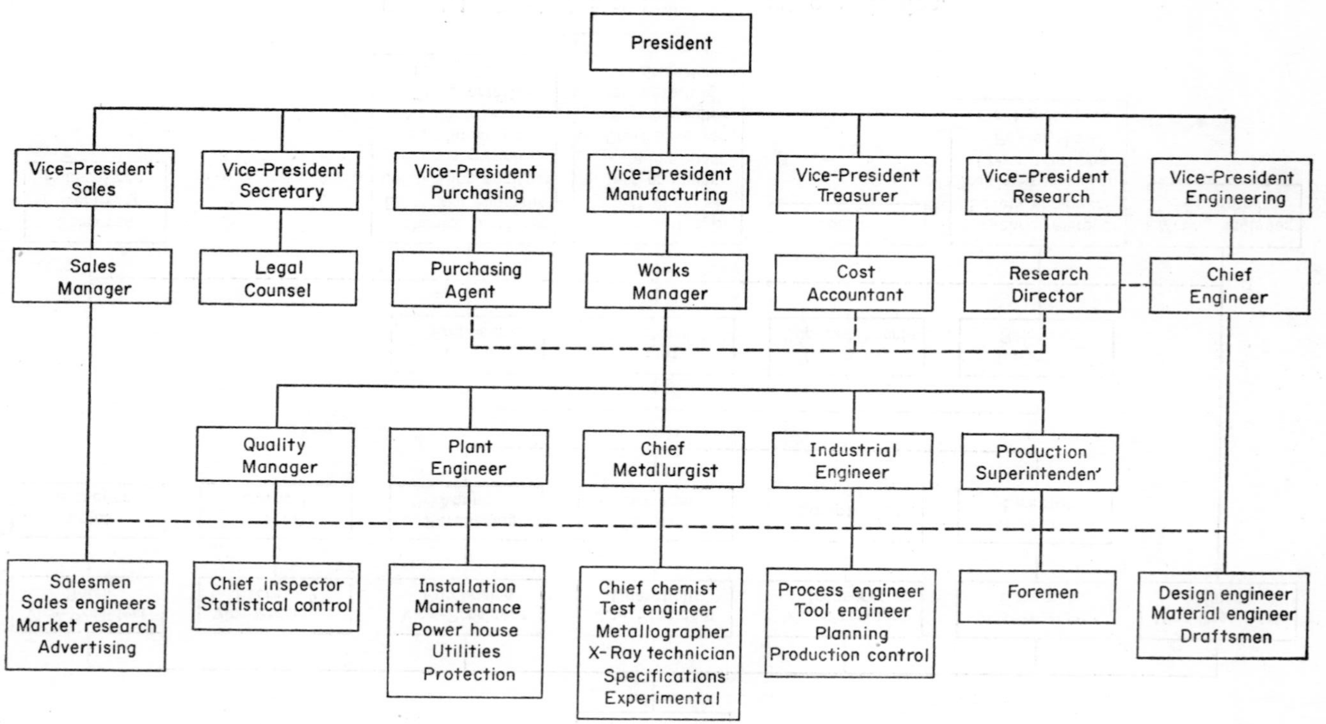

FIG. 1.2 Typical manufacturing plant organization.

panies within a given industry, vary to a large extent to meet local conditions.

The metallurgical department in a metallurgical plant is largely responsible for quality control and serves as a consultant to other departments within the plant and to the customers. In the absence of an established research department, the metallurgical department also may conduct research and follow experimental work through the plant.

Modern metallurgical control started about 1920 in response to consumer demands for consistent quality among heats of steel, and many industrial metallurgists received their early training in the chemical laboratory making routine chemical analyses. Metallurgical control expanded, however, from the chemical laboratory out into the plant in the form of temperature control of melting, casting and rolling, slag control, foundry sand control—in fact, control of any stage of the metallurgical process that would assure consistent results in the final product.

In the beginning, the metallurgical department struggled to enforce its recommendations against strong resistance by production departments. Progress was very slow. Production had been controlled for generations by the rule-of-thumb method with the emphasis too often placed on the "get it hot and get it out" philosophy. Today, as a result of an influx of scientifically trained men into the metallurgical industry and the ever-present pressure of metallurgically minded customers, the metallurgist has proved his value to management and gained the respect of the man in the shop.

The staff of a metallurgical department may consist of a single metallurgist in a small company or may include several hundred trained specialists, observers, and inspectors in a large steel plant. The steel plant's metallurgical department usually is supervised by a chief metallurgist and may include such key members as plant metallurgist, chief chemist, test engineer, metallographer, chief observer, chief inspector, specifications supervisor, statistical supervisor, sales metallurgists, etc. The plant metallurgist is charged with the broad supervision of metallurgical operations in the plant, such as casting, forging, heat-treating, etc., and is often consulted by the sales metallurgist and the customer regarding defects and service failures which may relate to plant operations. The chief chemist is responsible for the chemical analysis of the products and incoming raw materials. Much of the rapid control work of the chemical laboratory is being supplemented by spectographic analysis. The mechanical testing of metals is under the supervision of a test engineer who also acts as a consultant regarding specifications, failures, etc. The metallographer studies the structure of metals with the metallurgical microscope and assists in tracing the cause of service failures. He may also have cognizance of X-ray inspection for internal defects. The chief observer is in charge of a staff of metallurgical observers who record the complete history of every heat of steel from the initial charging of the furnace to the final hot-rolling operations, including processing temperatures, surface defects, scrap discard, etc. The chief inspector and his staff are responsible for dimensional inspection and surface finish resulting from mechanical operations, such as rolling, forging, casting, drawing, etc. A specifications supervisor checks customers' orders before they go to the shop to determine whether specifications can be met or whether an error may have been made by the customer. A statistical control supervisor organizes and analyzes the data reported by the observers and inspectors, arranging it in periodic summary reports which go to production heads and to management. These records also establish a source of complete information on every billet, ingot, and heat of steel for investigation of service complaints. The sales metallurgist acts in a liaison capacity with the sales department and the customer in disseminating metallurgical information, investigating complaints, reviewing specifications, etc.

The metallurgical department in a manufacturing plant is a service organization acting in an advisory capacity to the engineering, purchasing, production, inspection, sales, planning, and plant engineering departments. It is required to be more versatile than its counterpart in the metallurgical organization, since its activities cover a wide variety of nonmetallic materials and a wider range of both ferrous and nonferrous metals than is encountered in any one metallurgical plant.

An important part of the work of the metallurgical department in a manufacturing organization is concerned with making recommendations to the engineering department as to the best material for a given application and setting manufacturing specifications to be met by the production departments. It may also be called upon to establish purchasing specifications and to conduct acceptance tests on many items used in production and maintenance, including metals, refractories, fuel, paints, cutting and lubricating oils, rust-preventive compounds, etc. In periods of material shortages, it may be forced to recommend approval of substitute materials. When plant facilities involving chemical or metallurgical operations are to be installed, the metallurgical department works closely with the planning, production, and plant engineering departments in the selection of suitable equipment.

In a manufacturing organization, dimensional inspection is usually divorced from the metallurgical department, and customer liaison is often handled by sales engineers representing the sales department.

1.4 Metallurgy and the Engineer. Modern engineering applications demand a better fundamental knowledge of metallurgy than ever before. The engineer must understand the principles underlying metallurgical problems with which he will come in contact and must be able to present these problems to the metallurgist in a manner that can be interpreted and evaluated. In the absence of metallurgical talent, the engineer must be prepared to rely on his own resources for the intelligent solution of many such problems.

The interests of the engineer lie in the branches of physical and mechanical metallurgy. He not only must be able to determine the size and shape of a product from the standpoint of strength, but also must be able to determine from past experience and a knowledge of materials and processing methods that specific stress or deformation which can be allowed in certain parts.

Some of the factors that must be considered in selecting a material for a given product include:

1. Properties required.
2. Previous experience.
3. Availability.
4. Cost.
5. Processing method.

The processing method may strongly influence the actual selection of a material. Those factors that govern to a large degree the choice of the method of processing include:

1. Size and shape.
2. Properties of material.
3. Properties required.
4. Finish required.
5. Previous experience.
6. Availability.
7. Cost.

The properties required in a given part include strength, machinability, appearance, ability to be worked, etc. With a knowledge of the properties required, the engineer must call upon prior experience to determine those materials that have been satisfactory in similar applications in the past. Needless to say, the selection of a suitable material on the basis of properties and experience is not sufficient unless the material is currently available. Since engineering involves the economical application of science, the engineer's choice will be governed by the item of cost. In the final analysis, the advantages derived from the use of an expensive material will be balanced against the lower cost of an inferior material.

In the choice of a processing method, size and shape usually determine whether the part is to be cast, forged, rolled, drawn, extruded, or welded.

The inherent properties of the material also govern the process to be selected. Hard, brittle materials cannot be cold-worked successfully, and a material that is brittle at elevated temperatures cannot be hot-forged. The processing method will influence the properties that may be obtained in a given part. With most materials, there is a wide variation between the properties of castings and forgings. The various processes have commercial limitations which establish the final finish that can be obtained in a given process without resorting to subsequent finishing operations. Processing cost is largely a function of the quantity of required parts.

From the foregoing relationship of metallurgy to engineering, it should be self-evident why the engineer should receive training in the fundamentals of metallurgy with particular emphasis on applied physical metallurgy.

REFERENCES

Hoover and Hoover, *Georgius Agricola De Re Metallica,* London, 1912.

Howe, H. M., *The Metallurgy of Steel,* The Engineering and Mining Journal, 1st. ed., New York, 1890.

Rickard, *Man and Metals,* McGraw-Hill Book Co., New York, 1932.

Sauveur, Albert, *The Metallography and Heat Treatment of Iron and Steel,* 1st ed., University Press, Cambridge, Mass., 1912.

Smith and Gnudi, *Pirotechnia of Vannoccia Biringuccio,* American Institute of Mining and Metallurgical Engineers, New York, 1943.

Sullivan, *The Story of Metals,* American Society for Metals, Metals Park, Ohio, 1951.

QUESTIONS

1. About when in history was the first alloy cast by man? What was this alloy?
2. At approximately what period in history was iron ore first smelted?
3. What development made it possible to produce iron in large quantities? In what period of time did this development occur?
4. In what way is the development by James Watt of an improved steam engine related to progress in metallurgy?
5. Who was the inventor of the rolling mill, and at about what time?
6. What ferrous material was used predominantly for structures before the advent of steel on a large scale?
7. What developments made possible the production of large quantities of steel?
8. What individual was largely responsible for the application of the microscope to the study of metals?
9. Define metallurgy.
10. What are the three branches of metallurgy? Describe each.
11. Define metallography.
12. What is the general function of the metallurgical department in a metallurgical plant?

13. What is the function of the metallurgical department in a manufacturing plant?
14. Why is metallurgy of prime importance to the design engineer?
15. What factors must be considered in the selection of a material for a given product?
16. What factors govern the choice of the method of processing or manufacturing a given product?

II

THE STRUCTURE OF MATTER

2.1 The Atom. The most basic unit of matter is the atom, which in turn is a complex assemblage of particles. The distinctive difference between the elements stems from the difference in their atomic structures. An *atom* is an electronic structure having a diameter of approximately 15×10^{-8} cm with its mass concentrated in the nucleus. The *nucleus* of the atom is approximately 10^{-12} cm in diameter and is composed of a number of different particles. One of the particles in the nucleus is the *proton,* which has a mass of 1.66×10^{-24} g and a charge of $+ 4.80 \times 10^{-10}$ esu. The number of protons in the nucleus determines the charge on the nucleus and therefore determines the identity of the element. The *atomic number, Z,* indicates the number of protons in the nucleus.

Isotopes of elements may exist which have the same atomic number but different atomic weights. *Isobars* of elements also exist which have the same atomic weight but different atomic numbers.

The *mass number* of the nucleus A is the mass of the atom to the nearest whole number. This is based on the mass of the isotope of oxygen having a mass of 16.00. The atomic unit of mass is $\frac{1}{16}$ of the mass of the isotope of oxygen, which is 1.66×10^{-24} g.

The *neutron* is another particle within the nucleus and has an uncharged mass slightly greater than the proton. The number of neutrons is equal to the difference between the mass number A and the atomic number Z.

While the other particles present in the nucleus are of importance to the nuclear physicist, in general they are not of great consequence insofar as the usual properties are concerned.

Revolving around the nucleus in orbits are the *electrons* which have a mass of $1/1836 \times$ the mass of a proton and a charge of -4.8×10^{-10} esu. This charge is equal and opposite to the charge on a proton. The number of electrons is the same as the number of protons in a neutral atom.

2.2 Electronic Orbits. The first theory pertaining to the electronic state of the atom was applied to hydrogen. The assumption was made that the one electron in hydrogen occupied a circular orbit and the outward centrifugal force of the electrons was balanced against the inward attraction of the nucleus. The original theory as applied to hydrogen did not provide the reason for the production of spectral lines or the stability of the atom. The original classical theory would have permitted the electron to approach the nucleus with a continual emission of radiation of continually changing frequency.

The unsatisfactory situation resulting from the classical theory was corrected to a certain degree by Niels Bohr through the use of the *quantum theory*. This theory involved two principal assumptions. First, it was assumed that the electrons revolve about the nucleus in specified orbits or stationary states. Under this assumption, any emission or absorption of radiation would occur whenever an electron jumped from one stationary state to another stationary state. This meant that, when an electron jumped from one orbit to another, the atom would emit or absorb a quantum of radiation of frequency h which would be given by the expression

$$E_1 - E_2 = h$$

where h is Plank's constant $= 6.624 \times 10^{-27}$ erg sec.

The second principal assumption was that only stationary states that are stable are those for which the angular momentum is an integral multiple of $h/2\pi$. This means that an electronic orbit is concerned with the whole number which is called the quantum number. The possible orbits are $h/2\pi$, $2h/2\pi$, $3h/2\pi$, etc., or in general terms the angular momentum would be $nh/2\pi$ where n is the orbital quantum number. Later it was shown that the orbits need not be circular, and the Bohr theory was extended to include elliptical orbits and to take into consideration the variation of the mass of the electron in accordance with the theory of relativity.

The electron states have been characterized by two quantum numbers n and k. The *principal quantum number* is n, which is a measure of the total energy of the orbit. The major axes of the elliptical orbits are proportional to n^2; thus the orbits become larger with larger n. The quantum number k is designated as the *secondary quantum number,* and it is a measure of the angular momentum of the orbit. This is equal to $kh/2$. The ratio of minor to major axes of the elliptical orbit is given by the ratio k/n. The secondary quantum number k can have any value from 1 to, and including, n. Orbits for which $n = k$ are circular in character. There is precession of the elliptical orbits since the velocity, and therefore the mass, of the electron varies with distance from the nucleus.

The Bohr theory did not provide the complete story in view of the fact that the electrons were assumed to obey the laws of classical mechanics. Even the additions of quantum theory did not correct some of the deficiencies of the Bohr theory, which required considerable overhauling as far as the detail theory was concerned. However, the concept that electrons revolve around the nucleus in orbits that may be considered to lie in distinct shells at discreet distances from the nucleus seems to be justifiable and is probably sufficient for this discussion. After the corrections are made, it is not possible to describe a precise picture of the motion of electrons around the nucleus of the atom. The only picture that can be drawn is one that may represent the probability of electron density in the space surrounding the nucleus.

The electrons in an atom can occupy different energy states. Each of the electron states produces what may be considered to be an electron cloud pattern characterized by a definite energy. The electrons are divided among the different quantum shells or energy levels. The state of any electron in an atom may be described by four quantum numbers, which are n, l, m_l, and m_s; n has been previously described and is a measure of the energy of the electron in the state indicated; l is a measure of the angular momentum and may have values from 0 to $n - 1$. When $l = 0$ the electron is not at rest, but the motion does not give rise to an angular momentum. The quantum number m_l is a measure of the component of the angular momentum in a specified direction. This may have any value from $+l$ to $-l$ including 0. The quantum number m_s is an indication of the *electron spin* and may have a value of $\pm\frac{1}{2}$ depending upon the direction of the spin.

Electron states for which $l = 0$, 1, 2, and 3 are designated s, p, d, and f, respectively. The total quantum number n is indicated by a whole number; therefore if $n = 2$ and $l = 2$ the state would be designated by the symbol $2d$. The number of electrons in the subgroup is indicated by a superscript outside of the parentheses enclosing the symbol. For example, if there were two electrons in the state $2d$ then it would be written $(2d)^2$.

The most stable or normal state of any atom is that for which its energy is a minimum. When an atom is in a higher energy state, it is in an excited state. According to the *Pauli exclusion principle,* no two electrons in an atom can be in exactly the same state as defined by all four quantum numbers. This means that the quantum numbers n, l, and m may be the same for two electronic states of an atom, but one of them must have an m_s of $+\frac{1}{2}$ while the other must have an m_s of $-\frac{1}{2}$.

A consideration of the values that each of the quantum numbers may have shows that the maximum possible numbers of electrons with principal quantum numbers $n = 1, 2, 3, 4 \ldots n$ will be $2, 8, 18, 32 \ldots 2n^2$, respectively. When each of the quantum shells is completely filled to its stable configuration, one obtains an unusually stable element. There is provisional stability when the third and higher shells contain eight electrons each.

Table 2-I. ELECTRONIC CONFIGURATION OF ELEMENTS (Principal and Secondary Quantum Numbers).

Atomic Number	Element	1 0	2 0	2 1	3 0	3 1	3 2	4 0	4 1	4 2	4 3	5 0	5 1	5 2	5 3	6 0	6 1	6 2	6 3	7 0
1	H	1	..	..	..	..	..	..	..	..	..	..	..	..	..	..	..	..	..	..
2	He	2	..	..	..	..	..	..	..	..	..	..	..	..	..	..	..	..	..	..
3	Li	2	1	..	..	..	..	..	..	..	..	..	..	..	..	..	..	..	..	..
4	Be	2	2	..	..	..	..	..	..	..	..	..	..	..	..	..	..	..	..	..
5	B	2	2	1	..	..	..	..	..	..	..	..	..	..	..	..	..	..	..	..
6	C	2	2	2	..	..	..	..	..	..	..	..	..	..	..	..	..	..	..	..
7	N	2	2	3	..	..	..	..	..	..	..	..	..	..	..	..	..	..	..	..
8	O	2	2	4	..	..	..	..	..	..	..	..	..	..	..	..	..	..	..	..
9	F	2	2	5	..	..	..	..	..	..	..	..	..	..	..	..	..	..	..	..
10	Ne	2	2	6	..	..	..	..	..	..	..	..	..	..	..	..	..	..	..	..
11	Na	2	2	6	1	..	..	..	..	..	..	..	..	..	..	..	..	..	..	..
12	Mg	2	2	6	2	..	..	..	..	..	..	..	..	..	..	..	..	..	..	..
13	Al	2	2	6	2	1	..	..	..	..	..	..	..	..	..	..	..	..	..	..
14	Si	2	2	6	2	2	..	..	..	..	..	..	..	..	..	..	..	..	..	..
15	P	2	2	6	2	3	..	..	..	..	..	..	..	..	..	..	..	..	..	..
16	S	2	2	6	2	4	..	..	..	..	..	..	..	..	..	..	..	..	..	..
17	Cl	2	2	6	2	5	..	..	..	..	..	..	..	..	..	..	..	..	..	..
18	A	2	2	6	2	6	..	..	..	..	..	..	..	..	..	..	..	..	..	..
19	K	2	2	6	2	6	—	1	..	..	..	..	..	..	..	..	..	..	..	..
20	Ca	2	2	6	2	6	—	2	..	..	..	..	..	..	..	..	..	..	..	..
21	Sc	2	2	6	2	6	1	2	..	..	..	..	..	..	..	..	..	..	..	..
22	Ti	2	2	6	2	6	2	2	..	..	..	..	..	..	..	..	..	..	..	..
23	V	2	2	6	2	6	3	2	..	..	..	..	..	..	..	..	..	..	..	..
24	Cr	2	2	6	2	6	5	1	..	..	..	..	..	..	..	..	..	..	..	..
25	Mn	2	2	6	2	6	5	2	..	..	..	..	..	..	..	..	..	..	..	..
26	Fe	2	2	6	2	6	6	2	..	..	..	..	..	..	..	..	..	..	..	..
27	Co	2	2	6	2	6	7	2	..	..	..	..	..	..	..	..	..	..	..	..
28	Ni	2	2	6	2	6	8	2	..	..	..	..	..	..	..	..	..	..	..	..
29	Cu	2	2	6	2	6	10	1	..	..	..	..	..	..	..	..	..	..	..	..
30	Zn	2	2	6	2	6	10	2	..	..	..	..	..	..	..	..	..	..	..	..
31	Ga	2	2	6	2	6	10	2	1	..	..	..	..	..	..	..	..	..	..	..
32	Ge	2	2	6	2	6	10	2	2	..	..	..	..	..	..	..	..	..	..	..
33	As	2	2	6	2	6	10	2	3	..	..	..	..	..	..	..	..	..	..	..
34	Se	2	2	6	2	6	10	2	4	..	..	..	..	..	..	..	..	..	..	..
35	Br	2	2	6	2	6	10	2	5	..	..	..	..	..	..	..	..	..	..	..
36	Kr	2	2	6	2	6	10	2	6	..	..	..	..	..	..	..	..	..	..	..
37	Rb	2	2	6	2	6	10	2	6	—	—	1	..	..	..	..	..	..	..	..
38	Sr	2	2	6	2	6	10	2	6	—	—	2	..	..	..	..	..	..	..	..
39	Y	2	2	6	2	6	10	2	6	1	—	2	..	..	..	..	..	..	..	..
40	Zr	2	2	6	2	6	10	2	6	2	—	2	..	..	..	..	..	..	..	..
41	{Nb / Cb	2	2	6	2	6	10	2	6	4	—	1	..	..	..	..	..	..	..	..
42	Mo	2	2	6	2	6	10	2	6	5	—	1	..	..	..	..	..	..	..	..
43	Tc	2	2	6	2	6	10	2	6	6	—	1	..	..	..	..	..	..	..	..
44	Ru	2	2	6	2	6	10	2	6	7	—	1	..	..	..	..	..	..	..	..
45	Rh	2	2	6	2	6	10	2	6	8	—	1	..	..	..	..	..	..	..	..
46	Pd	2	2	6	2	6	10	2	6	10	—	—	..	..	..	..	..	..	..	..
47	Ag	2	2	6	2	6	10	2	6	10	—	1	..	..	..	..	..	..	..	..
48	Cd	2	2	6	2	6	10	2	6	10	—	2	..	..	..	..	..	..	..	..
49	In	2	2	6	2	6	10	2	6	10	—	2	1	..	..	..	..	..	..	..
50	Sn	2	2	6	2	6	10	2	6	10	—	2	2	..	..	..	..	..	..	..

Atomic Number	Element	1 0	2 0	2 1	3 0	3 1	3 2	4 0	4 1	4 2	4 3	5 0	5 1	5 2	5 3	6 0	6 1	6 2	6 3	7 0
51	Sb	2	2	6	2	6	10	2	6	10	—	2	3	..	..	..	..	..	..	..
52	Te	2	2	6	2	6	10	2	6	10	—	2	4	..	..	..	..	..	..	..
53	I	2	2	6	2	6	10	2	6	10	—	2	5	..	..	..	..	..	..	..
54	Xe	2	2	6	2	6	10	2	6	10	—	2	6	..	..	..	..	..	..	..
55	Cs	2	2	6	2	6	10	2	6	10	—	2	6	—	—	1	..	..	..	..
56	Ba	2	2	6	2	6	10	2	6	10	—	2	6	—	—	2	..	..	..	..
57	La	2	2	6	2	6	10	2	6	10	—	2	6	1	—	2	..	..	..	..
58	Ce	2	2	6	2	6	10	2	6	10	2	2	6	—	—	2	..	..	..	..
59	Pr	2	2	6	2	6	10	2	6	10	3	2	6	—	—	2	..	..	..	..
60	Nd	2	2	6	2	6	10	2	6	10	4	2	6	—	—	2	..	..	..	..
61	Pm	2	2	6	2	6	10	2	6	10	5	2	6	—	—	2	..	..	..	..
62	Sm	2	2	6	2	6	10	2	6	10	6	2	6	—	—	2	..	..	..	..
63	Eu	2	2	6	2	6	10	2	6	10	7	2	6	—	—	2	..	..	..	..
64	Gd	2	2	6	2	6	10	2	6	10	7	2	6	1	—	2	..	..	..	..
65	Tb	2	2	6	2	6	10	2	6	10	9	2	6	—	—	2	..	..	..	..
66	Dy	2	2	6	2	6	10	2	6	10	10	2	6	—	—	2	..	..	..	..
67	Ho	2	2	6	2	6	10	2	6	10	11	2	6	—	—	2	..	..	..	..
68	Er	2	2	6	2	6	10	2	6	10	12	2	6	—	—	2	..	..	..	..
69	Tm	2	2	6	2	6	10	2	6	10	13	2	6	—	—	2	..	..	..	..
70	Yb	2	2	6	2	6	10	2	6	10	14	2	6	—	—	2	..	..	..	..
71	Lu	2	2	6	2	6	10	2	6	10	14	2	6	1	—	2	..	..	..	..
72	Hf	2	2	6	2	6	10	2	6	10	14	2	6	2	—	2	..	..	..	..
73	Ta	2	2	6	2	6	10	2	6	10	14	2	6	3	—	2	..	..	..	..
74	W	2	2	6	2	6	10	2	6	10	14	2	6	4	—	2	..	..	..	..
75	Re	2	2	6	2	6	10	2	6	10	14	2	6	5	—	2	..	..	..	..
76	Os	2	2	6	2	6	10	2	6	10	14	2	6	6	—	2	..	..	..	..
77	Ir	2	2	6	2	6	10	2	6	10	14	2	6	7	—	2	..	..	..	..
78	Pt	2	2	6	2	6	10	2	6	10	14	2	6	9	—	1	..	..	..	..
79	Au	2	2	6	2	6	10	2	6	10	14	2	6	10	—	1	..	..	..	..
80	Hg	2	2	6	2	6	10	2	6	10	14	2	6	10	—	2	..	..	..	..
81	Tl	2	2	6	2	6	10	2	6	10	14	2	6	10	—	2	1	..	..	..
82	Pb	2	2	6	2	6	10	2	6	10	14	2	6	10	—	2	2	..	..	..
83	Bi	2	2	6	2	6	10	2	6	10	14	2	6	10	—	2	3	..	..	..
84	Po	2	2	6	2	6	10	2	6	10	14	2	6	10	—	2	4	..	..	..
85	At	2	2	6	2	6	10	2	6	10	14	2	6	10	—	2	5	..	..	..
86	Rn	2	2	6	2	6	10	2	6	10	14	2	6	10	—	2	6	..	..	..
87	Fr	2	2	6	2	6	10	2	6	10	14	2	6	10	—	2	6	—	—	1
88	Ra	2	2	6	2	6	10	2	6	10	14	2	6	10	—	2	6	—	—	2
89	Ac	2	2	6	2	6	10	2	6	10	14	2	6	10	—	2	6	1	—	2
90	Th	2	2	6	2	6	10	2	6	10	14	2	6	10	—	2	6	2	—	2
91	Pa	2	2	6	2	6	10	2	6	10	14	2	6	10	2	2	6	1	—	2
92	U	2	2	6	2	6	10	2	6	10	14	2	6	10	3	2	6	1	—	2
93	Np	2	2	6	2	6	10	2	6	10	14	2	6	10	5	2	6	—	—	2
94	Pu	2	2	6	2	6	10	2	6	10	14	2	6	10	6	2	6	—	—	2
95	Am	2	2	6	2	6	10	2	6	10	14	2	6	10	7	2	6	—	—	2
96	Cm	2	2	6	2	6	10	2	6	10	14	2	6	10	7	2	6	1	—	2
97	Bk	2	2	6	2	6	10	2	6	10	14	2	6	10	8	2	6	1	—	2
98	Cf	2	2	6	2	6	10	2	6	10	14	2	6	10	10	2	6	—	—	2
99	E	2	2	6	2	6	10	2	6	10	14	2	6	10	11	2	6	—	—	2
100	Fm	2	2	6	2	6	10	2	6	10	14	2	6	10	12	2	6	—	—	2
101	Mv	2	2	6	2	6	10	2	6	10	14	2	6	10	13	2	6	—	—	2

The outermost shell of electrons is called the *valency group* and the electrons in this group are called the *valence electrons.* The properties of the elements are closely related to the valence electrons. The manner in which the electrons are arranged in the various elements is indicated in Table 2-I. Because the properties of the atoms are largely determined by the electrons in the outer orbits, those that have similar structures should have similar properties. The elements that have one electron in the outer shell are listed below. The similarity of properties of these elements is well

Element	Atomic Number	Shells Occupied																
		$n = 1$	2		3			4				5			6			7
		$l = 0$	0	1	0	1	2	0	1	2	3	0	1	2	0	1	2	0
Lithium	3	2	1															
Sodium	11	2	2	6	1													
Potassium	19	2	2	6	2	6		1										
Rubidium	37	2	2	6	2	6	10	2	6			1						
Cesium	55	2	2	6	2	6	10	2	6	10		2	6		1			
Francium	87	2	2	6	2	6	10	2	6	10	14	2	6	10	2	6		1

known. The two outermost shells of the elements in this series, with the exception of lithium, have six electrons and one electron, respectively.

The inert gases are those elements in which the shells are just completely filled to form a stable element. Their electronic structures are indicated below:

Element	Atomic Number	Shells Occupied														
		$n = 1$	2		3			4				5			6	
		$l = 0$	0	1	0	1	2	0	1	2	3	0	1	2	0	1
Neon	10	2	2	6												
Argon	18	2	2	6	2	6										
Krypton	36	2	2	6	2	6	10	2	6							
Xenon	54	2	2	6	2	6	10	2	6	10		2	6			
Radon	86	2	2	6	2	6	10	2	6	10	14	2	6	10	2	6

All of these gases have the maximum allowable number of electrons in the next to the outer shell and six in the outer shell. All of these gases have similar characteristics.

The elements scandium (21) through nickel (28), yttrium (39) through palladium (46), lanthanum (57) through platinum (78), and actinium (89) through californium (98), have inner shells that are not filled in the regular order as is observed for the other elements. These elements are called the *transition elements.*

2.3 Periodic Table of Elements. In discussing the states of the electrons in the elements, it was shown that elements that have similar electronic states have similar properties. Mendelyeev, recognizing the periodicity of the elements, arranged them into a *periodic table.* The atomic structure of the elements provides a reason for this periodicity.

The elements are arranged in the periodic table vertically into *groups* and horizontally into *periods,* as shown in Fig. 2.1. The elements are ar-

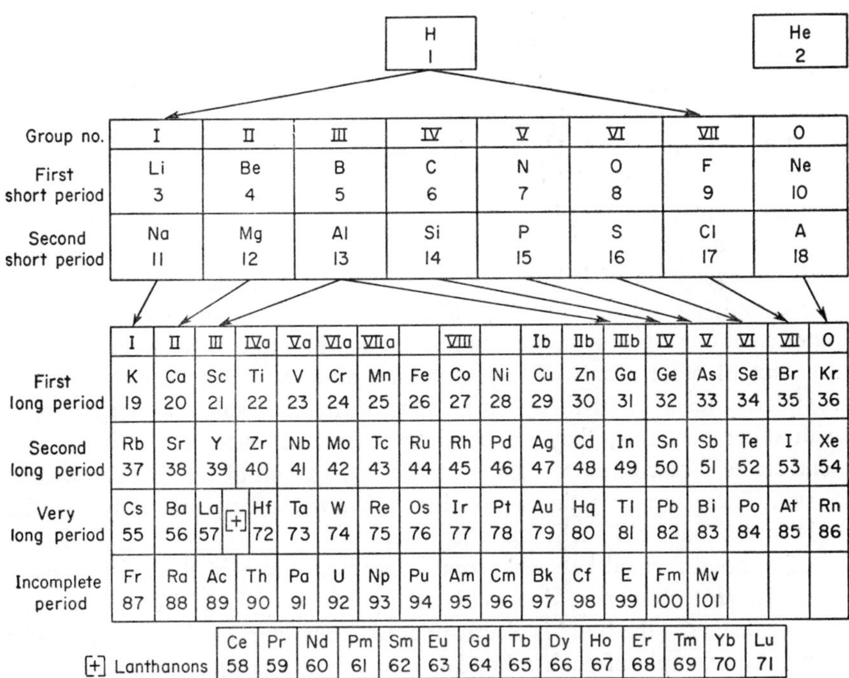

FIG 2.1 Periodic table of the elements.

ranged in the order of increasing atomic number from left to right in each of the periods. There are seven periods: one period containing hydrogen and helium, two short periods with eight elements each, two long periods with 18 elements each, a very long period containing thirty-two elements, and an incomplete period containing elements of atomic numbers 87 through 101. If one plots certain of the properties of the elements as a function of the atomic number, it becomes obvious that there is a systematic change and a definite periodicity of the elements.

The *transition elements* previously mentioned are those with atomic numbers 21 through 28 in the first long period and 39 through 46 in the second long period. The *rare earth metals,* or the *lanthanons,* are the elements with atomic numbers 57 through 71. The *actinons* are those elements with atomic numbers 90 through 98. The elements in the left and central portions of the periodic table are *metals,* while those elements in the right portion of the table are *nonmetals.* The transition from metals to nonmetals occurs in the elements beginning with group 3 and passing diagonally downward to the right—boron, silicon, germanium, arsenic, antimony, tellurium, and polonium. These elements have properties intermediate to the metals and nonmetals and are referred to as *metalloids.*

The periodic table of elements is extremely important to the metallurgist. It can be very helpful to the engineer in understanding the possible substitution of elements where scarcities may exist and in which substitutions may therefore resolve problems of economics or shortages.

2.4 Atomic Bond. Up to this point, consideration has been given to atoms alone. All materials are composed of a very large number of atoms that are bonded together in some manner. The cohesion between the atoms is dependent upon the character of the individual elements. The inert gases, with the exception of helium, are formed when a stable quantum shell is just completely filled. This illustrates the fact that eight electrons in the outer shell constitute a stable condition. Such an atom is unaffected by the presence of other atoms of the same kind or of any other kind. They are completely inert. Even though the atoms of the inert gases do not combine with other elements, they do bond together and exist in the liquid and solid states. The cohesion in this instance is the result of the *van der Waals' forces.* These are sometimes called polarization forces, and they are the only forces that hold such atoms together.

The bond formed between atoms other than those of the inert gases is of a different character, due to the difference in the electronic state. In many substances the bond is electrovalent in character and is called an *ionic bond.* An example of such a bond is found in the compound NaCl. Here, the valence electron of the sodium atom enters the one $2p$ state of the chlorine atom. The bond is electrostatic in character.

The formation of a stable state with the completion of the shell containing eight electrons is significant in another type of bond. Hydrogen, for example, has one electron. The first shell will be stable with two electrons. Therefore, if two atoms of hydrogen combine, they will share their electrons so that there is an equivalent of two electrons per atom. Nitrogen has a total of seven electrons; therefore, by each atom's sharing one electron, a stable octet is produced. This is called a *homopolar* or *covalent bond.*

Another type of bond is similar to the covalent type, but in this instance all of the atoms share each other's electrons, forming what may be called an electron gas or cloud. This sharing produces what is called a *metallic bond,* in which there is complete freedom of motion of the electrons when a group of atoms becomes associated. There is not a sufficient number of electrons in each atom to completely fill the outer shell, and the shared electrons are free to move from atom to atom. The only resistance to this movement is the result of collisions with other electrons. The relative free movement of electrons in metals accounts for electrical and thermal conductivity. The collisions which interfere with the flow of electrons accounts for the resistivity. The metallic bond which has been and is still under great discussion is not thoroughly understood.

There are many definitions for a metal, but this is a matter of individual preference. In some instances, *metals* have been defined as those substances that possess lustre, are opaque, exhibit plasticity, and conduct electricity. However, there are problems with this type of definition. Graphite, for example, is not a metal and yet it conducts electricity. In general, however, it is true that a metallic substance is one in which there are free electrons available by which a current may be conducted.

2.5 Crystal Structure. Since there is no direct linkage in the metallic bond, metal atoms try to surround themselves by as many similar atoms as possible. This leads to an attempt to obtain a close packing of atoms to form a structure. The atoms arrange themselves in a distinct pattern in space, called a *space lattice*. The spacing of the atoms is such that the electron shells of the different atoms do not interfere with each other, but there can still be exchanges of electrons from one atom to another. Each atom may be considered as a sphere having a definite diameter.

One finds that there are two ways in which spheres of equal size may be packed so as to have the closest possible packing. One method of packing is to start with a layer of spheres, with each sphere in contact with six adjacent spheres in the same plane. Another plane of spheres above the first is placed in such a manner that each sphere in the second layer rests in a pocket between the spheres of the first layer and touches three spheres in the layer below it. A third layer of spheres is placed so that each of the atoms is above the pockets of the first layer that were not filled by the second layer. Then there is a fourth layer of spheres which will be exactly above the spheres of the first layer. This continues on in the sequence "abc, abc," etc., where "a" is the first layer, "b" is the second layer, "c" is the third layer, and "a" repeats as the fourth layer. These layers are indicated as 1, 2, 3, and 4, respectively, in Fig. 2.2. This type of packing leads to the face-centered cubic lattice structure. This method of placing the spheres is illustrated in Fig. 2.2. The *face-centered cubic structure* appears

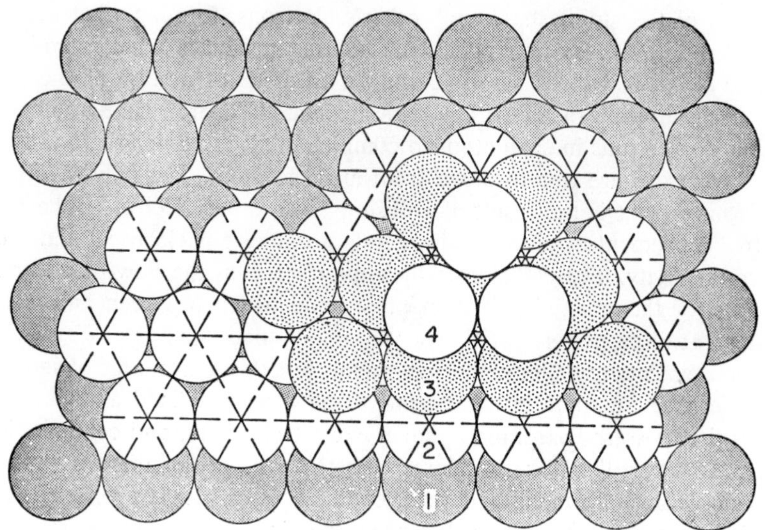

FIG. 2.2 Stacking of spheres to form face-centered cubic structure: (1) first (bottom) layer; (2) second layer; (3) third layer; (4) fourth layer.

when one cuts this packing through the plane indicated in the figure and the usual appearance of the face-centered cubic structure is shown in Fig. 2.3.

The other method of high-density packing of spheres consists of laying the first two layers as in the production of the face-centered cubic lattice, but the third layer of spheres is placed in such a way that the spheres will be directly over the spheres of the first layer. This structure will repeat in the form "ab, ab." This structure is called the *hexagonal close-packed lattice,* and the method of packing is shown in Fig. 2.4. The usual representation of the lattice is shown in Fig. 2.5.

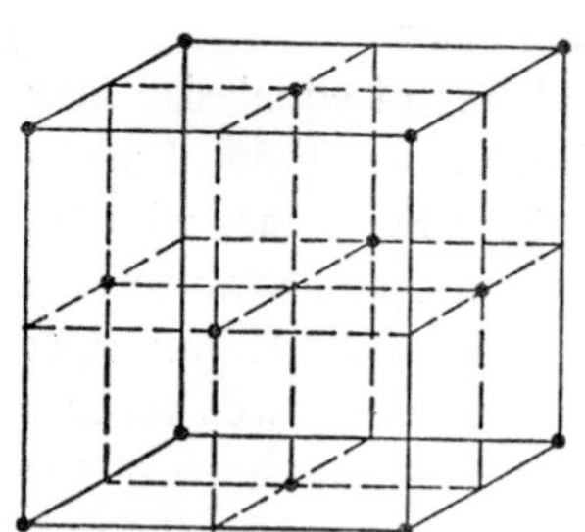

FIG. 2.3 Single lattice of face-centered cubic structure.

Metals are also found to form another structure which is not the closest possible packing, namely, the *body-centered cubic lattice,* as shown in Fig. 2.6. Each atom in this structure is in contact with only eight of its neighbors, in contrast to the other type of packing in which each atom was in contact with 12 neighbors. There are some instances in which metals form structures that are of the body-centered tetragonal structure shown in Fig. 2.7. In this instance, the atoms in one direction are slightly farther apart than in the other direction. There is a variety of lattice forms that are ob-

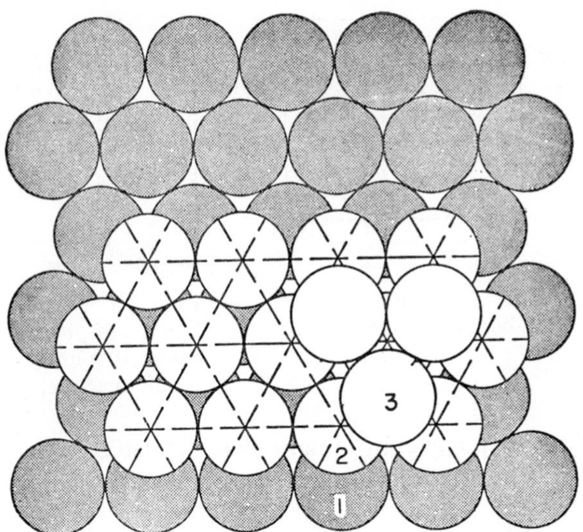

FIG. 2.4 Stacking of spheres to form hexagonal close-packed structure: (1) first (bottom) layer; (2) second layer; (3) third layer.

served in nature. However, the metals are found to form principally in the cubic or isometric, the hexagonal, and the tetragonal systems.

Space lattices are extremely small, being of the order of 4Å on an edge, approximately 100 millionth of an inch. The dimensions of the space lattices are characteristic of individual elements and vary from element to element.

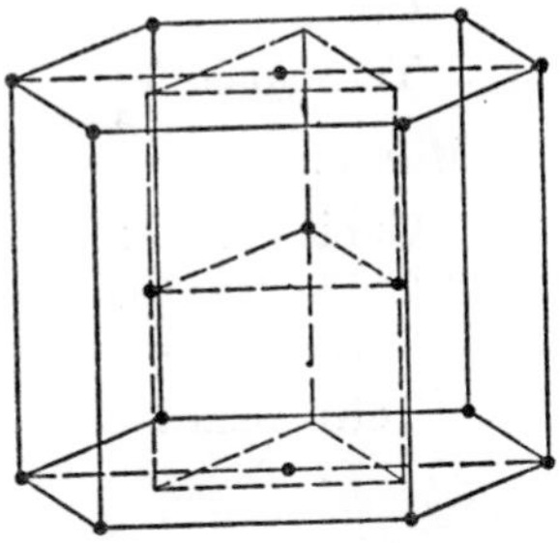

Some metals may exist in several lattice forms, depending upon the temperature. When a metal exists in more than one lattice form, it is said to be *allotropic* in character. Those metals that exist in only one form are:

Face-centered cubic
  Ca, Sr, Ni, Cu, Rh, Pd, Ag, Ir, Pt, Au, Pb, Al, Th
Body-centered cubic
  Li, Na, K, Rb, Cs, V, Mo, Ta, W, Ba, Cb
Tetragonal (face-centered)
  In
Hexagonal (close-packed)
  Be, Mg, Zr, Os, Zn, Cd, Te

FIG. 2.5 Single lattice of hexagonal close-packed structure.

Of the commercially important metals, iron, cobalt, tin, titanium, manga-

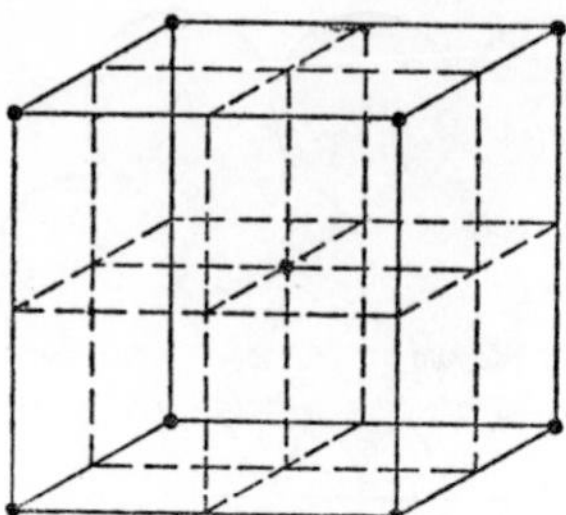

FIG. 2.6 Single lattice of body-centured cubic structure.

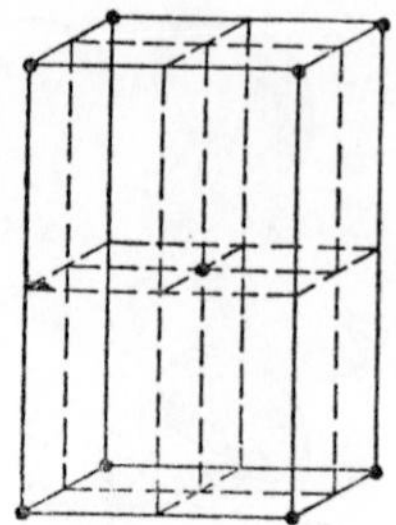

FIG. 2.7 Single lattice of body-centered tetragonal structure.

nese, and chromium may exist in more than one lattice form, i.e., they exhibit allotropy. The following elements have the allotropic forms indicated:

Iron

Body-centered cubic α	(below 1663 F)	
Face-centered cubic γ	(1663 F to 2557 F)	
Body-centered cubic δ	(2557 F to 2795 F)	(mp)

Cobalt

Close-packed hexagonal	(below 788 F)	
Face-centered cubic	(788 F to 2723 F)	(mp)

Tin

Cubic α	(gray below 55.8 F)	
Tetragonal β	(white 55.8 F to 450 F)	(mp)

Manganese

Cubic, 58 atoms per unit cell α	(below 1252 F)	
Cubic, 20 atoms per unit cell β	(1252 F to 2012 F)	
Tetragonal γ	(2012 F to 2080 F)	
Tetragonal δ	(2080 F to 2273 F)	(mp)

Chromium

Close-packed hexagonal β	(below 68 F)	
Body-centered cubic α	(68 F to 3270 F $\pm$ 90 F)	(mp)

Titanium

Close-packed hexagonal α	(below 1615 F)	
Body-centered cubic β	(1615 F to 3140 F)	(mp)

A combination of lattices produces a crystal. A *crystalline substance* is defined as one in which the atoms are arranged in a definite and repeating order. Single crystals are very rarely used in practice; however, they are employed in the investigation of the fundamental behavior of materials.

2.6 Crystallization of Metals. A metal in the liquid state is noncrystalline, and the atoms move freely among one another without regard to interspacial distances. The internal energy that these atoms possess prevents them from approaching one another closely enough to come under the control of the electrostatic fields of force. In many cases the atoms may be close enough to each other and, in fact, at times a particular atom may be surrounded by the proper number of atoms to form a lattice, but, since each atom is free to move, no structure is formed. The bonds between the atoms are easily broken and reformed, which is characteristic of the conditions of liquids.

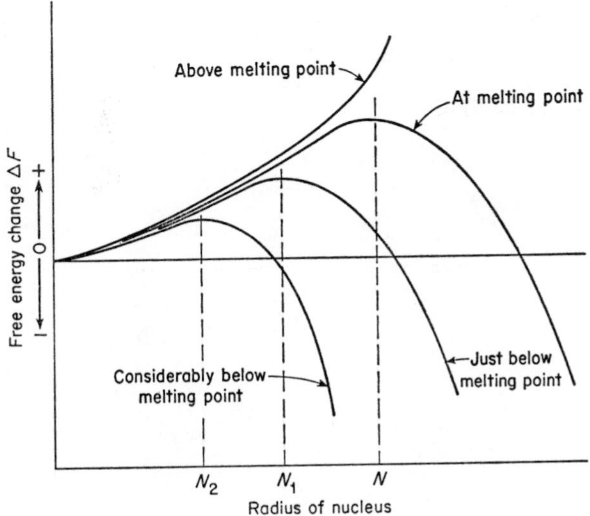

FIG. 2.8 Change of free energy vs. radius of nucleus for formation of a crystal lattice.

The solidification of a metal requires that a sufficient number of atoms shall exist in the proper arrangement to form a crystal that will grow. The minimum number of atoms required to form the nucleus of a crystal is not known with certainty. The number of atoms that will form a solid nucleus depends upon the decrease of energy when a liquid is replaced by solid, and the increase of energy resulting from the formation of an interface between liquid and solid phases. From thermodynamic considerations, the size of the nucleus required for solidification is indicated in Fig. 2.8. In this figure the change of energy of the system is plotted as a function of the radius of the nucleus. If a nucleus forms, there is an increase in energy; and if the nucleus is larger than that indicated by the maximum of each curve, the nucleus continues to grow. Under such conditions, nuclei should grow into finite crystals. If the temperature is just below the melting point, the radius of the nucleus can be smaller; if it is a size at least equal to "N_1," then it is certain that the crystal will grow. If the temperature is well below the melting point, the size of the nucleus can be very much smaller. It is obvious that the very large size of the nucleus demanded at the melting point is such that the formation of a solid at the true melting point might never occur. Nuclei of this critical size may form, but there is good likelihood that the atoms will disperse; hence solidification will never occur.

FIG. 2.9 Dendritic formation in magnesium.

The growth of a crystal possessing a cubic structure is in a direction normal to the cube planes in the three mutually perpendicular directions. This preferential growth leads to a skeletal type of development, which is called a *dendritic structure*. The dendritic pattern is not usually observable in a pure metal. Dendritic structures may be observed in alloys in which there is not complete solubility. Dendrites in some instances may be observed under certain growth conditions, as shown in Fig. 2.9.

When a pure metal is cooled to a temperature corresponding to the freezing point or slightly below, centers of crystallization consisting of space lattices appear. These lattices grow by the aggregation of more lattices about each center. This growth continues at the expense of the liquid. The lattice structures expand in the directions of the axes of the lattice until development is stopped by interference through contact with the vessel containing the liquid or by contact with adjacent growing grains. The resulting structure is made up of irregularly shaped *grains* in which the size and arrangement of the grains are influenced by the rate of cooling. Each of these grains is an individual single crystal, which is not permitted to develop its normal external symmetry because of interference from other grains or the vessel containing the metal.

A diagrammatic representation of the process of solidification, as originally given by Rosenhain, is shown in Fig. 2.10. In this diagram, the squares represent space lattices. In view (a), crystallization has begun at four cen-

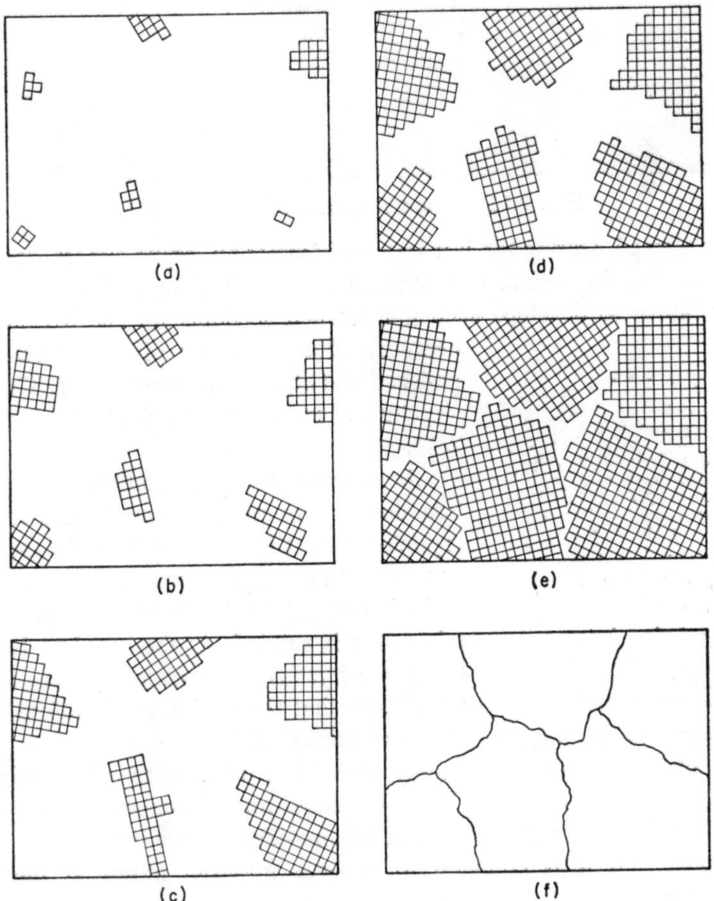

FIG. 2.10 Stages in the process of solidification of a metal. (After Rosenhain)

ters or nuclei. As crystallization continues, more centers appear and develop with space lattices of random orientation. Successive stages in the crystallization are shown in the illustration. Small crystals may join large ones, provided that they have about the same orientation, i.e., their axes are very

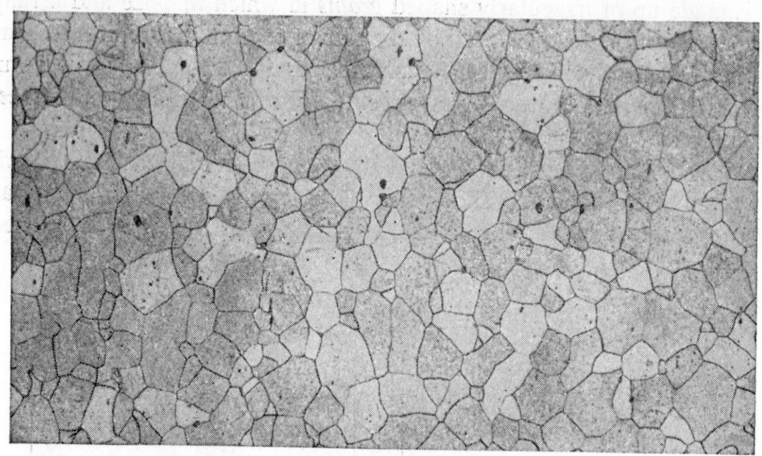

FIG. 2.11 Grain structure of iron (100×).

nearly aligned. During the last stages of formation, crystals meet, and there are places at the surface of intersection where it is impossible for another space lattice to develop. Such interference accounts for the irregular appearance of crystals in a piece of metal when polished and etched, as shown in (f). The grain structure of iron, for example, is shown in Fig. 2.11.

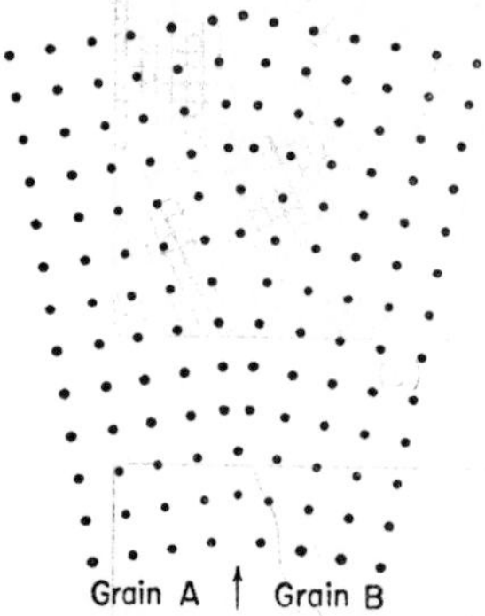

Grain A ↑ Grain B

FIG. 2.12 Grain boundary on lattice scale.

A schematic representation of a grain boundary, as it might appear in a metal that crystallizes in the cubic system, is shown in Fig. 2.12. In this figure, one of the axes of each grain is parallel to the other while one of the other axes is rotated. At the interface between these two crystals, the atomic distances are not correct, so that the atoms are therefore not in stable positions. These atoms are in a higher energy state. They will tend to adjust their positions to a condition of minimum energy. The lower density of atoms at the boundary provides a place for impurity atoms to reside, and it is common to observe impurity atoms at the grain boundaries.

If, during the growth process, the development of external features, such as regular faces, is prevented by interference from the growth of other nuclei, each unit is called a *grain* rather than a crystal. The term *crystal* is usually applied to a group of space lattices of the same orientation which shows symmetry by the development of regular faces. Each grain is essentially a single crystal. The size of the grain depends upon the temperature from which the metal is cast, the cooling rate, and the nature of the metal. Slow removal of heat at the solidification temperature will produce a few large grains, whereas the rapid removal of heat will produce many small grains.

Two parameters control the size of grains formed in the solidification of a metal: the rate of nucleation and the rate of growth. The first is the number of nuclei that form per unit of time and is dependent upon temperature. The second is the velocity of growth expressed in length per unit of time and is also dependent upon temperature. The rate of nucleation and the rate of growth as a function of temperature vary from metal to metal. In one example, this relationship may be shown in Fig. 2.13. By the slow

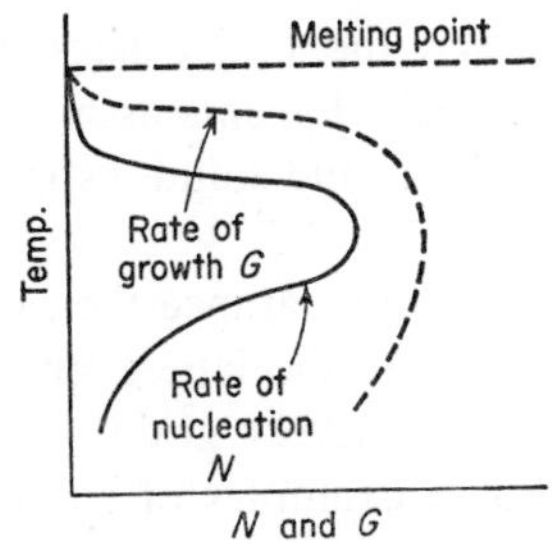

FIG. 2.13 Rate of growth and rate of nucleation vs. temperature, Case 1.

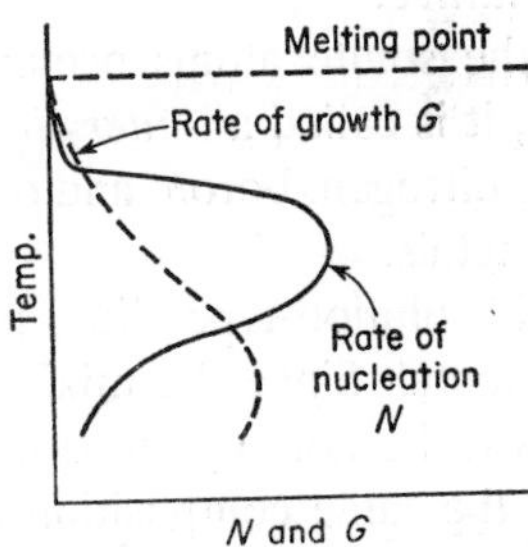

FIG. 2.14 Rate of growth and rate of nucleation vs. temperature, Case 2.

cooling of such a metal, crystallization will take place at a temperature near, but below, the equilibrium melting point. (At the melting point, equilibrium prevails, and the metal crystallizes and melts, since both solid and liquid can coexist.) In the metal illustrated, the grains will be small, since the rate of growth is not large compared with the rate of nucleation. If this metal is cooled more rapidly so that crystallization occurs at a temperature at which the rate of nucleation is very large compared with the rate of growth, the grain size may be very small. Much faster cooling in this event will produce a coarser grain size, since a point is reached where the rate of nucleation has decreased and the rate of growth has increased. The relation of the rate of nucleation and the rate of growth in another instance is illustrated in Fig. 2.14. In this instance, slow cooling will lead to a coarse grain, whereas more rapid cooling will yield finer grains. In most metals, the latter conditions prevail: rapid cooling leads to a finer grain structure.

2.7 Alloying. Metals in their pure form are seldom used in engineering applications. Most of the metallic materials used in engineering are combinations of metals known as alloys. An *alloy* is any combination of elements that results in a substance possessing metallic properties.

Elements may combine in different ways to form alloys. The components of alloys are usually completely soluble in the liquid state, i.e., they dissolve in each other in a manner similar to that of water and alcohol. In the solid state, however, the elements may form solid solutions, compounds, or mechanical mixtures.

2.8 Solid Solutions. Elements may combine to form alloys by completely dissolving in each other. The atoms of one element become a part of the space lattice of the other element, thus forming a solid solution. If the solute atoms take the place of some of the solvent atoms at lattice sites, it is called a *substitutional solid solution*. When nickel is added to copper, the nickel atoms will take the place of some of the copper atoms. If the solute atom takes a specific position in the lattice, then the structure is called a *superlattice*.

If the solute atoms occupy positions between the lattice sites of the solvent, it is called an *interstitial solid solution*. Such elements as hydrogen, carbon, nitrogen, boron, and oxygen form interstitial solid solutions in many of the metals.

Solid solution-type alloys present a microscopic appearance the same as a pure metal. Since the mixing of the elements is on the lattice, the alloy appears to be homogeneous on a microscopic scale. Every grain will have exactly the same composition and structure. Some examples of solid solution alloys are listed below:

1. Certain stainless steels (alloys of iron and chromium with or without nickel).
2. Monel metal (alloys of nickel and copper highly resistant to corrosion).
3. Sterling silver ($92\frac{1}{2}$ per cent silver, the remainder principally copper).

The solubility of one element in another is controlled, first, by the relative atomic diameters of solvent and solute atoms; second, by the relative electronegative valence of solvent and solute atoms; and third, by the relative valence of solvent and solute atoms. The formation of a solid solution is most favorable when the difference in the atomic diameters of the solute and solvent atoms is less than 14 or 15 per cent. If this difference is less than 14 or 15 per cent, extensive solubility may be expected; however, if the difference is greater, no solubility or, at most, limited solubility may be expected.

In addition to the size factor, one must consider the relative electronegative valence of solvent and solute atoms. The more electronegative the solute and the more electropositive the solvent, the greater is the tendency to produce a stable bond between solvent and solute atoms. When the size factor is favorable and one of the elements is strongly electronegative and the other strongly electropositive, it is most probable that a compound will be formed and not a continuous solid solution, or the solubility may be limited.

The extent to which one metal is soluble in another is related to the relative valence of the two elements. A metal of higher valence is more likely to be soluble in one of lower valence than the reverse. Under this condition, one would find that the solute atoms contribute toward an excess of valence electrons. An element of lower valence would leave a deficiency of electrons which would be unsatisfactory. With these criteria, one may, with fair surety, estimate the relative solubility of the elements.

In addition to these three fundamental rules of solubility, it is important to bear in mind that solubility tends to be favored when the crystal structure of the two metals is of the same type. This means that two metals, both having the face-centered cubic structure, are more likely to form a solid solution than two elements, one of which has a face-centered cubic and the other, a close-packed hexagonal structure. A few examples of alloys showing different degrees of solubility are indicated in Table 2-II.

The crystal structure of a solid solution is determined by the structure of the solvent. Some of the solvent atoms in the space lattice may be replaced by solute atoms. This is called a *substitutional solid solution.* However, the solute atoms may not replace any of the solvent atoms but may

TABLE 2-II. EXAMPLES OF SOLUBILITY IN ALLOYS.

System	Size Factor	Relative Electro-negativeness	Relative Valence	Structure	Relative Solubility
Ag-Au	Very favorable	Weak	Same	Both F.C.C.*	Continuous
Cu-Au	Borderline	Weak	Same	Both F.C.C.	Continuous (with a minimum)
Ag-Cu	Borderline	Weak	Same	Both F.C.C.	Restricted 8.8% Cu, 8.0% Ag
Ag-Zn	Favorable	Weak	Ag(1), Zn(2)	Ag(F.C.C.) Zn(C.P.Hex.)†	Restricted 37.8% Zn, 6.3% Ag
Ag-Sn	Favorable	Strong	Ag(1), Sn(4)	Ag(F.C.C.) Sn(Tetr.)	Restricted 12.2% Sn, 0.1% Ag; compounds formed
Cu-Cd	Unfavorable	Slight	Cu(1), Cd(2)	Ag(F.C.C.) Cd(C.P.Hex.)	Slight 1.7% Cd, 0.12% Cu; compounds formed

* Face-centered cubic.
† Close-packed hexagonal.

occupy positions between the solvent atoms, thus forming an *interstitial solid solution.* Both substitutional and interstitial solid solutions are found in common alloys. An example of an interstitial solid solution is the solid solution of carbon in gamma iron. As the number of solute atoms increases, the dimensions of the space lattice will be altered.

In general, the hardness or strength of a solid solution is greater than that of the solvent.

2.9 Metallic Compounds. Compounds formed between metals and nonmetals are commonplace and well known to all who are familiar with the fundamentals of chemistry. Examples of such compounds are salt (NaCl), iron oxide (Fe_2O_3), and silica (SiO_2). These chemical compounds are nonmetallic in character. Compounds formed between two metals and between metals and metalloids, however, possess metallic characteristics as shown by their great hardness and brittleness.

The metallic compounds differ in a remarkable respect from the ordinary chemical compounds in the apparent disregard of all rules of valence. For example, copper would appear to have a valence of three when combining with zinc, two-thirds when combining with tin, six when combining with aluminum. Some of the common, more important metallic compounds include: Cu_2Zn_3, which is the chief constituent of hard white brass; Cu_3Sn, which is a hard constituent of high-tin bronze; SnSb, a hard, wear-resisting crystal which is found in tin-base bearing metal.

Some of the compounds formed between metals and metalloids include Fe_3C, Mn_3C, Cr_4C, WC, W_2C, and Mg_2Si. The hard metallic compounds by themselves find some application where wear or abrasion resistance is desired. However, the presence of intermetallic compounds in alloys is usually of greatest importance when they are properly dispersed in the structure. When compounds are in the form of large particles in the structure, they are of relatively little value except when in sufficient abundance to improve abrasion resistance. The mechanical properties of many alloys can be altered markedly by changing the degree of dispersion of intermetallic compounds. This characteristic of the intermetallic compounds will be illustrated in the discussion of the heat treatment of steel and of precipitation hardening in later chapters.

The ratio of the number of free electrons to the number of atoms is the same in some compounds. These materials, called *electron compounds,* exhibit similar characteristics. Some of the electron compounds that possess similar characteristics are shown in Table 2-III. The compounds in each column have the same ratio of number of electrons to number of atoms. Some of the important intermetallic compounds include Cu_2Zn_3, which has already been cited.

TABLE 2-III. SOME ELECTRON COMPOUNDS

Electrons: Atoms = 3:2			Electrons: Atoms = 21:13	Electrons: Atoms = 7:4
Body-Centered Cubic	Complex Cubic	Close-Packed Hex.	"γ-Brass" Structure	Close-Packed Hex.
CuBe	Cu_5Si	Cu_3Ga	Cu_5Zn_8	$CuZn_3$
CuZn	AgHg	AgZn	Cu_9Al_4	$AgZn_3$
Cu_3Al	Ag_3Al	Ag_3Al	$Cu_{31}Si_8$	Ag_5Al_3
Cu_5Sn	$CoZn_3$	Ag_5Sn	Ag_5Cd_8	Au_3Sn
AgMg		Ag_7Sb	Au_5Zn_8	
AuMg		Au_5Sn	Mn_5Zn_{21}	
FeAl			Ni_5Zn_{21}	
NiIn			Pt_5Be_{21}	
			$Na_{31}Pb_8$	

2.10 Electron and Zone Theories. Metallic vapors are generally monatomic and are composed of free atoms with essentially no interaction between the atoms in the vapor. In the liquid and solid states, however, this is not the case. Here the atoms are bound together and therefore affect each other. The differences in the behavior of materials may be understood better if one has a knowledge of the characteristics of the electronic bonds in them. Many of the properties of metals cannot be adequately described in terms of the simple pairwise bonding forces.

The valence electrons in solids do not belong to particular atoms. They are shared by all the atoms of which the crystal is composed. An electron that belongs to a single atom has only certain permissible energy levels. When atoms are combined to form a solid, the number of permissible energy levels for shared electrons increases. The range of energy levels available in a crystal includes "allowed" and "forbidden" bands. Some aspects of this theory are presented here because the theory provides a qualitative explanation of some of the characteristics of crystals.

In some of the early theories, Drude and Lorentz considered that only the outer electrons are of importance in the metals, and therefore all of the atom, except the valence electrons, was neglected. The theory pictured a volume of electrons as forming an electron gas which is free to move through the entire volume of metal. The total energy of the electrons has been taken as the kinetic energy for an ideal gas, plus this potential energy in relation to a free electron. This energy, however, gives an over-estimate of the electronic specific heat and does not conform to experimental data. Furthermore, the relations developed give the wrong dependency of electrical conductivity on temperature. However, this theory did leave an important consequence, namely the concept of an electron cloud within the metal.

The work of Drude and Lorentz was modified by Sommerfeld who introduced the quantum mechanical concepts. The electrons in this theory were still assumed to be free in the electric field but with no interference from the ions present or from any interaction with each other. In other words, the concept was retained that a uniform potential exists within the metal. However, Sommerfeld used the Schroedinger equation to calculate the total energy. This led to the determination of the permissible energy values that the valence electrons might have. The number of possible states was found to increase rapidly with increasing energy. This may be represented by the relationship shown in Fig. 2.15.

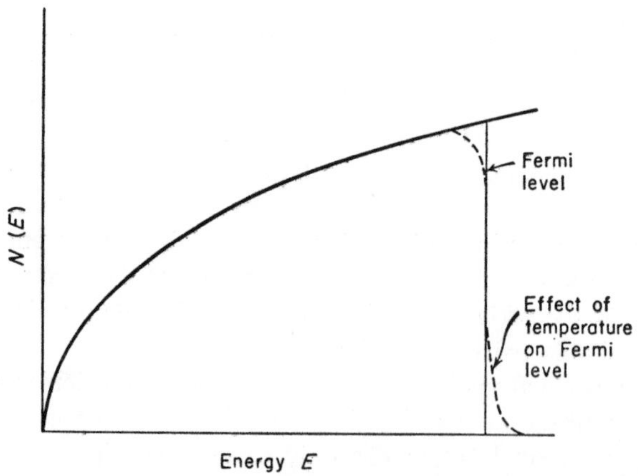

FIG. 2.15 Distribution of energy states as a function of the energy E.

The valence electrons always tend to occupy the lowest energy states that are available. The value of the energy level that divides the occupied from the unoccupied states is designated as the *Fermi level,* as indicated in Fig. 2.15. The type of relation shown in this figure holds for a temperature of 0 K. There is only a very slight effect of temperature on the Fermi level, as indicated by the dashed line, which is considerably exaggerated. The development of the theory to this extent allowed an explanation of the conduction of electricity through a metal. In the absence of an applied voltage the electrons move randomly. Any electrons moving in one direction are canceled by electrons moving in the opposite direction. Under an applied voltage, only electrons near the Fermi level experience a net change in energy; they then contribute to motion in a preferred direction which therefore constitutes a current. This theory does not account for nonconduction in some substances in which free electrons exist. This difficulty was removed when the concept of the periodic potential field in which the electrons move was introduced. This modification led to the *zone* or *band theory* of solids.

The Drude-Lorentz and the Sommerfeld theories neglected the effect of the ion cores. The potential energy for a one-dimensional array of ions is indicated in Fig. 2.16. Certainly the potential energy is 0 between the

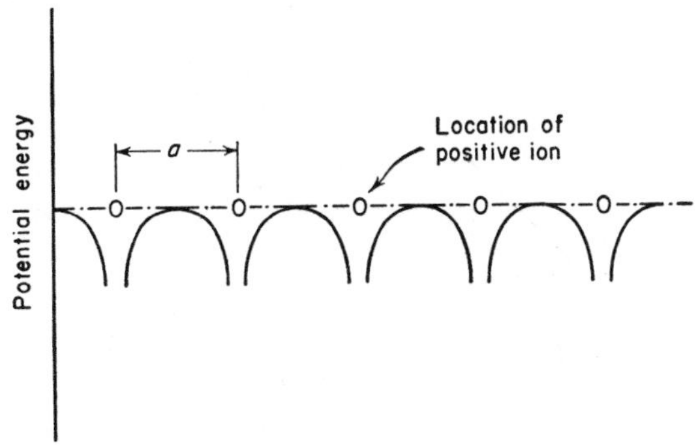

FIG. 2.16 Variation of potential energy of an electron moving on a one-dimensional array of positive ions.

ions. This was also true in the Sommerfeld theory, but around the ion the potential energy is very strongly negative. Again, the Schroedinger equation can be employed to obtain the energies of the electronic states under conditions of a periodic potential energy.

In view of the fact that the electrons have wave properties, it is possible for the electrons to be diffracted by a periodic grating such as the array of atoms in a crystal. Electrons of low energy possess a long wavelength compared with the atomic spacings. These electrons, of course, move through the crystal undisturbed. The electrons with high energy, however, have wavelengths very nearly equal to the atomic spacing; these electrons can be diffracted according to the *Bragg law,* which is

$$\lambda = 2a \sin \theta$$

where λ is the wavelength, a is the atomic spacing, and θ is the diffraction or Bragg angle. This expression can be put in terms of wave numbers where the wave number

$$k = \frac{2\pi}{\lambda}$$

This may be used in place of λ as a measure of energy. Then the relation would be

$$k = \frac{\pi}{a \sin \theta}$$

From the above relationship the necessary conditions for diffraction of electrons can be determined, and this is indicated in what is known as a *Brillouin zone*. If one considers a two-dimensional structure, i.e., a square lattice, one secures the energy levels in the Brillouin zones indicated in Fig. 2.17. In this figure, the shaded region is the first Brillouin zone which

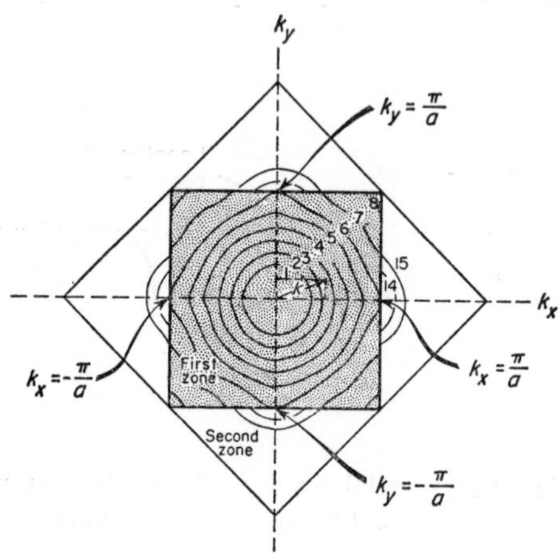

FIG. 2.17 Energy levels in the first Brillouin zone for a square lattice.

is considered to be filled, and the second zone is empty with two possible energy levels indicated. A point in this diagram represents an electron moving in the direction of the arrow and having an energy described by the wave number whose components are k_x and k_y.

The energy of free electrons considered in the Sommerfeld theory appears in the Brillouin zones as circles. However, in treating these energies by the zone theory, the energy contours become somewhat distorted circles, particularly in the vicinity of the boundaries of the Brillouin zones. At the boundary, there is a sudden increase as the boundary is crossed. This is most readily shown in a curve which presents the number of energy states, $N(E)$, as a function of the energy of the electronic states. This is illustrated in Fig. 2.18. The curve for the free electron theory is compared with the zone theory.

For a nonconductor, the first Brillouin zone is completely filled, as indicated in Fig. 2.19. This condition serves for nonconductors because it fulfills two conditions. First, there are just enough valence electrons to completely fill one or more of the zones. Generally, the number of states in each zone corresponds to two electrons per atom. It should be clear that any substance with an even number of valence electrons might be an insulator, although this is not always true. Second, higher unoccupied zones must always

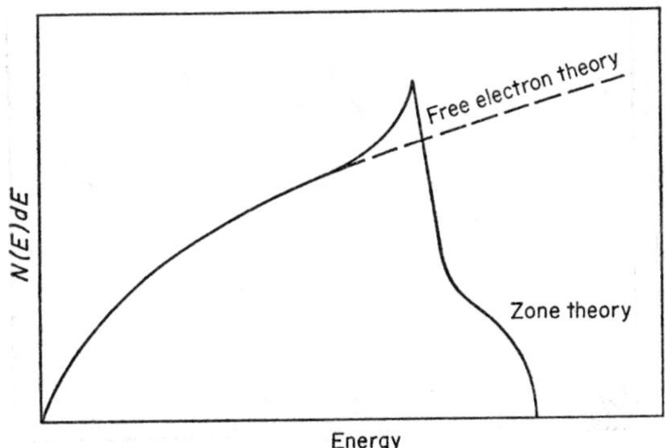

FIG. 2.18 Density of energy states vs. energy for the free electron theory and for the zone theory.

be separated from a lower filled zone by an energy gap. It is not possible, by the application of a voltage, to move an electron across this energy gap; therefore no current can flow. It should therefore be clear that the zone theory permits one to make a reasonable distinction between conductors and nonconductors.

In some instances the second Brillouin zone may overlap the first zone. In this case, higher energy electrons can be moved to other levels in the second zone and there will be conduction, as shown in Fig. 2.20.

There is a group of materials in which the energy gap between the filled and unfilled zones is sufficiently small so that electrons may be excited by thermal energy to move from the filled zones to empty zones, thus providing some current flow. Such materials are called *semiconductors.* At absolute zero temperature, a semiconductor would be a nonconductor since

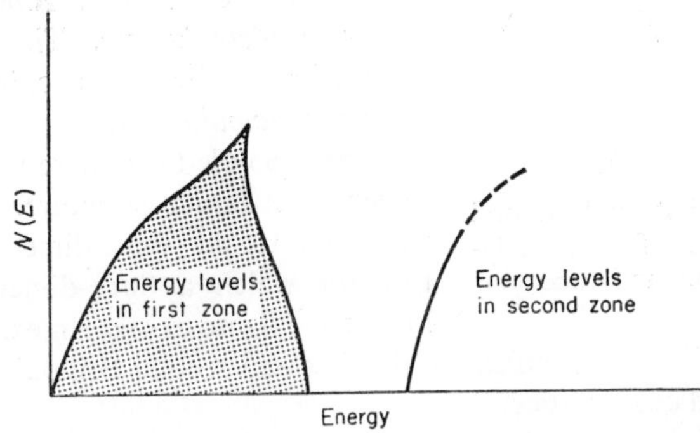

FIG. 2.19 Energy states vs. energy for an insulator corresponding to the Brillouin zones of Fig. 2.17. Shaded area indicates just sufficient valence electrons to fill the first zone.

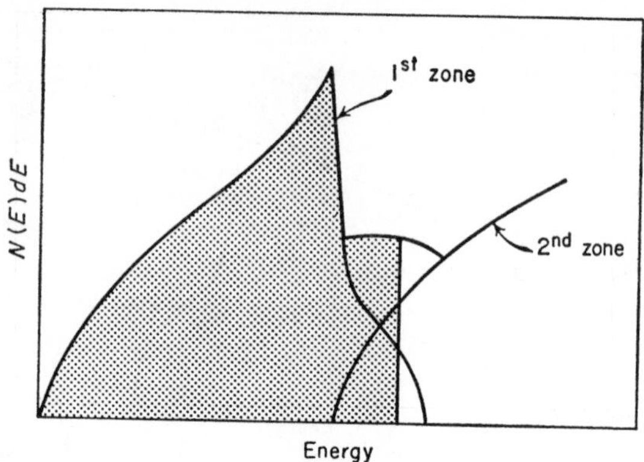

FIG. 2.20 Density of energy states vs. energy for a filled first zone overlapping a second zone.

there would be no thermal energy available to excite any electrons across the gap. There will be a few vacant states at the upper end of the normally filled zone, and a few electrons will occupy states near the lower end of the normally empty zone at finite temperatures. The vacant states behave as positive charges and are called "holes." Some impurity atoms may be introduced to provide extra electrons that enter the conduction zone, thus providing conduction by electrons only. Some impurities may hold electrons permitting their excitation across the gap; then conduction may occur by holes only.

A semiconductor in which current flows principally as a result of thermal excitation of electrons across the forbidden gap is called an *intrinsic semiconductor*. Semiconductors containing impurities to give conduction by electrons only in the conduction zone are called *n-type semiconductors*. Those in which conduction is by holes only are called *p-type semiconductors*.

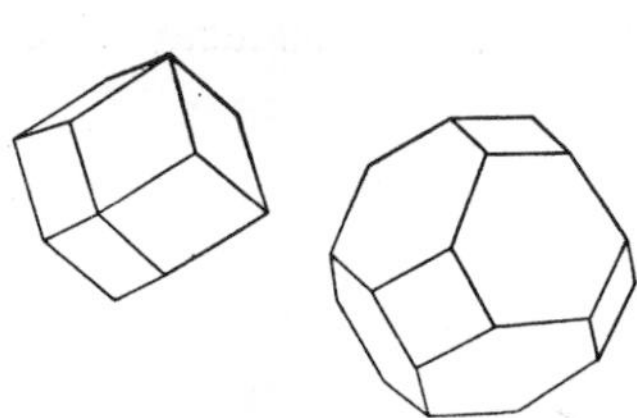

FIG. 2.21 First and second Brillouin zones for a body-centered cubic lattice.

The example taken here was for a two-dimensional lattice structure. This can be expanded to the three-dimensional structure for which a three-dimensional Brillouin zone figure exists. An example of this is shown in Fig. 2.21, indicating the first and second Brillouin zones for a body-centered cubic lattice.

The zone structures are approximate models of electronic energies. They provide a guide by means of which certain phenomena may be explained in solid state problems. Lattice imperfections or the presence of foreign atoms may modify the interpretations. The references should be consulted for further details on zone theory.

REFERENCES

Hume-Rothery, *Atomic Theory for Students of Metallurgy,* The Institute of Metals, London, 1960.

Hume-Rothery and Raynor, *The Structure of Metals and Alloys,* The Institute of Metals, London, 1954.

Jones, *The Theory of Brillouin Zones and Electronic States in Crystals,* North-Holland Publishing Co., Amsterdam, 1960.

Sinnott, *The Solid State for Engineers,* John Wiley and Sons, New York, 1958.

Van Vlack, *Elements of Materials Science,* Addison-Wesley Publishing Co., Reading, Mass., 1959.

QUESTIONS

1. What is an atom?
2. What are the principal particles of which an atom is composed?
3. What determines the charge on the nucleus of an atom and therefore determines the identity of the element?
4. What is the meaning of the atomic number of an atom?
5. What are isotopes?
6. What is the meaning of the mass number of a nucleus?
7. What is the significance of the electrons in an atom?
8. Describe the concept of electronic orbits in an atom.
9. What two quantum numbers characterize electronic states of atoms?
10. How is the state of any electron in an atom described?
11. What is the Pauli exclusion principle?
12. What is the significance of a valency group of electrons in an atom?
13. What is the special condition of the electrons in the inert gases?
14. What is the peculiar situation with respect to electronic states of the transition elements?
15. Describe the significance of the periodic table of elements.
16. Of what particular value is the periodic table to the materials engineer?
17. Name the four types of atomic bonds observed in materials.
18. Indicate the distinguishing features characterizing each of the four types of atomic bonds.
19. Define a metal.
20. What is a space lattice?
21. Explain the packing of atoms to form the hexagonal, close-packed, space lattice structure.
22. Explain the packing of atoms to form the face-centered cubic structure.
23. Describe the body-centered cubic lattice structure.
24. What is the meaning of "allotropic"?
25. Define a crystalline substance.
26. Describe the process of the crystallization of a metal.

27. What is a dendritic structure; how is it formed?
28. What is a grain?
29. Describe the process of solidification of a metal to form a polycrystalline structure.
30. What two parameters control the size of grains formed in the solidification of a metal?
31. Describe the conditions that will lead to coarse grains upon solidification.
32. What conditions will lead to a fine grain upon solidification?
33. Describe the atomic situation that exists at the grain boundaries of a metal.
34. Define an alloy.
35. What is a solid solution?
36. What two types of solid solutions may be found in metallic systems?
37. What factors control the solubility of one element in another?
38. What are the conditions most favorable to the formation of solid solutions?
39. What conditions are most favorable to the formation of compounds?
40. What condition determines the extent to which one metal is soluble in another?
41. What is a metallic compound? Give an example.
42. What are electron compounds?
43. What is the meaning and significance of the Fermi level?
44. Explain briefly the mechanism of electrical conductivity in metals.
45. How may the necessary conditions for defraction of electrons be determined?
46. What is the meaning and significance of the Brillouin zone?
47. What is the condition of the first Brillouin zone for a nonconductor?
48. Why does the zone theory of metals permit a distinction between conductors and nonconductors?
49. What is the situation with respect to the Brillouin zones for a conductor?
50. What is the condition of the Brillouin zones for semiconductors?
51. What is an intrinsic semiconductor?
52. What is an *n*-type semiconductor?

III

METALS AND ALLOY SYSTEMS

3.1 Metals. The features that distinguish the metallic elements from the nonmetallic elements have been discussed in Chapter II. In a qualitative sense, a *metal* is defined as a chemical element that is lustrous, hard, malleable, heavy, ductile, tenacious, and is a conductor of heat and electricity. A more rigorous identification of a metal was indicated in Chapter II.

Of the 98 elements, 73 are classified as metals. A list of the metallic elements, excluding the rare earth metals, is included in Table 3-I. Elements such as oxygen, chlorine, iodine, bromine, hydrogen, and the inert gases—helium, neon, argon, krypton, xenon, and radon—are considered *nonmetallic*. There is, however, a group of elements, including carbon, sulfur, silicon, and phosphorus, which is intermediate between the metals and nonmetals. Under certain circumstances, these elements portray the characteristics of nonmetals. They are referred to as *metalloids* and are included in Table 3-I.

The most widely used metallic elements include iron, copper, lead, zinc, aluminum, tin, nickel, and magnesium. Some of these are used extensively in the pure state, but by far the largest amount is consumed in the form of alloys. An *alloy* is a combination of elements, the combination of which exhibits the properties of a metal. Metals may be divided according to their characteristics as the *heavy metals* (iron, copper, nickel, etc.), *light metals* (aluminum and magnesium), *white metals* (lead, tin, zinc, cadmium, etc.), and *precious metals* (gold, platinum, palladium, silver, etc.).

The commercial importance of metals lies principally in their ability to combine with one another to produce substances having a wide range of properties that may not be attained with pure metals alone. There are many instances, however, in which the characteristics of a pure metal make it superior to an alloy for a specific application. An example is the use of pure copper for electrical conductors.

3.2 Systems, Phases, Structural Constituents. In considering the relation of the constituents that make up alloys at different temperatures, pres-

TABLE 3-I. METALLIC ELEMENTS AND METALLOIDS.

(Excluding rare earth metals)

Element	Symbol	Atomic No.	Atomic Wt.	Melting Point (°F)	Melting Point (°C)	Crystal Structure[a]	Lattice Constant (kX Units)[b]	Specific Gravity
Actinium	Ac	89	227.05	(2900)	(1600)	—	—	—
Aluminum	Al	13	26.97	1220.4	660.2	F.C.C.	a = 4.0408	2.699
*Americium	Am	95	241	—	—	—	—	—
Antimony	Sb	51	121.76	1166.9	630.5	Rhomb.	a = 4.4974 < = 57°6.5′	6.62
Arsenic	As	33	79.91	(1497)	(814)	Rhomb.	a = 4.151 < = 53°49′	5.73
***Barium	Ba	56	137.36	1300	704	B.C.C.	a = 5.015	3.5
Berkelium	Bk	97	—	—	—	—	—	—
***Beryllium	Be	4	9.02	2340	1280	C.P.Hex.	a = 2.2810 c = 3.5771	1.82
Bismuth	Bi	83	209.00	520.3	271.3	Rhomb.	a = 4.7361 < = 57°14.2′	9.80
Boron	B	5	10.82	4200	2300	Orth. (?)	a = 17.86 b = 8.93 c = 10.13	2.3
Cadmium	Cd	48	112.41	609.6	320.9	C.P.Hex.	a = 2.9727 c = 5.606	8.65
***Calcium	Ca	20	40.08	1560	850	F.C.C.	a = 5.56	1.55
Californium	Cf	98	—	—	—	—	—	—
Carbon	C	6	12.010	6700	3700	Hex.	a = 2.4565 c = 6.6906	2.22
**Cesium	Cs	55	132.91	82	28	B.C.C.	a = 6.05	1.9
Chromium	Cr	24	52.01	3430	1890	B.C.C.	a = 2.8787	7.19
Cobalt	Co	27	58.94	2723	1495	C.P.Hex.	a = 2.502 c = 4.061	8.9
Columbium	Cb	41	92.91	4380	2415	B.C.C.	a = 3.2941	8.57
Copper	Cu	29	63.54	1981.4	1083.0	F.C.C.	a = 3.6080	8.96
*Curium	Cm	96	242	—	—	—	—	—
**Francium	Fr	87	223	—	—	—	—	—
Gallium	Ga	31	69.72	85.60	29.78	Orth.- 1 F.C.	a = 4.517 b = 4.511 c = 7.645	5.91
Germanium	Ge	32	72.60	1760	958	Dia. Cube	a = 5.647	5.36
Gold	Au	79	197.2	1945.4	1063.0	F.C.C.	a = 4.0701	19.32
Hafnium	Hf	72	178.6	(3100)	(1700)	C.P.Hex.	a = 3.200 c = 5.077	11.4
Indium	In	49	114.76	313.5	156.4	F.C.T.	a = 4.585 c = 4.941	7.31
Iridium	Ir	77	193.1	4449	2454	F.C.C.	a = 3.8312	22.5
Iron	Fe	26	55.85	2802	1539	B.C.C.	a = 2.8606	7.87
Lanthanum	La	57	138.92	1519	826	C.P.Hex.	a = 3.754 c = 6.063	6.15
Lead	Pb	82	207.21	621.3	327.4	F.C.C.	a = 4.9395	11.34
**Lithium	Li	3	6.940	367	186	B.C.C.	a = 3.5019	0.53
***Magnesium	Mg	12	24.32	1202	650	C.P.Hex.	a = 3.2028 c = 5.1998	1.74
Manganese	Mn	25	54.93	2273	1245	Cub. Complex	a = 8.894	7.43
Mercury	Hg	80	200.61	−37.97	−38.87	Rhomb.	a = 2.999 < = 70°31.7′	13.55
Molybdenum	Mo	42	95.95	4760	2625	B.C.C.	a = 3.140	10.2
*Neptunium	Np	93	237	—	—	—	—	—
Nickel	Ni	28	58.69	2651	1455	F.C.C.	a = 3.5167	8.90
Osmium	Os	76	190.2	4900	2700	C.P.Hex.	a = 2.7298 c = 4.3104	22.5
Palladium	Pd	46	106.7	2829	1554	F.C.C.	a = 3.8824	12.0
Phosphorus	P	15	30.98	111.4	44.1	Cubic	a = 7.17	1.82
Platinum	Pt	78	195.23	3224.3	1773.5	F.C.C.	a = 3.9158	21.45
*Plutonium	Pu	94	239	—	—	—	—	—
Polonium	Po	84	210	(1100)	(600)	—	—	—
**Potassium	K	19	39.096	145	63	B.C.C.	a = 5.333	0.86
*Protactinium	Pa	91	231	(5400)	(3000)	—	—	—

TABLE 3-I. (*Continued*)

Element	Sym-bol	Atomic No.	Atomic Wt.	Melting Point (°F)	(°C)	Crystal Structure[a]	Lattice Constant (kX Units)[b]	Specific Gravity
***Radium	Ra	88	226.05	1300	700	—	—	5.0
Rhenium	Re	75	186.31	5740	3170	C.P.Hex.	$a = 2.7553$ $c = 4.493$	20
Rhodium	Rh	45	102.91	3571	1966	F.C.C.	$a = 3.7957$	12.44
**Rubidium	Rb	37	85.48	102	39	B.C.C.	$a = 5.62$	1.53
Ruthenium	Ru	44	101.7	4500	2500	C.P.Hex.	$a = 2.6984$ $c = 4.2730$	12.2
Scandium	Sc	21	45.10	2190	1200	F.C.C.	$a = 4.532$	2.5
Selenium	Se	34	78.96	428	220	Hex.	$a = 4.3552$ $c = 4.9494$	4.81
Silicon	Si	14	28.06	2605	1430	Dia. Cube	$a = 5.4173$	2.33
Silver	Ag	47	107.880	1760.9	960.5	F.C.C.	$a = 4.0774$	10.49
**Sodium	Na	11	22.997	207.9	97.7	B.C.C.	$a = 4.2820$	0.97
***Strontium	Sr	38	87.63	1420	770	F.C.C.	$a = 6.075$	2.6
Sulfur	S	16	32.066	(246.2)	(119.0)	F.C.Ortho.	$a = 10.48$ $b = 12.92$ $c = 24.55$	2.07
Tantalum	Ta	73	180.88	5425	2996	B.C.C.	$a = 3.2959$	16.6
Technetium	Tc	43	99	(4900)	(2700)	—	—	—
Tellurium	Te	52	127.61	840	450	Hex.	$a = 4.4469$ $c = 5.9149$	6.24
Thallium	Tl	81	204.39	572	300	C.P.Hex.	$a = 3.450$ $c = 5.514$	11.85
*Thorium	Th	90	232.12	3300	1800	F.C.C.	$a = 5.077$	11.5
Tin	Sn	50	118.70	449.4	231.9	B.C.T.	$a = 5.8194$ $c = 3.1753$	7.298
Titanium	Ti	22	47.90	3300	1820	C.P.Hex.	$a = 2.953$ $c = 4.729$	4.54
Tungsten	W	74	183.92	6170	3410	B.C.C.	$a = 3.1585$	19.3
*Uranium	U	92	238.07	2065	1130	Ortho.	$a = 2.852$ $b = 3.865$ $c = 4.945$	18.7
Vanadium	V	23	50.95	3150	1735	B.C.C.	$a = 3.033$	6.0
Yttrium	Y	39	88.92	2700	1490	C.P.Hex.	$a = 3.663$ $c = 5.841$	5.51
Zinc	Zn	30	65.38	787.03	419.46	C.P.Hex.	$a = 2.659$ $c = 4.935$	7.133
Zirconium	Zr	40	91.22	3200	1750	C.P.Hex.	$a = 3.223$ $c = 5.123$	6.5

[a] Crystal structure is for room temperature; other forms may exist at other temperatures.
[b] 1 kX unit = 1.00202 Å.
* Uranium metals.
** Alkaline metals.
*** Alkaline earth metals.

sures, and concentrations, one specifies the two or more metallic elements that are being considered and the variables to which these elements will be subjected. With these specifications, an alloy system is established. The term *system* has a wider application in that it refers to any portion of objective space within specified boundaries subject to specified variables. An *alloy system* is a combination of two or more elements forming alloys which are considered within a specified range of temperature, pressure, and concentration.

Systems are classified according to the number of components that constitute the system. A *component* is a unit of the composition variable of the system. If there are two components, it is a binary system; if three, a ternary;

if four, a quaternary system, etc. Even though compounds are formed in the system, the number of components remains the same. However, the components may be selected so that one is a compound, an example of which is the iron-iron carbide system.

In a particular alloy system, the components may combine within a certain temperature range to form two homogeneous coexisting portions. Each of these portions will be of different composition and will have different properties. However, each portion is homogeneous throughout, and, regardless of where a sample is taken in that portion, the composition will be the same. These homogeneous, physically distinct portions of the system are called *phases.* For example, a liquid may exist in equilibrium with a solid solution, or two solid solutions of different compositions may exist together. In each instance, two phases are present.

Many phases exist in some alloys. Gray cast iron, for example, is made up of the following phases in order of decreasing amounts: (1) a solid solution phase of iron containing some carbon, silicon, and phosphorus; (2) a soft metalloid, graphite (carbon); (3) a hard, brittle compound, Fe_3C (cementite); (4) a compound of the impurity sulfur, usually in the form MnS; (5) a compound of the impurity, phosphorus, present as Fe_3P and found with iron as a mixture called steadite; (6) a very small amount of nonmetallic compounds.

The phases in an alloy are not necessarily uniformly distributed throughout the structure. There are certain ways in which these phases may be associated to form the structure. The association of phases in a recognizably distinct fashion may be referred to as a *structural constituent* of the alloy. A *eutectic,* for example, is an intimate mechanical mixture of two or more phases having a definite melting point and a definite composition. An alloy may be composed of the eutectic as one structural constituent and one phase not intimately associated with the other phase as a second structural constituent. An example of a eutectic structure is shown in Fig. 3.1. The interrelation and distinction between phases and structural constituents will be more apparent in subsequent articles of this chapter.

3.3 Thermodynamic Considerations. One is usually concerned with the equilibrium between components in a system. Seldom is a system in a true state of equilibrium. However, it is important to know the limiting conditions that would prevail in the case of true equilibrium. In many instances the practical cases are so close to equilibrium that the conditions of ideal equilibrium may be utilized.

A system is in a state of *equilibrium* when the free energy of that system is at a minimum. The *free energy* may be defined in two ways. The first is called the Helmholtz free energy: it is given by the expression

$$F = U - TS$$

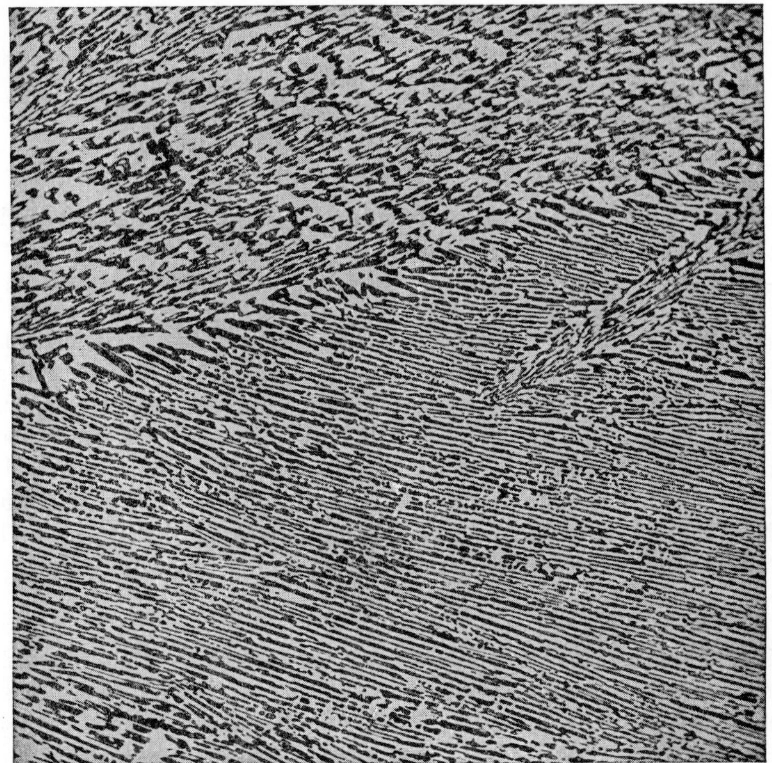

FIG. 3.1 Structure of eutectic of 40% cadmium and 60% bismuth (250×).

where U is the total energy of a system of volume V at pressure P and temperature T and S is the entropy of the system. Entropy is defined as $\int dQ/T$, where dQ is the heat transfer in an infinitesimal cycle and T is the absolute temperature. The second is the Gibbs free energy which is given by the relation

$$G = U + PV - TS$$

When one is concerned with the solid state, changes in volume are relatively small, and, usually, pressures are relatively low so that small changes in the Helmholtz free energy are about equal to small changes in the Gibbs free energy.

A relationship between the number of coexisting phases, the number of components making up the system, and the number of variables was developed by Willard Gibbs. The relation is a statement of the *phase rule,* which in its most general form is expressed as

$$p - c = V - F$$

where

p = the number of phases in equilibrium,
c = the number of components in the system,
V = the variables, temperature and pressure,
F = the number of degrees of freedom.

If the pressure is constant, as it is in most metallurgical considerations, then the phase rule may be stated as

$$p - c = 1 - F$$

The phase rule is of particular value in checking the results of experimental studies of phase equilibria. For example, in a system of lead and tin (two components), one may observe that the temperature remains constant at some period during the cooling of an alloy of a specified composition in this system. Under what conditions can this situation prevail? Since the temperature remains constant for a period of time during the removal of heat, the system certainly has no degree of freedom, that is, $F = 0$. The phase rule shows that in this case

$$p = 1 - F + c = 1 - 0 + 2 = 3$$

This means that in a two-component system in which there are no degrees of freedom, three phases will exist in equilibrium and the compositions of each of the three phases are fixed. This condition prevails when a eutectic is formed.

When two phases are in equilibrium in a two-component system,

$$F = 1 - p + c = 1$$

This means that the state of the system may be established by specifying one of the variables—the temperature or the composition of one of the two phases. If the composition of one of the phases is specified, then the temperature at which this phase is in equilibrium with the other phase is established. Furthermore, the composition of the second phase is established at the same time. The application of the phase rule in other sections of this chapter will help to clarify the use of the phase rule.

3.4 Equilibrium Diagram. The study of the interrelation of phases in an alloy system at different temperatures and for different alloy compositions is of importance in understanding the characteristics of alloys. It would indeed be a difficult and cumbersome task to tabulate the interrelation and composition of coexisting phases at all temperatures in an alloy system. Therefore, this information is given in the form of a diagram. In most cases, these relations are given for equilibrium conditions. The diagram shows the composition of the phases that are in equilibrium at all temperatures and is referred to as an *equilibrium* or *phase diagram.* Some prefer to employ the name *constitution diagram,* but it is probably well to apply this latter term to diagrams that show the phase relations under special circumstances of quasi equilibrium.

An *equilibrium diagram* may be defined as a plot of the composition of phases as a function of temperature in any alloy system under equilibrium conditions.

Equilibrium diagrams may be classified according to the relation of the components in the liquid and solid states as follows:

1. Components completely soluble in the liquid state.
   (a) Completely soluble in the solid state.
   (b) Partly soluble in the solid state.
   (c) Insoluble in the solid state.
2. Components partially soluble in the liquid state.
   (a) Completely soluble in the solid state.
   (b) Partly soluble in the solid state.
3. Components completely insoluble in the liquid state.
   (a) Completely insoluble in the solid state.

The first type of equilibrium diagram in which the components are completely soluble in the liquid state is most common in engineering alloys and will be given the most consideration in the following discussion.

3.5 Cooling Curves. One method of determining the temperatures at which phase changes occur in a system consists of following the temperature as a function of time as different alloys in the system are very slowly cooled. The data obtained in this manner form a cooling curve for each of the alloys. This method is particularly useful in studying the changes that occur during the solidification of alloys and, in some instances, may be used to advantage in determining transformations subsequent to solidification. However, in the latter case, the method may not be sufficiently sensitive; hence, other procedures may be employed.

If the equilibrium diagram of the antimony-bismuth system is to be determined, a series of alloys containing different amounts of antimony and bismuth will be made. Each sample is heated until molten and uniform in composition, and then it is allowed to cool very slowly. A thermocouple may be employed to determine the temperature as cooling proceeds.

Consider first a sample of pure antimony cooling slowly from the molten state. The cooling curve for this alloy is shown at (a) in Fig. 3.2. A *hold* occurs in the cooling curve beginning at x and ending at y. Examination of the metal in the crucible will indicate that solidification begins at the time corresponding to point x and is completed at the time corresponding to point y, i.e., solidification occurs at the constant temperature T_{Sb} (1167 F). The temperature remains constant during the process of solidification because of the liberation of heat of fusion.

Consider an alloy of antimony and bismuth containing 25 per cent bismuth and 75 per cent antimony. This homogeneous molten alloy is allowed to cool very slowly, and observations of temperature and time are made. The cooling curve for this alloy is shown at (b) in Fig. 3.2. Cooling continues in a uniform manner until the temperature reaches the value of 1095 F at which a *break* in the curve is encountered. At this temperature, the rate of cooling decreases. Examination will show that solidification begins at the break, or point x, and continues until the temperature corre-

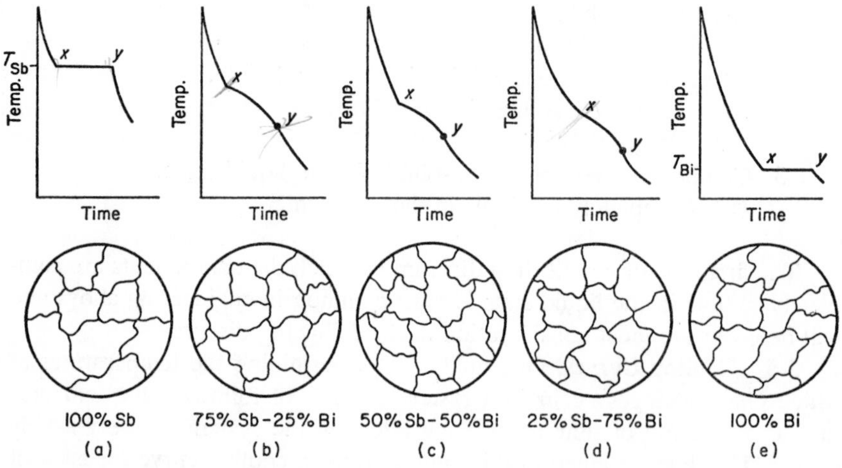

FIG. 3.2 Cooling curves for antimony-bismuth system.

sponding to point y, 800 F, is reached. At point y, the cooling rate changes again. This point y is designated as a *recovery*. Solidification occurs in this case over a range of temperature. The structure of the solid material at temperatures below 800 F consists of homogeneous grains, i.e., all grains are exactly of the same composition. It is a one-phase alloy, namely, a solid solution of antimony and bismuth.

Consider an alloy containing 50 per cent antimony and 50 per cent bismuth. These constituents are melted in a crucible and allowed to cool slowly. The cooling curve of this alloy is shown at (c) in Fig. 3.2. The break x (980 F), and the recovery, y (640 F), occur at lower temperatures than for the previous alloy. Examination of the structure of the solidified alloy will reveal that it consists of only one phase, namely, a solid solution of bismuth and antimony.

An alloy containing 75 per cent bismuth and 25 per cent antimony will produce a cooling curve as shown at (d) in Fig. 3.2. This curve has characteristics similar to those found in the preceding two cases. The break, *x* (800 F), and the recovery, *y* (560 F), occur at lower temperatures than in the preceding cases. The structure of the solid alloy will consist of homogeneous grains of solid solution.

The cooling curve of pure bismuth is shown at (e) in Fig. 3.2. In this case, a hold occurs in the cooling curve at a temperature of 520 F. Solidification begins at point *x* and is completed at point *y*. One may conclude from this discussion that pure metals solidify at constant temperature and their cooling curves exhibit only a hold, whereas the cooling curves of solid solutions exhibit a break and a recovery.

The results obtained from the cooling curves may now be plotted in a diagram of temperature as a function of composition called the equilibrium or phase diagram, Fig. 3.3. The temperature at which solidification begins,

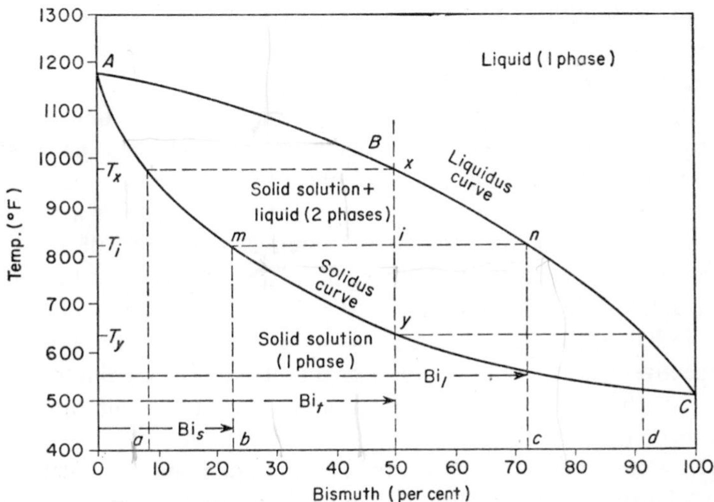

FIG. 3.3 Equilibrium diagram of antimony and bismuth.

as indicated by all points designated *x* in the cooling curves, will form the *liquidus curve*. The temperatures at which solidification is completed, namely, all points *y* of the cooling curves, form the *solidus curve* of the equilibrium or phase diagram.

The liquidus curve is, in reality, a plot of the composition of liquid that will be in equilibrium with solid at any given temperature. Similarly, the solidus curve indicates the composition of solid that will be in equilibrium with liquid at any given temperature. This is in accord with the phase rule as stated in Article 3.3. At a temperature of T_i in the equilibrium diagram of the antimony-bismuth system (Fig. 3.3), the composition of the liquid and the composition of the solid solution are determined by the intersection of this temperature horizontal with the liquidus and solidus curves, respectively. Any alloy that exists within the area enclosed by the liquidus and solidus curves will have one degree of freedom and will be considered as univariant.

3.6 Lever-Arm Principle. The equilibrium diagram indicates the composition of the phases that will be in equilibrium at any particular temperature. The diagram also makes it possible to determine the proportion of coexisting phases at any given temperature. Consider an alloy containing 50 per cent antimony and 50 per cent bismuth at temperature T_i in Fig. 3.3. At this temperature, the alloy is composed of the solid solution of composition *b* and the liquid of composition *c*. By considering the concentration of bismuth in the liquid, the solid, and the alloy as a whole, indicated by Bi_l, Bi_s, and Bi_t respectively, it can be shown that the amount of solid solution and liquid will be inversely proportional to the distances from the point representing the composition of the alloy to the points representing the composition of the phase in question.

In this particular alloy, the fraction of the total mass that is liquid is given by the relation $\overline{mi}/\overline{mn}$. The fraction of the total mass that is solid is given by the relation $\overline{in}/\overline{mn}$. The percentages of each phase can be secured by multiplying the fractions by 100. This is similar to the lever-arm principle of mechanics. The analogy may be illustrated as follows: If all of the liquid is placed at point *n* on the beam *mn*, all of the solid is placed at point *m*, and the fulcrum is considered to be at point *i*, the hypothetical beam *min* will be in balance if the amounts of liquid and solid placed at points *n* and *m*, respectively, are inversely proportional to the distances from the fulcrum. This lever-arm principle can be applied anywhere in the equilibrium diagram where two phases coexist.

3.7 Components Completely Soluble in Liquid and Solid States. A system that illustrates an equilibrium diagram in which there is complete solubility in the liquid and solid states is that of the antimony-bismuth system shown in Fig. 3.3. Consider an alloy containing 50 per cent antimony and 50 per cent bismuth. As cooling of this alloy occurs from a temperature greater than 980 F, solidification will begin when the temperature reaches the liquidus curve at point *x*. The composition of the first solid to form may be determined from the equilibrium diagram by passing a horizontal line from the point *x* through the two-phase region to the left until the first solid line, namely, the solidus curve, of the equilibrium diagram is reached. The intersection of the horizontal line corresponds to a solid solution containing 91 per cent antimony and 9 per cent bismuth. When the temperature has reached a value corresponding to point i on the equilibrium diagram, the composition of the solid that will be in equilibrium with the liquid can be found by extending a horizontal line at this temperature from the vertical line representing the composition of the alloy to the left until it intersects the solidus curve at m. The composition of the liquid at this same temperature can be determined by extending the same temperature horizontal to the right until it intersects the liquidus curve at n. The intersection points m and n correspond to the compositions b and c of the solid and liquid, respectively.

If equilibrium prevails, as it must in interpreting an equilibrium diagram, all of the solid solution must be of composition b and all of the liquid must be of composition c. This is different from the compositions that prevailed at higher temperatures. This means that there has been an adjustment in the composition of the solid and of the liquid during cooling from temperature T_x to temperature T_i. This adjustment in composition takes place by the process of diffusion. If cooling is too rapid, the required changes of composition cannot occur, and the compositions of the phases will not correspond to those indicated by the equilibrium diagram. The rate of cooling must be sufficiently slow to permit complete diffusion and readustment of the concentration.

As cooling continues from the temperature T_i to temperature T_y, solidification continues. Upon reaching temperature T_y, the composition of the liquid corresponds to the point d on the concentration axis, and the composition of the solid solution corresponds to that of the alloy being studied; hence, solidification is completed at this temperature, as shown by the application of the lever law.

In this system and, in general, in systems that possess this type of phase diagram, all alloys that exist below the solidus curve will consist of homogeneous solid solutions. Examples of binary alloys that form this type of diagram include nickel-copper, gold-silver, chromium-molybdenum, and tungsten-molybdenum.

3.8 Components Completely Soluble in Liquid State and Insoluble in Solid State. A system in which the components are completely soluble in the liquid state and insoluble in the solid state is illustrated by bismuth and cadmium.

The cooling curve for pure bismuth is shown at (a) in Fig. 3.4. Solidification begins at the temperature corresponding to x and is completed at the same temperature at point y. An alloy containing 20 per cent cadmium and 80 per cent bismuth will have the cooling curve shown at (b) in Fig. 3.4. This curve exhibits a break at point x at a temperature lower than the temperature at which solidification of pure bismuth occurred. The cooling curve then continues at a lower rate of cooling until point y is reached.

where there is a hold. The hold continues for a certain length of time until point z is reached, and solidification is completed. The structure of this solid alloy will consist of large grains of pure bismuth formed during the cooling from x to y and, in addition, an intimate mechanical mixture of crystals of bismuth and cadmium which formed at the constant temperature corresponding to the time interval from y to z. This latter combination is a eutectic.

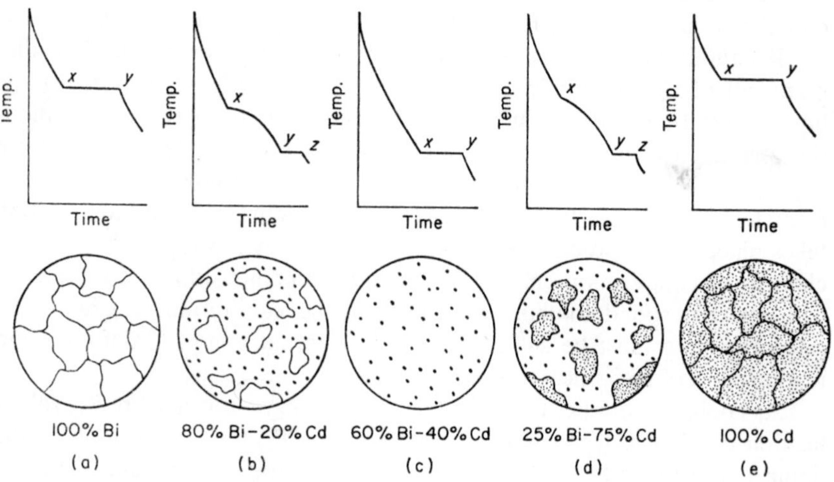

FIG. 3.4 Cooling curves for bismuth-cadmium system.

The cooling curve for an alloy containing 40 per cent cadmium and 60 per cent bismuth is shown at (c) in Fig. 3.4. This cooling curve is unique in that it has only a hold, occupying the time interval, x-y. This alloy in the solid state is composed entirely of an intimate mechanical mixture of grains of cadmium and grains of bismuth, forming the eutectic. This is the eutectic alloy of this system.

An alloy containing 75 per cent cadmium and 25 per cent bismuth will have a cooling curve as shown at (d) in Fig. 3.4. Solidification begins at the break x. A eutectic begins to form at point y, and solidification is completed at point z. The solid alloy will consist of large grains of cadmium and the eutectic.

The cooling curve of pure cadmium will exhibit only a hold shown at (e) in Fig. 3.4, solidification beginning at point x and being completed at point y. The temperatures at which solidification begins and ends for all of these alloys are plotted in the temperature-composition diagram to form the equilibrium diagram shown in Fig. 3.5. Had other alloys in this system been taken with less than 20 per cent cadmium or more than 75 per cent cadmium, the cooling curves for these alloys would also have had a horizontal line or a hold corresponding to the eutectic temperature. This indicates that no solubility exists between bismuth and cadmium in the solid state. It is important to indicate that there is probably no case in which there is not at least some slight solubility of one element in another. However, this solubility may be so slight that it cannot be readily detected. For

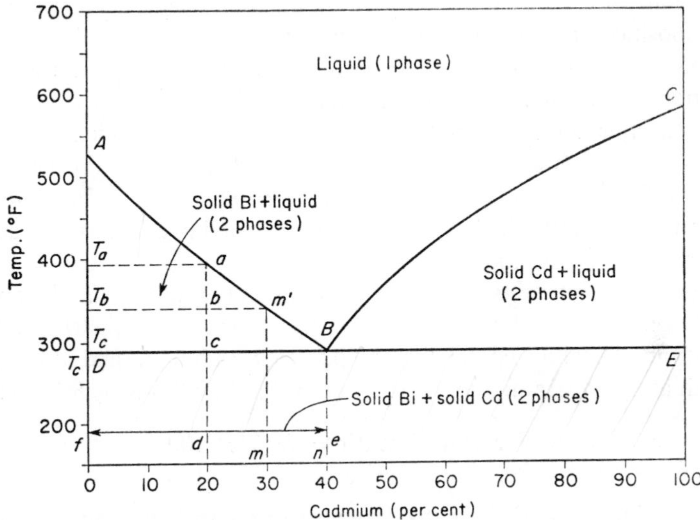

FIG. 3.5 Equilibrium diagram of bismuth and cadmium.

practical consideration, these alloys are considered as being insoluble in the solid state.

The phase regions of the bismuth-cadmium system are as follows:

Above curve *ABC*—homogeneous liquid solution—one phase.
Area *ABD*—solid bismuth + liquid—two phases.
Area *CBE*—solid cadmium + liquid—two phases.
Below *DBE*—solid bismuth + solid cadmium—two phases.

In the solid state, one may consider the regions of different structural constituents which should not be confused with the phase regions. All alloys below the line *DB* will consist of solid bismuth plus eutectic. All alloys below the line *BE* will consist of solid cadmium plus eutectic.

When an alloy containing 20 per cent cadmium and 80 per cent bismuth cools from above temperature T_a in Fig. 3.5, crystals of pure bismuth begin to form when temperature T_a is reached. Bismuth continues to solidify as the temperature changes from *a* to *b*. At temperature T_b, crystals of pure bismuth will be in equilibrium with a liquid of composition *m*, indicated by the intersection of this horizontal temperature line with the liquidus curve, *AB*, at point *m′*. Relative amounts of solid and liquid at this temperature may be determined by means of the lever-arm principle.

As cooling continues, more crystals of bismuth form. Just as the temperature reaches T_c, the alloy will be composed of large crystals of bismuth in equilibrium with a liquid of composition *n*. The cooling curve for an alloy of composition *n* (40 per cent cadmium and 60 per cent bismuth), *c* in Fig. 3.5, shows that solidification occurs at only one temperature with the formation of eutectic. Therefore, the liquid existing at temperature T_c will solidify at this temperature to form an intimate mechanical mixture of crystals of bismuth and crystals of cadmium which makes up the eutectic. The solidification of alloys to the right of point *n* may be analyzed in a similar manner and will be shown to consist of large grains of cadmium existing together with the eutectic.

The relative amounts of structural constituents can be found by applying the lever-arm principle; for example, the proportion of structural constituents in the alloy containing 20 per cent cadmium and 80 per cent bismuth at a temperature corresponding to *d* in Fig. 3.5 is as follows:

$$\text{Per cent bismuth} = \overline{de}/\overline{fe} \times 100$$
$$\text{Per cent eutectic} = \overline{fd}/\overline{fe} \times 100$$

The uniqueness of the eutectic with respect to temperature and composition is a further illustration of the application of the phase rule. At the eutectic temperature of the bismuth-cadmium system (Fig. 3.5), three

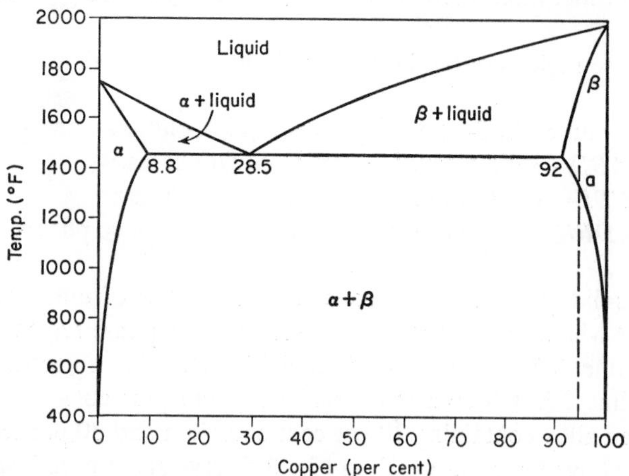

FIG. 3.6 Equilibrium diagram of silver and copper.

phases are in equilibrium, namely, liquid containing 40 per cent cadmium and 60 per cent bismuth, solid pure cadmium, and solid pure bismuth. In this state, there are no degrees of freedom, and the system is invariant. When one of the phases is eliminated, as by removing heat to complete solidification of the liquid, the temperature decreases. Two phases can coexist over a range of temperature.

3.9 Components Completely Soluble in Liquid State, Partially Soluble in Solid State. The most common and, therefore, the most important type of system is that in which there is partial solubility in the solid state. One example of a system of this type is silver-copper, Fig. 3.6. The changing

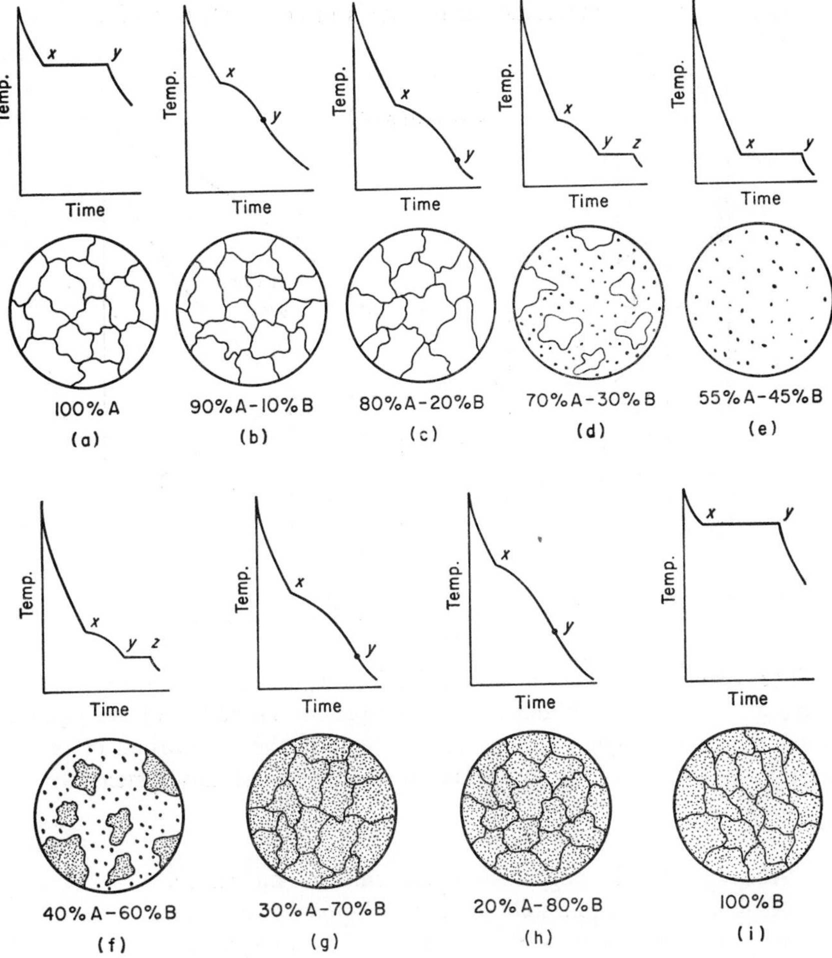

FIG. 3.7 Cooling curves for *A-B* system.

solubility in the solid state may be somewhat confusing to the beginner. Therefore, in order to simplify the discussion, a hypothetical system of elements *A* and *B* will be considered. The cooling curves for the pure components and a series of alloys in the system *A-B* are shown in Fig. 3.7. The cooling curves shown at (b), (c), (g), and (h) in Fig. 3.7 are of the type observed in Fig. 3.2, in which solid solutions were formed. It is, therefore, to be expected in this system *A-B* that solid solutions will form in these particular alloys. The cooling curves shown at (d) and (f) are similar to those found in the cooling curves of Fig. 3.4 in which a eutectic was formed. The cooling curve shown at (e) in Fig. 3.7 is characteristic of eutectic formation. It is important to note that the point *y* in (c) and (g) of Fig. 3.7 occurs at the same temperature as the holds in (d), (e), and (f).

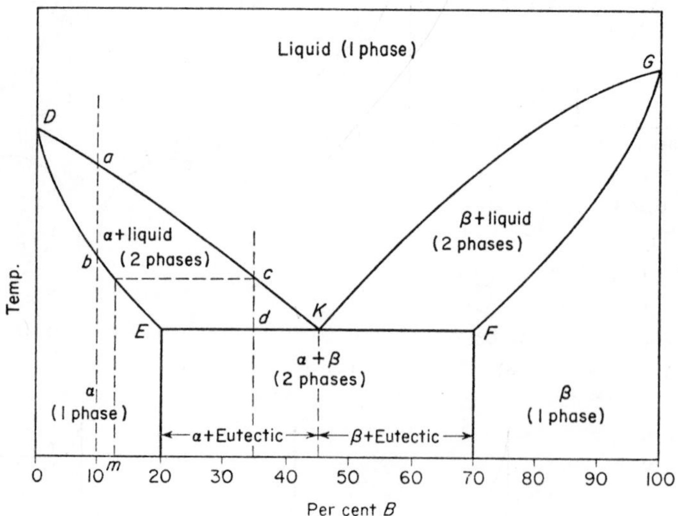

FIG. 3.8 Equilibrium diagram of *A* and *B*.

The temperatures at which solidification begins (all points *x*) and the temperatures at which solidification is completed (all points *y*) are plotted on the temperature-composition diagram to form the equilibrium or phase diagram as shown in Fig. 3.8. The phase regions of this system are as follows:

Above *DKG*—homogeneous liquid solution—one phase.
Region *DEK*—alpha solid solution + liquid—two phases.
Region *GKF*—beta solid solution + liquid—two phases.
Below *DE*—alpha solid solution—one phase.

Below *GF*—beta solid solution—one phase.

Below *EKF*—alpha solid solution + beta solid solution—two phases.

The solidification of any alloy containing 10 per cent *B* and 90 per cent *A* will begin at temperature *a* and will be completed at temperature *b*. In this case, the alloy will consist entirely of alpha solid solution. The solidification of an alloy containing 35 per cent *B* and 65 per cent *A* will begin at temperature *c* with the formation of alpha solid solution of composition *m*. The alpha solid solution will continue to form until the temperature *d* is reached. During the cooling from *c* to *d*, the composition of alpha solid solution will change from *m* and become richer in *B*. At temperature *d*, the alloy will be composed of alpha solid solution containing 20 per cent *B* and 80 per cent *A* in equilibrium with liquid of eutectic composition containing 45 per cent *B* and 55 per cent *A*. This liquid will solidify at constant temperature to form the eutectic consisting of grains of alpha solid solution and grains of beta solid solution. The regions in which the different structural constituents exist in the solid state are indicated in Fig. 3.8.

As indicated previously, most diagrams of this type are not as simple as that illustrated in Fig. 3.8. In most instances, there is a change in solubility in the solid state as indicated in the diagram for the silver-copper system, Fig. 3.6. In this system, the solubility of copper in silver and silver in copper changes as the temperature decreases below that of the eutectic temperature. An alloy containing 95 per cent copper and 5 per cent silver will be composed of beta solid solution at a temperature of 1400 F. Upon reaching the temperature indicated by point *a* on cooling, the solubility of silver in copper reaches its saturation value; in cooling below this, the solubility becomes less, and, therefore, alpha solid solution begins to precipitate from the beta solid solution. The phases present in the final alloy will be alpha and beta solid solutions. The curves bounding the alpha + beta region are called the *solvus* curves.

Other alloy systems that form equilibrium diagrams of this type include copper-tin, copper-zinc, copper-beryllium, aluminum-copper, magnesium-aluminum.

3.10 Systems Forming Compounds. It is common to find in certain systems the existence of compounds that may be represented by formulas of the type A_xB_y. Systems of this type include silver-lithium, silver-magnesium, aluminum-calcium, aluminum-cobalt, aluminum-copper, aluminum-iron, aluminum-lithium, aluminum-magnesium, aluminum-manganese, aluminum-nickel, arsenic-copper, arsenic-iron, gold-zinc, boron-iron, beryllium-copper, beryllium-nickel, bismuth-magnesium, iron-carbon, carbon-tungsten, calcium-magnesium, calcium-lead, columbium-iron, columbium-nickel, cadmium-copper, cadmium-antimony, copper-magnesium, copper-phosphorus, copper-antimony, copper-silicon, iron-phosphorus, iron-silicon, magnesium-silicon. A system in which a compound is formed is shown in Fig. 3.9. In this case, it is observed that a compound, Mg_2Si, is formed. This diagram can be interpreted by considering the two systems, Mg-Mg_2Si and Mg_2Si-Si. A system in which a compound is formed and some solubility exists in the solid state is shown in Fig. 3.10.

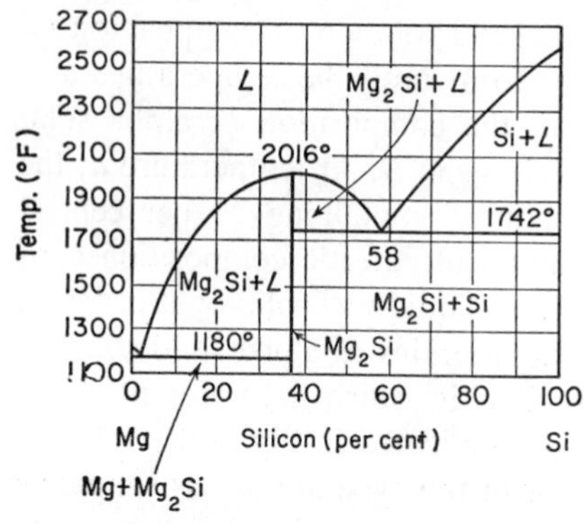

FIG. 3.9 Equilibrium diagram of magnesium and silicon.

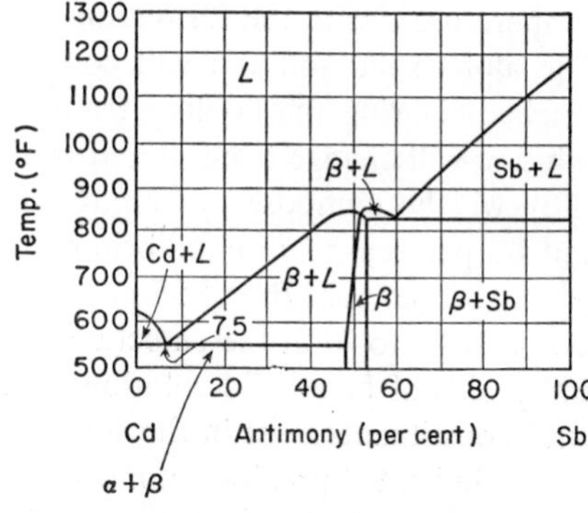

FIG. 3.10 Equilibrium diagram of cadmium and antimony.

In some instances, the compound dissociates before melting occurs. Alloys that contain compounds of this character are considered in the next section.

3.11 Peritectic Reaction. A *peritectic reaction* is essentially an inversion of a eutectic reaction. In the latter, there is a transformation from one phase to two phases on cooling. In a peritectic reaction, there is transformation from two phases to one phase on cooling. An example of a system in which a compound decomposes before melting is shown in Fig. 3.11. In this case, an alloy containing 34.64 per cent bismuth will transform on heating at a temperature of 700 F into gold and liquid. In the cooling of such an alloy, pure gold will crystallize from the liquid first; then at a temperature of 700 F, a reaction will occur to form the compound Au_2Bi. This would take place at constant temperature, forming a hold in the cooling curve. In this instance, cooling continues as soon as the gold and liquid react completely to form the compound Au_2Bi. Such reactions are usually slow. The compound Au_2Bi forms around the surface of each crystal of gold. The formation of a layer of Au_2Bi around every gold crystal prevents direct contact between the liquid and the gold crystals. Continued reaction will be very slow because it will depend upon diffusion of the atoms of the liquid through the compound to the gold crystals. This reaction is called a peritectic reaction because of this action, the Greek "peri" meaning "around."

A peritectic reaction may also occur when there is reaction between liquid and solid during the process of cooling to form a solid solution in the place of a chemical compound. This case may be illustrated by the higher temperature region of the iron-rich portion of the iron-carbon equilibrium diagram shown in Fig. 3.12. In this case, if an alloy containing 0.18 per cent carbon slowly cools from the liquid region, delta solid solution is

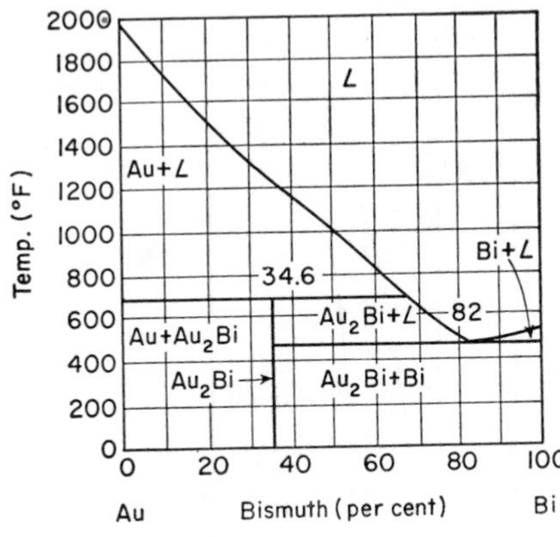

FIG. 3.11 Equilibrium diagram of gold and bismuth.

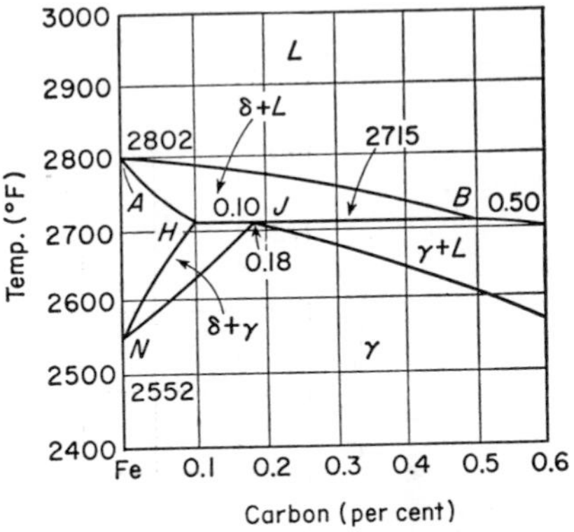

FIG. 3.12 Iron-rich portion of iron-iron carbide equilibrium diagram.

formed. When a temperature of 2715 F is reached, there is a reaction between the liquid and the delta solid solution to form gamma solid solution. The relative proportions of phases present before and after a peritectic reaction may be determined by the application of the lever-arm principle.

3.12 Eutectoid Transformations. In some systems, a transformation may occur within the solid state. A solid solution may transform to an intimate mechanical mixture of two or more phases. In a binary system, a mixture of two phases forms from the single solid solution. This will occur at a unique temperature, and the proportions of the two phases will be unique for the system under consideration. A structure of this type is called a *eutectoid,* an intimate mechanical mixture of two or more phases having a definite temperature of transformation in the solid state and a definite composition. An example of this is shown in a portion of the copper-aluminum system in Fig. 3.13. In this case, during the cooling of an alloy containing 11.8 per cent aluminum, the remainder copper, the beta solid solution transforms at a temperature of 1050 F to an intimate mechanical mixture of alpha solid solution and gamma solid solution. This forms a eutectoid. A more common example of a eutectoid of considerable commercial importance is found in the iron-iron carbide system, Fig. 7.1.

Systems may also possess a transformation that is similar to a peritectic reaction, but the reaction occurs within the solid state. These are called *peritectoid* reactions. An example of such a reaction is shown in the nickel-zinc system, Fig. 3.14. If an alloy containing 48 per cent zinc, the remainder nickel, slowly cools from a temperature above 1500 F, the beta prime phase forms from the alpha and beta phases. Since all of this reaction occurs in the solid state, it is called a peritectoid.

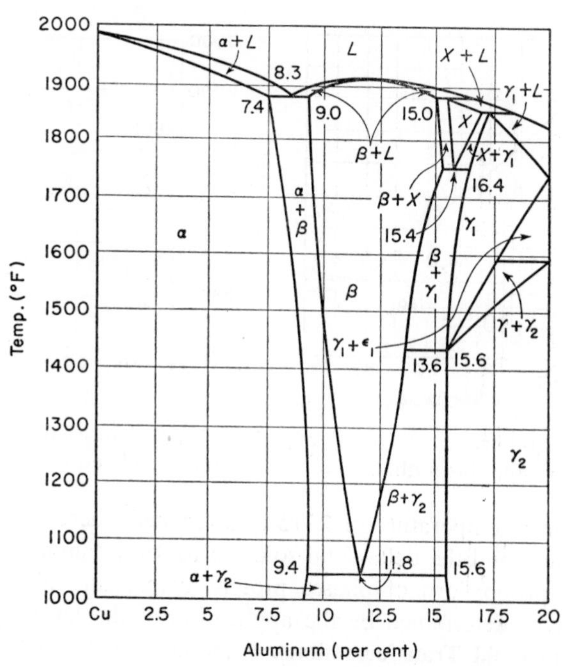

FIG. 3.13 Copper-rich portion of copper-aluminum equilibrium diagram.

Combinations of these transformations are usually found in the more complex diagrams. An example of one in which eutectics, partial solubility, eutectoids, peritectic and peritectoid transformations occur is illustrated in the copper-tin system, Fig. 17.9 of Chapter XVII.

3.13 Order/Disorder Transformations. Another type of transformation occurs within the solid state in some systems. Solid solutions are composed of atoms of different elements distributed in a random fashion on a space lattice. Sometimes the atoms in a solid solution within a specific range of composition may, at a certain temperature, assume a definite arrangement to form an *ordered structure*. An example of a system in which an ordered structure occurs on cooling is shown in Fig. 3.15 of the copper-gold system. Ordered structures occur at 25 and 50 atom per cent (50.9 and 75.5 wt. per cent) gold, at temperatures of 745 and 795 F, respectively.

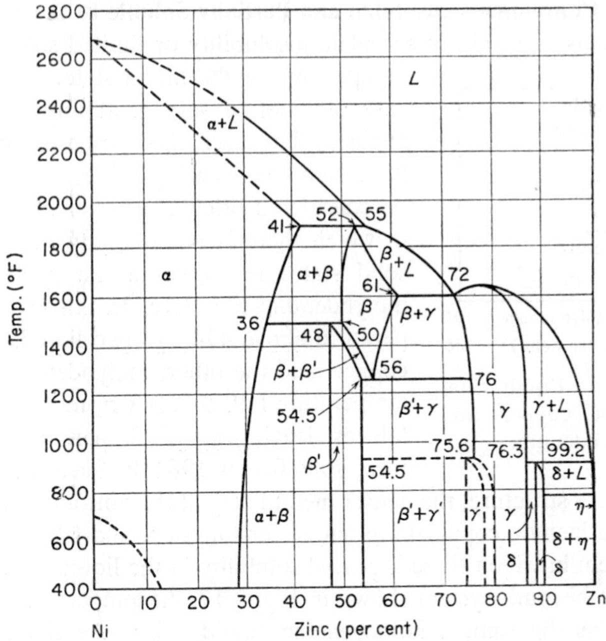

Fig. 3.14 Equilibrium diagram of nickel and zinc.

In each case, when the phase boundary is reached on cooling, the ordering occurs. However, there will be a two-phase region on each side of the α' and α'' areas. The interpretation of this feature of phase diagrams does not require any new principles.

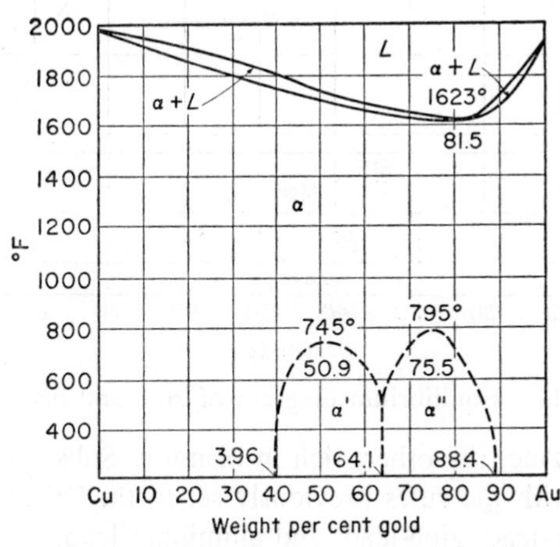

Fig. 3.15 Copper-gold system (order-disorder transformation).

3.14 Components Insoluble and Partially Soluble in the Liquid State. Some systems show either complete insolubility or limited solubility of the components in the liquid state. Many of these cases are not of very great importance to the engineer. However, there are a few cases which have some engineering value, and they should be considered briefly. An example of a system in which there is no solubility in the liquid or solid states is shown in the system, copper-molybdenum, Fig. 3.16. In this case, at a temperature above 4748 F, two liquids exist: one, copper, and the other, molybdenum. When the temperature falls to 4748 F, molybdenum crystallizes, leaving liquid copper. The copper, in turn, solidifies at 1981 F. Such a system gives rise to a layer structure, the denser metal being at the bottom. Other systems of this type include copper-tungsten, silver-tungsten, and silver-iron.

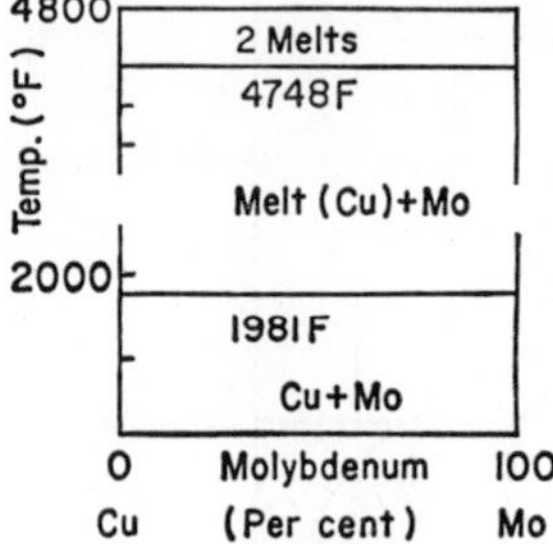

FIG. 3.16 Equilibrium diagram of copper and molybdenum.

A system in which there is partial solubility in the liquid state is typified by the zinc-bismuth system shown in Fig. 3.17. At temperatures above the dome in this diagram, a homogeneous liquid exists. As the temperature falls and reaches the boundary of the region of two melts, two liquid phases exist: one, rich in zinc, the other, rich in bismuth. Subsequent cooling will be in accordance with the rules previously set forth. Other systems of this type include copper-lead, zinc-lead, and aluminum-lead.

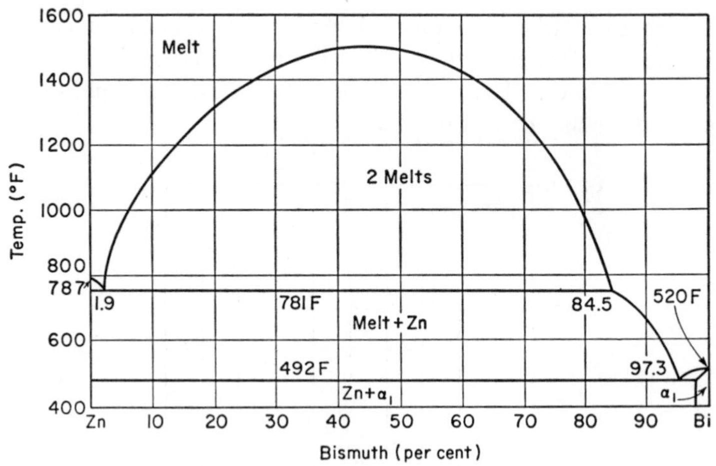

FIG. 3.17 Equilibrium diagram of zinc and bismuth.

In the system iron-copper, there are two distinct liquids at high temperatures, but, as the temperature decreases, a homogeneous liquid region is formed in the equilibrium diagram.

It has been possible to make homogeneous alloys of those metals which are not soluble in the liquid state by the methods of powder metallurgy. These materials are reduced to powders, compacted, and heated to temperatures below the melting range. This treatment produces a homogeneous alloy which may have properties of engineering importance.

3.15 Ternary Phase Diagrams. The simple binary systems are seldom utilized in engineering applications. The majority of alloys contain more than two components. Representing the phase relations that exist over a range of temperatures in systems containing more than two components presents a real problem. In alloys containing three components (ternary

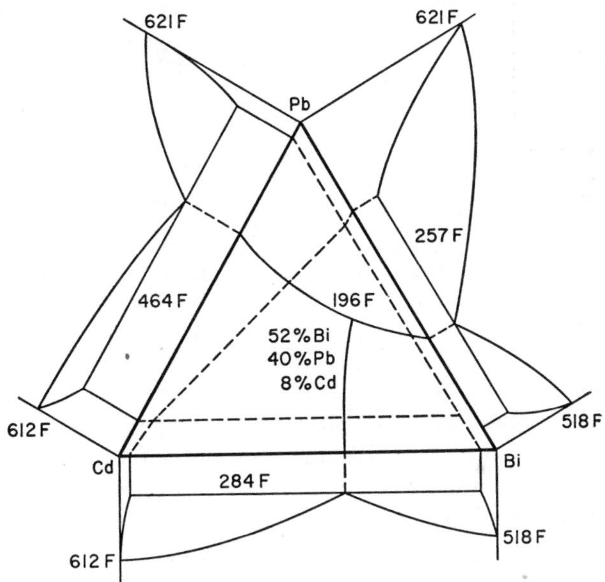

FIG. 3.18 Ternary equilibrium diagram of bismuth, cadmium, and lead.

systems), the phase diagram consists of a three-dimensional figure, the base of which is an equilateral-composition triangle, and the temperature is plotted vertically. In the place of curves to indicate phase boundaries, surfaces are encountered. These diagrams will have liquidus surfaces, solidus surfaces, etc. An example of a ternary diagram of very simple character is illustrated in Fig. 3.18 for the Cd-Pb-Bi system. The representation of phases in equilibrium for systems containing more than three components usually requires sections in which the proportions of certain of the components are held constant. The details of these more complex systems are beyond the scope of this text.

3.16 Transformation Diagrams. While the equilibrium diagram presents the composition and proportion of phases that will be in equilibrium at any temperature, it does not present a graphical picture of the changes in the proportions of phases. It is, therefore, convenient to construct a *phase transformation diagram.* Such a diagram is a plot of temperature as a function of the percentage of phases present. This diagram provides a means by which the formation of phases and structural constituents can be visualized. It is of particular value in studying alloys in which complex transformations occur, particularly within the solid state. The transformation diagram consists of a rectangular figure in which the temperature is represented on the ordinate and the percentage of phases or structural constituents is represented on the abscissa. If any phase or structural constituent occupies 100 per cent of the structure, it is represented by the full width of the transformation diagram. As an example, consider an alloy of 10 per cent B and 90 per cent A in the equilibrium diagram of Fig. 3.8. At temperature a, solidification of the alloy begins with the formation of the solid solution alpha. This gives point a in the transformation diagram shown in Fig. 3.19. As cooling continues, the amount of alpha solid solution increases until temperature b is reached. At this point, the entire structure is composed of grains of alpha solid solution, as indicated at b in the transformation diagram.

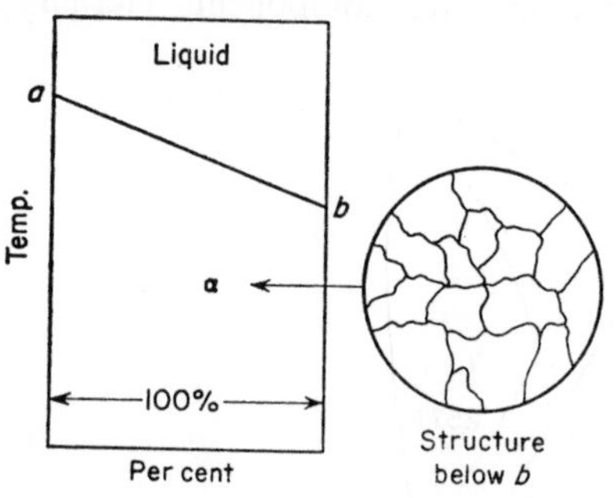

FIG. 3.19 Phase transformation diagram for 90% A–10% B alloy.

If one considers an alloy containing 35 per cent B and 65 per cent A, solidification begins at temperature c with the formation of grains of alpha solid solution. The beginning of solidification is indicated at c in the transformation diagram of Fig. 3.20. As cooling continues, solidification of the alpha solid solution proceeds until temperature d is reached. Upon reaching this temperature, the structure consists of the following proportion of phases:

$$\frac{45 - 35}{45 - 20} \times 100 = \frac{10}{25} \times 100 = 40\% \text{ alpha}$$

$$\frac{35 - 20}{45 - 20} \times 100 = \frac{15}{25} \times 100 = 60\% \text{ liquid}$$

Therefore, at d in Fig. 3.20, 40 per cent of the full width is occupied by the alpha solid solution (proeutectic α), and 60 per cent by liquid. The liquid at temperature d is of eutectic composition and transforms to the eutectic, composed of alpha and beta solid solutions. A vertical line is drawn in the transformation diagram to separate the eutectic from the alpha solid solution that formed in cooling from *c* to *d*. A horizontal line is drawn in the transformation diagram at temperature *d*. This horizintal line must be

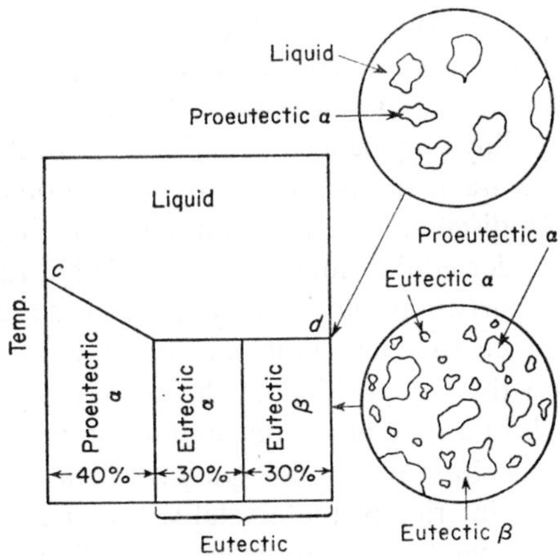

FIG. 3.20 Phase transformation diagram for 65% *A*–35% *B* alloy.

divided in accordance with the proportion of alpha and beta in the eutectic. This is as follows:

$$\text{alpha} = \frac{70 - 45}{70 - 20} \times 100 = \frac{25}{50} \times 100 = 50\%$$

$$\text{beta} = \frac{45 - 20}{70 - 20} \times 100 = \frac{25}{50} \times 100 = 50\%$$

A vertical line is drawn at the point delineating the two phases in the eutectic. The percentage of the entire structure occupied by the eutectic alpha, i.e., the alpha solid solution forming a part of the eutectic, may be computed by taking 50 per cent of the amount of liquid of eutectic composition existing at temperature *d*, or $60 \times 0.50 = 30$ per cent. The result of these computations is presented in Fig. 3.20. The principles given here may be utilized in the analysis of equilibrium diagrams of a more complex character.

3.17 Dendritic Structure in Alloys. The formation of dendrites during the crystallization of pure metals was discussed in Article 2.6. There the statement was made that a dendritic structure could not be detected readily because of the high purity of the material.

In alloys in which two phases crystallize from the liquid state, the dendritic structure is usually very easily detected. If a section of a cast two-phase alloy is machined to a smooth surface and etched with a suitable reagent, a tree-like pattern may be revealed. This condition is, in reality, a segregation, and may be called dendritic segregation.

If one looks at this structure under the microscope, it is observed that the phase that solidified first forms the main stem and branches of the dendrite, while the other phase fills the interstices. The interstices of the dendrite also contain impurities which are not readily soluble in the phase that first solidified. The dendritic structure in alloys can be destroyed most easily by hot working and, in some cases, by heat treatment. Dendritic segregation in a cast alloy may be a source of weakness. Attempts are made to eliminate this structure as far as possible by forging or by heat treatment.

3.18 Coring. In discussing equilibrium diagrams, it was assumed that cooling took place under equilibrium conditions. If the rate of diffusion of the components of a particular alloy is very low and the rate of cooling too rapid, there will not be sufficient time for equilibrium to be established. Under this condition the composition of the phases will not be as represented by the equilibrium diagram. Nonequilibrium cooling may lead to a variation in this composition of a solid solution which crystallizes from the melt. In this case, the interior of the crystal will be richer in the principal constituent than the other layers, producing a condition known as *coring*.

3.19 Properties of Alloys. The equilibrium diagrams give some indication of the relative properties of alloys in a given system. It is difficult, and in most cases impossible, to predict the absolute values of the properties of a series of alloys.

For a system in which there is complete solubility in the solid state, as in Fig. 3.3, the hardness, density, electrical conductivity, and corrosion resistance may be expected to vary as shown in Fig. 3.21. The relation of these properties to composition for a system in which there is complete insolubility in the solid state is illustrated in Fig. 3.22. In a system in which there is partial solubility, the relation of composition to properties is a combination of the two preceding cases as shown in Fig. 3.23.

The mechanical properties of an alloy depend upon two factors: (1) the properties of the phase or phases of which it is composed, and (2) the manner in which the several phases are associated to form the structure. The second factor is possibly the more important. In two-phase alloys in which one phase surrounds the grains of the other, the alloy as a whole usually has the properties of the continuous phase.

The mechanism of plastic deformation and the strength of crystals of pure metals are considered in Chapter V. Since the strength or hardness of any crystalline substance depends upon the resistance to slip on certain crystallographic planes, it is obvious that any mechanism or factor that tends to prevent slip within a crystal will bring about an increase of strength or hardness.

Metals and alloys of engineering importance consist of an aggregate of grains of varying orientation. The interference offered by the grain boundaries precludes the possibility of one continuous slip throughout the whole mass of the metal or alloy. Those metals or alloys that have a complex space lattice may be expected to have higher strength and greater hardness than those with a simpler lattice. An illustration of this fact may be found in the intermetallic compounds which have complex space lattices and are harder than the pure metals of which they are composed.

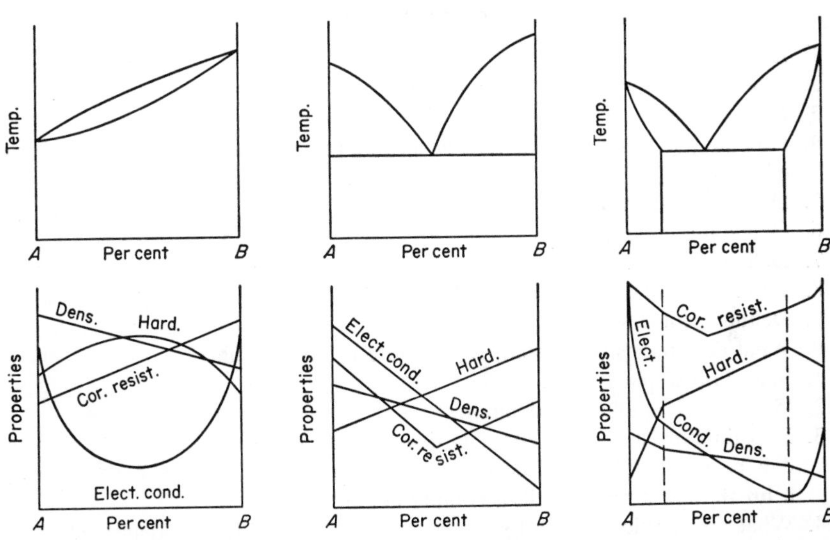

FIG. 3.21 Relation of properties to composition for system with complete solid solubility.

FIG. 3.22 Relation of properties to composition for system with solid insolubility.

FIG. 3.23 Relation of properties to composition for system with partial solid solubility.

In a given metal, resistance to slip may be increased by (1) an alteration of the space lattice through the formation of a compound or a solid solution; (2) a distortion of the lattice structure by plastic deformation (strain hardening); and (3) a precipitation of another phase within the lattice structure (precipitation hardening).

REFERENCES

Barrett, *Structure of Metals,* McGraw-Hill Book Co., New York, 1943.

Clark, *Engineering Materials and Processes,* International Textbook Co., Scranton, Pa., 1959.

Findlay, *The Phase Rule and Its Application,* Longmans, Green & Co., New York, 1935.

Guy, *Elements of Physical Metallurgy,* Addison-Wesley Publishing Co., Reading, Mass., 1959.

Hume-Rothery and Raynor, *The Structure of Metals and Alloys,* The Institute of Metals, London, 1954.

Hume-Rothery, Christian, and Pearson, *Metallurgical Equilibrium Diagrams,* The Institute of Physics, London, 1952.

Metals Handbook, *American Society for Metals,* Metals Park, Ohio, 1961.

Rhines, *Phase Diagrams in Metallurgy,* McGraw-Hill Book Co., 1956.

QUESTIONS

1. Define a metal.
2. What is a metalloid? Give an example.
3. Define an alloy.
4. How may the metals be divided according to characteristics? Give examples of each.
5. Define a system.
6. Define a component.
7. Define a phase.
8. What is a structural constituent?
9. Define a eutectic.
10. What determines if a system is in a state of equilibrium?
11. State Gibbs' phase rule.
12. What is an equilibrium diagram?
13. Explain how a cooling curve is determined. What significant information can be secured from a cooling curve? Explain the significance of a hold, a break, and a recovery in a cooling curve.
14. What are the liquidus and solidus curves in an equilibrium diagram?
15. Describe the phase changes that occur during the very slow cooling of an alloy containing 60 per cent antimony and 40 per cent bismuth from a temperature of 1100 F to 500 F.
16. Derive the lever law as applied to equilibrium diagrams.
17. Determine the percentage of phases present in an alloy of 30 per cent antimony and 70 per cent bismuth at a temperature of 700 F.
18. Describe the phase changes that occur in the very slow cooling of an alloy consisting of 80 per cent bismuth and 20 per cent cadmium from a temperature of 600 F to 200 F.
19. Draw the cooling curve for an alloy of 80 per cent bismuth and 20 per cent cadmium.
20. Explain by means of the phase rule why there is a hold in the cooling curve of an alloy of 60 per cent bismuth and 40 per cent cadmium.
21. How many degrees of freedom does an alloy containing 30 per cent cadmium and 70 per cent bismuth have at a temperature of 300 F?
22. Describe the phase changes that occur in an alloy containing 90 per cent *B* and 10 per cent *A*, according to Fig. 3.8, in slow cooling from a temperature above the liquidus to below the solidus curve.

23. Describe the phase changes that occur in the slow cooling of an alloy containing 60 per cent *B* and 40 per cent *A*, according to the equilibrium diagram in Fig. 3.8, in cooling from a temperature above the liquidus to below the eutectic temperature.
24. What is a solvus curve?
25. What is a peritectic reaction?
26. Draw an equilibrium diagram that illustrates the existence of a peritectic reaction.
27. Define a eutectoid.
28. What is a peritectoid reaction?
29. Explain an order-disorder transformation.
30. Draw an equilibrium diagram for an alloy system that shows partial solubility in the liquid state.
31. Draw the phase transformation diagram for an alloy containing 30 per cent *B* and 70 per cent *A*, according to the diagram in Fig. 3.8.
32. Draw the phase transformation diagram for an alloy containing 65 per cent zinc and 35 per cent nickel, according to Fig. 3.14.
33. What is coring?
34. Show by sketches how the density, hardness, corrosion resistance, and electrical conductivity may be expected to vary for an alloy in which: (a) there is complete solubility in the solid state; (b) there is complete insolubility in the solid state; (c) there is incomplete solubility in the solid state.
35. Upon what two factors do the mechanical properties of alloys depend?
36. In what ways may resistance to slip in crystals be increased?

IV

PHYSICAL PROPERTIES

4.1 Properties of Materials. A material is completely described by specification of its properties. These properties can be divided into three classes: physical, mechanical, and chemical properties. Those characteristics that are employed to describe a material under conditions in which external forces are not concerned are usually called *physical properties.* These characteristics are, in most instances, structure-insensitive or intrinsic properties of the material. This means that any variations in the structure have very little, if any, effect upon the particular property. The *mechanical properties* include those characteristics of a material that describe its behavior under the action of external forces. These are, for example, the strength characteristics of the material; they are definitely structure-sensitive and are seriously affected by small degrees of imperfections. The *chemical properties* describe the combining tendencies, corrosion characteristics, reactivity, solubilities, etc., of substance.

This chapter will be devoted to a brief discussion of some of the physical properties of metals in the light of the fundamental discussions that have been presented in previous chapters.

4.2 Specific Heat. The specific heat of a substance is the amount of heat that is required to raise a gram-mole of a substance 1 K at constant pressure (C_p), or at constant volume (C_v). In the case of most solids the difference between the specific heats for constant volume and constant pressure is negligible.

The *law of Dulong and Petit* is an empirical relationship stating that the gram atomic specific heat ($C_v \times$ atomic weight) of all elements is constant. This gram atomic specific heat is equal to three times the value of the universal gas constant (1.986 cal/C/mole). This law holds only at or near room temperature. As the temperature approaches absolute zero, the specific heat approaches zero.

The increase of specific heat with rising temperature to around room temperature is not accounted for by this empirical relationship. When the methods of classical physics are employed, the same relation results. Therefore, the methods of quantum physics are required to provide an explanation for the variation of specific heat with temperature.

The quantum theory was applied to this problem by Einstein, but his solution did not fit the experimental data in the vicinity of absolute zero. However, an expression given by Debye provided reasonable agreement at high temperatures in which the value of the specific heat approached $3R$ and at the low temperatures approached zero as the cube of temperature. Hence this relationship is in close agreement with the experimental data. According to the Debye theory, the following expression is given for the atomic specific heat:

$$\frac{dE}{dT} = C = 9R\left[4\left(\frac{T}{\theta}\right)^3 \int_0^{\theta/T} \frac{x}{e^x - 1}\,dx - \frac{\theta}{T}\frac{1}{(e^{\theta/T} - 1)}\right]$$

In this expression $x = h\gamma/kT$ is dimensionless, $\theta = h\gamma_n/kT$ is the Debye temperature, T is the temperature in degrees K, R is the universal gas constant, and γ_n is the maximum frequency of atomic vibration. In the neighborhood of room temperature, the Debye equation approaches $3R$ while at the low temperature the expression becomes

$$C_v = \frac{234\,RT^3}{\theta^3}$$

which approaches zero as absolute zero is reached. The Debye relation is reasonably satisfactory, but it is not exact in all instances and seems to be the closest approach to the experimental evidence at present.

The theories of specific heat have been worked out primarily for the elemental solids and not for any combinations of elements, such as alloys. A satisfactory relationship has not been established for these combinations because of the tremendous complexities involved.

4.3 Thermal Expansion. When thermal energy is added to a substance, a change is produced in the balance between the attractive and repulsive forces in the structure and therefore a change in the parameters occurs. Any change in the parameters of the structure means a change in dimension; this becomes evident in the coefficient of thermal expansion. A large number of substances have crystal structures that are anisotropic. These materials will exhibit anisotropy in the expansion behavior.

The potential energy of a pair of atoms in a structure may be indicated by the expression

$$E = -\frac{a}{r^m} + \frac{b}{r^n}$$

where a, b, m, and n are constants with n greater than m, and r is the atomic distance. This relationship gives a plot of the energy of a crystal in terms

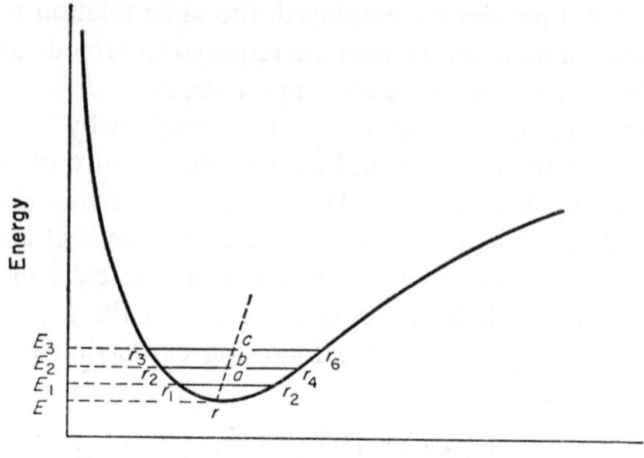

FIG. 4.1 Energy of a crystal as a function of the interatomic spacing of the atoms.

of the interatomic spacing of atoms of the form shown in Fig. 4.1. The shape of the minimum portion of the curve determines the degree of expansion. Covalent solids exhibit a rather steep curve on each side of the minimum, and these substances have small expansions. The same is true of

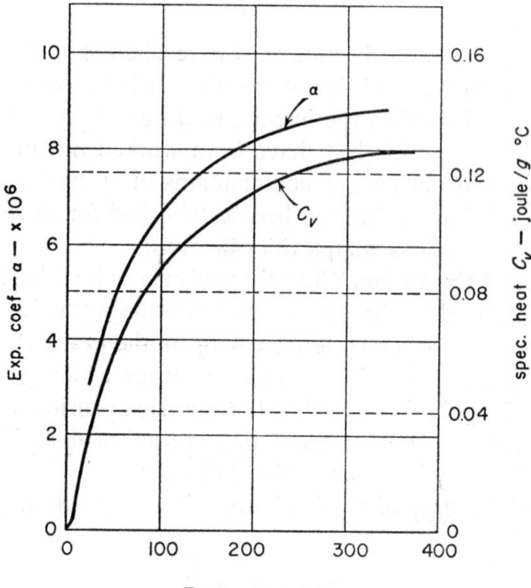

FIG. 4.2 Similarity of temperature dependency of specific heat, C_v, and expansion coefficient, α, of platinum. (Zwikker, *Physical Properties of Solid Materials,* Pergamon Press, London, 1954)

materials with the ionic bond. Those materials that possess the metallic bond exhibit greater expansion; those materials that are molecular solids have the greatest expansion of all substances. Therefore, by examining the curve of energy vs. atomic distance, one may gain some idea of the order of magnitude of expansion that may be encountered in that substance.

The coefficient of expansion is similar in its dependence upon temperature as specific heat. The similarity even includes the variation with the cube of the temperature near absolute zero. An example of the similarity

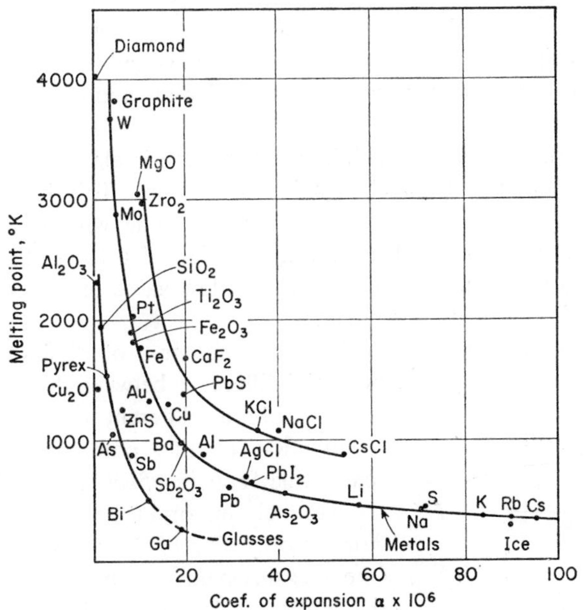

FIG. 4.3 Relation between melting point and coefficient of thermal expansion. (Zwikker, *Physical Properties of Solid Materials*, Pergamon Press, London, 1954)

of the variation of specific heat and coefficient of expansion with temperature for one material is indicated in Fig. 4.2.

There is an approximate correlation between the melting point and the expansion coefficients, which is illustrated in Fig. 4.3. This indicates that the higher the melting point, the lower the coefficient of expansion. Sometimes this is a very useful relationship.

4.4 Melting Point. Materials in the liquid state do not possess a geometric arrangement of the atoms. The atoms are free and are not bonded to each other. When solidification occurs, bonding forces become effective to form a solid in which the atoms form a geometrical pattern. The melting point is related to the bonding forces in solids. The covalent bonded solids have the highest melting points. Ionic bonded solids in which the bond is not quite as strong as in the covalent type have somewhat lower melting points. Materials having the metallic-type bond are next in order of melting point. The materials with the lowest melting points are the molecular bonded materials. Diamond is an excellent example of the perfect covalent bond, and it has the highest melting point of any solid.

The melting point of metals is related to the electronic energies that determine the cohesive forces between the atoms. There is a wide variation in these cohesive forces from type of atom to type of atom. For example, the alkali metals have low melting points and large interatomic distances. The alkali earth metals have higher melting points and decreased distances. The transition metals have very high melting points and even smaller interatomic distances.

4.5 Thermal Elastic Effect. Whenever there is elastic deformation in a substance there will be a change in temperature. This is a very small effect and is usually inconsequential. Work is done whenever there is a change in volume resulting from the application of a force; this work appears in the form of heat. In metals with a low value of the coefficient of expansion, this is a negligible effect.

4.6 Thermal Electric Effect. Two conductors of dissimilar material joined together will produce a potential difference between the two junctions if they are at different temperatures. The production of this potential difference is called the *Seebeck effect*. The Seebeck effect is extremely important because it is the basis for the operation of thermocouples, used for the measurement of temperature. The Seebeck effect is illustrated schematically in Fig. 4.4. If one places a third conductor in the circuit, the potential difference between the two initial junctions will not be affected as long as the new junctions remain at the same temperature. Usually, anything that occurs between the junctions will not influence the potential difference. This law was stated by *Magnus*. However, it is only correct when the conductors are homogeneous and isotropic—this means that it is only correct for cubic materials in which strain and defects are absent.

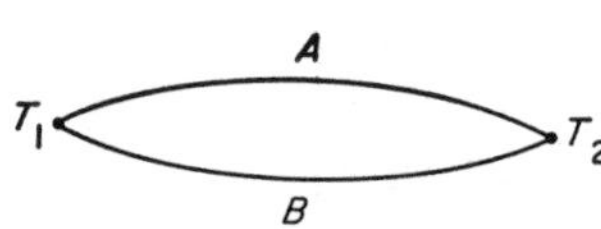

FIG. 4.4 Temperature difference (T_1 and T_2) at junction of dissimilar metals (A and B) produces potential difference (Seebeck effect).

It is of interest to note that, if two noncubic crystals of the same substance are put together but are oriented differently, a potential will be produced at the junctions. A potential difference can also be produced with two wires of the same material with one wire annealed and the other cold-drawn. These characteristics are important to keep in mind when using thermocouples.

The *thermal electric power* of a substance is equal to the electromotive force produced by a temperature difference of 1 C between the hot and cold junctions. Values of thermoelectric power for different materials are presented in the handbooks.

If an external electromotive force is impressed upon a thermocouple circuit so as to cause a current to flow in the circuit, heat will be evolved at one junction and absorbed at the other. A reversal of the current will reverse the heat effect. This effect is called the *Peltier effect* and is illustrated in Fig. 4.5.

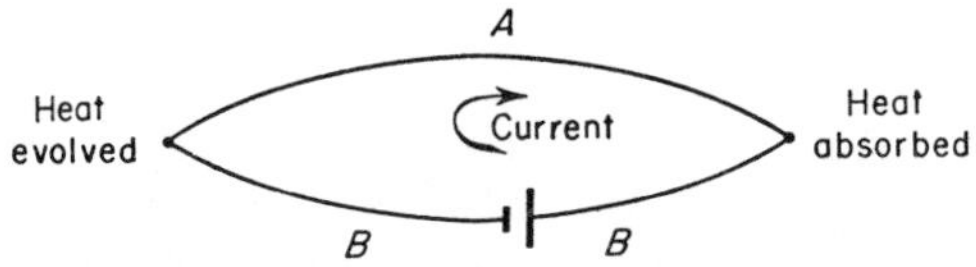

FIG. 4.5 Potential impressed on couple producing current flow with heat evolved at one junction and heat absorbed at other junction (Peltier effect).

If a current is passed through a bar that is conducting heat, heat will be either absorbed or evolved depending upon the material and the direction of current flow. This is the *Thompson effect,* which is illustrated in Fig. 4.6. In the case of iron, if the current flows from the cool end to the hot end, the bar becomes warmer. A reversal of the current will cause the bar to cool. In copper, the effect is opposite; in lead, the effect is practically nil.

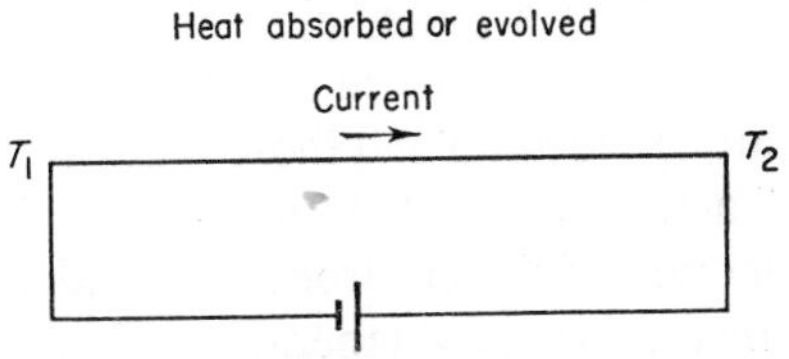

FIG. 4.6 Electric current through wire with temperature difference (conducting heat) causes heating or cooling of wire (Thompson effect).

4.7 Thermal Conductivity. There are three mechanisms by which the transfer of thermal energy in solids may be explained. The first is by the free electrons that exist in the metals. The second is by molecules that may occur in the organic solids, and the third by lattice vibration. In the metals, thermal conduction occurs primarily through electron transfer but the lattice vibration contributes to the transfer.

4.8 Thermal Diffusivity. The thermal diffusivity of a substance is defined by the relation

$$\frac{k}{C_p\rho}$$

where k is the thermal conductivity, C_p is the specific heat at constant pressure, and ρ is the density of the material. *Diffusivity* is a measure of the rate of change in temperature of a solid under a given set of conditions. In general, the greater the specific heat, the smaller will be the change in temperature across a solid for a given set of transfer conditions.

4.9 Vapor Pressure. The vapor pressure of a substance is a measure of its escaping tendency. Most solids have very low vapor pressures at ordinary temperatures. The higher the vapor pressure, the greater the escaping tendency which varies over a wide range with the bonding forces, as illustrated in Table 4-I.

TABLE 4-I. VAPOR PRESSURES OF SOME METALS AT TEMPERATURES BELOW THE MELTING POINT.

Cadmium	At 385 F (307 C)	0.0001 atmos (mp = 609.6 F (320.9 C))
Chromium	At 2588 F (1420 C)	0.0001 atmos (mp = 3430 F (1890 C))
Zinc	At 750 F (399 C)	0.0001 atmos (mp = 787 F (419 C))
Magnesium	At 1126 F (608 C)	0.001 atmos (mp = 1202 F (650 C))

4.10 Electrical Conductivity. One of the significant characteristics of the metals is their ability to conduct electricity. This property is attributable to the existence of free electrons in the structure. Conduction occurs because the Brillouin zone or zones are not completely filled, thus allowing electrons to be moved into higher energy states.

When a potential is applied to a conductor, electrons that are always in continual motion are acted upon, so that those moving in one direction are speeded up and those moving in the opposite direction are slowed down. There is a net flow of electrons in a given direction. A resistance to the motion of the electrons is provided by the positive ions of the metal lattice. Increasing the temperature of the material increases the vibrational energy of the ions and therefore provides more interference and hence greater resistance with increasing temperature.

The mechanism of electrical conductance can be explained on the basis of the zone or band theory. Only electrons near the maximum energy state can be pushed up to a higher energy state. This maximum energy condition must be adjacent to vacant electron levels or they cannot be moved into higher energy states. If this maximum is too close to the zone boundary, then electrons will be hindered from moving into the higher states. This situation accounts for the higher conductivity of the monovalent metals. The first Brillouin zone of this type of metal is only half filled; therefore metals such as sodium, potassium, copper, gold, and silver have very good conductivity. The addition of any alloying element tends to fill the empty levels and therefore tends to restrict the conductivity, because there will be fewer levels into which the electrons can be raised.

The transition elements exhibit lower electrical conductivity than most of the other metals. This situation is probably related to the higher density of the energy states and the transitions between energy states of different bands. In a similar manner, the very low conductivity of bismuth is accounted for on the basis of its zone structure.

4.11 Ferromagnetism. Ferromagnetic materials are those that behave like iron when placed in a magnetic field. Elongated samples of these materials align their long axes parallel to a strong magnetic field and become strongly magnetized. The only elements that behave as ferromagnetic substances are alpha iron, cobalt, nickel, gadolinium, and possibly dypsprosium. These elements are in the transition group; they have unfilled *d* shells of electrons. The ferromagnetism of these elements is associated with the alignment of the spin of the electrons. Many alloys of these elements also possess ferromagnetic characteristics. There are some alloys in which the components of the alloy are not ferromagnetic, but the alloy is ferromagnetic. This is true of the Heusler-type alloys which contain Cu, Mn, Al, or Ag, Mn, or Cu, Mn, In. There are some nonmetallic materials that exhibit ferromagnetic properties. These are called *ferrites,* which are cubic spinels and have a structure that is similar to that of magnetite, Fe_3O_4.

Each electron behaves as a magnet and, in the presence of a magnetic field, aligns itself with the field in accordance with the spin direction. A material that is magnetized must have more electrons aligned in one direction than in the other, otherwise there would be balancing effects. The alignment of electrons of certain spins in ferromagnetic materials is irreversible. The alignment in ferromagnetic materials occurs even in the absence of an external field, but these alignments occur in what are called domains, and when they are so arranged there is a net zero effect. A *domain* is a region in which the magnetic moments are aligned in the same direction. These domains have been observed through special techniques. The references should be consulted for a more detailed consideration of ferromagnetism.

The ferromagnetic behavior of materials decreases with increasing temperature. That temperature at which the magnetization reaches zero is called the *Curie temperature,* which for iron is 1430 F (777 C), for nickel 690 F (366 C), and for cobalt 2150 F (1177 C).

A phenomenon associated with ferromagnetic materials is that of *magnetostriction.* When a ferromagnetic material is magnetized, it experiences an elastic deformation. When nickel is magnetized, there is a contraction in all of the principal crystallographic directions. An alloy of 80 per cent nickel and 20 per cent iron, called *permalloy,* exhibits an expansion in all crystallographic directions upon magnetization. When a stress is applied to these substances, a change in magnetization occurs.

4.12 Paramagnetism and Diamagnetism. Materials that exhibit weak permeability to the extent that they may be considered nonmagnetic materials, but yet have greater permeability than a vacuum, are referred to as *paramagnetic substances*. The paramagnetic materials are most likely to be those elements that have an odd number of electrons, because under this condition the electron spins may not be balanced. However, even atoms in which there is an even number of electrons may also be paramagnetic under conditions in which the inner electron shells are incomplete, as is found in the transition elements.

Atoms that possess a symmetrical electronic structure may not have any permanent magnetic moments. Such materials are said to be *diamagnetic*. In this instance the permeability is less than that of a vacuum. Diamagnetic materials include copper, silver, bismuth, etc.

REFERENCES

Cottrell, *Theoretical Structural Metallurgy,* Longmans, Green & Co., New York, 1948.

Magnetic Properties of Metals and Alloys, Symposium, American Society for Metals, Metals Park, Ohio, 1959.

Raynor, *An Introduction to the Electron Theory of Metals,* The Institute of Metals, London, 1949.

Sinnott, *The Solid State for Engineers,* John Wiley & Sons, New York, 1958.

Zwikker, *Physical Properties of Solid Materials,* Pergamon Press, London, 1954.

QUESTIONS

1. What are the three classes of properties of materials? Define each.
2. Define "specific heat."
3. Compute the specific heat of aluminum at room temperature by means of the law of Dulong and Petit.
4. Is the law of Dulong and Petit applicable at temperatures other than room temperature?
5. How is the shape of the curve of energy vs. interatomic spacing related to the expansion characteristics of an element?
6. How is the coefficient of expansion related to the melting point?
7. What is the relationship of the melting point to the type of bond in solids?
8. What is the thermal elastic effect? How is this related to the coefficient of expansion?
9. What is the Seebeck effect?
10. What is the thermal electric power of a substance?
11. What is the Peltier effect?
12. What is the Thompson effect?
13. By what three mechanisms may the transfer of thermal energy in solids be explained?
14. By what mechanism does thermal conduction in metals primarily occur?

15. What is thermal diffusivity?
16. What is the significance of vapor pressure in metals?
17. Explain the mechanism of electrical conduction in a metal.
18. What are the electron characteristics of ferromagnetic materials?
19. What is a magnetic domain?
20. What is the Curie temperature?
21. What is magnetostriction?
22. What are paramagnetic substances?
23. What are diamagnetic substances?

V

MECHANICAL PROPERTIES

5.1 Introduction. The engineer is most interested in the way in which metals will respond to the application of external forces. The reaction of materials to the action of external forces is indicated by the *mechanical properties of materials*. These properties relate primarily to the elastic and plastic behavior, as well as the over-all strength and fracture characteristics of materials.

Since the metals used in engineering are made up of a large number of crystals, thus making them polycrystalline in character, it is important to understand the mechanical behavior of single crystals. With some basic understanding of the behavior of single crystals, one will have a better appreciation of the mechanical behavior of polycrystalline materials. Therefore, the behavior of single crystals will be discussed first, and then some of the characteristics of the polycrystalline metals will be considered.

This discussion of mechanical properties is not intended to be exhaustive but to show the relationship to the structure of matter presented in an earlier chapter. The mechanical properties, for the most part, are structure-sensitive characteristics of the material. This is in distinct contrast to most of the physical properties, which are structure-insensitive properties.

5.2 Elasticity. Most engineering materials are utilized under what is known as elastic conditions. This means that when a material is deformed the deformation is not permanent. When a force is applied to deform a single crystal of a metal, there is a deformation of the lattice structure. If the force is applied in only one direction of the single crystal, such as in compression, the distance between atoms will be decreased in that direction and increased in directions 90° to this. The ratio between the strain in the two directions is called *Poisson's ratio*. The atomic distances are slightly altered, but the atoms exert a resisting force which tends to restore them to their normal positions. The resisting force constitutes a stress. A *stress* may be defined as the internally distributed forces that tend to resist deformation. When the externally applied force producing the deformation is removed, the lattice returns to its normal dimensions provided the deformation has not been great enough to produce permanent deformation.

The relationship between the forces acting on a material and the deformation is very significant. The force applied to a simple cube can be resolved into forces perpendicular to the three axes of the cube. Each of these forces can be resolved into tension or compression forces perpendicular to the cube faces and into a shear force in the plane of the cube faces. This provides, then, three principal stresses, which are perpendicular to the cube faces, and six shear stresses. These are illustrated in Fig. 5.1. The

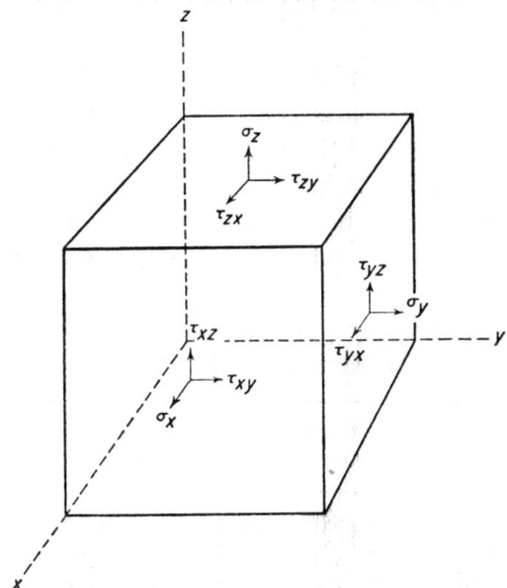

FIG. 5.1 Shear and normal stresses on a simple cube.

conditions for equilibrium require that three of the shear stresses be balanced by the other three shear stresses. The three principal stresses are designated as σ_x, σ_y, and σ_z, which are the stresses in the x, y, and z directions, respectively. The shear stresses τ_{yx}, τ_{zx}, and τ_{yz} are equal to τ_{xy}, τ_{xz}, and τ_{zy}, respectively. The strains that correspond to these stresses are ϵ_x, ϵ_y, ϵ_z, γ_{xy}, γ_{yz}, and γ_{xz}. The strains γ_{xy}, γ_{yz}, and γ_{xz} are equal to the shear strains γ_{xy}, γ_{yz}, and γ_{xz}, respectively.

According to *Hooke's Law*, strain is a linear function of stress for small values of strain. This condition may be expressed as follows:

$$
\begin{aligned}
\epsilon_x &= S_{11}\sigma_x + S_{12}\sigma_y + S_{13}\sigma_z + S_{14}\tau_{xy} + S_{15}\tau_{xz} + S_{16}\tau_{zy} \\
\epsilon_y &= S_{21}\sigma_x + S_{22}\sigma_y + S_{23}\sigma_z + S_{24}\tau_{xy} + S_{25}\tau_{xz} + S_{26}\tau_{zy} \\
\epsilon_z &= S_{31}\sigma_x + S_{32}\sigma_y + S_{33}\sigma_z + S_{34}\tau_{xy} + S_{35}\tau_{xz} + S_{36}\tau_{zy} \\
\epsilon_{xy} &= S_{41}\sigma_x + S_{42}\sigma_y + S_{43}\sigma_z + S_{44}\tau_{xy} + S_{45}\tau_{xz} + S_{46}\tau_{zy} \\
\epsilon_{xz} &= S_{51}\sigma_x + S_{52}\sigma_y + S_{53}\sigma_z + S_{54}\tau_{xy} + S_{55}\tau_{xz} + S_{56}\tau_{zy} \\
\epsilon_{yz} &= S_{61}\sigma_x + S_{62}\sigma_y + S_{63}\sigma_z + S_{64}\tau_{xy} + S_{65}\tau_{xz} + S_{66}\tau_{zy}
\end{aligned}
$$

In these relations S_{ij} is called the *compliance constant* and is one of the elastic constants. The components of stress can be expressed mathematically in terms of strain in the following relations:

$$\begin{aligned}
\sigma_x &= C_{11}\epsilon_x + C_{12}\epsilon_y + C_{13}\epsilon_z + C_{14}\epsilon_{xy} + C_{15}\epsilon_{yz} + C_{16}\epsilon_{zy} \\
\sigma_y &= C_{21}\epsilon_x + C_{22}\epsilon_y + C_{23}\epsilon_z + C_{24}\epsilon_{xy} + C_{25}\epsilon_{yz} + C_{26}\epsilon_{zy} \\
\sigma_z &= C_{31}\epsilon_x + C_{32}\epsilon_y + C_{33}\epsilon_z + C_{34}\epsilon_{xy} + C_{35}\epsilon_{yz} + C_{36}\epsilon_{zy} \\
\tau_{xy} &= C_{41}\epsilon_x + C_{42}\epsilon_y + C_{43}\epsilon_z + C_{44}\epsilon_{xy} + C_{45}\epsilon_{yz} + C_{46}\epsilon_{zy} \\
\tau_{zy} &= C_{51}\epsilon_x + C_{52}\epsilon_y + C_{53}\epsilon_z + C_{54}\epsilon_{xy} + C_{55}\epsilon_{yz} + C_{56}\epsilon_{zy} \\
\tau_{xz} &= C_{61}\epsilon_x + C_{62}\epsilon_y + C_{63}\epsilon_z + C_{64}\epsilon_{xy} + C_{65}\epsilon_{yz} + C_{66}\epsilon_{zy}
\end{aligned}$$

The values of C_{ij} are taken as the moduli, which are also referred to as *elastic constants.*

The constant $C_{ij} = C_{ji}$ and $S_{ij} = S_{ji}$ so that only 21 of these constants are required for the most general case, which is a solution for the elastic constants of a triclinic crystal. In the cubic system it is necessary only to evaluate C_{11}, C_{12}, and C_{44}. In the hexagonal system the constants C_{11}, C_{12}, C_{13}, C_{33}, and C_{44} are required. The compliance constants and the moduli for a few of the metals are given in Table 5-I. It will be obvious from the

TABLE 5-I. COMPLIANCE CONSTANTS AND ELASTIC MODULI FOR SOME METALS.

Metal	10^{-13} cm²/dyn			10^{12} dyn/cm²		
	S_{11}	S_{12}	S_{44}	C_{11}	C_{12}	C_{44}
Aluminum	15.9	−5.8	35.2	1.08	0.622	0.284
Copper	14.9	−6.2	13.3	1.70	1.23	0.75
α-Iron	7.57	−2.82	8.62	2.37	1.41	1.160
Tungsten	2.573	−0.729	6.604	5.01	1.98	1.514

From Boos and MacKenzie, *Progress in Metal Physics*, Vol. II, Pergamon Press, London, 1950.

values in this table that metallic single crystals are anisotropic; that is, they possess different properties in different directions. While a single metallic crystal is anisotropic, the polycrystalline materials with which the engineer is usually concerned are usually isotropic, since the grains are very small and take essentially all possible orientations. For isotropic materials, the relations are simplified.

The elasticity of a material is somewhat of an idealized condition. In general, the strains must be very small in order to have Hooke's law strictly apply. Under normal engineering conditions, the deviation from Hooke's law is relatively small and may be neglected, but, from a fundamental point of view, it is important to recognize that there are deviations. The relationship between the stress and the strain is controlled by the intensity of the atomic bond, and the differences in moduli are actually associated with the bond characteristics of the different elements. Since the details of the microscopic character of the theory of elasticity cannot be covered in this text, complete works on this subject should be consulted for more information.

5.3 Plasticity. Plasticity is the property of a material by virtue of which it may be permanently deformed when it has been subjected to an externally applied force great enough to exceed the elastic limit. The subject of plasticity is of great importance to an engineer for it is this property that, in most cases, enables him to shape metals in the solid state.

The minimum stress that should cause permanent deformation can be computed from a knowledge of the bond strength. The result of such computations gives values that are from 100 to 1000 times the stress required to initiate plastic deformation of a crystal as determined by testing.

5.4 Slip Process. The examination of a single crystal after it has been subjected to plastic deformation reveals that deformation takes place by a gliding or a shearing action on certain planes within the crystal. This gliding or shearing action is discontinuous, and it occurs by distinct movement on certain planes. This is called a slip process. In some materials this movement can be heard, as in the case of what is called "tin cry." Deformation may also take place by a twinning process.

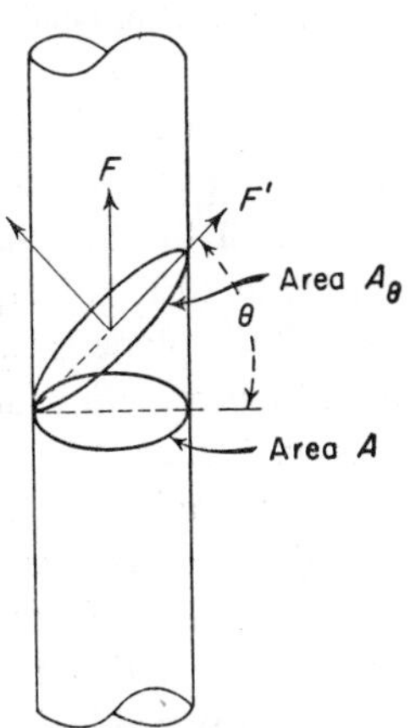

FIG. 5.2 Shear stress in relation to normal stress.

Slip will occur most readily in certain directions on certain specified planes. The stress that will just produce the first plastic deformation in the direction of easiest slip and on the plane of easiest slip is called the *critical shearing stress.* If a force F is applied to a simple cylindrical crystal, as shown in Fig. 5.2, the component of the force that will produce motion on the plane will be in the plane and in the direction of easiest slip. The component of this force is

$$F' = F \sin \theta$$

acting on the plane of area

$$A' = A/\cos \theta$$

Then the shear stress will be

$$\sigma_s = \frac{F \sin \theta}{A/\cos \theta} = \sigma \sin \theta \cos \theta$$

Some typical values of the critical shear stress for slip in crystals are shown in Table 5-II.

TABLE 5-II. CRITICAL SHEAR STRESS, PLANES, AND DIRECTION OF SLIP ON SOME METALS AT 20 C.

Metal	Impurity Content Per Cent	Slip Plane	Direction of Slip	Critical Shear Stress kg/mm²
Copper	0.1	(111)	$[10\bar{1}]$	0.10
Silver	0.01	(111)	$[10\bar{1}]$	0.060
Gold	0.01	(111)	$[10\bar{1}]$	0.092
Nickel	0.2	(111)	$[10\bar{1}]$	0.58
Magnesium	0.05	(0001)	$[11\bar{2}0]$	0.083
Zinc	0.04	(0001)	$[11\bar{2}0]$	0.094

From Schmid, International Conference on Physics, Vol. II, *The Solid State of Matter*, Phys. Soc. London, 1935.

The planes of easiest slip are those planes having the greatest atomic density. In face-centered cubic lattices, the (111) planes will slip most readily. In the body-centered cubic structures, the slip planes are (110), (112), or (123). For those metals that crystallize in the hexagonal system, the (0001) plane is the plane of easiest slip. In a similar manner, the direction of easiest slip is also the direction corresponding to the greatest atomic density. Hence, for the face-centered cubic structure, it is $[10\bar{1}]$. In the body-centered cubic structure it is [111]. In the hexagonal structure it is $[11\bar{2}0]$. These planes and directions are also shown in Table 5-II.

The ability of any metal to plastically deform is largely related to the electronic cloud surrounding each of the atoms. The metallic bond is largely responsible for the plastic behavior of the metals. In other words, atoms may be displaced with respect to each other without permanently breaking the bond. This is not true in covalent- and ionic-type structures.

The critical shear stress decreases as the temperature increases. The melting point of a metal is that temperature at which the thermal energy is just adequate to cancel the bonding forces such that the atoms are no longer held together. At the melting point, then, the critical shear stress is zero. This condition is not approached asymptotically as the melting point is approached, but at temperatures just below the melting point there is a finite critical shear stress and at the melting point there is an abrupt decrease to zero.

The influence of foreign atoms in the structure of a single crystal may have a pronounced effect on its strength characteristics. The importance of this effect is shown very forcefully in Fig. 5.3. In general, insoluble impurities do not influence the critical shear strength of single crystals. This is because these atoms cannot occupy any position on the space lattice of the crystal. In polycrystalline substances, such foreign atoms may have a distinct effect.

The critical shear strength of a single crystal of a metal is increased by plastic deformation. This work hardening effect is illustrated in the case of a single crystal of zinc in Fig. 5.4. As the amount of elongation increases, the critical shear stress increases. Aging of the material that has been plastically deformed may have a marked influence, as illustrated in Fig. 5.5. The mechanisms by which these effects occur in most instances are quite complex, and more specialized texts should be consulted for these discussions.

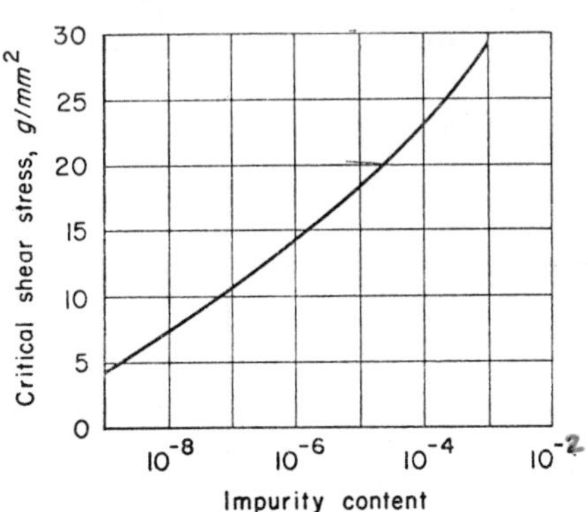

FIG. 5.3 Critical shear stress of mercury single crystals vs. impurity content. (Greenland, *Proc. Roy. Soc. A163*, **52** (1937))

5.5 Dislocations. A previous section of this chapter mentioned the fact that the calculated stress to initiate plastic deformation does not agree with experiments. According to calculations, the stress to produce an elastic deformation of a crystal of the order of 10 per cent should be from 100 to 1600 kg/mm^2. Such values are several thousand fold greater than the observed stress. Attempts have been made over a long period of time to explain this discrepancy. One theory, proposed by Griffith, considered that there are submicroscopic cracks in crystals which localize the applied stress or cause stress concentration. A network of these cracks could be responsible for the discrepancy between theoretical and experimental critical shear stress. The Griffith theory has not yielded completely satisfactory results. Smekel, on the other hand, has associated the discrepancy with a network or *mosaic structure* of defects. This theory has not proved to be completely satisfactory.

Originally it was believed that, by the slip process, whole planes of atoms in a lattice moved with respect to the atoms on the other side of the plane. Computation of the force required to accomplish this mass movement of a plane gives a value much greater than is observed experimentally. Hence a search was made for a completely different mechanism by which plastic deformation could occur. A theory was devised through the efforts of Orowan, Taylor, and Polanyi. The theory was later refined and advanced by such individuals as Reed, Frank, Burgers, and Shockley. Modifications of these theories have been made by many other investigators and have led to what is now called the dislocation theory of plastic deformation.

A *dislocation* is an imperfection in the space lattice. The literature on dislocation theory has now become quite voluminous. Some of the less complex aspects of dislocations will be discussed here. More details will be found in the references.

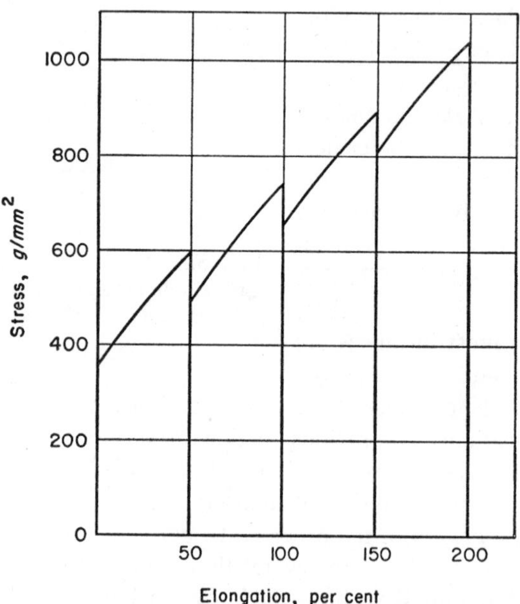

FIG. 5.4 Stress-strain diagram for single crystals of zinc with repeated cycles. (Haase and Schmid, *Zei. f. Physik* **33,** 413 (1925))

The simplest type of dislocation is called an *edge dislocation.* It is characterized by an extra plane of atoms above or below the dislocation line, as indicated in Fig. 5.6. If the extra plane of atoms is above the dislocation line, it is arbitrarily called a *positive* edge dislocation. If the extra plane is below the line, it is called a *negative* edge dislocation. An edge dislocation could also be defined as a structure in which one plane of atoms above or below the dislocation line is absent. An edge dislocation runs through the lattice perpendicular to the plane of the paper for the section shown.

The real significance of this model lies in the mobility of the dislocation. The dislocation can be displaced to either the left or the right by the application of a suitable force. The critical shear stress required to move a dislocation is of the order of 100th to 1000th of the stress calculated from lattice binding force. The movement of an edge dislocation accomplishes

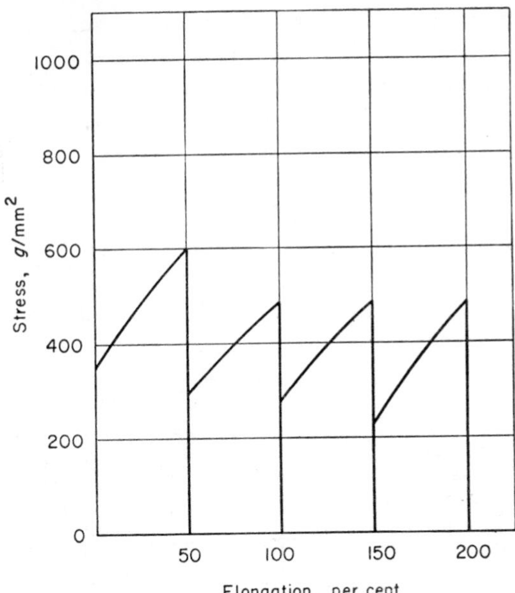

FIG. 5.5 Stress-strain diagram of a single crystal of zinc with repeated cycles interrupted by one-day intervals. (Haase and Schmid, *Zei. f. Physik* **33,** 413 (1925))

slip on the plane and the final result is as though the entire plane of atoms moved.

A dislocation can be described quantitatively by means of the *Burgers vector b*. The magnitude of this vector can be determined by tracing a closed

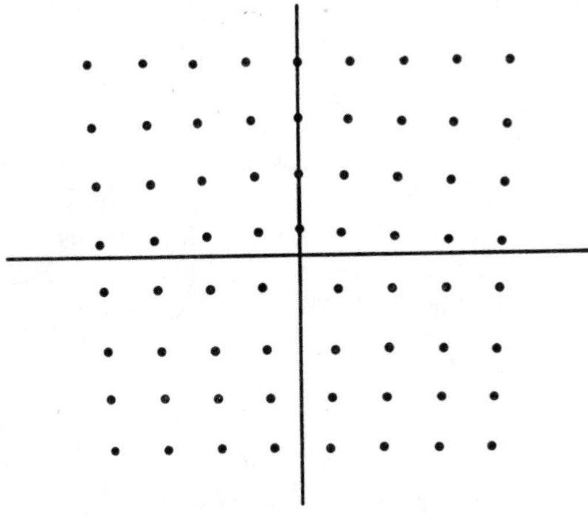

FIG. 5.6 Edge dislocation (positive).

loop in a dislocation free part of the lattice. The vector needed to close the loop is the Burgers vector. The vector is always given in terms of the lattice parameters of the crystal. An edge dislocation can never end inside the crystal: it can end at a grain boundary or at the surface of the crystal or in another kind of dislocation.

Another type of dislocation is called a *screw dislocation.* In this instance a helical surface is formed by the atomic planes around what is called the screw dislocation line. This is illustrated in Fig. 5.7. A *dislocation loop* may be formed within a crystal by the combination of edge dislocations and screw dislocations. An edge dislocation can end in a screw dislocation, and vice versa.

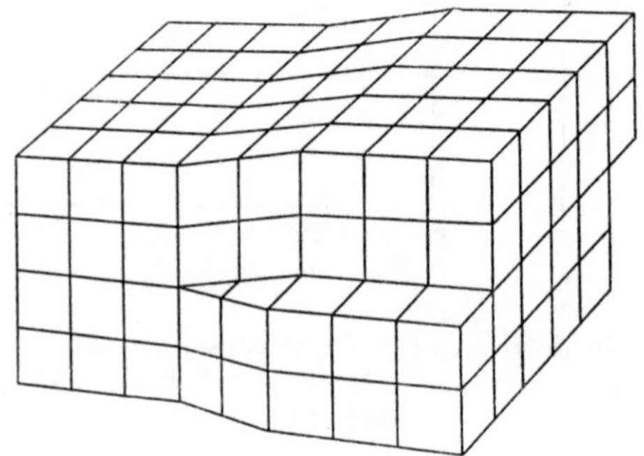

FIG. 5.7 Screw dislocation.

Another characteristic of dislocations is their ability to multiply. A process has been conceived by which dislocations can be generated from existing dislocations. This is believed to occur during the initial stages of the yielding phenomenon in certain types of metals. The action of dislocations can be modified by the presence of foreign elements in the lattice structure. Impurities such as carbon or nitrogen in iron may restrain the motion of a dislocation so that higher stresses may be required to produce deformation.

The dislocation theory was conceived to explain the mechanism of plastic deformation so that theory and experiment would be in agreement. One may be concerned about the proof that dislocations actually exist. By special techniques it has been shown that defects of this character do actually exist. By suitable etching techniques the location of groups of dislocations have been revealed and shown to move. These dislocations account for plastic deformation.

Unusual strength may be developed in metals grown in the form of whiskers. The strength of these very minute, almost invisible, crystals is attributed to their extreme purity and perfection. The absence of defects in the lattice makes plastic deformation more difficult; therefore, very high stresses which are more nearly those required theoretically are necessary to initiate plastic deformation.

5.6 Twinning. Deformation of crystals was stated to occur by either one of two processes, by a slip mechanism or by twinning. *Twinning* is a process distinctly different from that of slip. In twinning, the rows of atoms parallel to what is called the twinning plane are displaced along the plane a distance that is proportional to the distance from the twinning plane. This results in the formation of a structure that is a mirror image of the lattice on the other side of the twinning plane. The process of deformation by twinning is illustrated in Fig. 5.8.

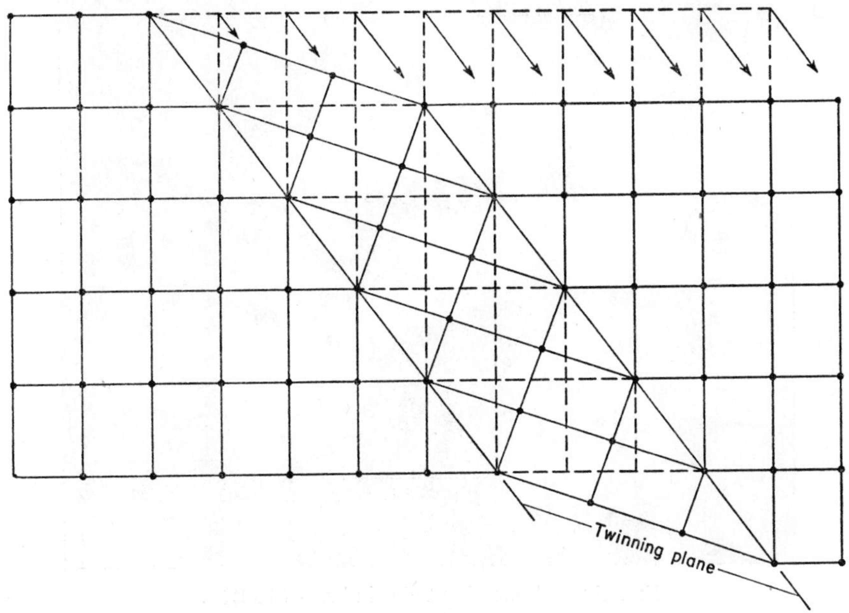

FIG. 5.8 Deformation by twinning.

Plastic deformation by twinning occurs principally in body-centered cubic and hexagonal metals. In the body-centered cubic lattices, the twinning plane is [112], and in the hexagonal close-packed structure the twinning plane is [102]. Deformation by twinning in the face-centered cubic metals does not seem to occur.

Metals that are plastically deformed by mechanical twinning alone or in a combination with slip include tin, zinc, magnesium, bismuth, and antimony. Twins may appear in alpha iron and alloy ferrite as *Neumann bands* resulting from impact deformation. Similar bands have been observed in copper and its alloys.

Twins may be formed by annealing of some cold-worked metals of the face-centered cubic type, including austenitic steels, silver, gold, copper, and nickel; that is, heating to a proper temperature after plastic deformation. These are called *annealing twins*. The formation of annealing twins is possibly caused by the crystalline growth of twinned nuclei produced by the previous cold work. An example of annealing twins in brass is shown in Fig. 5.9. The dark and light areas show the twinned structure. There is

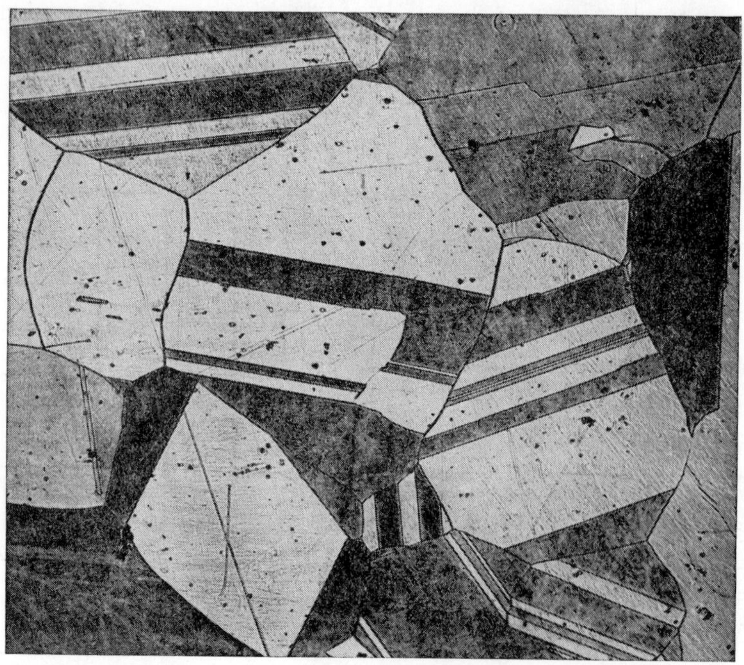

Fig. 5.9 Annealing twins in brass (100×).

some evidence to indicate that twinning also involves the motion of dislocations through the lattice structure, and not a mass movement of blocks of atoms.

5.7 Influence of Grain Boundaries. The metals utilized in engineering are composed of a multitude of single crystals of all possible orientations. The material is polycrystalline. The previous sections of this chapter have indicated the characteristics of single crystals. Their directional properties have been emphasized. The mechanism of deformation has been explained on the basis of the dislocation theory. All of these fundamental concepts are applicable to polycrystalline substances, but in this instance one must recognize that the grain boundaries may have an influence on the mechanical behavior of the polycrystalline aggregate. If all of the crystals were oriented the same in a polycrystalline metal, definite and distinct directional properties would be experienced, because of dislocation interactions. The orientation of the crystals (grains) in an iron-silicon alloy to produce directional magnetic behavior is an example of this effect.

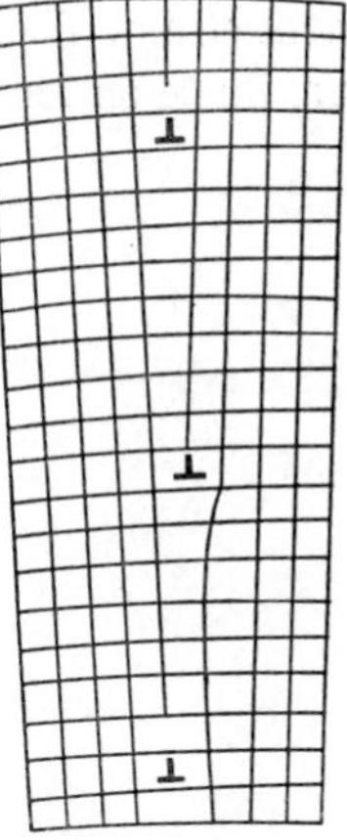

FIG. 5.10 Grain boundary formed by edge dislocation.

A *grain boundary* may be considered as being formed by edge dislocations, as indicated in Fig. 5.10, provided the boundary angle does not exceed approximately 26°. Grain boundaries are regions of misfits in the material and usually consist of an array of dislocations. They serve as a source and a sink for dislocations. Thus the population of dislocations in polycrystalline materials is much greater than in a single crystal. This increased density of dislocations may contribute to the increased strength in comparison to a single crystal. The widely varied orientation of the crystals making up the polycrystalline substance tends to produce a homogeneous isotropic mass. This indeed is fortunate because then it is possible to employ the relatively simple methods of engineering mechanics.

5.8 Elasticity of Polycrystalline Metals. The elasticity of single crystals was discussed briefly in Art. 5.2. Because of the polycrystalline character of the metals in practical use, provided the orientations are completely random, the three principal constants can be utilized in calculations of the stress-strain relations. However, if by some method preferred orientations occur, serious errors may result in the use of the three basic constants, namely, the tension modulus, the shear modulus, and Poisson's ratio.

The elastic constants for metals cannot as yet be derived because of the assumptions that are made in the fundamental relations and do not hold in the case of materials with the metallic type of bond.

5.9 Strength of Polycrystalline Metals. The meaning of the term "strength of a metal" is not always clear. The word "strength," to some individuals, means the resistance to plastic deformation and hence the stress that will just produce permanent deformation. To others, strength means the stress required to bring about fracture. Therefore it is desirable to establish clearly the significance of certain points on the curve of stress vs. strain of a metal, such as shown in Fig. 5.11. The first recognizable deviation from the linear portion of the stress-strain curve occurs at the *proportional limit*. This is defined as the maximum stress to which a material can be subjected without any deviation from the proportionality of stress and strain. In general, the proportional limit is less with the increasing accuracy of the measurement of both stress and strain. A term that is closely related to the proportional limit is the *elastic limit,* which is defined as the maximum stress to which a material can be subjected without causing permanent strain. This depends upon the accuracy with which the measurements are made and tends to become lower as the accuracy improves.

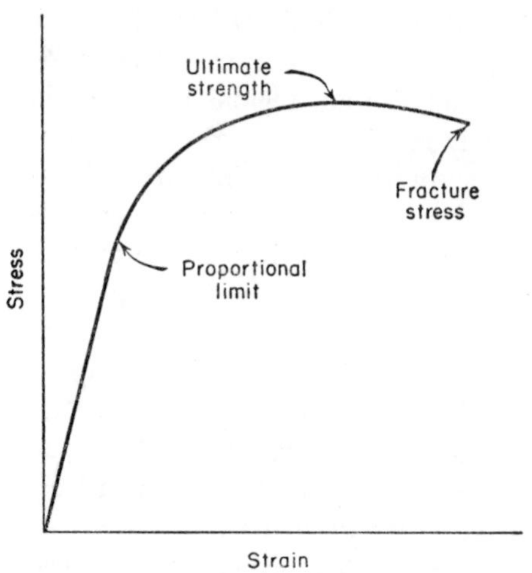

Fig. 5.11 Engineering stress-strain diagram.

In view of the fact that the proportional limit and the elastic limit may be a little difficult to determine, the engineer utilizes an arbitrarily defined point on the stress-strain curve which is designated as the yield strength. The *yield strength* is defined as the stress at which the material exhibits a specified limiting permanent deformation. In most cases this limiting permanent deformation is 0.2 per cent. In some materials the stress-strain diagram exhibits a discontinuity at stresses slightly above the proportional limit, as indicated in the stress-strain diagram for a low-carbon steel shown in Fig. 5.12. This material exhibits what is called a yield point. The *upper yield point* is that stress at which there occurs a sudden strain with an accompanying decrease in the resisting stress. The reduced value of the stress is then called the *lower yield point.* The material is never in a state of strain intermediate between the strain corresponding to the upper yield point and the strain corresponding to the lower yield point. The yielding is a slip process which occurs as a result of the pile-up of dislocations. When these dislocations break through the grain boundaries, there is cataclysmic slip which gives rise to the yield phenomenon. This behavior occurs only in body-centered cubic and hexagonal close-packed structures. The yield-point phenomenon is associated with the presence of a very small amount of a solute atom in the structure. In iron the solute is carbon or nitrogen or both. Such solute atoms occupy a position below the extra half-plane at an edge dislocation and on a line parallel to the dislocation. This line of atoms is referred to as an atmosphere. The atmosphere tends to restrain the motion of the dislocation. If the force tending to move the dislocation becomes great enough, yielding may be initiated.

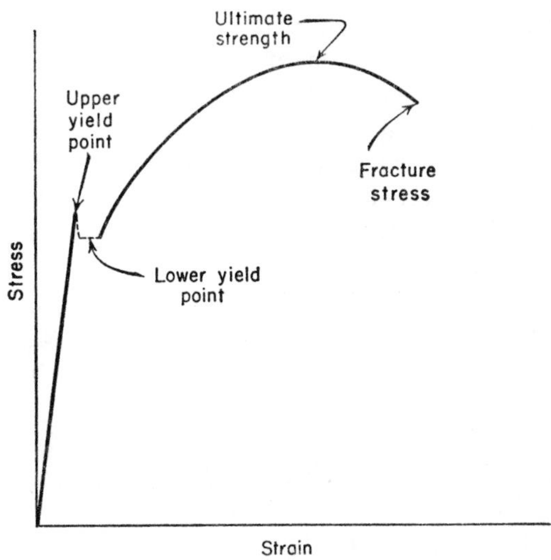

FIG. 5.12 Engineering stress-strain diagram for a low-carbon steel.

After yielding, the material will not exhibit a distinct yield point upon immediate retest. However, aging at room temperature or above allows the solute atoms to return to their favorable sites by diffusion. The recovery time is a function of temperature. The reappearance of the yield point is called *strain-aging*.

The maximum stress to which a material can be subjected in terms of the engineering stress, namely the force divided by the original cross-sectional area on which the force acts, is designated as the *ultimate strength* of the material. The stress at which fracture occurs is called the *fracture stress*. The ultimate strength and the fracture stress are indicated in Figs. 5.11 and 5.12.

5.10 Hardness. Hardness to the engineer refers to resistance to penetration. In another sense, hardness is a measure of resistance to permanent deformation and is related to the bond in lattice structure. There are a variety of methods by which hardness may be determined which are discussed in detail in texts on materials testing.

5.11 Other Mechanical Properties. There are many other mechanical properties that are of interest in specialized cases. For example, time is frequently an important factor in the behavior of materials; the interrelationship between stress, strain, and time is considered to be in the field of *rheology*.

While creep is a time effect, it is a plastic deformation that is of greatest consequence in metals utilized at elevated temperatures. The proportional limit or yield strength of a material is usually markedly affected by increasing temperature. The proportional limit may be so low that it would be almost impossible to use the material at elevated temperature without having a very large mass of the material. Therefore, under these circumstances, stresses are permitted which will produce plastic deformation over a period of time. So long as this is realized and provision is made for it, creep may be tolerated for a certain period of time. *Creep* has been the subject of very extensive investigations and forms a very important part of the study of the strength of materials.

Another property that is of considerable importance in certain applications is the response of materials to dynamic loading. Many materials behave quite differently when they are subjected to an impact or rapidly applied stress. The marked decrease of energy-absorbing capacity which occurs in some materials, notably low-carbon steel with decreasing temperature, is significant in many applications. The temperature at which this marked decrease of energy occurs is called the *transition temperature*. Much work has been done on the behavior of materials under these conditions.

A large portion of the metals in service for machine parts fail by fatigue. In most applications in engineering, materials are subjected to alternate loading and unloading. This type of stress application may cause failure at stresses very much below that which would cause failure under static conditions.

Another property that is of interest to those concerned with the fundamental behavior of materials is anelasticity. *Anelasticity* refers to the nonelastic properties of solids. In general, the anelasticity is inversely proportional to the bonding forces in a polycrystalline substance. Anelasticity pertains to the deformation of a material which upon rapid removal of the load would appear to be a permanent deformation, but over a period of time this deformation disappears.

5.12 Effect of Radiation on Mechanical Properties. During developments in the field of atomic energy, it has been shown that structural materials of reactors that are exposed to neutron bombardment may experience changes in mechanical properties. While this is a somewhat specialized field, it is important to recognize that radiations may have some effect on the mechanical properties of materials. The influence of radiation upon the properties has unfortunately been called radiation damage. Damage usually means a decrease in the material's usefulness. This is not always the case. It is better that this discussion be thought of as the effect of radiation upon properties.

Irradiation by neutron bombardment has very little influence on the modulus of elasticity of materials. Some investigators have reported slight effects, but in general these have been so slight as to be considered inconsequential.

Irradiation increases the yield strength of zirconium, some aluminum alloys, steel, and stainless steel. Increases in yield point vary from material to material but an annealed aluminum 1100 has shown an increase of 151 per cent in yield strength. Type 301 austenitic steel has shown an increase of about 126 per cent. Annealing after irradiation tends to counteract the effect.

Irradiation has about the same effect on tensile strength as it has on yield strength. The percentage change, however, is not usually as great. In 1100 aluminum the increase in tensile strength is about 91 per cent. The 301 stainless steel has an increase of about 15 per cent.

Irradiation appears to cause some decrease in the percentage elongation of materials. In view of the increased strength, this effect is not surprising.

Some results have indicated that irradiation causes an increase of the ductile-to-brittle transition temperature. This is a characteristic of the plain carbon steels and may be of some consequence in the design of pressure vessels for reactors. More details on the effects of radiation are presented in Chapter XXII, which is concerned with the metallurgy of nuclear engineering.

REFERENCES

Cottrell, *Dislocations and Plastic Flow in Crystals,* Oxford University Press, London, 1953.

Sinnott, *The Solid State for Engineers,* John Wiley and Sons, New York, 1958.

QUESTIONS

1. Define the mechanical properties of materials.
2. What is meant by the elasticity of a material?
3. What is Poisson's ratio?
4. Define stress.
5. What are the stress conditions for equilibrium?
6. What is Hooke's law?
7. What controls the relationship between stress and strain in a material?
8. What is plasticity?
9. What is the critical shearing stress of a material?
10. To what atomic characteristic is plasticity related?

11. What is the effect of foreign atoms in the structure of a single crystal?
12. What is the effect of plastic deformation upon the critical shear strength of a single crystal?
13. Why is the mass slip process not a satisfactory explanation of the mechanism of plastic deformation?
14. What is a dislocation?
15. Explain the mechanism of deformation by means of an edge dislocation.
16. What is the Burgers vector?
17. What is a screw dislocation?
18. What is a dislocation loop?
19. What is twinning?
20. What are Neumann bands? How are they produced?
21. What are annealing twins? How are they produced?
22. Describe the grain boundary of a crystal in relation to dislocations.
23. Under what conditions may the elastic constants of single crystals be used in polycrystalline materials?
24. What is the proportional limit of a material?
25. Define elastic limit.
26. Define yield strength.
27. Define yield point.
28. Distinguish between the upper and lower yield points.
29. How is the yield mechanism accounted for on the basis of dislocation theory?
30. What is strain-aging?
31. Define ultimate strength.
32. Define hardness in an engineering sense.
33. What is rheology?
34. What is creep?
35. What is anelasticity?
36. What is the general effect of radiation upon the mechanical properties of materials?

VI

PRECIPITATION HARDENING AND RECRYSTALLIZATION

6.1 Discovery and Significance. For many years, the properties of the nonferrous alloys could not be altered except by cold working and annealing to produce recrystallization. In 1906, Dr. Alfred Wilm discovered that the hardness of a quenched aluminum alloy containing 4 per cent copper and 0.5 per cent magnesium changed upon aging at room temperature. Dr. Wilm reported this phenomenon in 1911, but he was not able to provide an explanation.

The work of Merica, Waltenberg, and Scott in 1919 indicated that the essential condition for age hardening discovered by Wilm, is the precipitation of a phase from a solid solution in which solubility decreases with decreasing temperature. While their explanation was overly simple and conceived of precipitate particles providing mechanical interference to slip, it provided an impetus for the discovery of many alloys that exhibit age or precipitation hardening.

6.2 Precipitation Hardening. Precipitation hardening can be well illustrated by considering certain alloys in the aluminum-copper system. The aluminum-rich end of the aluminum-copper system is shown in Fig. 6.1. The maximum solubility of copper in aluminum is about 5.6 per cent of copper by weight at 1018 F (548 C). At room temperature, less than about 0.25 per cent copper is soluble in aluminum.

If an alloy of this system containing about 4 per cent copper is heated to a temperature of 950 or 1000 F (510 or 540 C), copper will be completely dissolved in the aluminum. If the alloy is then slowly cooled to room temperature, a new phase will appear at the grain boundaries of the alpha solid solution. This is the theta solid solution, which is characterized structurally as the compound $CuAl_2$. Although the designation of this phase as the compound $CuAl_2$ is not precise, it may be convenient. After this treatment, the alloy will have a tensile strength of approximately 35,000 $lb/in.^2$

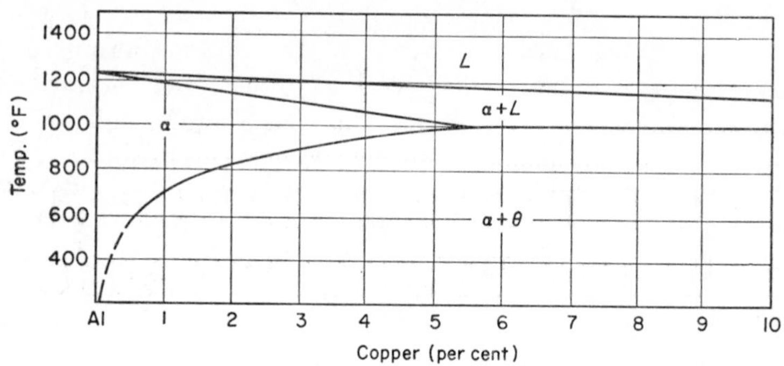

FIG. 6.1 Aluminum-rich end of the aluminum-copper equilibrium diagram.

If the same alloy is given a solution treatment by heating to a temperature between 970 and 1000 F (520 and 540 C) so that the copper is taken into solution and it is then rapidly cooled, the alloy will have a tensile strength of approximately 45,000 lb/in.[2] with a Brinell hardness of approximately 55.

If the hardness of this alloy is determined at frequent intervals, a change will be noted. After about ten days, the hardness will have reached a value of about 104 Brinell, and the tensile strength will be between 55,000 and 60,000 lb/in.[2] Since the change in properties occurs at room temperature, the phenomenon is referred to as natural age hardening. It has been found

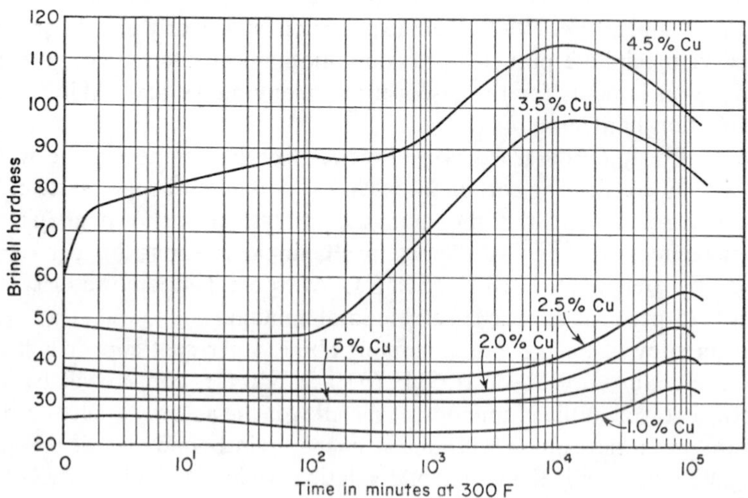

FIG. 6.2 Precipitation hardening curves for aluminum-copper alloys quenched in water at 212 F and aged at 300 F. (After Hunsicker, *Symposium on the Age Hardening of Metals*)

that any alloy system which shows limited or decreasing solubility in the solid state may be susceptible to age hardening or to precipitation hardening. In some alloys, aging at room temperature does not produce increased hardness and strength in a reasonable time; but by aging at a somewhat higher temperature, precipitation hardening occurs. This treatment is called *artificial aging,* and its effect is illustrated in Fig. 6.2.

6.3 Mechanism of Precipitation Hardening. Any process that increases the resistance to slip or the motion of dislocations in the lattice structure of crystals will effect an increase in the resistance to deformation or an increase of strength. This may be accomplished by the formation of a solid solution (solution-strengthening), or by controlled allotropic transformation, or by cold work, or by controlled precipitation from a solid solution.

In the case of alloys of the type discussed in the preceding article, the aluminum-copper alloy is typical. The as-quenched alloy may be softer than the alloy that is slowly cooled. The solid solution of the two components which exists at the elevated temperature is retained at room temperature by a quenching treatment. However, at room temperature, the solid solution is supersaturated, and there is a strong tendency for the solute atoms to aggregate and coalesce into particles of the precipitating phase. The application of the term *precipitation hardening* to this behavior stems from the early explanation of the mechanism that was believed to prevail. The condition of supersaturation and the fact that under equilibrium conditions two phases should exist at room temperature led early investigators to believe that the increased strength experienced on aging a quenched alloy of this type was the result of the formation of a very finely dispersed precipitate of the second phase. This first explanation was given by Merica, Waltenberg, and Scott in 1919.

While the precipitate idea presented by Merica, *et al.,* was basically sound, it did not provide the complete explanation for some of the phenomena observed subsequently. Jeffries and Archer proposed in 1922 that the optimum hardening corresponded to the formation of a critical particle size of the precipitate. However, changes in electrical conductivity and lattice parameter did not agree with the theoretical aspects that had been proposed. Furthermore, X-ray diffraction studies revealed structural changes to occur before hardening was observed. Merica in 1932 proposed that the hardening was caused by the formation of "knots" in the structure during a pre-precipitation stage of the aging. In 1935 Wasserman and Weerts presented the possibility of the existence of a transition lattice before the actual precipitate formed. Then in 1938, Guinier and Preston independently concluded from X-ray diffraction investigations that minute aggregates of solute atoms formed within the supersaturated solvent matrix. These aggregates are now known as *Guinier-Preston zones*. The actual precipitate is nucleated from the G-P zone stage and grows to a complete particle which is microscopically recognizable. In this state, however, the maximum strength has been passed and the alloy is in an over-aged state.

While the theory seems to be reasonable, although not simple, there remains much to be investigated and understood. Whatever may be the more complex theory of precipitation hardening, one can rightfully think of the phenomenon as involving the decomposition of a supersaturated solid solution. During some state in this decomposition, the distribution of the atoms produces an increase of hardness and strength. It is well established that the hardening effect takes place on aging prior to the formation of physically distinct, microscopic particles of a precipitating phase. As finite particles of the precipitate form and subsequently coalesce into larger particles, the hardness and strength decrease. This behavior is in many respects similar to the softening of hardened steel on tempering in which hardness decreases as the particles of cementite and ferrite become larger. At room temperature, most alloys do not exhibit any softening with aging because the mobility of the atoms is highly restricted, thus preventing coalescence.

Alloys, which exhibit decreasing solid solubility and do not harden after quenching and aging at room temperature, may be hardened by heating after the quenching treatment. An example of this is shown in Fig.

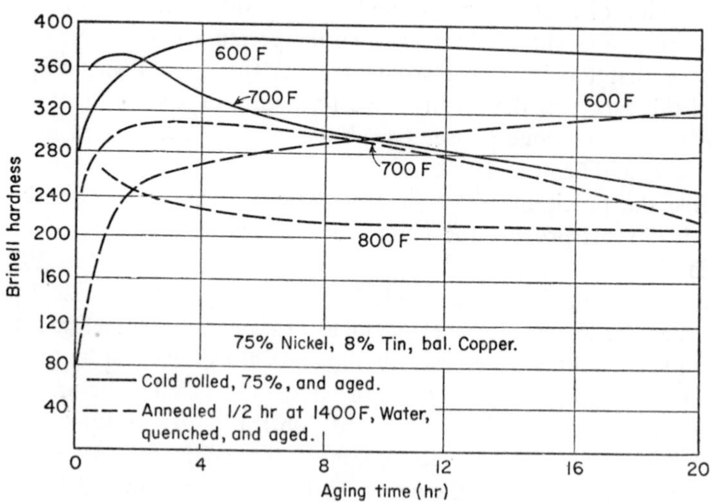

FIG. 6.3 Brinell hardness of nickel-copper-tin alloy aged at various times and temperatures. (Wise and Eash, *Trans. AIME*, Inst. of Metals Div., Vol. III, 1934, p. 218)

6.3 for a nickel-copper-tin alloy containing 75 per cent nickel, 8 per cent tin, with the remainder copper. The maximum hardness attainable through precipitation alone occurs by heating to approximately 600 F (315 C) for about 20 hr. By heating at a higher temperature, such as 700 F (370 C), a maximum hardness is attained in about 4 hr. This example illustrates that precipitation hardening involves both time and temperature.

The above discussion has been concerned with the formation of the precipitate and does not account for an increase of hardness and strength. The dislocation theory provides a means of understanding why a precipitate produces this effect. A particle of a second phase such as a precipitate can act as an obstruction to the motion of a dislocation. Any such obstruction provides increased resistance to deformation and hence increased strength and hardness. A dislocation can pass through an array of solute atoms in the manner indicated in Fig. 6.4. The greater the spacing of the particles, the easier the dislocations can pass through the array. This condition accounts for the fact that as a precipitate becomes coalesced the strengthening effect is decreased. As the dislocation line bends to pass around the obstructions, a dislocation loop is formed at each of the obstructions.

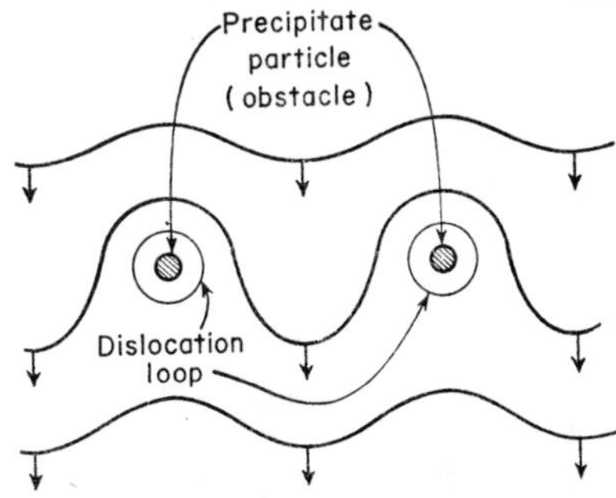

FIG. 6.4 Movement of a line of dislocations in a lattice containing obstacles in the form of precipitate particles.

6.4 Treatment of Precipitation-Hardenable Alloys. The treatment of alloys for precipitation hardening requires very careful control, because in most instances the temperature at which the homogeneous solid solution is attainable lies within a very narrow range. Heating above this temperature may result in burning, embrittlement, or actual fusion. Too low a temperature, on the other hand, will permit the complete separation of a portion of the second phase in the grain boundaries, thus preventing the attainment of maximum hardening. Proper time must be allowed for complete homogenization at the proper temperature. This process is known as the *solution treatment.* The time element will depend upon the type of alloy being treated and the section of the part. The time may vary from 10 to 15 min up to a matter of several days. With some alloys, it is necessary to transfer the parts from the heating furnace to the quenching bath in a very short time. Delay in this transfer may permit gross precipitation of the phase that is to bring about hardening. In some alloys, such as nickel-silicon and iron-copper, etc., precipitation does not occur very rapidly; hence, air cooling may be sufficiently rapid so that maximum hardening can be attained later by artificial aging.

The combination of aging temperature and aging time, known as the *precipitation treatment,* is rather critical for some alloys. As a general rule, a decrease of the aging temperature requires a considerable increase of the time of aging. In some alloys, lower aging temperatures bring about appreciable increase in strength without materially decreasing the ductility. This is evident in some aluminum-copper alloys.

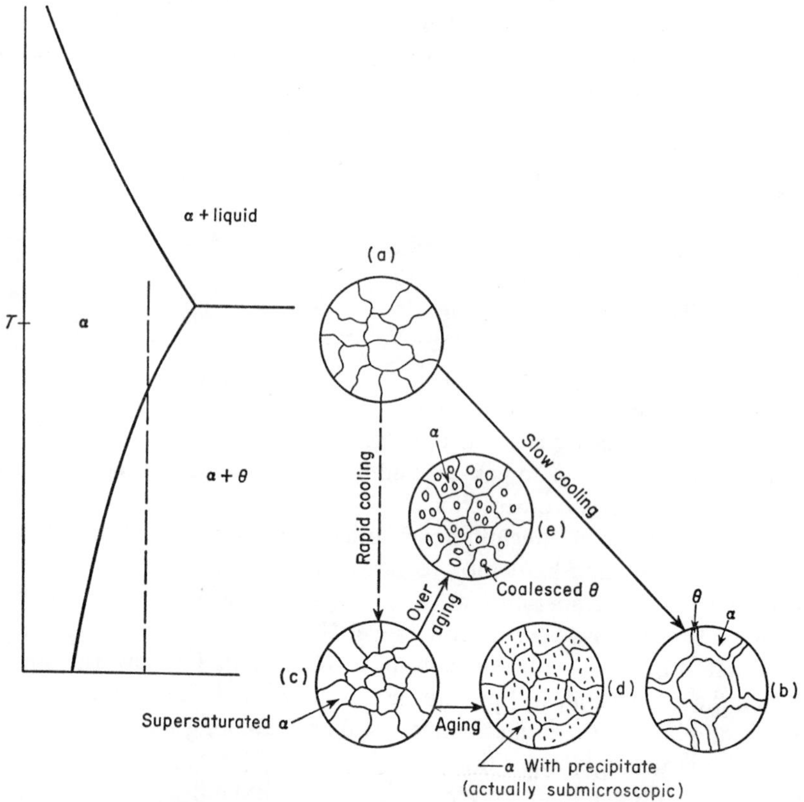

FIG. 6.5 Schematic representation of structures produced in different treatments of precipitation-hardenable alloys.

The steps involved in the precipitation hardening of alloys may be illustrated schematically as shown in Fig. 6.5. A portion of a typical equilibrium diagram in which alloys are susceptible to precipitation hardening is shown in the left portion of the figure. The alloy being used for the example is represented by the dotted line on the equilibrium diagram. The treatment and structures are shown schematically at the right of the figure. At temperature *T*, the structure consists of homogeneous grains of solid solution alpha as shown at (a). If the alloy is slowly cooled from temperature *T* to room temperature, the structure will consist of grains of alpha solid solution surrounded by the theta phase as shown at (b). In this event, precipitation and coalescence are complete; sufficient time has been given for the complete separation of the theta phase. If the alloy is rapidly cooled, as by quenching in water, from temperature *T*, the solid solution alpha is retained at room temperature as indicated at (c). When the alloy in this structural condition is aged at room temperature or slightly above, precipitation occurs in the manner that has been indicated. The structure of the alloy in this condition has been indicated at (d). Although this sketch indicates the

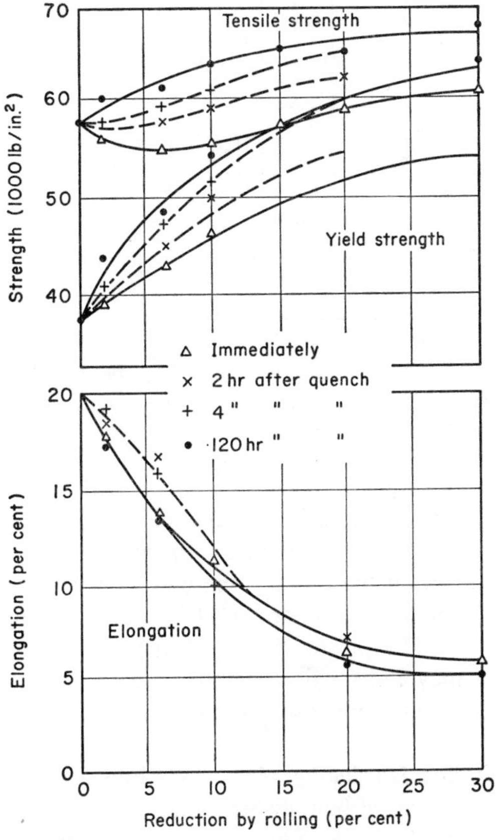

FIG. 6.6 Effect of plastic deformation at different stages of aging on mechanical properties of Duralumin sheet. (Sachs and Van Horn, *Practical Metallurgy*)

presence of a visible precipitate, this is only for convenience in representation; actually a precipitate cannot be observed after an aging time that will produce the maximum hardening. If the alloy is aged at a higher temperature for a sufficiently long time, the theta phase will physically precipitate and coalesce as indicated at (e).

Any plastic deformation that is to be applied to precipitation-hardenable alloys is usually done prior to the precipitation treatment when the metal is in the soft or annealed state. As an example, aluminum alloy rivets, which harden by natural aging, are usually driven before precipitation occurs. Precipitation hardening in alloys of this type can be prevented by holding them at a temperature of about 32 F (0 C) after the solution treatment. Hence, rivets of such a material can be held in readiness by keeping

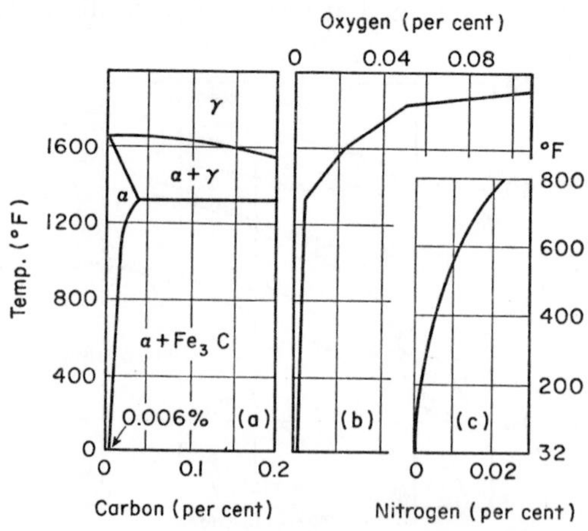

FIG. 6.7 Solubility of carbon, oxygen, and nitrogen in fairly pure iron.

them at this temperature before being driven; then when in place at room temperature, precipitation will occur.

The relation between strength, elongation, and the amount of plastic deformation at different stages of aging in an aluminum-copper alloy is shown in Fig. 6.6. From these data, one may conclude that, for this particular alloy, it is preferable to apply plastic deformation subsequent to the aging treatment. The plastic deformation of an alloy in the solution-treated condition may reduce the effective over-all strength of the alloy. This effect may be due to some precipitation caused by the plastic deformation.

6.5 Precipitation Hardening of Low-Carbon Steel. The solubility of certain elements, such as carbon, oxygen, and nitrogen, in alpha iron, commonly referred to as ferrite, decreases with decreasing temperature. This is shown in Fig. 6.7.

The solubility curves of carbon, oxygen, and nitrogen in iron have the characteristic shape of the precipitation-hardenable alloys. In the very low-carbon steels, precipitation hardening may be attained by quenching and then aging. The temperature of aging influences the maximum increase of hardness attainable and the time required to bring about this change. The hardening is probably due to precipitation of iron carbide, iron nitride, etc. Oxygen is found to have a pronounced effect upon this precipitation. Thoroughly deoxidized steels do not usually show the aging characteristics. The effect of aging very low-carbon steel at different temperatures is shown in Fig. 6.8. The precipitation in very low-carbon steel is not of practical con

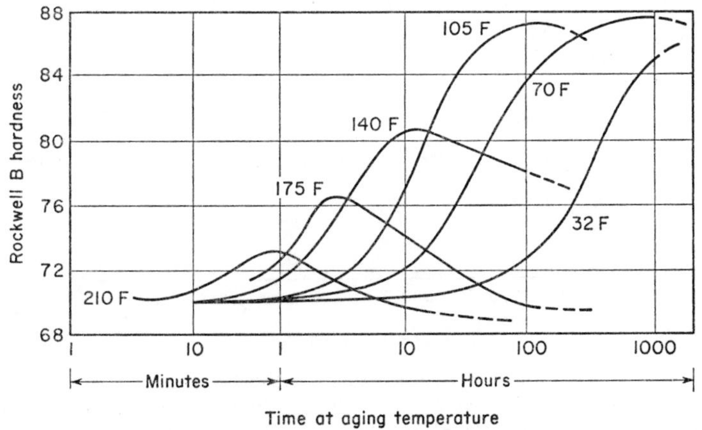

FIG. 6.8 Effect of aging 0.06% carbon steel after quenching from 1325 F. (Davenport and Bain, *Trans. ASM* **23,** 1935, p. 1047)

sequence because of the minor effect and because it occurs in the extremely low-carbon steels which are seldom used commercially.

6.6 Precipitation Hardening in Other Alloys. There is a large number of other alloys in which precipitation hardening is observed and utilized in practice. One of these is a stainless steel described in greater detail in Chapter XIV. Specifically, this material is designated 17-7 PH stainless steel and contains 0.07% C, 17% Cr, 7% Ni, and 1.2% Al. The properties in both the annealed state and in the precipitated state are indicated in the following table:

	Annealed	*Hardened*
Ultimate strength lb/in.^{2}	129,000	230,000
Yield strength lb/in.^{2}	47,400	216,600
Elongation in 2 in.	38.6	7.2
Hardness	B86	C46

Some of the alloys for use at elevated temperatures are strengthened markedly by precipitation hardening. Another example of the influence of precipitation is illustrated in the case of certain steels when heated after quenching (tempered). The effect is clearly shown in Fig. 9.19 for the influence of molybdenum on secondary hardening. The effect of tempering on high speed steel is shown in Fig. 6.9. The greater hardness secured by tempering at a temperature of about 1200 F results from the formation of a carbide precipitate.

There are also the beryllium-copper alloys, copper-silicon alloys, etc., which are also examples of the multitude of alloys in which precipitation hardening is important.

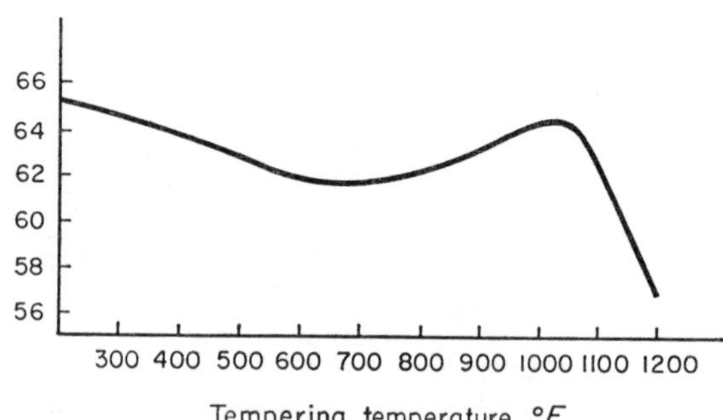

FIG. 6.9 Effect of tempering temperature on the hardness of an 18-4-1 high-speed steel, oil-quenched from 2350 F. (Palmer and Luerssen, *Tool Steel Simplified,* Carpenter Steel Co., Reading, Pa., 1948)

6.7 Recrystallization. The plastic deformation of any metal produces internal stresses and alters the electrical conductivity as well as the magnetic permeability of the material. However, heating of a plastically deformed metal will effect a relief of the internal stresses and a recovery of electrical conductivity and magnetic permeability. The effect of heating to different temperatures on the room temperature properties and the change that occurs in the grain structure are shown in Fig. 6.10. As the temperature is first increased, no change in the structure can be detected and the strength properties are not affected. However, the electrical conductivity and magnetic permeability increase during the early stages of heating. Furthermore, the internal stress is markedly decreased during this period. This stage is called the *recovery period.* At the temperature corresponding to point (a), nuclei of stress-free grains begin to form in the slip planes, undoubtedly as a result of the residual distortion energy in the crystal lattice. These nuclei will generally form in the slip planes, at grain boundaries, or at twinning planes, wherever there is a high energy concentration. As heating continues, the nuclei grow to visible size as new, equiaxed, stress-free grains, the original plastically deformed grains gradually disappearing. The severely deformed grains are, in one sense, eaten away by the stress-free grains during their process of growth. At a temperature corresponding to (b) in Fig. 6.10, the metal is essentially free of internal stress, and the structure consists entirely of very small equiaxed, stress-free grains. The temperature interval between points (a) and (b) is a zone of recrystallization during which nuclei form and grow into the equiaxed, stress-free grains.

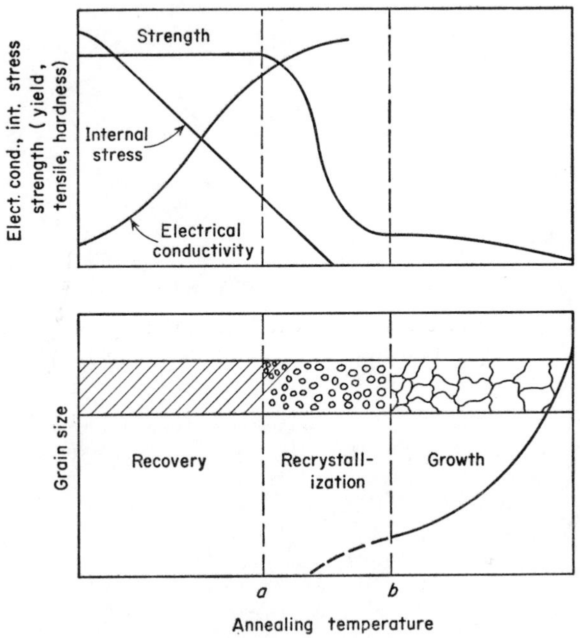

FIG. 6.10 Effect of heating a plastically deformed metal on grain size and properties.

The change of structure that occurs on heating an alloy containing 80 per cent copper and 20 per cent zinc (brass) after plastic deformation is shown in Fig. 6.11. The structure of the alloy after plastic deformation by reducing the section area by 60 per cent is shown in (a) where it can be noted that the grains are severely distorted. Heating of the plastically deformed alloy to a temperature of 750 F (400 C) produces a refinement of the grains and removes the distortion as shown in (b). Heating to a temperature of 1000 F (540 C) causes some grain growth as indicated in (c). By heating to a temperature of 1500 F (815 C), marked grain growth occurs as shown in (d).

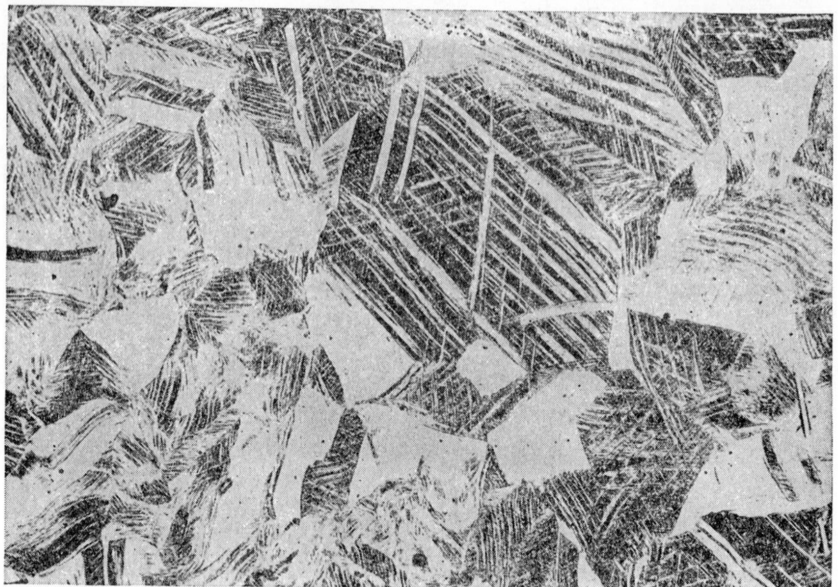

FIG. 6.11a Structure of brass (80% Cu 20% Zn) as plastically deformed by 60% reduction of section area (100×).

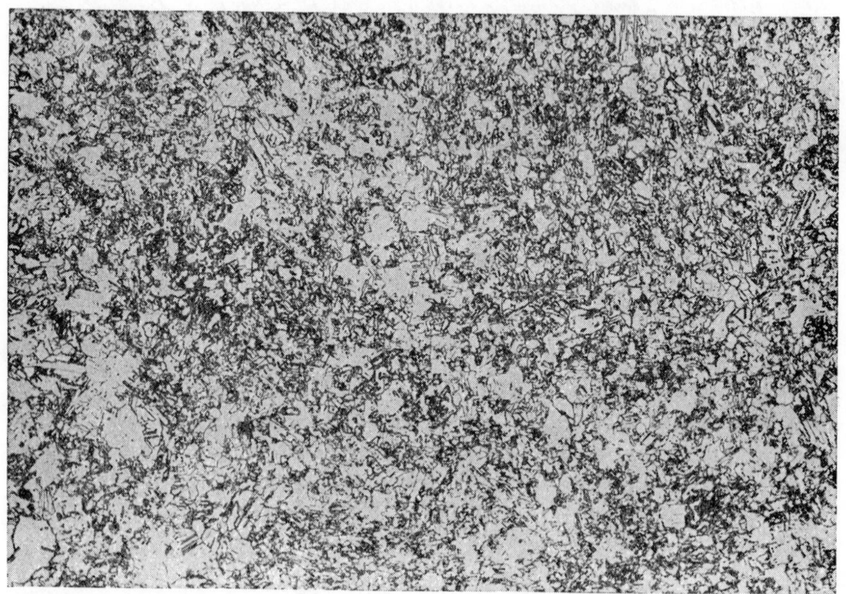

FIG. 6.11b Structure of brass (80% Cu 20% Zn) plastically deformed by 60% reduction of section area and heated to 750 F (100×).

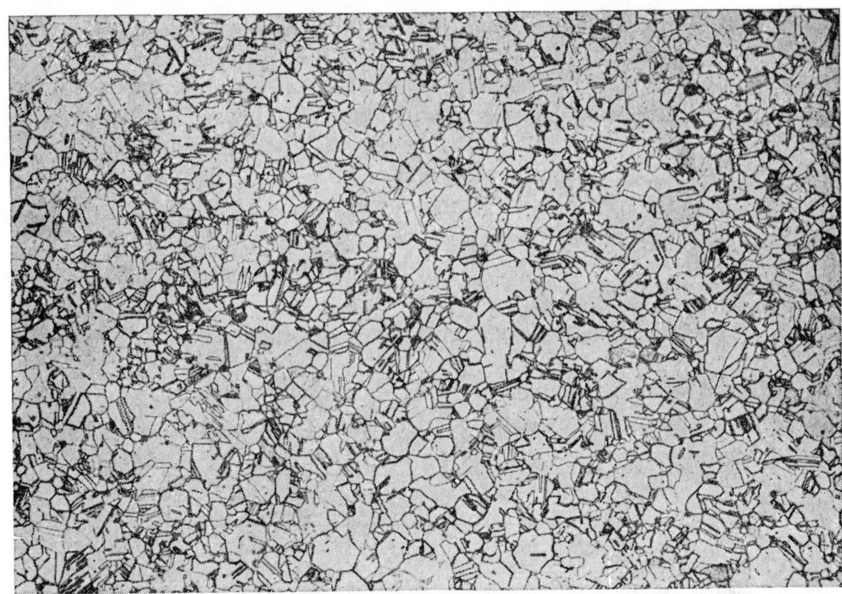

FIG. 6.11c Structure of brass (80% Cu 20% Zn) plastically deformed by 60% reduction of section area and heated to 1000 F (100×).

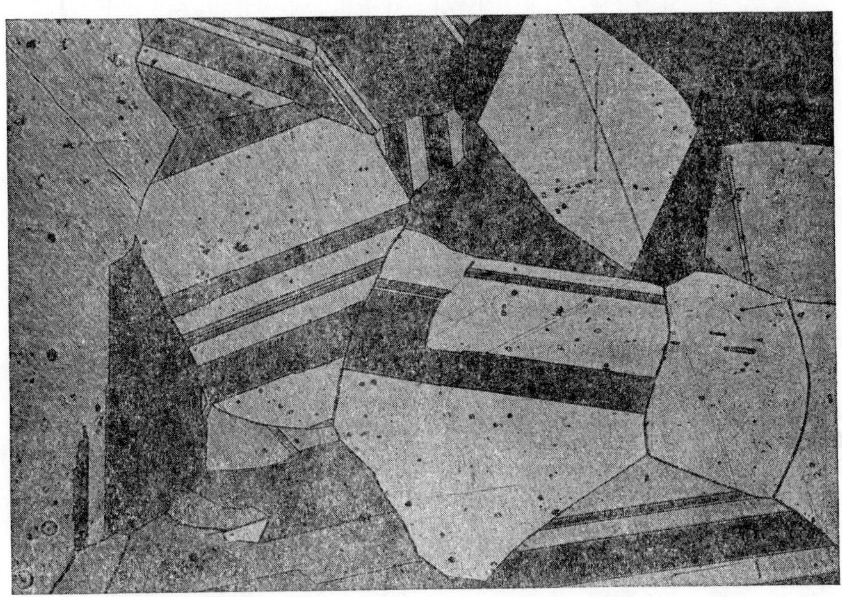

FIG. 6.11d Structure of brass (80% Cu 20% Zn) plastically deformed by 60% reduction of section area and heated to 1500 F (100×).

The *recrystallization temperature* is defined as the lowest temperature at which equiaxed, stress-free grains appear in the structure of a previously plastically deformed metal. The recrystallization temperature depends upon several factors, the principal ones of which are as follows:

1. The severity of plastic deformation.
2. The grain size prior to plastic deformation.
3. The temperature at which plastic deformation occurs.
4. The time for which the plastically deformed metal is heated to attain recrystallization.
5. The presence of dissolved or undissolved elements.

The extent to which the degree of plastic deformation influences the recrystallization temperature of electrolytic iron, low-carbon steel, and cartridge

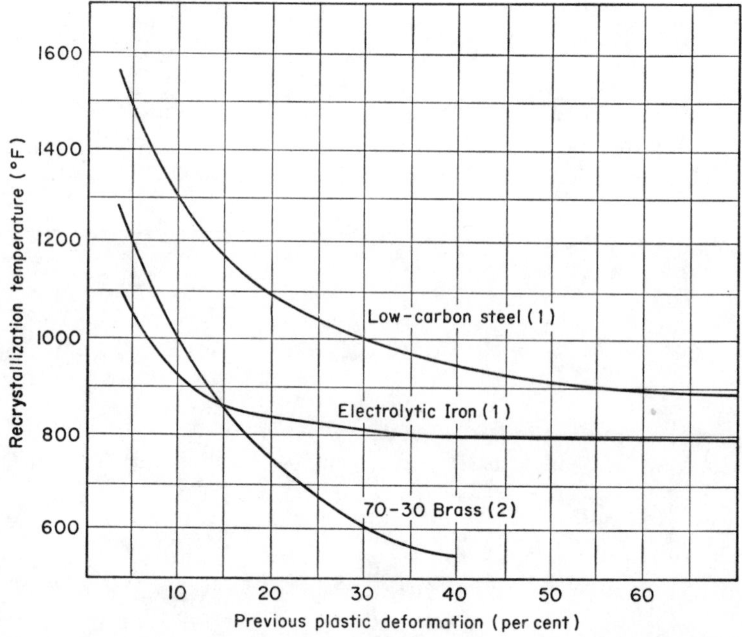

Fig. 6.12 Effect of amount of plastic deformation on the recrystallization of iron, steel, and brass. (1) After Kaiser and Taylor, *Trans. ASM* **27,** 1939, p. 227. (2) After Mathewson and Phillips, *Trans. AIME* **54,** 1916, p. 608.

brass is shown in Fig. 6.12. This shows that the greater the amount of plastic deformation, the lower the recrystallization temperature. The finer the grain size prior to the plastic deformation, the lower will be the recrystallization temperature. The lower the temperature at which plastic deformation occurs, the lower the temperature of recrystallization. Longer times at the temperature to which a plastically deformed metal is heated will give evidence of a lower recrystallization temperature. If, for example, a metal is plastically deformed a certain amount and heated to a particular temperature for 15 min, no evidence of recrystallization can be observed. If it is heated for 2 or 3 hr at the same temperature, recrystallization can be noted. On the other hand, if this same metal, plastically deformed the same amount, is heated to a somewhat higher temperature, complete recrystallization can be observed in 15 min. This illustrates the effect of temperature and time at temperature upon the recrystallization temperature. Soluble impurities and alloy elements generally increase the temperature of recrystallization. Insoluble impurities have little effect except to inhibit subsequent grain growth. The recrystallization temperatures for some metallic elements after severe plastic deformation are given in Table 6-I.

TABLE 6-I. APPROXIMATE MINIMUM RECRYSTALLIZATION TEMPERATURES.

Lead	Below room temp.		Iron	840 F	450 C
Tin	Below room temp.		Platinum	840 F	450 C
Cadmium	Room temp.		Nickel	1110 F	600 C
Zinc	Room temp.		Titanium	1200 F	650 C(?)
Magnesium	300 F	150 C	Beryllium	1290 F	700 C
Aluminum	300 F	150 C	Molybdenum	1650 F	900 C
Copper	390 F	200 C	Tantalum	1830 F	1000 C
Gold	390 F	200 C	Tungsten	2190 F	1200 C
Silver	390 F	200 C			

It is important to recognize that plastic deformation followed by recrystallization furnishes a means of securing grain refinement in metals and alloys that do not exhibit allotropic transformation.

6.8 Grain Growth after Recrystallization. When a plastically deformed metal is heated to temperatures above the recrystallization temperature, the small, recrystallized grains grow larger. This increase in size of the grains occurs by a process of coalescence or reorientation of adjoining grains and is a function of time and temperature. As the grain size increases, the strength and hardness continue to decrease as indicated in Fig. 6.10. If the metal has been uniformly plastically deformed prior to recrystallization, all grains become uniformly larger as they are heated above the recrystallization temperature. If, however, the metal has not been uniformly plastically deformed or the composition is not the same throughout the structure, an abnormally large and nonuniform grain size may result. The production of abnormally large and nonuniform grains is called *germina-*

tion. Germination may also be produced by very slow heating after mild deformation.

The production of coarse grains by growth after recrystallization is not generally desirable. Furthermore, germination resulting from heating after plastic deformation is extremely detrimental, particularly if the germination is carried to the point of the formation of extremely large grains. General ductility of the metal is greatly impaired and serious consequences may occur if a metal in this condition is put into service. Nonuniform heating or the existence of temperature gradients in a plastically deformed metal may lead to germination. The presence of certain dispersed phases in the structure of an alloy may assist in the germination process. Therefore, if a plastically deformed metal is to be subjected to localized heating as by welding, the structure and properties may be adversely affected. Any localized heating of a plastically deformed metal will usually impair the metal for the use for which it is intended.

REFERENCES

Barrett, *Structure of Metals,* McGraw-Hill Book Co., New York, 1943.

Hardy and Heal, *Report on Precipitation, Progress in Metal Physics,* Vol. 5, Interscience Publishers, New York, 1954.

Jeffries and Archer, *The Science of Metals,* McGraw-Hill Book Co., New York, 1924.

Merica, Waltenberg, and Scott, *Heat Treatment and Constitution of Duralumin, Trans. AIME* **64,** 1920, pp. 41-77.

Metals Handbook, American Society for Metals, Metals Park, Ohio, 1961.

Precipitation from Solid Solution, American Society for Metals, Metals Park, Ohio, 1959.

Sachs and Van Horn, *Practical Metallurgy,* American Society for Metals, Metals Park, Ohio, 1940.

Symposium on the Age Hardening of Metals, American Society for Metals, Metals Park, Ohio, 1940.

QUESTIONS

1. Describe the general process of precipitation hardening, using an aluminum-copper alloy as an example.
2. Describe the structural changes that occur in the precipitation hardening of an alloy.
3. What part does the solution treatment play in the treatment of a precipitation-hardenable alloy?
4. What type of alloys are most likely to exhibit precipitation hardening?
5. Draw typical curves of hardness versus aging time at different temperatures for a hypothetical alloy that exhibits precipitation hardening.
6. What part do the Guinier-Preston zones play in the mechanism of precipitation hardening?

7. Explain why precipitation produces an increase of strength in terms of the dislocation theory.
8. Show by means of a sketch the steps involved and the structural changes that occur in the precipitation hardening of alloys.
9. Name some typical alloys in which precipitation hardening is observed.
10. Describe the changes that occur during the heating of a plastically deformed metal to successively higher temperatures.
11. Define recrystallization temperature.
12. Upon what factors does the recrystallization temperature depend?
13. What benefits are derived through recrystallization?
14. What occurs to the structure of a metal when heated to temperatures above the recrystallization temperature?
15. What is germination, and how is it produced?
16. What is the effect of germination upon the properties of a metal?
17. What factors contribute to germination?

VII

IRON-CARBON ALLOYS

7.1 Iron. The metal iron serves as the base for some of the most important engineering alloys. A relatively small amount of iron is used in the pure state in comparison with that used in alloys. The analysis of several forms of iron is given in Table 7-I. The most readily available form of commercially pure iron is that which is made in the open-hearth furnace by the Armco Steel Corporation and is commonly referred to as *ingot iron.* It is available in the form of sheet, strip, plate, rod, wire, and castings. This material, under the trade name of Armco Ingot Iron, is used extensively for drainage culverts, roofing, ducts, grave vaults, and as a base for porcelain enamel as used in refrigerator cabinets, stoves, sign panels, etc.

Commercially pure iron is relatively soft and is readily cold-worked. Typical mechanical properties of annealed Armco iron are as follows:

Tensile strength	45,000 lb/in.^{2}
Yield point	30,000 lb/in.^{2}
Elongation in 2 in.	35-40%
Reduction of area	75%

The strength of iron can be raised by strain hardening. The magnetic characteristics of pure iron, namely, high permeability and low remanence, make its use in direct-current magnetic circuits very desirable.

There are three allotropic forms of iron, namely, delta (δ), gamma (γ), and alpha (α), in the order named from the melting point to room temperature. The structures of these forms are, respectively, body-centered cubic, face-centered cubic, and body-centered cubic. Magnetically, there are two forms of alpha iron. It is paramagnetic in the temperature range of 1414 F to 1663 F (768 to 906 C) and ferromagnetic at temperatures below 1414 F (768 C). The temperature at which this magnetic transformation occurs is called the *Curie* temperature.

TABLE 7-I. ANALYSIS (%) OF SEVERAL FORMS OF IRON.
(After *Metals Handbook*)

Designation	C	Mn	P	S	Si
Armco ingot iron	0.012	0.017	0.005	0.025	Trace
Electrolytic iron	0.006	—	0.005	0.004	0.005
Hydrogen purified iron	0.005	0.028	0.004	0.003	0.0012
Carbonyl iron	0.0007 0.00016	—	—	—	—

7.2 Wrought Iron. Prior to the extensive use of steel, wrought iron was the principal metal for construction purposes. *Wrought iron* is a mechanical mixture of very pure iron and a silicate slag. From about 2 to 4 per cent slag is mechanically mixed in the iron. The older puddling process of producing wrought iron has been largely replaced by a more modern method by which the quality can be more carefully controlled.

A typical percentage analysis of wrought iron follows:

Carbon	0.02
Manganese	0.03
Phosphorus	0.12
Sulfur	0.02
Silicon	0.15
Slag	3.00

The average properties of wrought iron are summarized in the following table:

Modulus of elasticity	29×10^6 lb/in.^{2}
Shear strength	38,000 to 40,000 lb/in.^{2}
Rockwell hardness	B55
Tensile strength	48,000 to 52,000 lb/in.^{2}
Yield point	30,000 lb/in.^{2}
Elongation in 8 in.	25% minimum
Reduction of area	40-48%
Charpy impact	40-44 ft-lb

As in the case of pure iron, the ultimate strength of wrought iron can be increased considerably by cold working followed by a period of aging. Wrought iron is noted for its high ductility and for the ease with which it can be forged and welded. Its adaptability to welding is no doubt due to the intimate mixture of slag which acts as a flux. It has been claimed that the slag also acts to prevent the inroads of corrosion.

Wrought iron is never cast. All shaping is accomplished by hammering, pressing, or forging. Examples of its use are for lap- and butt-welded pipe, flat and corrugated sheet metal, ornamental fences and grills, and a wide variety of iron-bar products.

7.3 Binary Alloys of Iron. A very large number of elements form alloys with iron. Some of the elements that combine with iron to form commercially important alloys include carbon, chromium, nickel, silicon, and manganese. The reader is referred to the *Metals Handbook* for a detailed discussion of the phase relations and properties of these many alloys.

7.4 Iron-Carbon Alloys. Carbon is the most important alloying element with iron. Iron and carbon form an intermetallic compound, represented by the formula Fe_3C, which is very hard and brittle. Iron carbide contains 6.67 per cent carbon by weight and is commonly referred to as *cementite.*

Iron carbide in the pure form dissociates readily into iron and graphite on heating to above 2012 F (1100 C). The properties of iron are greatly changed by the combination with carbon. The relative solubility of carbon in the different allotropic forms of iron contributes very markedly to the characteristics of the iron-carbon alloys. The iron-carbon alloys containing from a trace to 1.7 per cent carbon with only minor amounts of other elements are referred to as *plain carbon steels.* Those alloys which contain

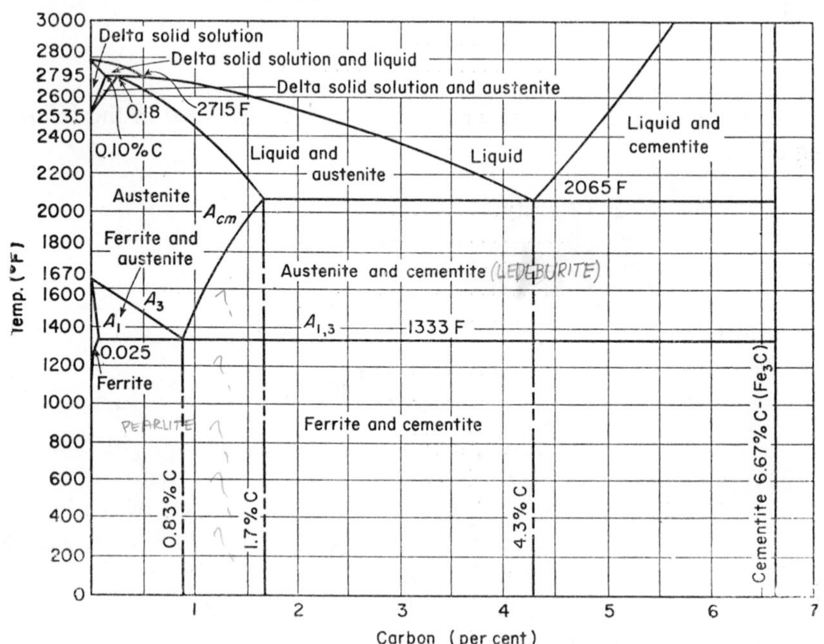

FIG. 7.1 Equilibrium diagram of iron and iron carbide.

from 1.7 to 6.67 per cent carbon and only small amounts of other elements are referred to as the *white cast irons*. Since under equilibrium conditions the carbon in these alloys is in the form of the compound iron carbide, the system may be referred to as the iron-iron carbide system.

7.5 Iron-Iron Carbide Equilibrium Diagram. The equilibrium diagram of iron and iron carbide is shown in Fig. 7.1. In this system, solid solutions of carbon in delta, gamma, and alpha iron exist, namely, delta solid solution, austenite, and ferrite, respectively. A eutectic (ledeburite) occurs at a composition of 4.3 per cent carbon and consists of austenite (1.7 per cent carbon) and iron carbide (cementite). There is a eutectoid (pearlite) containing 0.83 per cent carbon composed of ferrite and cementite. It is customary to refer to steels containing less than 0.83 per cent carbon as *hypoeutectoid steels,* and to those containing more than 0.83 per cent carbon as *hypereutectoid steels.* A summary of these phases and structural constituents follows:

Name	*Description*
Delta solid solution.......	Solid solution of carbon in body-centered cubic or delta (δ) iron.
Austenite...............	Solid solution of carbon in face-centered cubic or gamma (γ) iron.
Ferrite..................	Solid solution of carbon in body-centered cubic or alpha (α) iron.
Cementite...............	Intermetallic compound Fe_3C.
Pearlite..................	Eutectoid of ferrite and cementite.
Ledeburite..............	Eutectic mixture of austenite and cementite.

In determining the cooling curves for alloys containing less than 1.7 per cent carbon, arrests are experienced when the transformations occur. The temperatures at which the transformations in the solid state take place are called *critical temperatures* or *critical points.* The loci of these critical temperatures are given symbols. The curve separating the phase regions, austenite and austenite plus ferrite, is designated by A_3. The temperature of the magnetic transformation is designated as A_2. The eutectoid temperature is denoted by A_1 for alloys to the left of the eutectoid, and by $A_{3,1}$ for alloys to the right of the eutectoid. The curve separating the phase regions, austenite and austenite plus cementite, is designated as A_{cm}. The interval between A_1 and A_3 is called the *critical range.*

An understanding of the equilibrium diagram is best obtained by following the process of solidification and subsequent cooling of different alloys in this system.

7.6 Transformation in the Range 0 to 1.7 Per Cent Carbon. Consider the slow cooling of an alloy containing 0.3 per cent carbon under equilibrium conditions from a temperature of 2800 F (1538 C) to atmospheric temperature. In describing the transformations that occur in this alloy upon cooling, it is helpful to construct a transformation diagram of the type described in Chapter III. For convenience, a portion of the iron-iron carbide equilibrium diagram is reproduced in Fig. 7.2. At temperature *a* in Fig. 7.2, solidification begins with the formation of crystals of delta solid

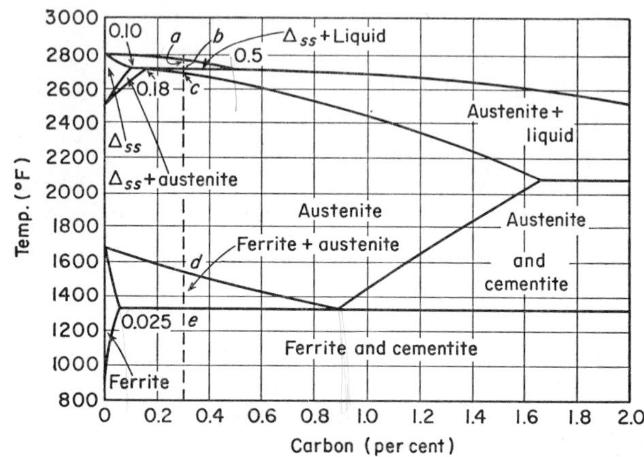

FIG. 7.2 Portion of equilibrium diagram of iron and iron carbide.

solution (Δ*ss*). With continued cooling, delta solid solution continues to form until temperature *b* is reached. At this instant, the alloy will contain the following proportions of delta solid solution and liquid:

$$\text{Per cent } \Delta ss = \frac{0.5 - 0.3}{0.5 - 0.1} \times 100 = \frac{0.2}{0.4} \times 100 = 50\%$$

$$\text{Per cent liquid} = \frac{0.3 - 0.1}{0.5 - 0.1} \times 100 = \frac{0.2}{0.4} \times 100 = 50\%$$

This relation is shown in the transformation diagram of Fig. 7.3. At this temperature, the peritectic reaction occurs. When this reaction is completed, the phases will consist of austenite and liquid in the following proportions:

$$\text{Per cent austenite} = \frac{0.50 - 0.30}{0.50 - 0.18} \times 100 = \frac{0.20}{0.32} \times 100 = 62.5\%$$

$$\text{Per cent liquid} = \frac{0.30 - 0.18}{0.50 - 0.18} \times 100 = \frac{0.12}{0.32} \times 100 = 37.5\%$$

These proportions are shown in the transformation diagram of Fig. 7.3.

As cooling continues, more austenite forms until temperature *c* (Fig. 7.2) is reached, at which solidification is completed. At this temperature, the mass will be composed entirely of grains of austenite as indicated in the transformation diagram of Fig. 7.3. This phase persists with continued cooling until a temperature corresponding to *d* (Fig. 7.2) is reached. At

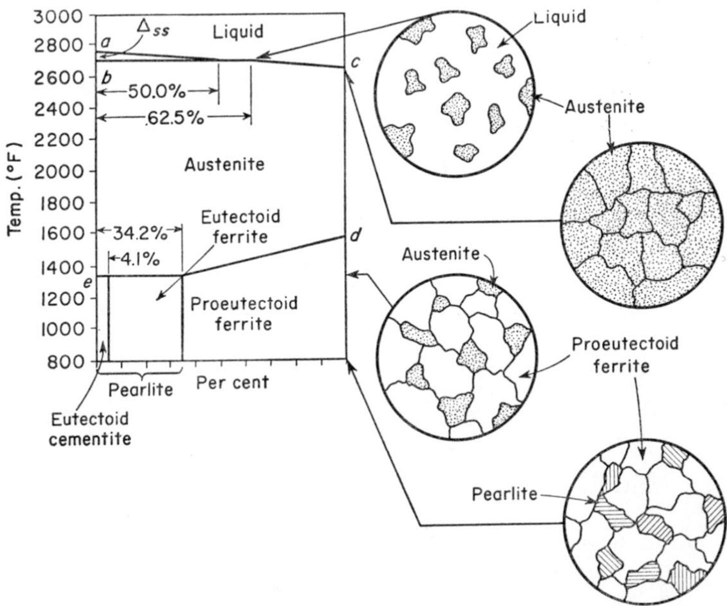

FIG. 7.3 Phase transformation diagram for 0.3% carbon-iron alloy.

this temperature, the A_3 transformation curve is intercepted, and ferrite begins to form. Ferrite continues to form with decreasing temperature until *e* (1333 F (723 C)) is reached. As the temperature just reaches *e*, the alloy is composed of the following proportions of phases:

$$\text{Per cent ferrite} = \frac{0.83 - 0.30}{0.83 - 0.025} \times 100 = \frac{0.53}{0.805} \times 100 = 65.8\%$$

$$\text{Per cent austenite} = \frac{0.30 - 0.025}{0.83 - 0.025} \times 100 = \frac{0.275}{0.805} \times 100 = 34.2\%$$

The austenite will contain 0.83 per cent carbon, corresponding to the eutectoid of this system. It will transform at this temperature to form the eutectoid called pearlite. The pearlite will be composed of the following proportions of ferrite and cementite:

$$\text{Per cent ferrite} = \frac{6.67 - 0.83}{6.67 - 0.025} \times 100 = \frac{5.84}{6.645} \times 100 = 87.9\%$$

$$\text{Per cent cementite} = \frac{0.83 - 0.025}{6.67 - 0.025} \times 100 = \frac{0.805}{6.645} \times 100 = 12.1\%$$

The ferrite that forms prior to the formation of the pearlite is called *proeutectoid ferrite.* The ferrite that is incorporated in the pearlite is called *eutectoid ferrite.* In similar manner, the cementite present in the pearlite is called *eutectoid cementite.* In the over-all structure, the two phases, ferrite and cementite, are distributed in the following proportions:

Proeutectoid ferrite = 65.8% }
Eutectoid ferrite 34.2 × 0.879 = 30.1% } .... Total ferrite = 95.9%
Eutectoid cementite .... 34.2 × 0.121 = 4.1%

These proportions are shown in the transformation diagram of Fig. 7.3.

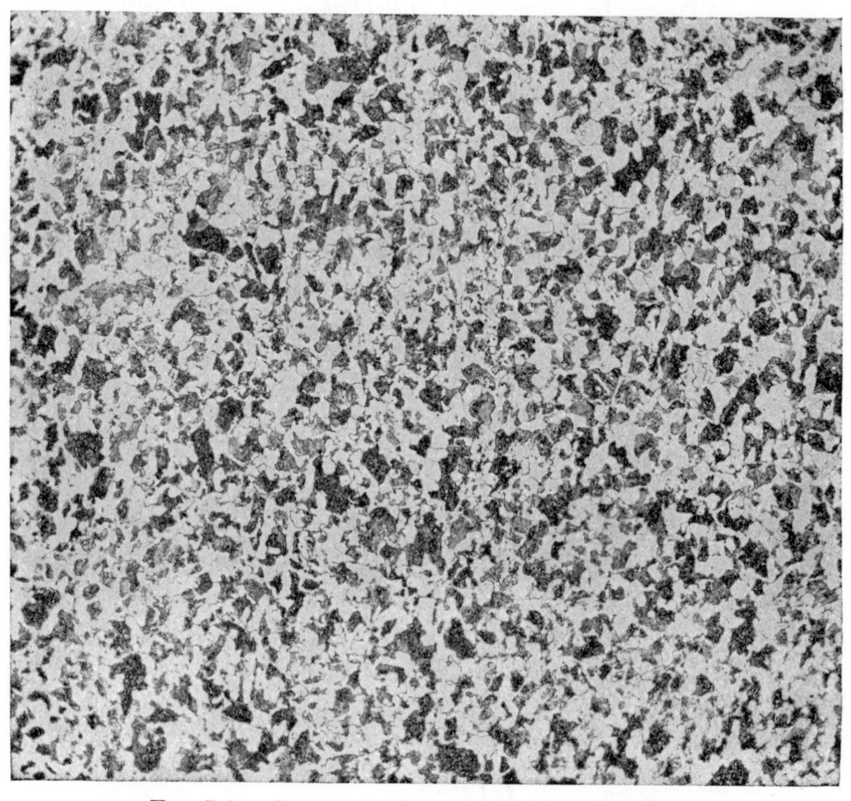

FIG. 7.4a Structure of hypoeutectoid steel (150×).

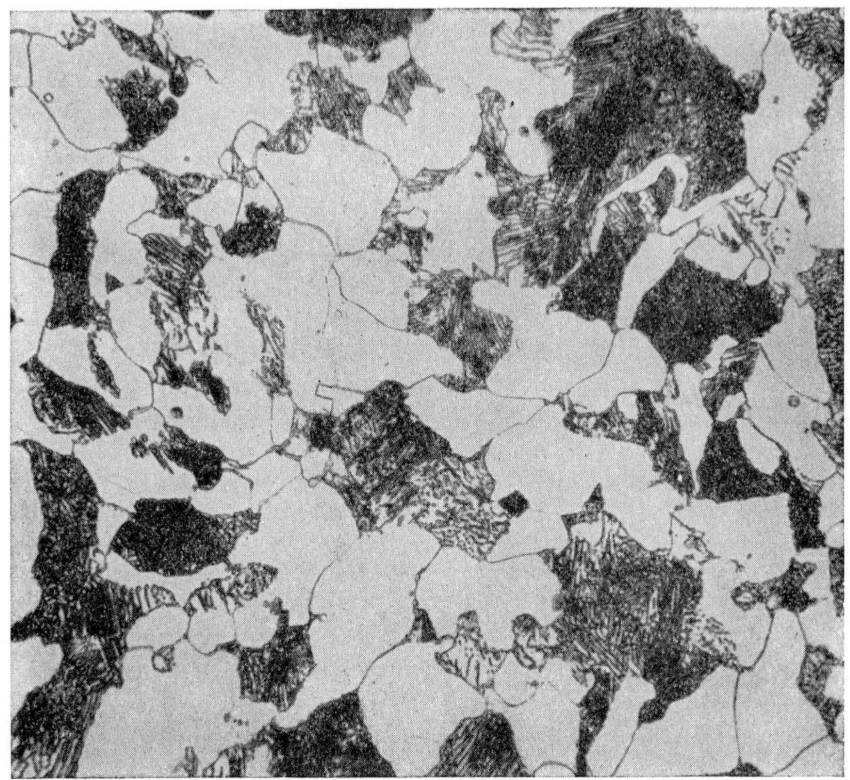

FIG. 7.4b Structure of hypoeutectoid steel (1000×).

With further cooling of this alloy, a small amount of cementite will precipitate from the ferrite because of the decreasing solubility of carbon in alpha iron with the decreasing temperature. For most purposes, this decreasing solubility may be neglected.

Structurally, as indicated by the transformation diagram, the alloy will be composed of ferrite and pearlite. The structure of such an iron-carbon alloy is shown in the photomicrographs of Fig. 7.4.

The transformation of other alloys in the range 0 to 1.7 per cent carbon may be analyzed in a similar manner. The structure of a steel of eutectoid composition is shown in Fig. 7.5 and consists entirely of pearlite. However, it is important to recognize that alloys containing more than 0.83 per cent carbon will form proeutectoid cementite in the place of proeutectoid ferrite. The structure of an alloy containing 0.9 per cent carbon is shown in Fig. 7.6. The white network is cementite, and the grains consist of pearlite.

7.7 Transformation in the Range 1.7 to 6.67 Per Cent Carbon. An alloy containing 3 per cent carbon, when cooled under equilibrium conditions from a temperature of 2500 F (1371 C), will begin to solidify with the formation of grains of austenite at *a* in Fig. 7.7. Austenite will continue to solidify until the temperature reaches *b*, 2065 F (1129 C). Just as this temperature is reached, the alloy will be composed of 50 per cent austenite and 50 per cent liquid of eutectic composition in accordance with the following relations:

$$\text{Austenite} = \frac{4.3 - 3.0}{4.3 - 1.7} \times 100 = \frac{1.3}{2.6} \times 100 = 50\%$$

$$\text{Liquid} = \frac{3.0 - 1.7}{4.3 - 1.7} \times 100 = \frac{1.3}{2.6} \times 100 = 50\%$$

This distribution is shown in the transformation diagram of Fig. 7.8. The austenite that forms from the liquid prior to the formation of the eutectic is called *proeutectic austenite*. For the moment, disregard the remaining

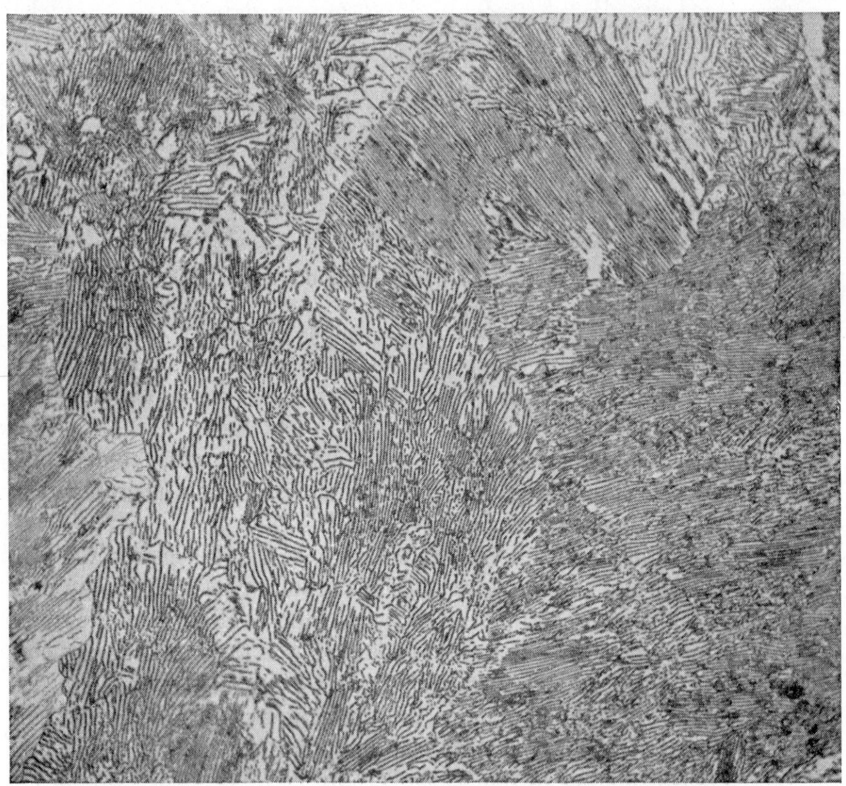

FIG. 7.5a Structure of eutectoid steel (500×). (Pearlite)

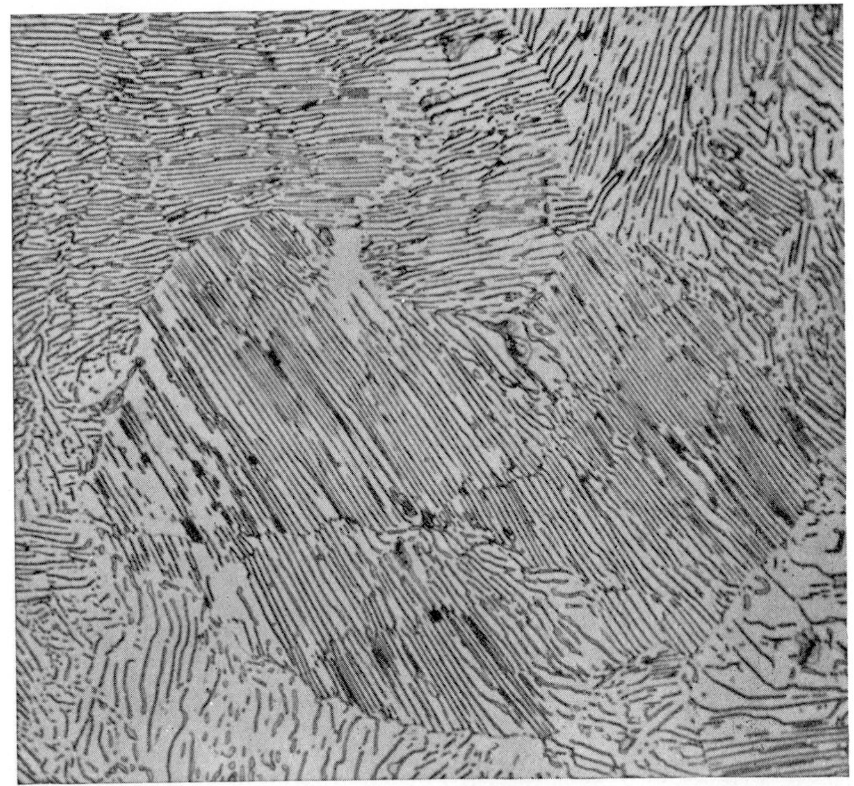

FIG. 7.5b Structure of eutectoid steel (1000×). (Pearlite)

liquid, and concentrate upon the proeutectic austenite at temperature *b* (2065 F (1129 C)). This austenite contains 1.7 per cent carbon. An alloy of this carbon content, in cooling from a temperature of 2065 F to 1333 F (1129 to 723 C), will precipitate cementite from the austenite because of the decreasing solubility of carbon in the austenite. Just as the temperature reaches 1333 F, this alloy which contains 1.7 per cent carbon will be composed of the following proportions of austenite of eutectoid composition and cementite:

$$\text{Austenite} = \frac{6.67 - 1.70}{6.67 - 0.83} \times 100 = \frac{4.97}{5.84} \times 100 = 85.0\%$$

$$\text{Proeutectoid cementite} = \frac{1.70 - 0.83}{6.67 - 0.83} \times 100 = \frac{0.87}{5.84} \times 100 = 15.0\%$$

These phases compose 50 per cent of the total weight since they are derived from the proeutectic austenite which existed at 2065 F (1129 C). The proportion of the whole will then be as follows:

$$\text{Austenite} = 50 \times 0.85 = 42.5\%$$
$$\text{Proeutectic proeutectoid cementite} = 50 \times 0.15 = 7.5\%$$

These proportions are represented at *c* in the transformation diagram. The cementite that precipitated from the austenite prior to the formation of the eutectoid is called, as previously, *proeutectoid cementite.* Since this constituent was derived from the proeutectic austenite, it is designated as *proeutectic proeutectoid cementite* in order to distinguish it from another cementite that may form from the liquid not being considered at the moment.

At 1333 F (723 C), the austenite of eutectoid composition will trans-

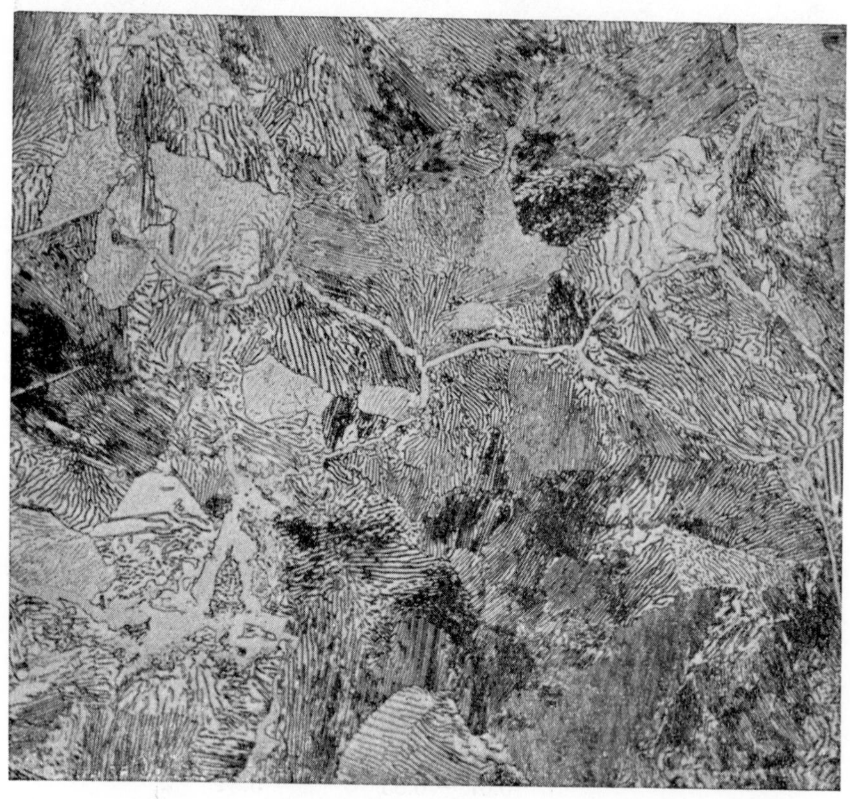

FIG. 7.6a Structure of hypereutectoid steel (500×).

FIG. 7.6b Structure of hypereutectoid steel (1000×).

form to pearlite, consisting of ferrite and cementite in the following proportion:

$$\text{Ferrite} = \frac{6.67 - 0.83}{6.67 - 0.025} \times 100 = \frac{5.84}{6.645} \times 100 = 87.9\%$$

$$\text{Cementite} = \frac{0.83 - 0.025}{6.67 - 0.025} \times 100 = \frac{0.805}{6.645} \times 100 = 12.1\%$$

Hence, the proportion of the whole will be as follows:

$$\text{Eutectoid ferrite} = 42.5 \times 0.879 = 37.4\%$$
$$\text{Eutectoid cementite} = 42.5 \times 0.121 = 5.1\%$$

These proportions are shown in the transformation diagram.

Returning now to the liquid of eutectic composition (4.3% C) at a

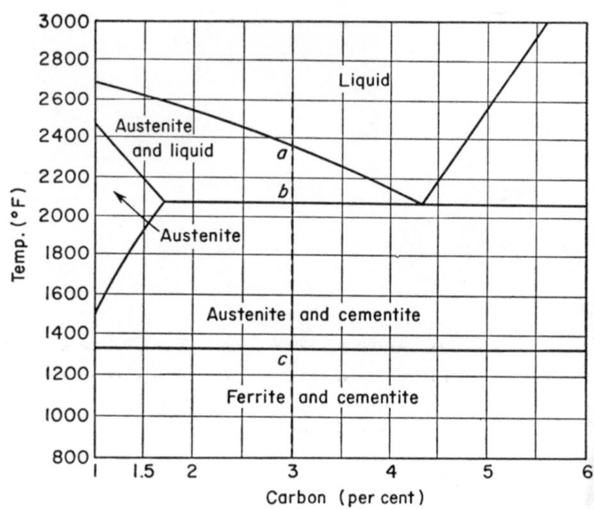

FIG. 7.7 Portion of equilibrium diagram of iron and iron carbide.

temperature of 2065 F (1129 C), this liquid will transform to produce the eutectic composed of austenite and cementite in the following proportions:

$$\text{Austenite} = \frac{6.67 - 4.3}{6.67 - 1.70} \times 100 = \frac{2.37}{4.97} \times 100 = 47.7\%$$

$$\text{Cementite} = \frac{4.3 - 1.70}{6.67 - 1.70} \times 100 = \frac{2.6}{4.97} \times 100 = 52.3\%$$

These phases compose 50 per cent of the total weight, and their proportion of the total is as follows:

$$\text{Eutectic austenite} = 50 \times 0.478 = 23.9\%$$
$$\text{Eutectic cementite} = 50 \times 0.522 = 26.1\%$$

The eutectic austenite contains 1.7 per cent carbon. This constituent is of the same carbon content as the previously considered proeutectic austenite. Hence, the cooling of the eutectic austenite will produce the same transformations as the proeutectic austenite. Proeutectoid cementite precipitates from the austenite until the temperature reaches 1333 F (723 C), or temperature c in the transformation diagram. This proeutectoid cementite is designated as *eutectic proeutectoid cementite,* since it came from the eutectic constituent. The proportion of the total is as follows:

$$\text{Eutectic proeutectoid cementite} = 23.9 \times 0.15 = 3.58\%$$

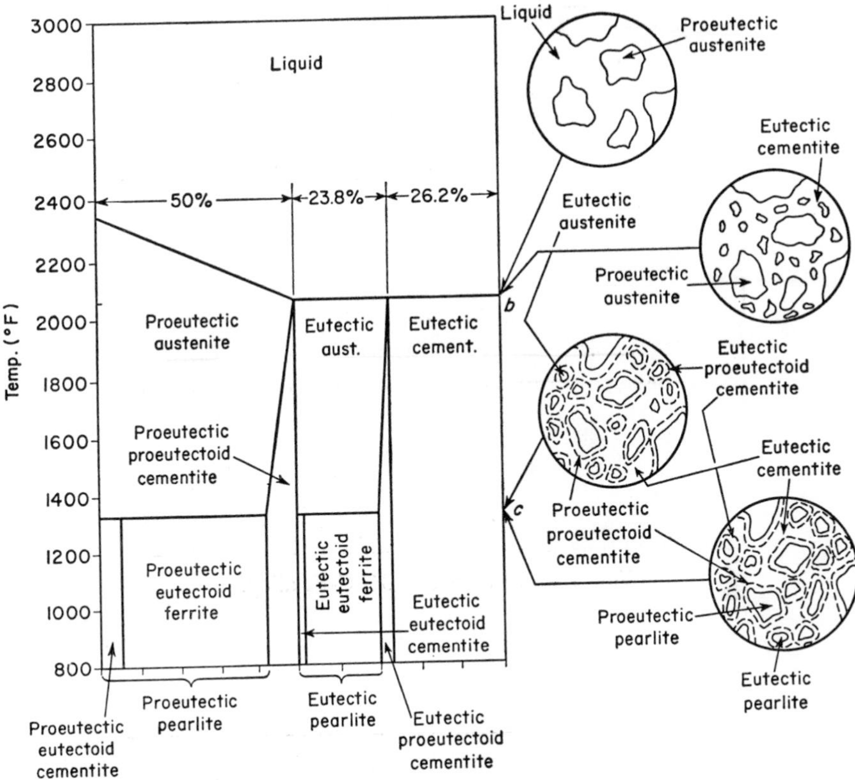

FIG. 7.8 Phase transformation diagram for a 3% carbon-iron alloy.

The remaining austenite of eutectoid composition ($23.9 \times 0.85 = 20.3\%$) transforms to pearlite in the manner previously described and in the same proportions of ferrite and cementite, namely 87.9 per cent ferrite and 12.1 per cent cementite. The word *eutectic* is prefixed to the eutectoid ferrite and eutectoid cementite to distinguish these constituents from the eutectoid ferrite and cementite that came from the proeutectic constituents. The proportion of these constituents of the whole is as follows:

$$\text{Eutectic eutectoid ferrite} = 20.3 \times 0.879 = 17.8\%$$
$$\text{Eutectic eutectoid cementite} = 20.3 \times 0.121 = 2.5\%$$

The cementite that is part of the eutectic at a temperature of 2065 F (1129 C) is called eutectic cementite and does not change with further cooling. The change in structure in this particular alloy is shown schematically at the right of the transformation diagram (Fig. 7.8). This should be

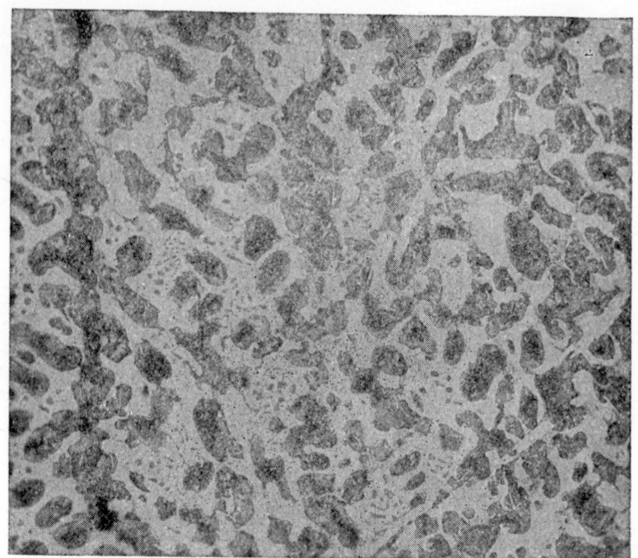

FIG. 7.9 Structure of 3% C-Fe alloy.

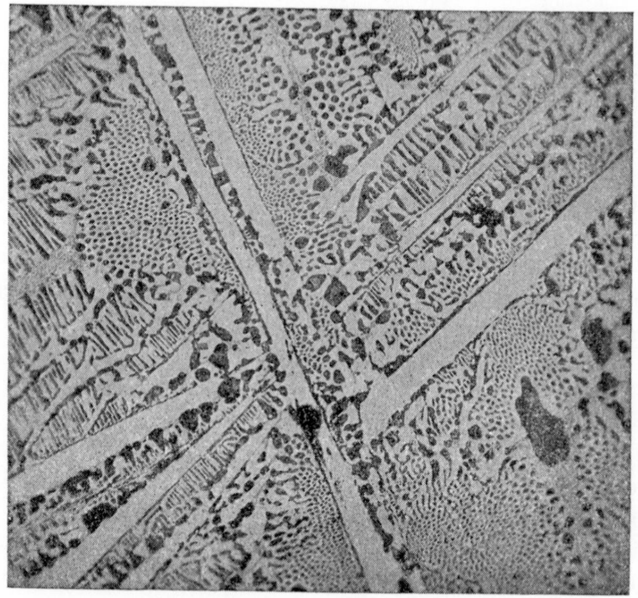

FIG. 7.10 Structure of 4.8% C-Fe alloy.

compared with the photomicrograph of a 3 per cent carbon-iron alloy shown in Fig. 7.9.

The terms associated with the structures formed in the transformation of iron-carbon alloys containing more than 1.7 per cent carbon are summarized as follows:

Name	*Description*
Proeutectic austenite	Austenite formed directly from liquid in cooling over a range of temperature prior to the formation of eutectic.
Eutectic austenite	Austenite formed on solidification of liquid of eutectic composition.
Proeutectic cementite	Cementite formed directly from liquid in cooling over a range of temperature prior to the formation of eutectic.
Eutectic cementite	Cementite formed on solidification of liquid of eutectic composition.
Proeutectoid cementite	Cementite precipitated from austenite in cooling from eutectic to eutectoid temperature.
Eutectoid cementite	Cementite formed on transformation of austenite of eutectoid composition.
Eutectoid ferrite	Ferrite formed on transformation of austenite of eutectoid composition.
Proeutectic proeutectoid cementite	Proeutectoid cementite derived from proeutectic austenite.
Eutectic proeutectoid cementite	Proeutectoid cementite derived from eutectic austenite.
Eutectic pearlite	Pearlite derived from eutectic austenite.
Proeutectic pearlite	Pearlite derived from proeutectic austenite.

The structure of an iron-carbon alloy containing 4.8 per cent carbon is shown in the photomicrograph of Fig. 7.10. By constructing the transformation diagram for this alloy, it should be possible for the student to identify the source of each portion of the structure. The white needles in the structure are proeutectic cementite.

7.8 Properties of Iron-Iron Carbide Alloys. In Chapter III, it was stated that the properties of an alloy depend upon the properties of the phases and the manner in which those phases are arranged to make up the structure. For the iron-iron carbide alloys under equilibrium conditions, the phases are always ferrite and cementite at room temperature. Ferrite is a relatively soft material having a tensile strength of approximately 40,000 lb/in.^{2} and is very ductile, as evidenced by an elongation of about 40 per cent in 2 in. Cementite is very hard and brittle with a probable tensile strength of 5000 lb/in.^{2}

The combination of these two phases in the form of a eutectoid (pearlite) produces an alloy of much greater tensile strength than that of either of the phases. The tensile strength of pearlite is of the order of 115,000 lb/in.^{2}, and the elongation is approximately 10 per cent in 2 in. The tensile strength and Brinell hardness of alloys increase with increasing carbon content up to about 0.83 per cent carbon. The ductility, as expressed by elongation and reduction of area, decreases with increasing carbon content. When the carbon content exceeds eutectoid proportions, the increase of tensile strength is relatively small. However, the Brinell hardness continues to increase because the structure contains a greater proportion of hard cementite. With more than 1.7 per cent carbon, the tensile strength gradually decreases, and the reduction of area and percentage elongation become practically zero. These relations are shown quantitatively in Fig. 7.11. The

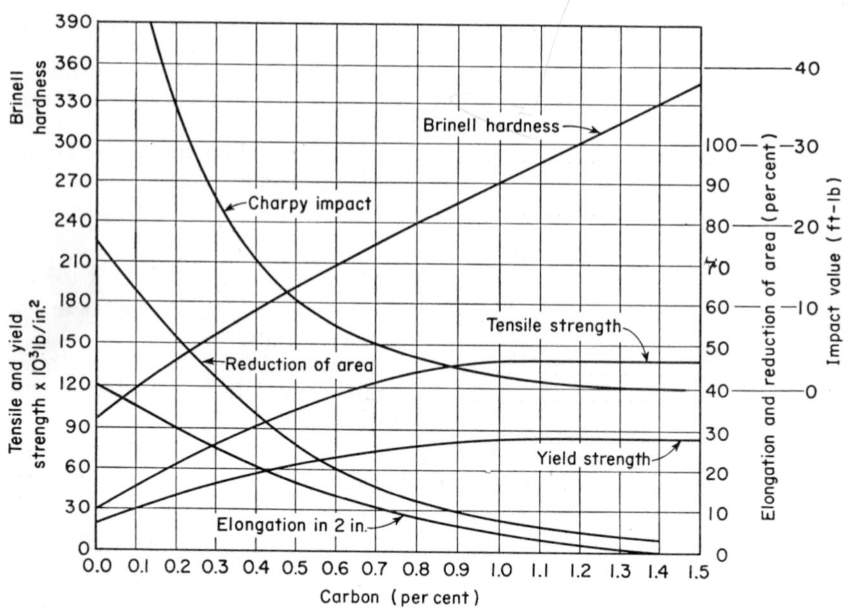

FIG. 7.11 Effect of carbon on mechanical properties of hot-worked steel. (By permission from *Alloys of Iron and Carbon* by Sisco, copyright 1937, McGraw-Hill Book Co., Inc.)

properties and uses of steel and white cast iron will be discussed in more detail in a later chapter.

7.9 Effect of Small Quantities of Other Elements. In the discussion of the equilibrium diagram of iron and iron carbide, it has been assumed that only pure substances are involved. Under practical conditions other elements may be present, and one must understand the effect of each of these elements upon the characteristics of these alloys. The elements to be considered here, although in relatively small amounts, are not necessarily to be considered as impurities. Many of them have beneficial effects when properly used, although, under certain conditions, their presence may be harmful. The elements sulfur, phosphorus, silicon, and manganese will be considered. Other elements which are intentionally added for the purpose of modifying the properties of the iron-iron carbide alloys are considered in Chapter IX.

Sulfur: Sulfur combines with iron to form a sulfide, FeS, and with manganese, if it is present, to form a sulfide, MnS. Iron sulfide freezes out of the melt principally at the grain boundaries in the structure, thus reducing the hot-working properties of the alloy because of its lower melting point. The presence of sulfur in the form of iron sulfide renders steel brittle, or *hot-short,* at elevated temperatures. Since manganese has a strong affinity for sulfur, the ill effects of the latter may be reduced by the addition of manganese to form the insoluble manganese sulfide, which either passes into the slag or is found as well-distributed inclusions throughout the structure. It is usually recommended that the ratio of manganese to sulfur be in the range of 3 to 1 to about 8 to 1. The sulfur content in most steels is maintained at less than 0.05 per cent. Sulfur in the range of 0.075 to 0.15 per cent contributes machinability as a result of the presence of sulfide inclusions and, therefore, is used for screw stock.

Phosphorus: Phosphorus in small amounts dissolves in the ferrite and increases the strength and hardness. Most steels contain less than 0.05 per cent phosphorus. A larger percentage of phosphorus increases the strength with decrease in ductility and dynamic properties, rendering the steel *cold-short.*

Silicon: When the amount of silicon in the steel is less than approximately 0.2 per cent, it is entirely dissolved in the ferrite. In this amount, its effect is very slight. In amounts greater than 0.2 per cent and not over approximately 0.4 per cent, silicon raises the elastic limit and ultimate strength of the steel without greatly reducing the ductility. With more than about 0.4 per cent silicon, a marked decrease of ductility is observed in the plain carbon steels. Silicon acts as a deoxidizing agent in the melting and refining of steel, therefore tending to make for greater soundness in casting. In cast iron, silicon has a marked effect in decreasing the stability of cementite. In such alloys with more than about one per cent silicon, the cementite may be rendered so unstable that it dissociates into iron and graphite. Details of the effect of silicon in cast iron will be found in Chapter XVI.

Manganese: Manganese is a very important element in the iron-iron carbide alloys, and the benefits derived from its combination with sulfur have already been indicated in the discussion of the effect of sulfur. Manganese will form a solid solution with ferrite in carbon-free iron. In the presence of sufficient carbon, the compound Mn_3C tends to form. The combination of cementite and manganese carbide has a marked hardening effect on steel. With more than 2.0 per cent manganese, the ductility of the steel is appreciably impaired. Manganese may be employed in certain steels in amounts in the range of 10 to 14 per cent for special alloying purposes.

REFERENCES

Aston and Story, *Wrought Iron,* A. M. Byers Co., Pittsburgh, Pa., 1941.

Clark, *Engineering Materials and Processes,* International Textbook Co., Scranton, Pa. 1959.

Cleaves and Thompson, *The Metal Iron,* McGraw-Hill Book Co., New York, 1935.

Epstein, *Alloys of Iron and Carbon,* McGraw-Hill Book Co., New York, 1936.

Metals Handbook, American Society for Metals, Metals Park, Ohio, 1961.

QUESTIONS

1. What are the characteristics of commercially pure iron? What are some of its uses?
2. Name the three allotropic forms of iron and indicate the lattice structure of each.
3. What is the Curie point in iron?
4. What is wrought iron? What are its characteristics and uses?
5. What is cementite?
6. What is a plain carbon steel?
7. What is white cast iron?
8. Draw the equilibrium diagram of the iron-iron carbide system.
9. What is the eutectic in the iron-iron carbide system called?
10. What name is given to the eutectoid in the iron-iron carbide system? What is the carbon content?
11. Distinguish between hypoeutectoid and hypereutectoid steels.
12. What is austenite?
13. How are the critical temperatures in steel designated? What is the critical range?
14. Explain by means of a phase transformation diagram what occurs during the slow cooling of a steel containing 0.06 per cent carbon.
15. Explain by means of a phase transformation diagram what occurs during the slow cooling of a steel containing 1.0 per cent carbon. Compute the percentage of phases and structural constituents present in this alloy at room temperature.
16. Describe by means of a phase transformation diagram the transformation of an alloy containing 4.8 per cent carbon during slow cooling. Compute

the amount of phases and structural constituents present in this alloy at room temperature.

17. Describe by means of a phase transformation diagram what occurs upon slow cooling of an alloy containing 1.3 per cent carbon. Compute the per cent of phases and structural constituents present at room temperature.
18. Show by means of a graph the approximate relationship between tensile strength and percentage carbon in hot-worked steels.
19. What is the effect of sulfur upon the properties of steel?
20. What is the effect of phosphorus on the properties of steel?
21. What is the effect of silicon upon the properties of steel?
22. What is the effect of manganese upon the properties of steel?
23. What is the usual recommended ratio of manganese to sulfur in steel? Why is this ratio maintained?

VIII

HEAT TREATMENT OF STEEL

8.1 Heat Treatment. In the general sense, heat treatment may be defined as an operation or combination of operations involving the heating and cooling of a metal or alloy in the solid state for the purpose of obtaining certain desirable conditions or properties. The usefulness of steel is due largely to the relative ease with which its properties may be altered by properly controlling the manner in which it is heated and cooled. The changes that occur in the properties of steel are directly related to changes in the structural make-up of the steel.

While the steels employed in industry are not pure iron-iron carbide alloys, the equilibrium diagram of these pure alloys serves admirably in the study of the heat treatment of steel and of the structural changes that take place therein. It must be remembered that the presence of other elements will alter the diagram, although small quantities of them will have very little effect. In any event, the basic principles which may be gained by a study of the diagram of the pure alloys can be applied to the commercially available alloys. Since heat treatment involves the heating and cooling of alloys, the transformations that occur during heating and cooling will be considered separately.

8.2 Critical Temperatures on Heating and Cooling. In Chapter VII the critical temperatures were designated as A_3, A_1, and $A_{3,1}$. These would be the indications of transformation under strictly equilibrium conditions. In practice, however, strict equilibrium conditions do not prevail. If a hypoeutectoid steel is slowly heated, transformation will not occur at the A_1 and A_3 temperatures, but at somewhat higher temperatures. The slower the heating, the closer will these temperatures approach the equilibrium critical points. These critical temperatures observed on heating are designated as Ac points, the letter c being taken from the French *chauffage*—"heating." Hence, the critical points on heating are designated as Ac_1, $Ac_{3,1}$, Ac_3.

If the steel is slowly cooled, the critical temperatures will be found to be lower than the equilibrium values and are designated by the symbol *Ar*, the letter *r* being taken from the French *refroidissement*—"cooling." The critical points on cooling are, therefore, designated as Ar_3, $Ar_{3,1}$, Ar_1.

The difference between *Ac* and *Ar* temperatures is illustrated in Fig. 8.1, in which the change of length that occurs during the heating and cooling of a steel is recorded. When transformation begins on heating, Ac_1, the specimen contracts while the temperature increases up to the end of transformation, Ac_3, whereupon the specimen continues to expand. Upon cooling, the reverse effect is observed. The dilatometric method, illustrated by

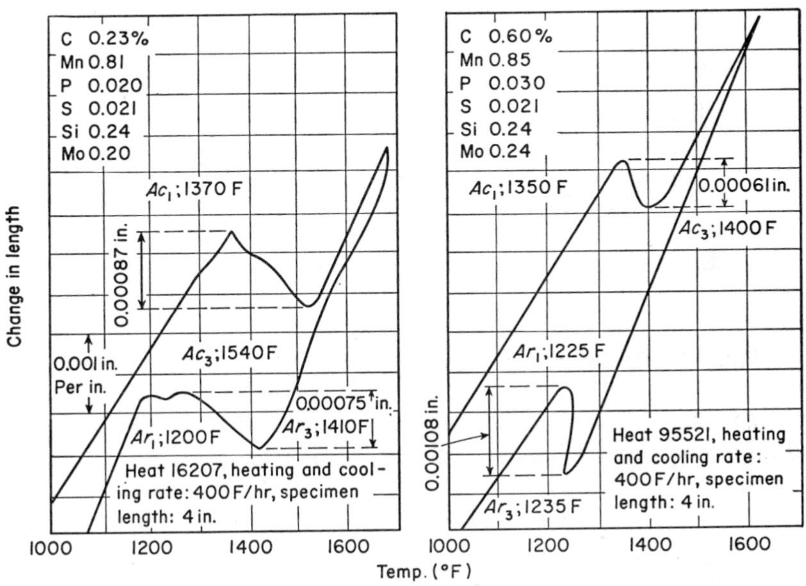

FIG. 8.1 Dilation vs. temperature curves for two C-Mo steels. (After *ASM Metals Progress Data Sheet* **34,** 1946)

the curves of Fig. 8.1, is utilized for the determination of the critical temperatures. The values obtained are dependent upon the rate of heating and cooling.

8.3 Transformations on Heating. Consider a hypoeutectoid steel containing 0.30 per cent carbon which has been slowly cooled under near-equilibrium conditions and, therefore, consists, at room temperature, of grains of proeutectoid ferrite and grains of pearlite as shown in Fig. 7.4. Referring to the equilibrium diagram, Fig. 8.2, if this steel is heated to any temperature below the Ac_1 point and then cooled in any desired manner, no alteration of structure will occur. However, if the steel is heated to just above the Ac_1 temperature, the pearlite will transform to austenite.

The actual mechanism of this transformation is one involving nucleation and growth. It is believed that eutectoid ferrite and cementite react at their interface to form nuclei of austenite. Growth of these nuclei takes place with the absorption of additional ferrite and cementite until each pearlite grain has transformed or become austenitized. Time and temperature play an important part in the austenitizing process. The allotropic change of

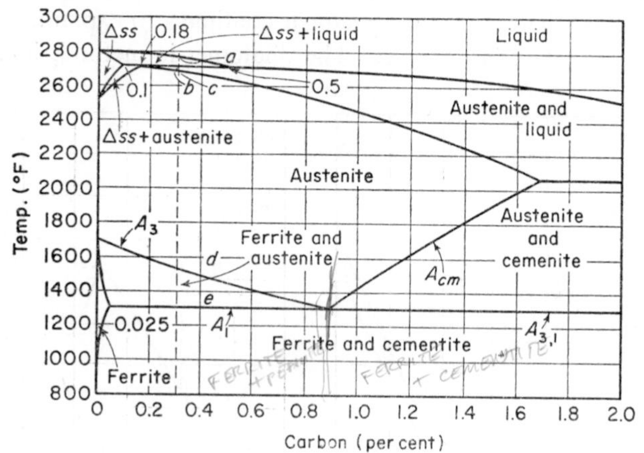

FIG. 8.2 Portion of equilibrium diagram of iron and iron carbide.

alpha to gamma iron is rapid compared to the time required for solution and diffusion of cementite, as illustrated by the relation

$$\text{Ferrite} + \text{Fe}_3\text{C} \leftrightharpoons \text{Austenite}$$

A coarse-grained structure and the presence of certain alloying elements greatly increase the time and temperature required for complete solution and diffusion.

At a temperature just above Ac_1, the structure of the steel being considered will be composed of grains of proeutectoid ferrite and grains of austenite. As the temperature is raised above Ac_1, the ferrite will be transformed to austenite gradually until, at the Ac_3 temperature, the transformation will be completed.

In hypereutectoid steels, the transformation on heating begins at the $Ac_{3,1}$ temperature by the transformation of pearlite to austenite. Heating the steel to temperatures above $Ac_{3,1}$ will bring about the solution of proeutectoid cementite in the austenite. At the A_{cm} temperature, the cementite is completely dissolved in the austenite. Any steel will be entirely austenitic under conditions of equilibrium when the temperature is above the Ac_3 or A_{cm} curves.

This discussion assumes that equilibrium conditions prevail, i.e., that heating is very slow. During any heating process in which a transformation occurs, the diffusion of one element into another requires time, as has been noted previously. If heating is too rapid, complete uniformity of the structure will not be attained.

According to the equilibrium diagram, if a steel that has been heated into the austenite range above the upper critical temperature is heated to a still higher temperature, no additional phase change occurs. Changes may occur, however, in grain size, as discussed in Article 8.9.

8.4 Transformations on Cooling. If the steel containing 0.30 per cent carbon is slowly cooled from above the Ac_3 temperature, proeutectoid ferrite will begin to precipitate from the austenite at the Ar_3 point. This transformation of austenite to ferrite is caused by allotropic change of gamma to alpha iron. Since alpha iron dissolves only a very small amount of cementite, the concentration of carbon in the austenite becomes greater until the eutectoid composition is reached. At the eutectoid temperature, the remaining austenite of eutectoid composition transforms to pearlite.

The mechanism of pearlite formation is again one of nucleation and growth. Upon cooling to Ar_1 or slightly below, nuclei of cementite form at the austenite grain boundary. Transformation of gamma to alpha iron takes place rapidly compared to the time required for diffusion of carbon to form cementite, as represented by

$$\text{Austenite} \leftrightarrows \text{Ferrite} + \text{Fe}_3\text{C}$$

Accordingly, edgewise growth of alternate plates of ferrite and cementite takes place into the austenite grain as times goes on, until under equilibrium conditions the entire grain is transformed. The method of growth from primary and secondary nuclei has been represented schematically in Fig. 8.3. The structure of pearlite was shown in Fig. 7.5.

Upon slowly cooling a hypereutectoid steel, proeutectoid cementite will begin to precipitate at the A_{cm} temperature. As the cementite is precipitated, it will migrate to the grain boundaries of the austenite and form a fine network structure which can be observed at room temperature after the steel has completed its transformation. The austenite-pearlite transformation in the high-carbon steels will occur at the $Ar_{3,1}$ temperature.

It should by now be recognized that, upon slow cooling, the transformation of austenite of eutectoid composition involves the simultaneous transformation of gamma to alpha iron and the precipitation of cementite due to the low solubility of carbon in alpha iron. With slow cooling, the cementite will be in lamellar form interspaced with ferrite to form pearlite. Since this lamellar mixture of ferrite and cementite is derived by the mutual divorcement of ferrite and cementite, migration of these phases is involved in the transformation. Migration or diffusion out of the austenite grains requires time. If the steel is cooled more rapidly, diffusion will be completed at some temperature lower than that for equilibrium transformation. At the lower temperature, migration is more difficult; hence, the interlamellar spacing will be smaller. This finer distribution of ferrite and cementite will result in a modification of the mechanical properties.

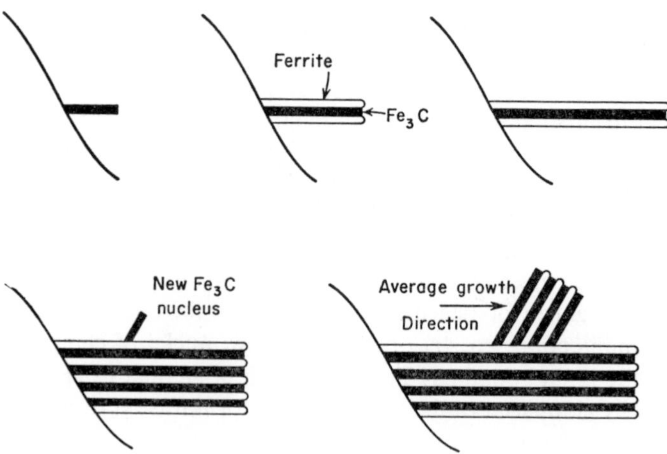

FIG. 8.3 Nucleation and growth of pearlite lamellae. (After Mehl, *Trans. ASM* **29,** 1941, p. 813)

The alteration of the mechanical properties of steel by heat treatment depends upon the control of the dispersion of ferrite and cementite. Proper control of this dispersion in hypoeutectoid and eutectoid steel requires that the structure be completely austenitic prior to cooling. If the steel is heated to a temperature between the Ac_3 and Ac_1 points, all of the ferrite will not be dissolved. Regardless of the cooling rate, this ferrite will be retained, thus reducing the hardness that otherwise might be obtained.

8.5 Austenite Transformation at Constant Temperature. The character of the structure that forms from austenite upon transformation depends upon the composition and grain size of the austenite, and particularly upon the temperature at which the transformation occurs. Therefore, a basic concept of the transformation of austenite can be secured by determining the time required for the transformation to begin and end at a series of constant subcritical temperatures. By making the transformation occur under isothermal conditions, the rate of nucleation and the rate of grain growth remain constant, thus eliminating two variables that might confuse the study.

To make this study, several small pieces of the steel being investigated are heated to above the upper critical temperature long enough to form homogeneous austenite. One by one the pieces are cooled instantaneously to the subcritical temperature for which the transformation time is to be determined. Each piece of the steel is held at the subcritical temperature for a different length of time before being quenched into water. Then, each piece is prepared for microscopic examination to reveal the extent to which transformation has occurred as a function of time. The same procedure is followed for a series of subcritical temperatures.

To illustrate this method of studying the transformation of austenite, consider a steel containing 0.89 per cent carbon and 0.29 per cent manganese. A group of small pieces of this steel is heated to a temperature of 1625 F (885 C) and then plunged into a bath of suitable liquid maintained at a temperature of 1300 F (704 C). Each piece is allowed to remain in the bath for a specified length of time, such as 100 sec, 1000 sec, 10,000 sec, etc., and then quenched in water. The microstructure of the piece that remained in the bath at a temperature of 1300 F (704 C) for 10^4 sec is found to consist of a few very small areas of pearlite with the remaining area composed of martensite. The pieces that were held for increasingly longer times at this temperature will consist of progressively more pearlite, whereas the piece that was held for 10^5 sec will be composed of practically all pearlite. Hence, at a temperature of 1300 F (704 C), transformation of this steel begins in 10^4 sec and is completed in 10^5 sec.

By making a similar experiment, but at a temperature of 1100 F (593 C), the transformation is observed to begin in 1 sec and to be completed in 7 sec. Experiments at other subcritical temperatures down to about 450 F (232 C) show different times for the beginning and completion of transformation. At temperatures below about 450 F (232 C), the reaction appears to be independent of time and to be only a function of temperature. These experimental data are plotted in the manner shown in Fig. 8.4 to form the isothermal transformation diagram, consisting of a curve for the time at which transformation begins and one for the time of completion of transformation. The temperatures at which the time-independent reaction starts and finishes are designated as the M_s and M_f temperatures, respectively. The equilibrium transformation temperature is designated as *Ae*. The curves forming this diagram have been, in some instances, called "S" curves because their shapes are sometimes similar to the letter "S"; frequently, however, the diagram is called a "TTT" diagram (time-temperature-transformation).

The structure that is formed by the transformation of austenite at subcritical temperatures is dependent upon the temperature of transformation. If transformation occurs just below the equilibrium temperature (*Ae*), the structure will be composed of ferrite and cementite, arranged in a coarse lamellar pattern known as coarse pearlite with a representative hardness of about Rockwell C 15. Transformations at lower temperatures down to about 1000 F (538 C) will produce lamellar structures, but of a much finer texture, which may be called fine pearlite with a hardness of about Rockwell C 45. The decrease of interlamellar spacing of the pearlite with decreasing temperature of transformation is continuous; therefore, the difference between coarse and fine pearlite is one of degree only and is not characterized by a sharp demarcation.

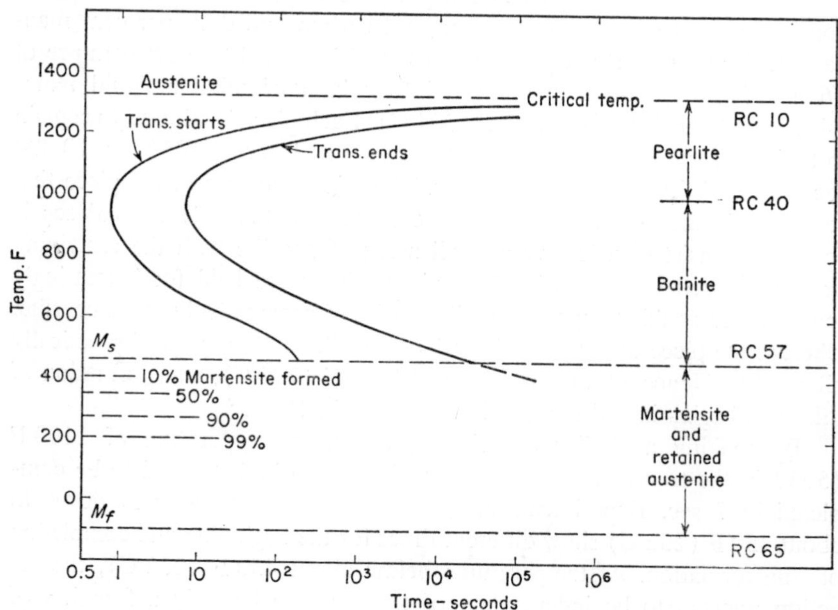

FIG. 8.4 Isothermal transformation diagram for eutectoid steel.

The transformation of austenite at temperatures below about 1000 F (538 C) down to about 400 F (204 C) produces a structure in which the ferrite and cementite are not arranged in lamellar form but in a feathery or acicular pattern in which the ferrite and cementite are not resolved under the microscope. This structure, known as *bainite,* has a feathery appearance when the transformation occurs in the upper portion of the temperature range of 1000 (538 C) to about 400 F (204 C), blending into a more acicular appearance when transformation occurs in the lower portion of this range. The hardness of bainite will vary over a range of approximately Rockwell C 45-60. The structure of partially formed bainite of the feathery type is shown in Fig. 8.5. The mechanism by which bainite forms is different from that involved in the transformation of austenite to pearlite. Pearlite is nucleated by cementite, and the formation of the cementite is accompanied by the formation of ferrite, whereas bainite is nucleated by ferrite which leads to the precipitation of cementite. This change in the mechanism of transformation accounts for the difference in the appearance of these two structures.

FIG. 8.5 Fine pearlite (dark), bainite (feathery), and martensite (white with slight structure) (2000×).

When austenite transforms at temperatures in excess of about 400 F, the face-centered cubic structure is completely transformed to the body-centered cubic structure of ferrite, and the ferrite is interspersed with cementite. Transformation at lower temperatures occurs as a function of temperature level independent of time and takes place by a process which appears to involve not nucleation and growth but instead a shear mechanism that produces a body-centered tetragonal structure in which carbon is dissolved. This solid solution of carbon in a body-centered tetragonal structure of iron is in an unstable, highly stressed state and has a hardness of Rockwell C 65 or somewhat less, depending upon the carbon content as shown in Fig. 9.1. This structure, which is typical of that found in a fully hardened steel, is called *martensite* and has the appearance of an assembly of needles as shown in Fig. 8.6. Since martensite occupies a greater volume than austenite, a resultant dimensional expansion takes place.

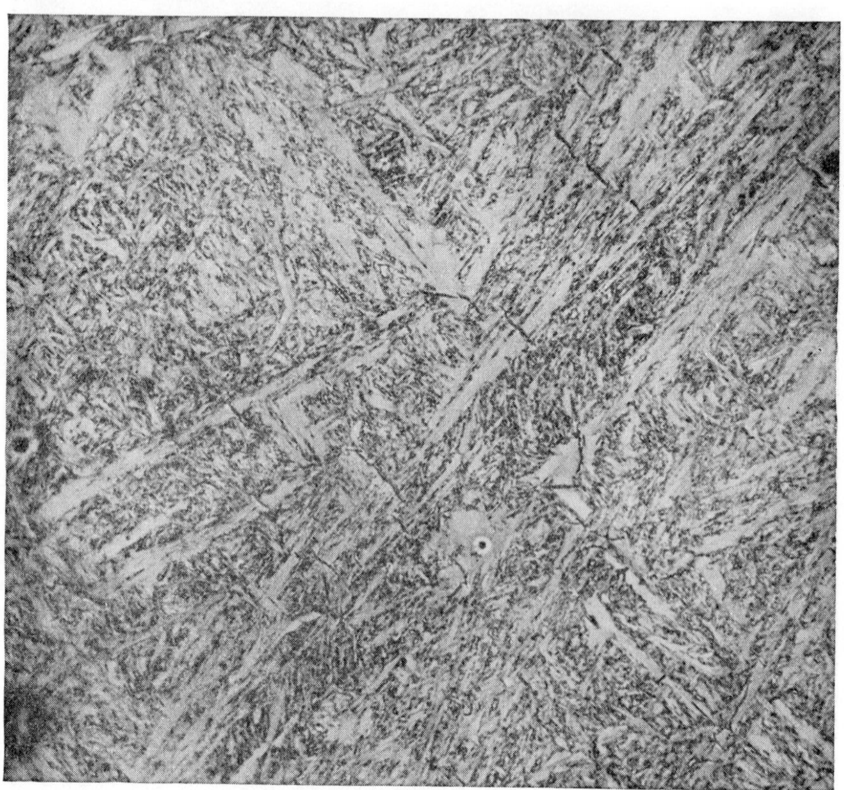

FIG. 8.6a Martensite (500×).

One may visualize the transformation of austenite to martensite as a transition stage in the change from the face-centered cubic structure of austenite to the body-centered cubic structure of ferrite. The low temperature at which martensite forms greatly decreases the chance for carbon atoms to diffuse out of the lattice; hence, they remain in solution in the transition lattice. In many instances, all of the austenite does not transform to martensite; thus, the martensite needles are observed to exist in a matrix of austenite called *retained austenite*. The austenite is relatively soft; therefore, its presence detracts from the hardness usually desired in a steel requiring full hardening.

The austenite-martensite transformation starts at some temperature designated as M_s and continues with respect to decreasing temperature down to some temperature referred to as M_f, at which the reaction ceases even though retained austenite may still exist. These temperatures are influenced by the composition of the steel and particularly by the carbon content as

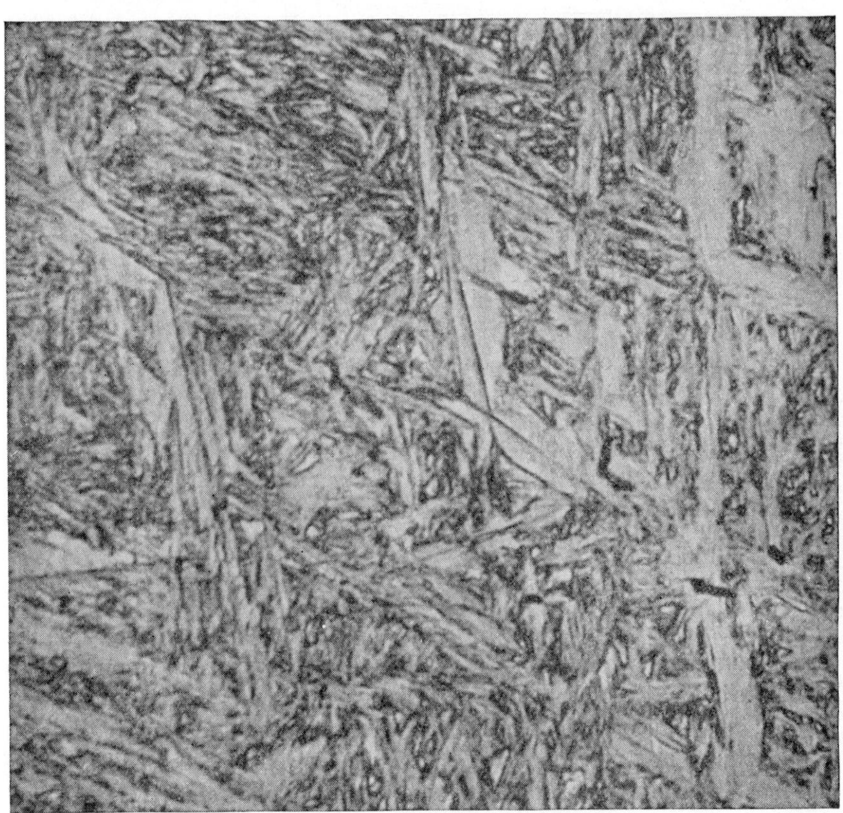

FIG. 8.6b Martensite (2000×).

shown in Fig. 8.7. The existence of retained austenite at or below the M_f temperature may be caused by a stabilization of the retained austenite, rendering it sluggish in its response to conversion. In the quenching of steels subject to retained austenite, it is important that continuous cooling down to the M_f temperature take place since martensite forms in the range M_s—M_f only with decreasing temperature. An appreciable delay at room temperature tends to stabilize retained austenite against further conversion.

The M_f temperature of some alloy steels is below room temperature. More complete transformation of the austenite may be attained by cooling to the M_f temperature, followed by reheating to room temperature and then cooling to the M_f temperature a second and a third time. With each cooling cycle a smaller percentage of retained austenite is converted to martensite.

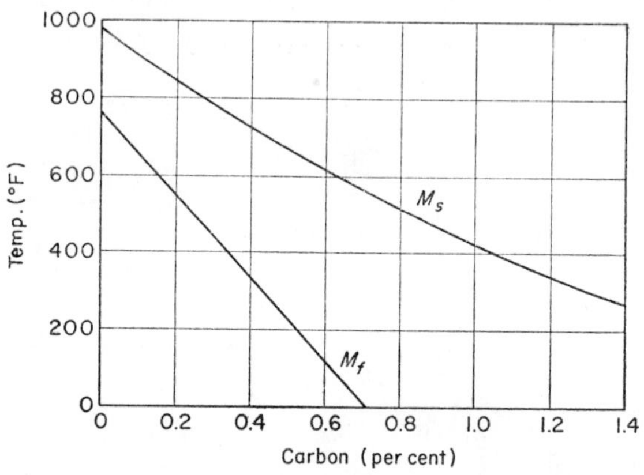

FIG. 8.7 Influence of carbon on M_s and M_f temperatures. (Troiano and Greninger, *Metal Progress* **50,** 1946, p. 303)

The M_s temperatures of many alloy steels have been determined by direct measurement. They may be approximated for most applications by the use of an empirical formula such as that of Grange and Stewart[1] which is given as

$$M_s(°\text{F}) = 1000 - 650 \times \%\text{C} - 70 \times \%\text{Mn} - 35 \times \%\text{Ni} - 70 \times \%\text{Cr} - 50 \times \%\text{Mo}$$

This formula will apply when all carbides are dissolved in the austenite, the carbon content is within the range of 0.20—0.85 per cent, the molybdenum content is below 1 per cent, and the chromium content is below 1.5 per cent.

The importance of complete austenitization or solution of carbides prior to quenching should be apparent, since undissolved carbides will result in higher M_s and M_f temperatures than would be indicated by the actual chemical composition.

Data on the M_f temperatures is of equal importance wherever it is necessary to minimize retained austenite in quenched steels. These temperatures will generally range from 325 to 475 F (163 to 246 C) below the corresponding M_s temperatures. The values of M_f temperatures are not generally available because of the difficulty of measuring small amounts of retained austenite of the order of 5 per cent or less. The influence of carbon

[1] *The Temperature Range of Martensite Formation, Trans. AIME* **167,** 1946, p. 467.

content on the proportion of martensite formed at given temperatures below the M_s is shown in Fig. 8.8, as derived from experimental data.

Isothermal transformation diagrams for hypoeutectoid and hypereutectoid steels include, in addition to the curves for the beginning and end of the transformation of austenite, a curve indicating the beginning of the

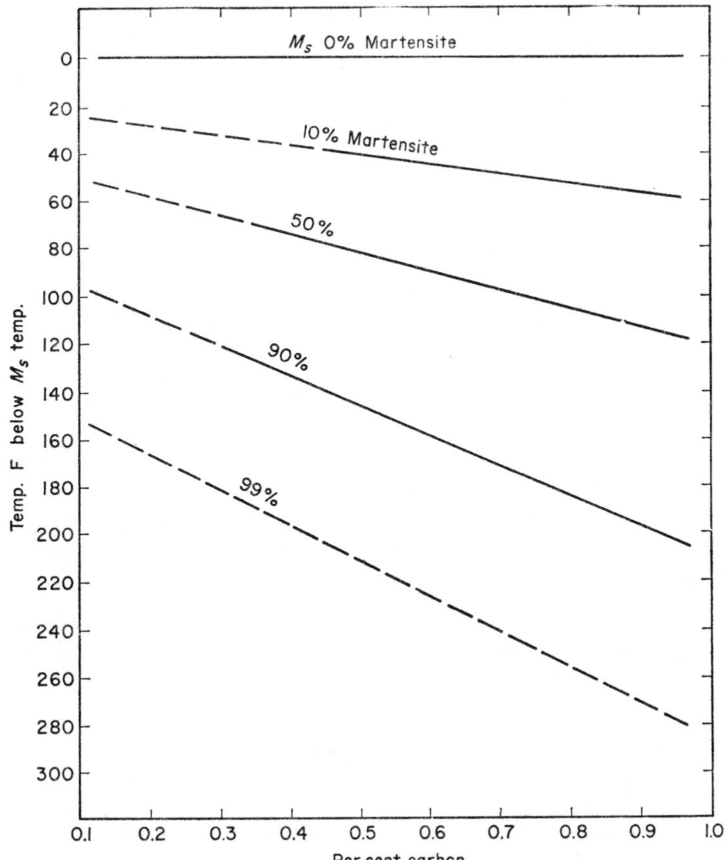

FIG. 8.8 Relationship between per cent carbon in steel and proportion of martensite formed at a given temperature. (After Grange and Stewart, *Trans. AIME* **167,** 1946, p. 467)

formation of the proeutectoid constituent, ferrite or cementite. These diagrams have not only an *Ae* temperature, but also an Ae_3 or Ae_{cm} temperature. Typical examples of diagrams of this type are shown in Fig. 8.9.

8.6 Austenite Transformation on Continuous Cooling. The most common method of hardening steel does not involve isothermal transformation,

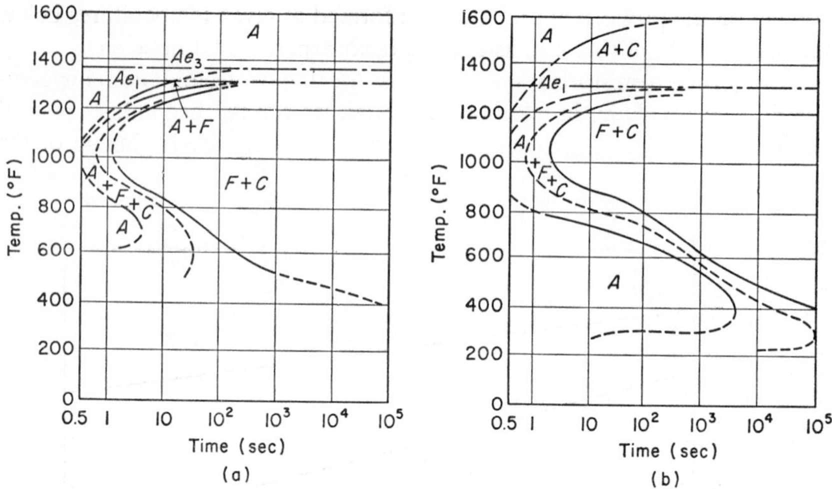

FIG. 8.9 Isothermal transformation diagram. (After *Atlas of Isothermal Transformation Diagrams,* U. S. Steel Corp.) (a) 0.54% C, 0.46% Mn steel austenitized at 1670 F; grain size 7-8. (b) 1.13% C, 0.30% Mn steel austenitized at 1670 F; grain size 7-8.

but rather the quenching of the steel into some medium that rapidly abstracts heat from the piece. The rate of cooling depends upon the cooling medium employed as shown in Fig. 8.10. Curves indicating the beginning and end of transformation are indicated in Fig. 8.10, and, although they resemble similar curves of the isothermal transformation diagram, they are displaced to lower temperatures and longer times. These curves are the loci of the beginning and ending of transformation for the indicated cooling curves.

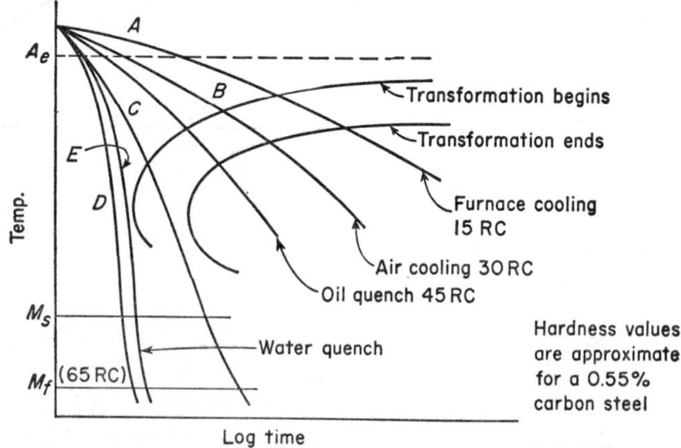

FIG. 8.10 Typical continuous cooling transformation curves.

Curve *A* represents slow cooling as in a furnace with transformation taking place at temperatures near the *Ae* temperature, thus forming coarse lamellar pearlite. The cooling condition indicated by curve *B* represents what may be expected with air cooling with the formation of a finer pearlite than formed by cooling condition *A*. Cooling of the character shown by curve *C* initiates the formation of fine pearlite, but insufficient time in the

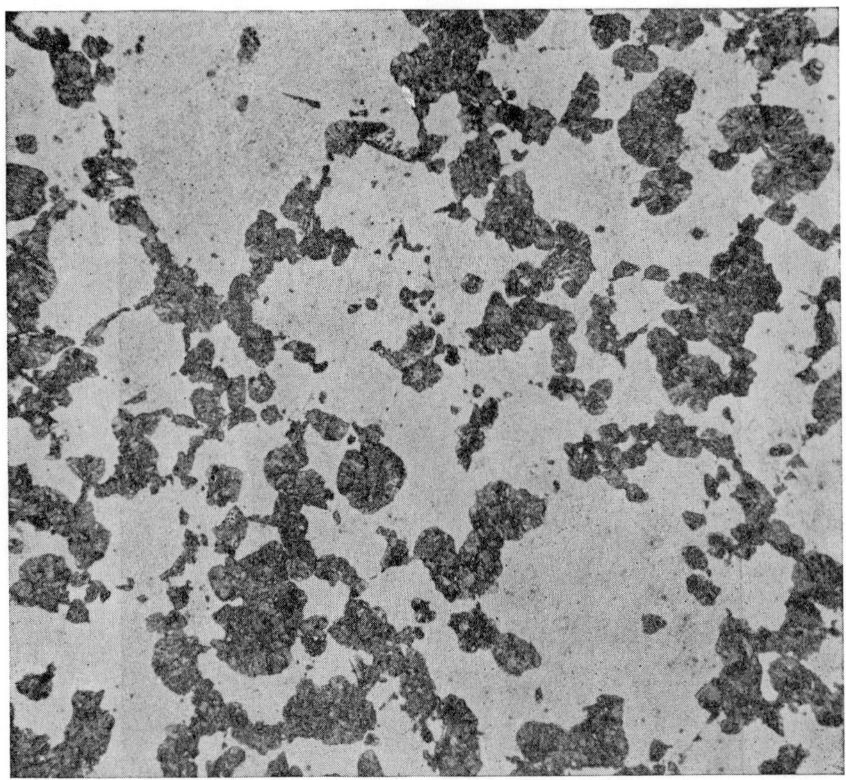

FIG. 8.11 Structure of fine pearlite (dark nodules) and martensite (white) produced by split transformation (250×).

upper range of temperature prevents the completion of this transformation, and some of the austenite transforms to martensite at the lower temperature. Under this condition, the transformation occurs to a certain degree in two stages and is called a *split transformation*. The structure of a steel that has transformed in this manner is shown in Fig. 8.11; it consists of a mixture of fine pearlite and martensite. Structures may also be obtained in which fine pearlite, bainite, and martensite are present as shown in Fig. 8.12.

With a sufficiently fast rate of cooling, such as indicated by curve *D* of

Fig. 8.10, transformation does not occur until a low temperature is reached, and then martensite forms. There will be some cooling condition, such as indicated by curve E, that will just produce a fully hardened steel; the cooling rate that will accomplish this is called the *critical cooling rate*. A cooling rate slower than the critical will allow the formation of some fine pearlite and, therefore, will prevent full hardening.

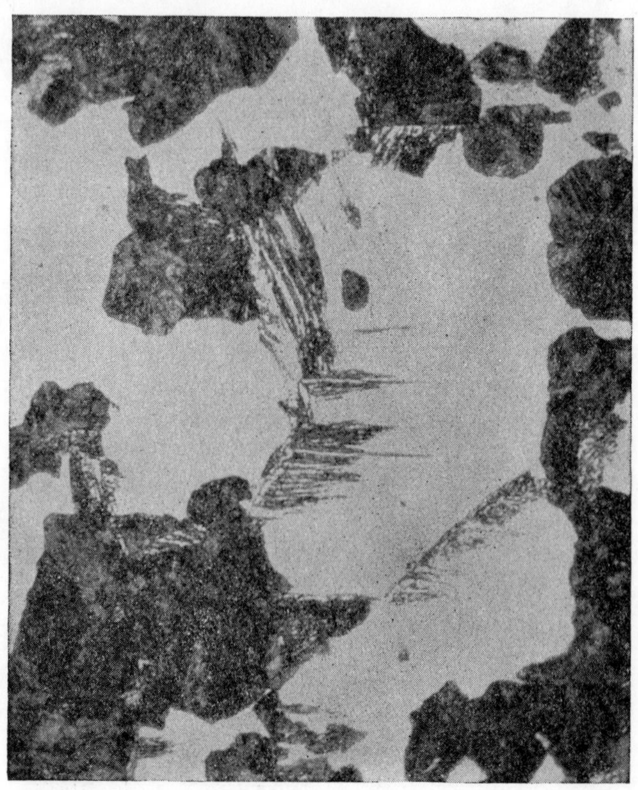

FIG. 8.12 Fine pearlite with some bainite (feathers) and the remainder martensite (white) (1000×).

8.7 Annealing. In general, the term *annealing* refers to any heating and cooling operation that is usually applied to induce softening. More specifically, however, annealing may be divided into two operations, namely, full annealing and process annealing. In *full annealing,* the steel is heated usually to about 100 F (40 C) above the upper critical temperature and held for the desired length of time, followed by very slow cooling as in the furnace. The purpose of full annealing is threefold: to soften the steel and improve ductility; to relieve internal stresses caused by previous treatment; and to refine the grain.

In *process annealing,* the steel is heated to a temperature below or close to the lower critical temperature, followed by any desired rate of cooling. Its principal purposes are to soften the steel partially and to obtain release of internal stresses. In this treatment, grain refinement by phase transformation is not accomplished as it is in full annealing. Process annealing is used extensively in the treatment of sheet and wire. For this purpose, the steel is heated to a temperature between 1020 and 1200 F (549 and 649 C).

In this text, the term *annealing,* when used in connection with the treatment of steel, will mean full annealing. When annealing is applied for the purpose of refining grain, the steel is heated to a temperature 50 to 100 F (10 to 40 C) above the upper critical point. This has been considered in Article 8.9 on grain size. Even though the heating temperature may be considerably above the upper critical point, the treatment is still called "annealing" if the steel is slowly cooled from this temperature.

In special cases, annealing may be accomplished by isothermal transformation at temperatures above the pearlite nose of the isothermal transformation diagram. This process is called *isothermal* or *cyclic annealing* and decreases the total time required for an annealing operation, provided adequate facilities are available.

8.8 Normalizing. In general, the term *normalizing* refers to the heating of steel to approximately 100 F (40 C) above the upper critical temperature, followed by cooling in still air. The purpose of normalizing may be to refine grain structure prior to hardening the steel, to harden the steel slightly, or to reduce segregation in castings or forgings. It is common practice to normalize a steel at a temperature of about 100 F (40 C) above the upper critical point for the purpose of refining the grain.

A hypoeutectoid steel consisting of a structure of ferrite and coarse pearlite may be made easier to machine if the ferrite and cementite are more finely distributed. A very soft steel has a tendency to tear in machining; therefore, the slight increase in hardness obtained by normalizing leads to a more brittle chip and thus improves machinability.

When a steel casting solidifies and cools in the mold, a certain amount of segregation occurs, and the austenite grains are rather coarse. This structure is not conducive to the best properties. The situation, however, may be altered by heat treatment. If heated to just above the critical range, grain refinement will be very slow and adjustment of composition will be extremely slow. At higher temperatures, the reduction of segregation will be much more rapid because the rate of diffusion is higher. Therefore, the steel is normalized at a temperature well above the upper critical point. At this temperature, although the grain size may be larger than it was just above the critical temperature, some refinement will have taken place. The rate of cooling from this higher temperature will be somewhat immaterial since the steel will be given a further treatment to improve the grain size later.

In some cases, the properties obtained by normalizing (air cooling) from just above the critical temperature satisfy the design requirements. In this event, the normalizing treatment is employed as a mild quench.

The use of normalizing in the treatment of forgings and castings will be considered in the chapters dealing with those subjects.

8.9 Grain Size. The properties of any alloy are affected not only by the character of phases present, but also by the size of grains that are present in the structure. In an annealed steel, the size of the ferrite and pearlite grains has an influence on the properties, and their grain size is controlled to a certain degree by the size of the austenite grains from which they form on cooling. The grain size of the austenite which exists at temperatures above the upper critical (Ac_3) also has a marked effect upon the rate of transformation during continuous cooling or under isothermal conditions and, hence, upon the hardening characteristics of steel.

When a steel is heated, the austenitic grain size is a minimum at the temperature at which the austenite forms; then, by continued heating, the size of the grains increases primarily as a function of temperature and secondarily as a function of time. The size of the austenite grains in a hypoeutectoid steel attained by heating to a particular temperature can be determined by cooling the steel slowly from this temperature to a temperature slightly below the Ar_3 point and then cooling it rapidly as by quenching in water. By slow cooling to a temperature a little below the Ar_3 point and above the Ar_1 point, proeutectoid ferrite is expelled from the austenite grains into the grain boundaries, forming a network of ferrite which outlines the austenite grains. Quenching from this temperature prevents further ferrite separation and transforms the remaining austenite into martensite. The size of the austenite grains that existed at the temperature from which cooling began can be determined by microscopic observation of the structure produced by this treatment. The austenitic grain size of a hypoeutectoid steel as revealed by proeutectoid ferrite is shown in Fig. 8.13.

The grain size is usually reported as a number which can be calculated from the relation, $n = 2^{N-1}$, where n is the number of grains per square inch at a magnification of 100×, and N is the grain size number commonly called the ASTM grain size. This method of designating grain size is recommended by the American Society for Testing Materials in their Standard E19-46. Instead of measuring the number of grains in a known area, the grain size can be determined by comparing the structure at the proper magnification with the ASTM standard grain-size chart. The ASTM grain-size numbers, together with their equivalents, are given in Table 8-I.

The grain size of a eutectoid steel can be determined in two ways. In one method, the steel is heated to the desired temperature and is then quenched in a manner to cool the steel at a rate somewhat less than the critical cooling rate. Since the transformation of austenite to pearlite begins at the surface of each austenite grain, this treatment will permit a small

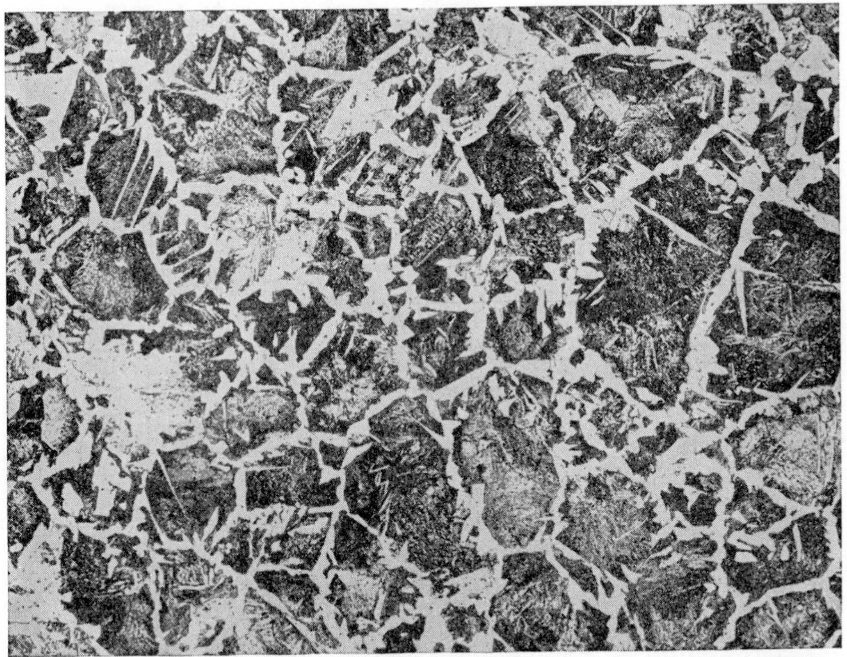

FIG. 8.13 Proeutectoid ferrite indicating former austenite grain boundaries (100×).

amount of fine pearlite to form at the austenite grain boundaries and the remainder of the grain to transform to martensite. The network of fine pearlite, such as shown in Fig. 8.14, allows the grain size to be determined microscopically. In the other method, the steel is cooled at a rate greater than the critical cooling rate, then reheated to a temperature of 600 F (315 C) for approximately an hour, and then quenched in water. After this treatment, the polished section of the steel is etched with a solution of picric and hydrochloric acids in ethyl alcohol which reveals the size of the former austenite grains by contrast, as shown in Fig. 8.15.

TABLE 8-I. ASTM GRAIN SIZE.

	ASTM No.	Mean Number of Grains per in.^{2} at 100×	Grains per mm^2
	−3	0.06	1
	−2	0.12	2
	−1	0.25	4
	0	0.5	8
Usual range	1	1	16
	2	2	32
	3	4	64
	4	8	128
	5	16	256
	6	32	512
	7	64	1024
	8	128	2048
	9	256	4096
	10	512	8200
	11	1024	16400
	12	2048	32800

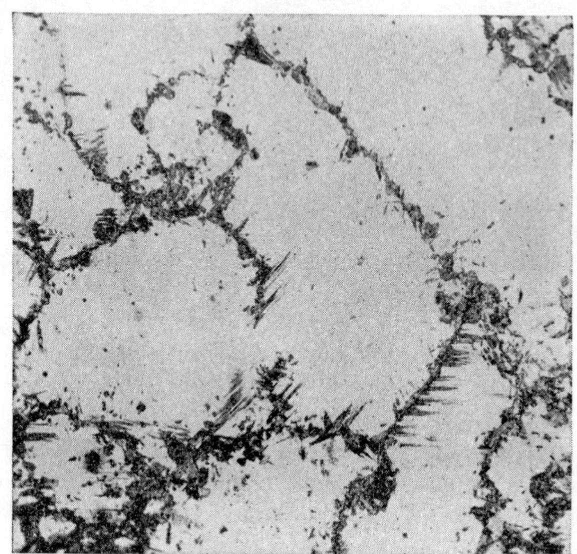

FIG. 8.14 Fine pearlite and bainite indicating former austenite grain boundaries (100×).

The austenitic grain size of hypereutectoid steels can be determined by slowly cooling the steel from the desired temperature to below the $Ar_{3,1}$ temperature. In this treatment, the proeutectoid cementite forms at the austenite grain boundaries, producing a network around the remaining austenite which transforms to pearlite. An example of this structure is shown in Fig. 8.16.

FIG. 8.15 Austenitic grain size revealed by quenching to produce martensite, followed by tempering at 600 F (315 C): Vilella etch (100×).

The grain size of tool steels is frequently determined by a method developed by Shepherd in the United States and by Jernkontoret in Sweden in which the fracture of the steel is compared with a set of standard fractures. The appearance of a fracture is directly related to the grain size of the steel; therefore, a set of standard fractures numbered from 1 to 10 has been established that compares favorably with the ASTM grain-size numbers. This method of determining grain size by comparing fractures is advantageous because it does not require microscopic observation, but at present it is restricted to use for tool steels.

The relation between austenitic grain size and temperature may be determined by heating samples of the steel to different temperatures above

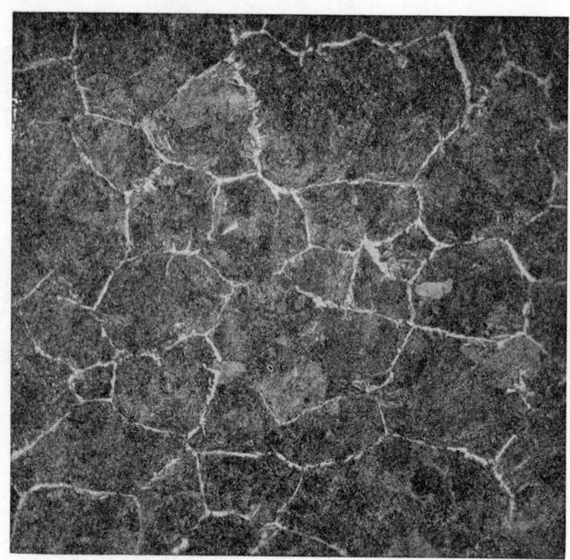

FIG. 8.16 Cementite indicating former austenite grain boundaries (100×).

the Ac_3 point and treating them by one of the procedures described. Fig. 8.17 shows the character of the grain size-temperature relationship for two different steels. One of the steels begins to coarsen almost immediately upon heating above the upper critical temperature, whereas the other steel does not coarsen appreciably until it is heated to a temperature considerably above the critical, whereupon the grain size increases very markedly. A steel that exhibits delayed coarsening is called a *fine-grained* steel; whereas one that begins to coarsen directly upon heating above the critical temperature is called a *coarse-grained* steel. The difference in the grain-growth tendency of different steels has been attributed to the manner in which the steel has been deoxidized in its manufacture. Those steels that are deoxidized with aluminum tend to have fine-grained characteristics; whereas those that are deoxidized with silicon are usually of the coarse-grained character. The addition of certain grain-growth inhibitors, such as vanadium, titanium, zirconium, etc., also promotes fine-grained characteristics in steels.

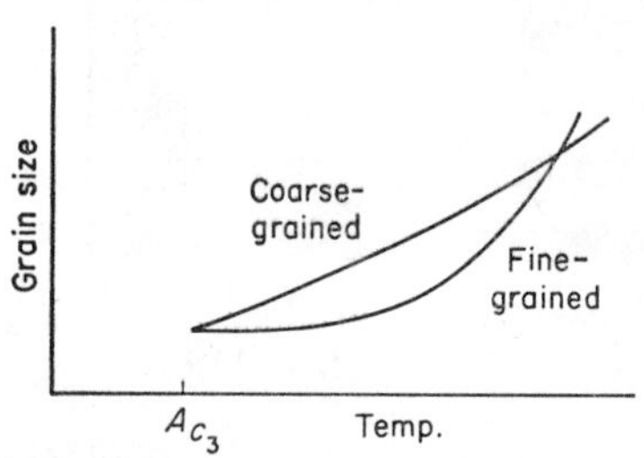

FIG. 8.17 Grain size vs. temperature for two steels, fine-grained and coarse-grained.

8.10 Effect of Austenitic Grain Size on Properties. The austenitic grain size of a steel is extremely important in controlling the properties.

This influence is attributable to the effect of the austenitic grain size on the reaction characteristics during transformation. The relation between the grain size and properties is tabulated below:

PROPERTY	AUSTENITIC GRAIN SIZE	
	Fine	Coarse
Depth of hardening	Shallower	Deeper
Retained austenite	Less	More
Warpage in quenching	Less	More
Penetration in carburizing	Shallower	Deeper
Possibility of quenching cracks and checking in grinding	Less	More
Possibility of soft spots in quenching	More	Less
Internal stress after quenching	Lower	Higher
Embrittlement by cold working	Less	More
Toughness	More	Less
Machinability after normalizing	Inferior	Better

The basis for this summary compilation will be apparent in Article 8.12.

8.11 Relation of Austenite to Ferrite Grain Size. Although the austenitic grain size cannot be evaluated by observing the size of the ferrite grains, the two are definitely related. The coarse austenite grains yield coarse ferrite grains, and the fine austenite grains yield fine ferrite grains. The grain size of austenite has a marked effect on the distribution of ferrite and pearlite with certain rates of cooling. In cooling a hypoeutectoid steel in the critical range, proeutectoid ferrite is expelled from the austenite grain, migrating toward the grain boundary along certain planes of the grain. With slow cooling, all the proeutectoid ferrite reaches the boundary; with a faster rate of cooling, some of the ferrite may remain in the transit position when the remaining austenite transforms to pearlite. A structure of this character is called a *Widmanstätten structure*. The formation of this structure is typical in coarse-grained steels with certain rates of cooling. It is relatively common in steel castings and welds.

8.12 Effect of Grain Size on Rate of Transformation. The discussion in the preceding article on ferrite separation indicates that coarse-grained steels should be more sluggish in transforming than the fine-grained steels. This is true as shown by comparing the isothermal transformation diagrams of a fine-grained and a coarse-grained steel of the same composition as in Fig. 8.18. A slower rate of cooling may be employed in the coarse-grained steel than in a fine-grained steel in order to obtain complete hardening. The transformation curve for the coarse-grained steel is displaced somewhat to the right of that for the fine-grained steel.

Three factors influence the transformation of austenite to pearlite: namely, the size of the austenite grains; the rate of nucleation; and the rate

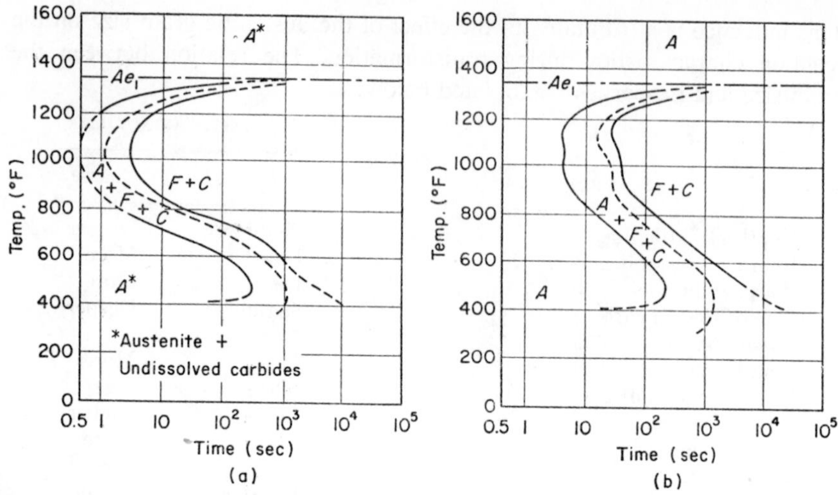

FIG. 8.18 Isothermal transformation diagram. (After *Atlas of Isothermal Transformation Diagrams,* U. S. Steel Corp.) (a) 0.87% C, 0.30% Mn, 0.27% V steel, austenitized at 1500 F (815 C); grain size 11. (b) 0.87% C, 0.30% Mn, 0.27% V steel, austenitized at 1925 F (1052 C); grain size 2-3.

of growth. The influence of the grain size is shown schematically in Fig. 8.19. Nucleation occurs at the austenite grain boundary. If it is assumed that the rate of nucleation and the rate of growth are constant, then a longer time will be required to complete the transformation to pearlite in the large grain than in the small grain. This results in a lower rate of transformation of the entire structure. As a consequence, the steel with the coarser grains of austenite will have a lower critical cooling rate than the steel with finer austenite grains. From this analysis, many of the properties compared in Article 8.10 can be justified.

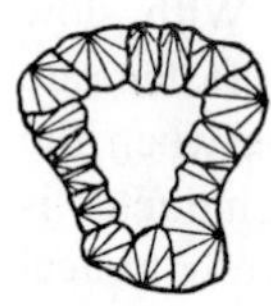

FIG. 8.19 Effect of grain size on time of transformation to pearlite with rate of growth constant. (After Mehl)

8.13 Hardening by Quenching. By controlling the manner in which the transformation of austenite occurs, it is possible to increase greatly the strength and hardness of steel. Maximum hardness may be attained by cooling the material at a rate equal to, or greater than, the critical cooling rate. The increase in hardness attainable with different plain carbon steels is shown in Fig. 8.20. These curves show that the maximum per cent increase of hardness by quenching is obtained in steel containing between 0.35 and 0.70 per cent carbon. Above about 0.55 per cent carbon, there is a small increase in the hardness value obtainable by quenching. This may be explained on the basis of the amount of eutectoid that is present. The existence of proeutectoid ferrite (ferrite formed between Ar_3 and Ar_1), which does not contribute to the hardness of the steel, would naturally account for the lower hardness obtained in the lower-carbon steels; whereas the existence of proeutectoid cementite in hypereutectoid steels generally will not contribute greatly to increased hardness.

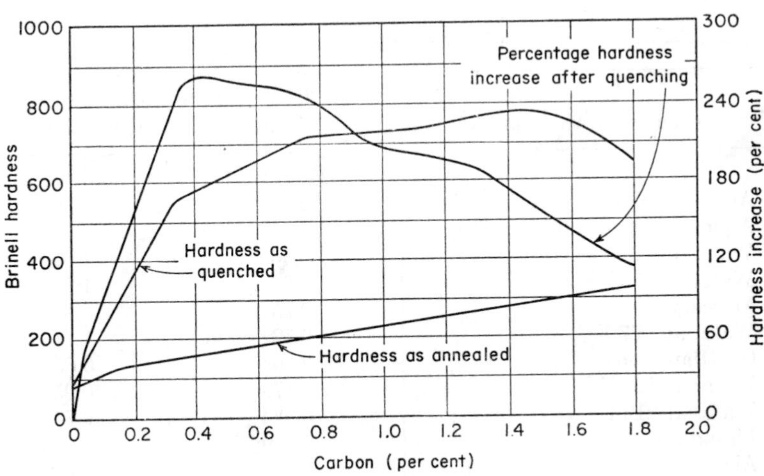

FIG. 8.20 Relation between Brinell hardness and carbon content. (Epstein, *The Alloys of Iron and Carbon,* after Esser and Eilender. By permission of the publishers, McGraw-Hill Book Co., New York, copyright 1936)

The term *hardening* as used in the heat treatment of steel refers to that process of cooling by which the steel is made hard. The steel is quenched in some medium, whereby heat is removed from the steel at the desired rate. It is not always possible or convenient to employ a quenching medium that will produce that rate of cooling which will give the steel a required hardness. In hardening, it is usually customary to quench the steel into certain liquids which are readily available.

The cooling rate for hardening is controlled by the selection of the proper quenching medium. Quenching media include water, dilute sodium hydroxide solution, sodium chloride brine, oil, oil-water emulsion, and air. Quenching characteristics of liquid coolants are controlled by the following factors:

1. Temperature of the medium.
2. The specific heat: the amount of heat necessary to raise the temperature one degree per unit weight.

3. The heat of vaporization: the amount of heat necessary to vaporize unit weight.
4. Thermal conductivity of the medium.
5. Viscosity.
6. Agitation: the rate of movement of piece or flow of coolant.

TABLE 8-II. RELATIVE COOLING RATES FOR SOME QUENCHING MEDIA.*
(After *Metals Handbook*)

Quenching Medium	Cooling Rate Relative to That of Water at 65 F†	
	(1328 − 1022 F)	(At 392 F)
Aqueous solution 10% NaOH	2.06	1.36
Aqueous solution 10% H_2SO_4	1.22	1.49
Water at 32 F	1.06	1.02
Water at 65 F	1.00	1.00
Mercury	0.78	0.62
Water at 77 F	0.72	1.11
Oil (Rapeseed)	0.30	0.055
Water at 122 F	0.17	0.95
Water at 212 F	0.044	0.71
Liquid air	0.039	0.033
Air	0.028	0.007

* Determined at center of a 4-mm diameter Nichrome ball through the temperature ranges indicated during quenching from 1580 F.

† Cooling rates for water at 65 F are 3260 F per sec through range 1328 to 1022 F, and 810 F per sec at 392 F.

The relative cooling rates that may be obtained with some quenching media are shown in Table 8-II. As can be seen in the table, there is a wide range of cooling rates available. Where extremely rapid cooling is required, water spray may be used. It is general practice to quench machinery steel in oil. Water quenching of steel containing more than 0.35 per cent carbon may be dangerous because of the liability of cracking and warping. When austenite transforms, there is an increase in volume, as indicated previously in Fig. 8.1. If the transformation occurs at a very low temperature, as in the formation of martensite, this increase in volume will set up very high stresses in the steel which may be of sufficient magnitude to cause the development of microscopic cracks or visible rupture.

Certain steels may not develop maximum potential hardness even though they are cooled at a rate equal to, or greater than, the critical cooling rate. This situation may be attributable to the presence of retained austenite in the quenched steel. When the carbon content exceeds about 0.65

per cent, the M_f temperature will be at or below room temperature, as previously shown in Fig. 8.7. This condition requires some method of *cold treatment* or *cold stabilization* to bring about the conversion of retained austenite to martensite at subatmospheric temperatures. Such methods include the use of mixtures of dry ice and alcohol for temperatures down to about −100 F (−73 C), mechanical refrigerators for temperatures down to −150 F (−100 C), and liquid nitrogen for temperatures as low as −320 F (−195 C). It is seldom necessary to cool below −120 F (−85 C) except in cases where maximum dimensional stability is required, as in the case of gauges or machine parts requiring close tolerance. The linear expansion resulting from the transformation of 1 per cent retained austenite is approximately 140 microinches per inch. Cold treatment has been employed to advantage to produce:

1. Increased hardness and wear resistance.
2. Dimensional stability of tools, gauges, and machine parts.
3. Elimination of grinding cracks.
4. Increased cutter tool life.
5. Improved magnetic properties.
6. Growth of undersized dies, etc., for salvage.

Since carbon content has the greatest influence in lowering the M_s and M_f temperatures, the greatest value of cold treatment is found in the treatment of high-carbon or carburized steels containing austenite stabilizing elements such as manganese, chromium, and nickel.

8.14 Interrupted Quench. Large sections of certain steels in which high surface hardness is desired require water quenching. However, the high stresses induced by such a rapid quench may be objectionable. This objection may be overcome in some cases by time quenching. The procedure is to quench the steel into water for a predetermined length of time after which it is transferred to an oil-quenching bath for completion of the transformation. The results of the application of this treatment to an SAE 52100 steel are given in the following discussion: Bars 2½ in. in diameter and 6 in. long were heated to 1550 F (843 C). Each bar was quenched into water for a specified length of time and then transferred to the oil-quenching tank. The results of this treatment are shown in Fig. 8.21.

By the usual quenching procedure where the cooling rate is greater than the critical, transformation to martensite occurs at relatively low temperatures. In a heavy section, this transformation will not take place simultaneously throughout the section. This may lead to serious internal stresses, and possibly cracking of the piece. This difficulty, however, may be eliminated or the danger reduced by a procedure called *martempering.*

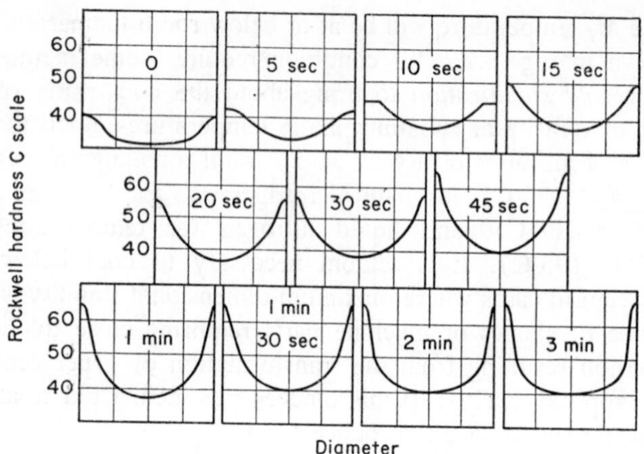

FIG. 8.21 Hardness distribution obtained in time-quenching bars of SAE 52100 steel. Quenched in water for times indicated and then into oil. (Grossmann, *Principles of Heat Treatment*)

In this treatment the steel is cooled rapidly to a temperature just above the M_s and held until the temperature becomes uniform across the section. Then the piece is allowed to cool in air. In this way, the transformation tends to occur more uniformly throughout the section. The steps in this treatment are illustrated in Fig. 8.22.

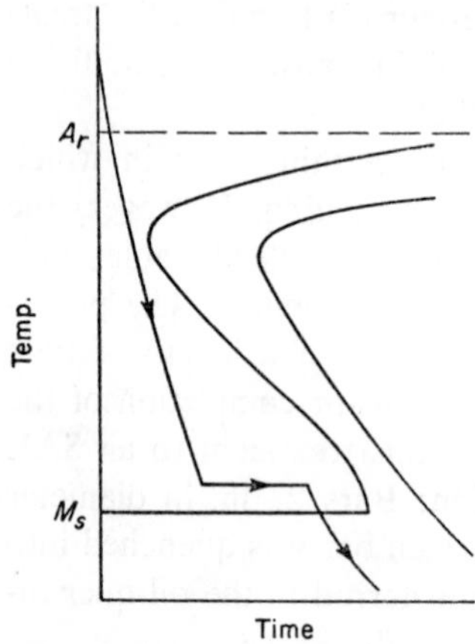

FIG. 8.22 Cooling conditions for martempering in relation to continuous cooling transformation curves.

8.15 Tempering. *Tempering* is the process of heating a **hardened** steel to any temperature below the lower critical temperature, and then cooling at a desired rate. The object of tempering is to reduce the hardness and to relieve the internal stresses of a quenched steel in order to obtain greater ductility than is associated with the high hardness of the quenched steel. If retained austenite is present in the hardened structure, tempering may serve to cause isothermal conversion to bainite. If a fully hardened steel, that is, one in which the structure consists entirely of martensite, is heated to successively higher temperatures above room temperature, submicroscopic particles of cementite will be expelled from the body-centered tetragonal lattice, and the lattice will become body-centered cubic. With continued heating, the particles of cementite grow to microscopic size by a process of coalescence and growth. This coalescence will be more pronounced the higher the tempering temperature. The structure of a quenched and tempered steel is shown in Fig. 8.23. This process is one that involves both time and temperature. Since the hardness of the

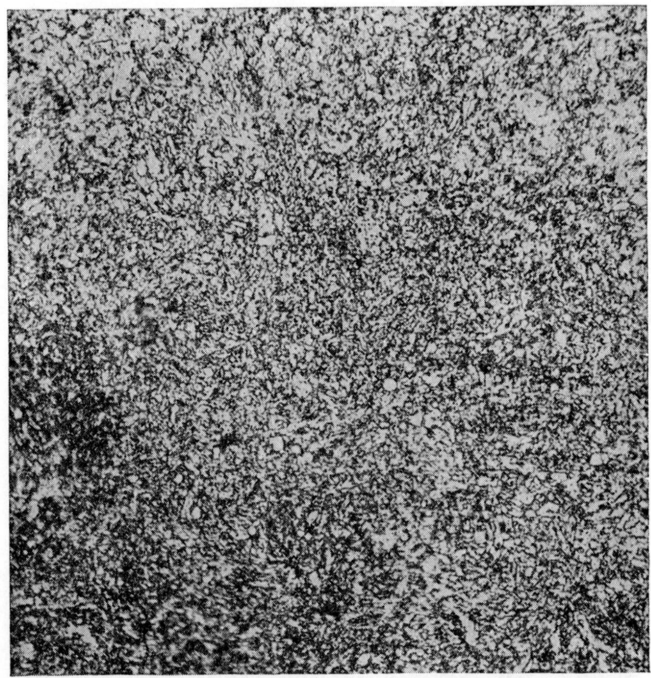

FIG. 8.23 SAE 1080 steel quenched in water from above upper critical temperature and tempered at 1200 F (500×).

steel is due to the fineness of the distribution of the two phases present, an increase in the size of these particles will cause a softening of the steel. The decrease of hardness and strength and the increase in ductility with increasing tempering temperature for one particular steel are shown in Fig. 8.24. Dimensionally, tempering generally brings about an over-all contraction. The dimensional change may be influenced by conversion of any retained austenite which tends to transform isothermally to bainite as a function of tempering time.

If the steel is tempered at a temperature just below the Ac_1 point, the cementite will collect into small spheroids surrounded by ferrite. This treatment is known as *spheroidizing*. It is employed as a toughening treatment, and in high-carbon steels machinability may be improved. The structure of

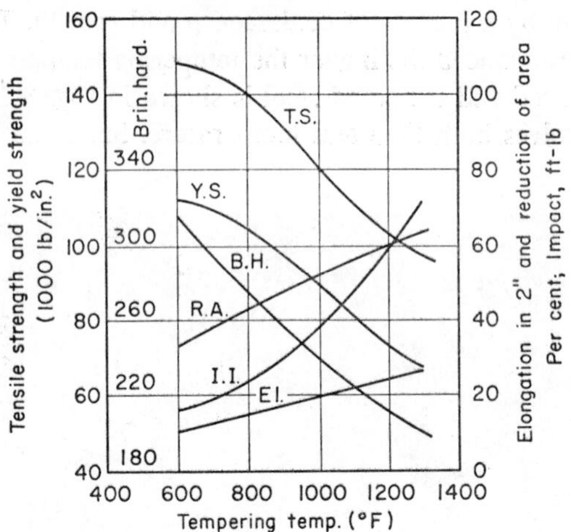

FIG. 8.24 Properties of SAE 1045 steel quenched in water from 1475-1525 F and tempered at temperatures indicated.

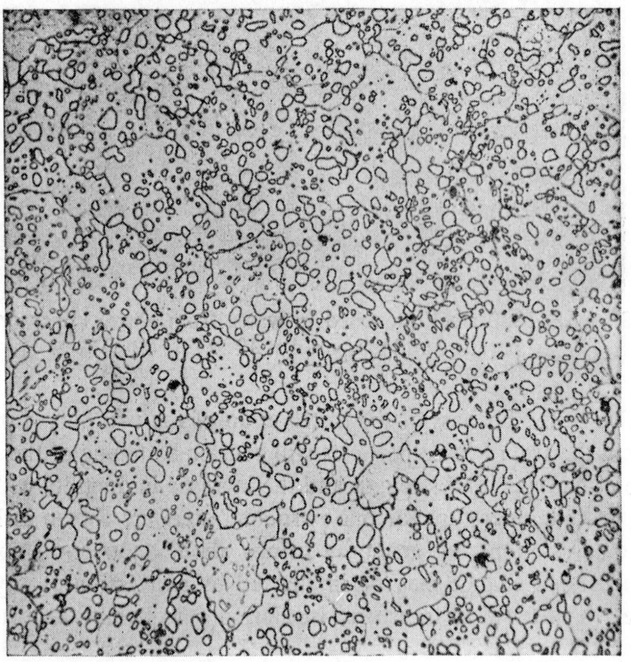

FIG. 8.25 Structure of a spheroidized steel (500×).

a spheroidized steel is shown in Fig. 8.25. A comparison of the mechanical properties of annealed and spheroidized steel is given below:

Property	Pearlitic (1% Carbon)	Spheroidized (1% Carbon)
Yield point (lb/in.²)	85,000	40,000
Tensile strength (lb/in.²)	150,000	78,000
Reduction of area (%)	12-15	57
Brinell hardness	300	156

(After Bain, *Alloying Elements in Steel*, ASM, Metals Park, Ohio, 1939.)

The most common method of treating a steel to obtain particular mechanical properties is to quench the steel in the proper medium from just above the Ac_3 temperature, followed by tempering at the proper temperature. The plain carbon steels containing less than about 0.35 per cent carbon are usually quenched in water; those containing 0.35 to 0.55 per cent carbon, in oil or water; those containing in excess of 0.55 per cent carbon, in oil. The choice may be influenced by other factors, such as section size and shape. A noted exception is with high-carbon, low-manganese steels which may also be water-quenched.

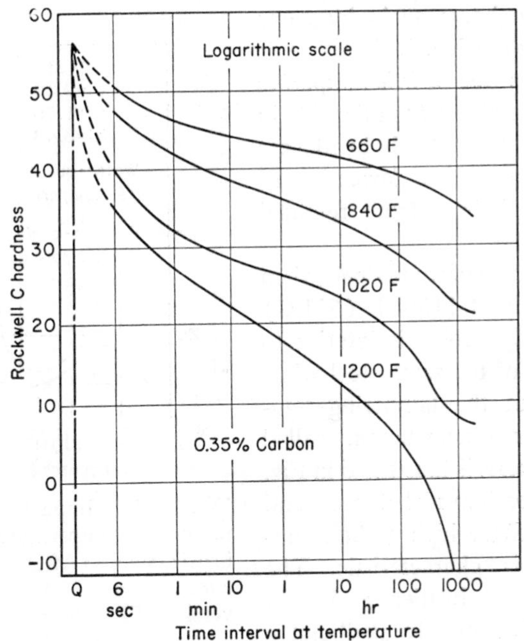

FIG. 8.26 Hardness vs. time for tempering a 0.35% carbon steel at four temperatures. (After Bain, *Alloying Elements in Steel*)

Since the process of tempering involves the diffusion, coalescence, and growth of cementite particles throughout the structure, time will be required for the softening to occur at any given temperature. This effect is illustrated in Fig. 8.26 which is a plot of hardness vs. time for four tempering temperatures. The time required at the tempering temperature is usually some 20 to 30 min. However, at the lower temperatures, a somewhat longer time may be necessary.

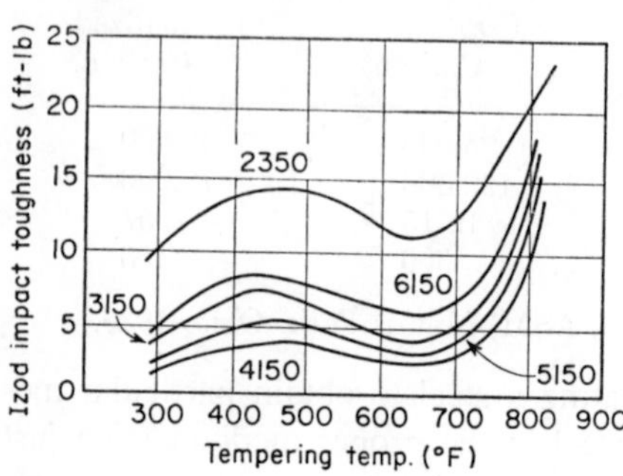

FIG. 8.27 Impact toughness vs. tempering temperature for five alloy steels fully hardened before tempering. (Grossmann, *Principles of Heat Treatment*)

Fig. 8.27 shows the variation of the toughness of a steel with tempering temperature as indicated by the Izod impact values for different tempering temperatures. The toughness gradually increases with increase of tempering temperature, but between 450 and 650 F (232 and 343 C), the Izod value decreases very noticeably. Above this temperature range, the toughness again increases to a very high value. This range is referred to as the *brittle-tempering range* and is particularly noticeable in certain alloy steels. Such brittleness is probably due to the formation of martensite or bainite from retained austenite.

The toughness of some steels when tempered to about 1150 F (621 C) and very slowly cooled is inferior to the toughness of the same steel when quenched from the tempering temperature. This phenomenon, called *temper brittleness,* is illustrated in Fig. 8.28. The chromium-nickel steels and others of relatively high carbon content show this characteristic. The cause of temper brittleness is not clear, but it is probably attributable to the precipitation of some phase from ferrite. Slow cooling from the temperature range of 1000 to 1200 F (538 to 649 C) permits this precipitation, while quenching retains this

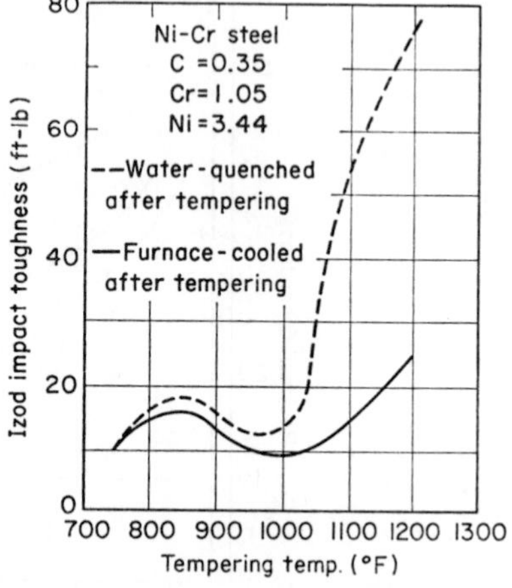

FIG. 8.28 Impact toughness vs. tempering temperature for a temper-brittle steel. (After Nagasawa, *Science Report,* Tohoku Imperial University, First Series, October 1936, pp. 1078-1087)

phase in solution. Temper brittleness can be removed by reheating the steel to the tempering temperature followed by quenching. Molybdenum tends to decrease the effect of temper brittleness.

8.16 Austempering. The quench and temper method of hardening steel to a desired hardness value has the inherent disadvantage of the possibility of producing cracks. In order to obtain the desired hardness, it may be necessary to quench the steel in oil since air cooling would render the steel softer than the desired hardness. After quenching in oil, the steel may be too hard, and it must be tempered. If one could obtain a quenching medium that would cool the steel at the desired rate, there might be less tendency to crack. However, this procedure is not practical in most instances, so another method may be followed for certain steels.

By referring to the isothermal transformation diagram, it will be noted that the desired properties can be secured by transformation at some specific temperature. The steel must be cooled at a rate that will avoid passing through the nose of the continuous cooling transformation curve with resultant formation of fine pearlite. Cooling must be stopped and the tempera-

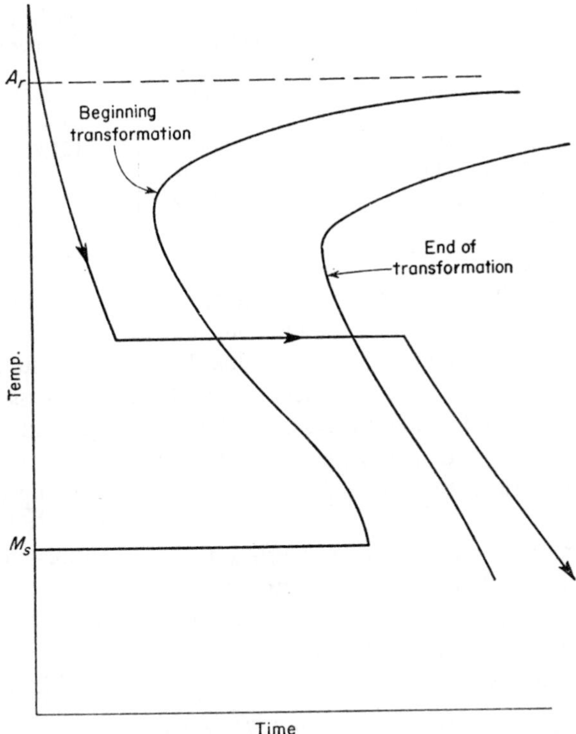

FIG. 8.29 Cooling conditions for austempering in relation to isothermal transformation curves.

ture held at that value for the proper time to produce the desired structure. In this way, none of the austenite transforms to martensite; hence, the formation of microscopic cracks is prevented. This process, called *austempering,* is illustrated in Fig. 8.29. Only those steels for which the initial cooling rate can be fast enough in the desired section to prevent the formation of fine pearlite can be treated in this manner.

A comparison of the properties obtained by the austemper and the quench and temper methods is given below for a steel containing 0.74 per cent carbon as reported by Davenport, Roff, and Bain[2] for an 0.180-in. diameter rod. The austemper treatment in this example consisted of heating the steel to 1450 F (788 C), quenching into a lead-alloy bath at 580 F (304 C) for 15 min, and then quenching into water. The hardness with this treatment was Rockwell C 50.4. The quench and temper treatment was done by heating to 1450 F (788 C), quenching into oil at 70 F (21 C), and tempering for 30 min at 600 F (316 C). By this treatment, the steel had a hardness of Rockwell C 50.2. The properties obtained by each treatment are given in the following table:

	Austemper	Quenched and Tempered
Hardness, Rockwell C	50.4	50.2
Ultimate strength (lb/in.^{2})	282,700	246,700
Yield strength (lb/in.^{2})	151,300	121,700
Elongation in 6 in. (%)	1.9	0.3
Reduction in area (%)	34.5	0.7
Impact (ft-lb)	35.3	2.9

There is very little difference in the hardness, strength, and yield strength of the steel treated by each of these methods, but there is a large difference in the reduction of area and the impact value. From these results, it may be concluded that, for this particular steel, the austemper process is far superior to the quench and temper process. The lower ductility and impact value of the quenched and tempered piece may be due to the formation of the submicroscopic cracks to which reference has been made in a previous paragraph.

Improvement of the dynamic properties by the use of austempering cannot be attained in all steels. Its application will depend in a large degree on the character of the isothermal transformation curve, upon the size of the pieces treated, and upon the final hardness desired.

8.17 Hardenability. The maximum hardness obtainable after quench-

[2] *Microscopic Cracks in Hardened Steel, Their Effects and Elimination, Trans. ASM* **22,** 1934, pp. 289-310.

ing steels of different carbon content was given in Article 8.13. In that discussion, it was assumed that the entire section of the piece was transformed to the fully hardened state. With small sections, such hardnesses will be obtained; but with larger pieces, the center of a bar may not become as hard, because transformation will occur at a higher temperature. The effect of increasing the size of bar on the uniformity of hardening can be shown by quenching a series of bars of different diameters into the same quenching medium from the same temperature. Each bar is sectioned, and the variation of hardness is measured from surface to surface along a diameter. The results are then plotted as shown in Fig. 8.30. This shows that even in the one-half-inch diameter bar, the center of the piece is somewhat softer than the outside surface. The one-inch diameter piece shows almost the same surface hardness, but the center hardness is very much less. The larger-diameter pieces have a much lower hardness at the surface and in the interior. The effect of size on the hardening of steel is related to the term *hardenability*. Hardenability may be defined as susceptibility to hardening by quenching. A material that has higher hardenability is said to harden more uniformly throughout than one that has lower hardenability. This may be clearly shown by comparing the hardness distribution of the 0.45 per cent carbon steel shown in Fig. 8.30 with the hardness distribution of a 0.40 per cent C alloy steel shown in Fig. 8.31. The surface hardness of the alloy steel remains at much higher values than in the plain carbon steel as the size of the piece becomes larger.

In times of critical material shortages, an understanding of hardenability has made possible the direct substitution of an available steel having a hardenability curve closely approaching that of the steel previously found satisfactory for a given part.

Hardenability is directly related to the isothermal transformation diagrams. A steel with a low critical cooling rate will harden deeper on quenching than a steel with a high critical cooling rate. Even though the center of a large bar cools slower than the surface, that rate may be equal to or greater than the critical rate; hence, full hardening will result. Any condition that shifts the isothermal transformation curve to the right will improve hardenability. A steel can be made to harden deeper by coarsening the austenitic grain, since such treatment will shift the beginning of transformation to longer times. In doing this, one must remember that the other characteristics typical of a steel composed of coarse grains will be obtained which may not be to the advantage of the design. It may be better to select another steel, some alloy steel, which even though composed of fine grains may harden deeper by virtue of the inherent effect of the alloying element on the rate of transformation.

The hardenability of a steel may be obtained by determining the diameter of bar that will just harden through to the center. This is accomplished by heating round bars of different diameter to the austenitizing temperature and quenching into oil or water. The length of each bar is at least five times the diameter. After quenching, the bars are cut in half, and a hardness survey is made along a diameter or etched to reveal the depth of hardening. In this method, devised by M. A. Grossmann and his associates,[3] that portion of the section which is occupied by a structure composed of more than 50 per cent martensite is taken as the hardened region. That portion in

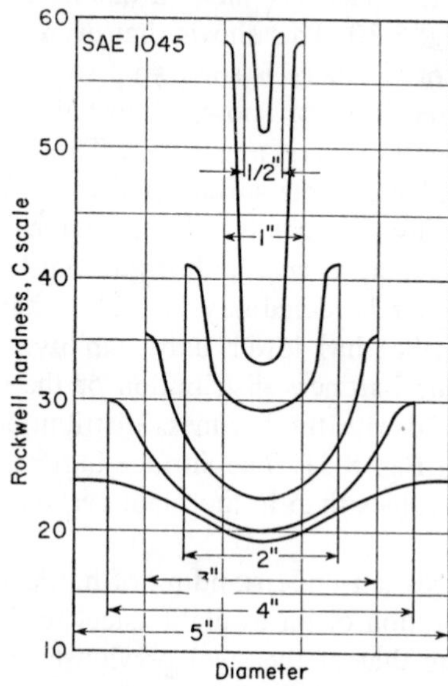

FIG. 8.30 Hardness distribution in SAE 1045 steel for various diameter round bars quenched in water. (Grossmann, *Principles of Heat Treatment*)

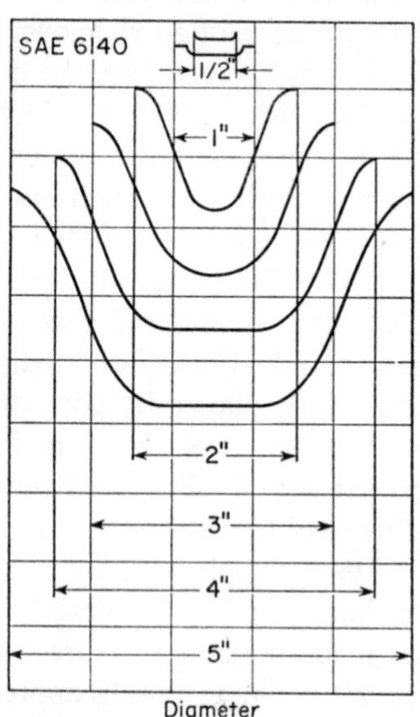

FIG. 8.31 Hardness distribution in SAE 6140 steel for various diameter round bars quenched in water. (Grossmann, *Principles of Heat Treatment*)

which the structure consists of less than 50 per cent martensite is considered as the unhardened region. In the round bar, an area in the center which contains less than 50 per cent martensite is called the unhardened core. The zone of 50 per cent martensite is made quite distinct by etching the section and provides a simple method of delineating the depth of hardening. Sometimes the inflection point in the hardness distribution curve along the diameter indicates the location of the 50 per cent martensite zone. How-

[3] Grossmann, Asimow, and Urban, *Hardenability, Its Relation to Quenching, and Some Quantitative Data, Hardenability Symposium,* American Society for Metals, Metals Park, Ohio, 1938, pp. 124-196.

ever, this is not always reliable because the remaining 50 per cent of the structure may contain different products of transformation, depending upon the grade of steel. It would be better if the zone in which 100 per cent martensite is just produced could be used to delineate the unhardened core, but this structure is not easily located. The size of bar in which the zone of 50 per cent martensite occurs at the center is taken as the *critical diameter.* This is a measure of the hardenability of the steel for the particular quenching medium employed. It is preferable to express the hardenability of the steel in such a way that the quenching medium is not a qualifying factor. This can be done if some standard quenching condition is taken for reference. The quenching condition can be expressed quantitatively by means of thermodynamic considerations involved in heat transfer. The *severity of quench* is indicated by the heat transfer equivalent H which is given by the relation

$$H = \frac{f}{K}$$

where f = heat transfer factor in Btu/in.^{2} sec° F, K = thermal conductivity in Btu/in. sec° F, thus making the units of H in.$^{-1}$

The most rapid cooling rate possible would be with a severity of quench, H, of infinity. Therefore, the inherent hardenability of a steel can be expressed as the diameter of bar that will form a structure composed of 50 per cent martensite at the center when quenched with $H = \infty$. This is called the *ideal critical diameter* and is designated D_I. The relation between critical diameter, ideal critical diameter, and severity of quench can be determined from thermodynamic considerations. These relations are presented in Figs. 8.32 and 8.33. These relations provide a method of obtaining the value of D_I from the quenching procedure used to determine the critical diameter, provided H is known. The severity of quench can be determined experimentally, and typical values for different quenching conditions are given in the following table.[4]

AGITATION OF QUENCHING MEDIUM	MOVEMENT OF PIECE	SEVERITY OF QUENCH			
		Air	Oil	Water	Brine
None	None	0.02	0.3	1.0	2.2
None	Moderate	—	0.4-0.6	1.5- 3.0	—
None	Violent	—	0.6-0.8	3.0- 6.0	7.5
Violent or spray	—	—	1.0-1.7	6.0-12.0	—

[4] Grossmann and Asimow, *Hardening on Quenching, Iron Age* **145,** 1940, April 25, pp. 25-29; May 2, pp. 39-45.

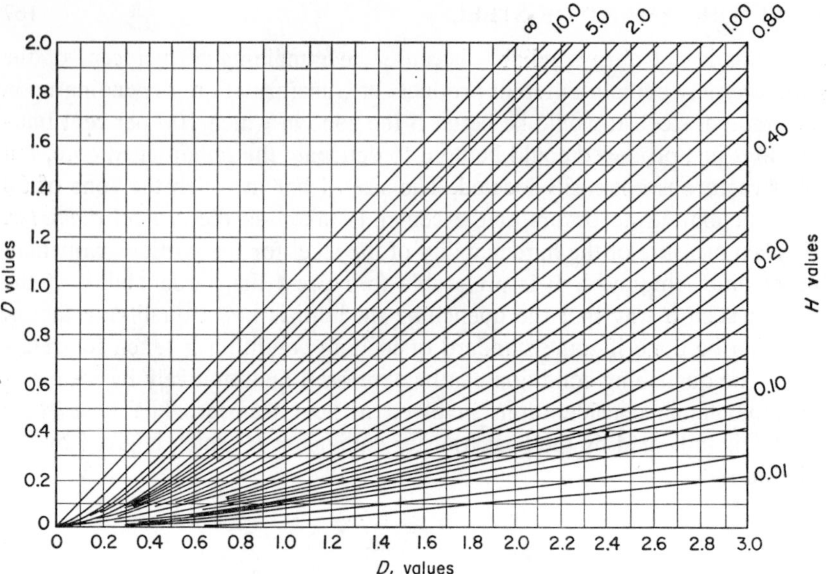

FIG. 8.32 Relation between critical diameter (D), severity of quench (H), and ideal critical diameter (D_I) for bars less than 3 in. in diameter. (Grossmann, Asimow, and Urban, *Hardenability, Its Relation to Quenching, and Some Quantitative Data, Hardenability Symposium,* American Society for Metals, Metals Park, Ohio, 1938, pp. 124-196)

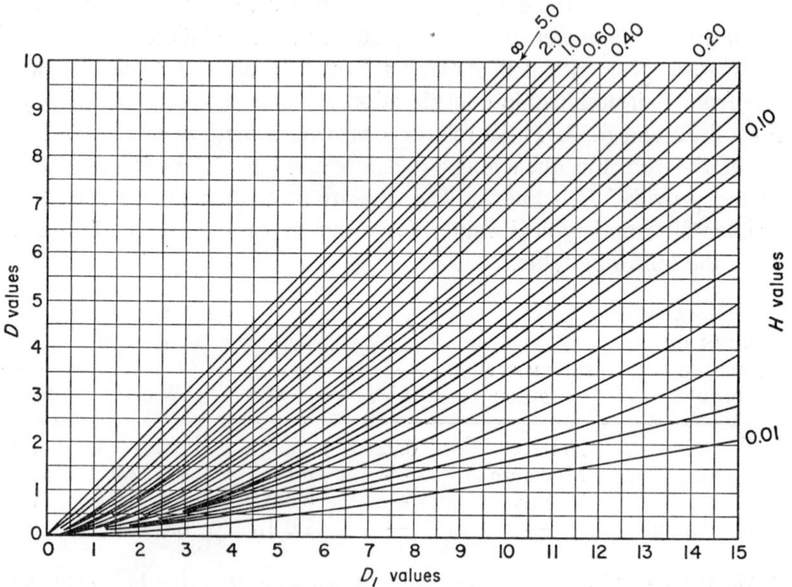

FIG. 8.33 Relation between critical diameter (D), severity of quench (H), and ideal critical diameter (D_I) for bars with diameter up to 15 in. (Grossmann, Asimow, and Urban, *Hardenability, Its Relation to Quenching, and Some Quantitative Data, Hardenability Symposium,* American Society for Metals, Metals Park, Ohio, 1938, pp. 124-196)

Probably the most common method of determining the hardenability of a steel is the *end-quench test,* sometimes called the Jominy test. A bar of the steel is machined to the dimensions given in Fig. 8.34. The sample is heated to the proper austenitizing temperature. After being heated for the proper length of time, the specimen is quickly placed in a fixture, also shown in Fig. 8.34. A quick-acting valve admits water at constant pressure through an orifice of standard size. The water impinges on one end of the specimen.

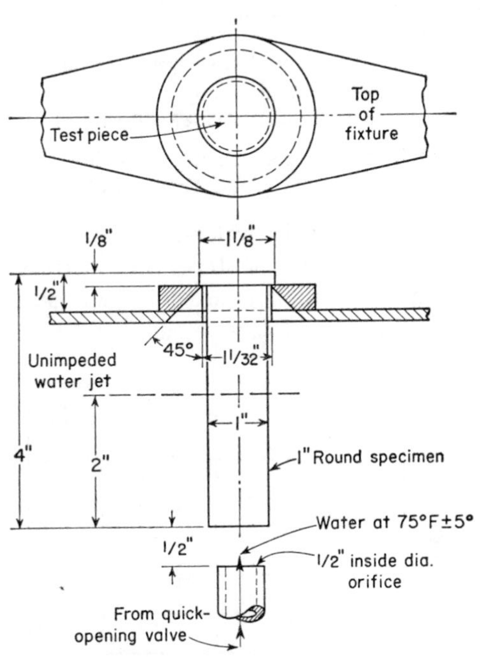

FIG. 8.34 Specimen and fixture for end-quench test.

The end quenching continues until the bar is cooled nearly to atmospheric temperature. A flat is ground on the surface of the specimen 0.015 in. deep. Rockwell hardness values are determined every $\frac{1}{16}$ in. along the length of the specimen from the quenched end and are plotted as shown in Fig. 8.35.

The end-quench test does not provide directly a single figure indicating the hardenability of the steel. However, it is possible to calculate the ideal critical diameter from the end-quench results. This may be accomplished by

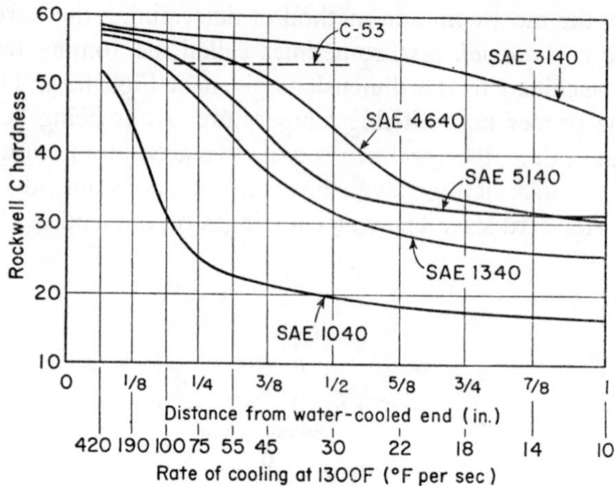

FIG. 8.35 End-quench curves for carbon and low-alloy steels containing 0.40% carbon. (Jominy, *Metal Progress* **38,** 1940, p. 685)

determining the distance from the water-quenched end of the end-quench bar at which a structure consisting of 50 per cent martensite is attained. The diameter of round bar that will produce this same structure at its center when quenched with a severity of $H = \infty$ can then be determined by means

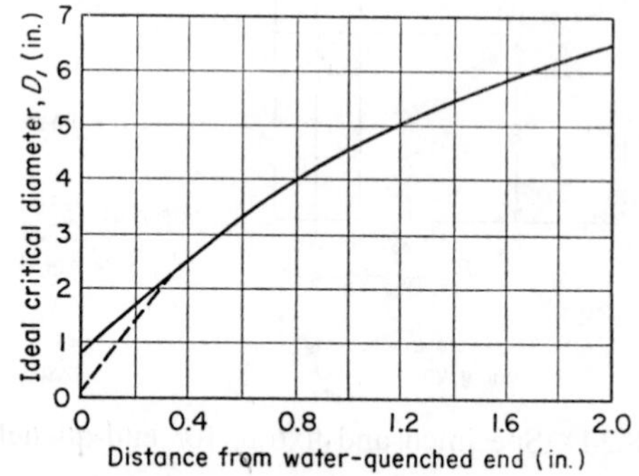

FIG. 8.36 Relation between ideal critical diameter (D_I) and distance from water-quenched end of end-quench bar to position of 50% martensite. (Asimow, Craig, and Grossmann, modified by Hodge and Orehoski[5])

of Fig. 8.36. The value so obtained is the ideal critical diameter. Another method of determining the ideal critical diameter from the end-quench

[5] Asimow, Craig, and Grossmann, *Correlation between Jominy Test and Quenched Round Bars, Trans. SAE* **49,** 1941, pp. 283-292. Hodge and Orehoski, *Hardenability Effects in Relation to the Percentage of Martensite, Trans. AIME* **167,** 1946, pp. 502-512.

curve will be explained by example in Chapter IX. A direct comparison of the end-quench curves may be used for a comparison of hardenability. The position of the inflection point in terms of the distance from the quenched end may be used in some instances for comparison of hardenability. The distance from the quenched end at which a certain hardness is observed may be used as an indication of hardenability. However, one must bear in mind that the absolute value of hardness is not necessarily a matter of hardenability. The maximum hardness attainable in a fully hardened steel depends primarily on the carbon content. It is customary to express the distance at which a certain hardness is obtained in the end-quench test by means of a code. If a hardness of Rockwell C 50 is obtained at a distance of $\frac{5}{16}$ in. from the quenched end, it is written $J_{50} = 5$. In specifying the hardenability of a steel to be used for a particular application, a minimum and a maximum hardness may be given for a certain distance from the quenched end, such as $J_{45/55} = 7$. This means that a minimum of Rockwell C 45 and a maximum of Rockwell C 55 must occur at a distance of $\frac{7}{16}$ in. from the quenched end. Hardenability limits for many of the popular grades of alloy steel have been adopted as specification standards by the Society of Automotive Engineers, under the title *hardenability bands*. The hardenability bands for these steels are given in the appendix, and their use is discussed in Chapter X.

The proper size of bar from which the standard end-quench specimen can be made may not always be available. Provision is made for such circumstances in ASTM tentative standards A255-48T.

In order to make use of the results of end-quench tests, a correlation must be established between the cooling conditions at the different positions along the end-quench bar and at different positions in any quenched bar. To a close approximation it can be said that if the hardness at different positions in two bars of the same steel is the same, then the cooling conditions at those positions are the same. From thermodynamic considerations, one should consider half-temperature cooling times. However, the cooling rate at a certain temperature is a satisfactory criterion of cooling conditions. A given cooling rate in a given steel will produce a certain hardness value. The scale at the top of the end-quench diagram gives the cooling rates for the distances from the quenched end. Cooling rates have been determined for different positions in round bars for oil and water quenching. By these data, the positions in round bars can be correlated with positions in the end-quench specimen. These data are shown graphically in Fig. 8.37.

The desirability of securing complete hardening of a steel on quenching is important with respect to the properties of a steel in the tempered condition. A superior combination of strength and ductility can be secured by

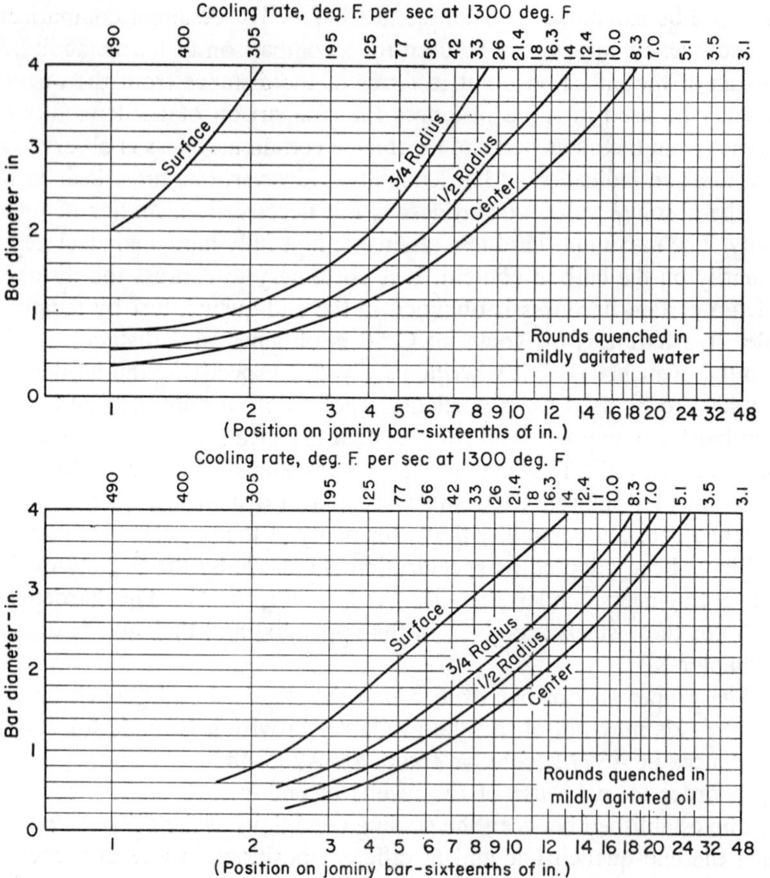

FIG. 8.37 Correlation of surface, center, and intermediate position cooling rates, bar sizes, and distance from quenched end of end-quench test bar, for water and oil quenching round bars. (Boegehold, *SAE Jour.* **52,** Oct. 1944, pp. 472-485)

quenching to produce martensite throughout the section, followed by tempering to restore toughness for the desired strength.

A marked difference in the ratio of yield strength to tensile strength is observed between a steel when tempered from a fully martensitic structure and when tempered from an incompletely hardened structure. This effect has been shown by Crafts and Lamont as presented in Fig. 8.38. These investigators determined the yield ratio for representative carbon and low-alloy steels with a wide range of carbon and alloy content. Cylindrical sections varying from $\frac{1}{2}$ to 4 in. in diameter were quenched in oil and tempered at 1000 F (538 C), 1100 F (593 C), and 1200 F (649 C). They expressed incomplete hardening as the difference between the as-quenched hardness (R_Q) and the maximum hardness (R_M) that could be obtained for a steel of that composition. The result of their study shows the definite decrease of yield ratio with greater incompleteness of hardening.

The relation between the as-quenched structure and the ductility as expressed as reduction of area when tempered to give a tensile strength of 125,000 lb/in.^{2} has been determined by Hollomon, Jaffe, McCarthy, and Norton.[6] With a quenched structure entirely martensitic, the reduction of area was 68 per cent; with initially a mixture of martensite and bainite, the reduction of area was 63 per cent; and with martensite and pearlite, it was 58 per cent.

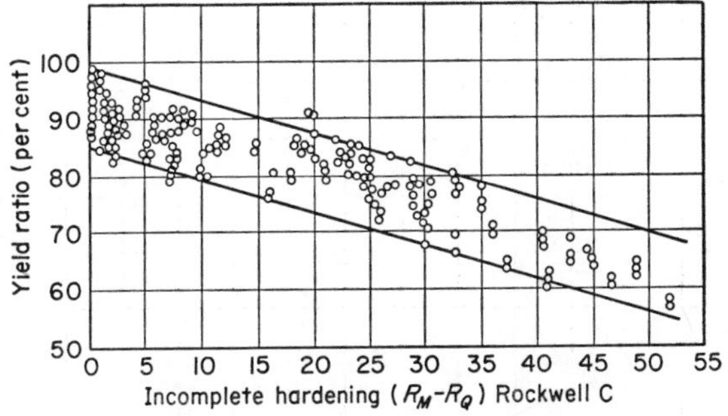

FIG. 8.38 Effect of incomplete hardening on the ratio of yield strength to tensile strength. (Crafts and Lamont, *Hardenability and Steel Selection*, Pitman Publishing Corp.)

The hardness distribution obtained across the section after tempering to a given temperature depends not only on the temperature, but also on the structure that is produced on quenching. This is illustrated by tempering an end-quenched bar to different temperatures. Such an example is shown in Fig. 8.39. Curves of this type are particularly helpful in selecting the proper tempering temperature to secure the desired properties in a given part.

8.18 Factors in Heat Treatment. The heat treatment of steel is a relatively simple procedure, provided the heat-treater thoroughly understands all of the factors that are involved and the fundamentals underlying the transformations. In order to heat-treat a piece of steel properly, three factors must always be specified: first, the temperature to which the steel is to be heated; second, the time that the steel is to be held at this particular temperature; and third, the rate at which the steel is to be cooled from this temperature.

[6] *Effects of Microstructure on the Mechanical Properties of Steel, Trans. ASM* **38,** 1947, pp. 807-847.

The specifications of these three details is determined by several factors: first, the analysis of the steel; second, the past history of the steel, mechanical and thermal; third, the size of the piece being treated; fourth, the shape of the piece being treated; and fifth, the properties desired and the purpose of the heat treatment.

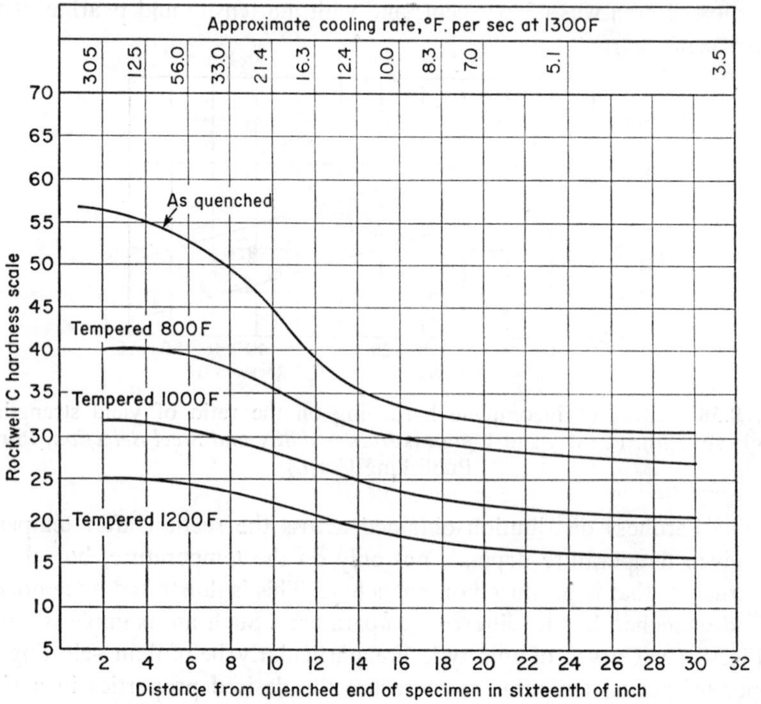

FIG. 8.39 Typical end-quench and tempered end-quench curves. (*Courtesy of Joseph T. Ryerson and Son, Inc.*)

The temperature to which the steel must be heated in any treatment is controlled by the analysis of the steel, its past history, and the purpose of the heat treatment. The time at which the steel is to be held at temperature is controlled by all five factors; and the rate at which the steel is to be cooled is controlled by the analysis of the steel, the size and shape of the piece, and the properties desired. Each of these factors will be discussed in detail.

The analysis of the steel is the principal factor governing the temperature to which the steel must be heated. This is determined by the iron-iron carbide equilibrium diagram, Fig. 8.2. In hardening operations, the temperature must be above the Ac_3 point for the hypoeutectoid steels and above the $Ac_{3,1}$ point for the hypereutectoid steels. In tempering operations, the tempering temperature depends upon the properties required and the characteristics of the steel as shown by the tempering curve. The analysis of the steel also controls the time the steel must be held at temperature. Some constituents do not dissolve or diffuse as readily as others; therefore, a longer time is required to attain structural uniformity.

The past history dictates the treatment that is to be given to the steel and, therefore, dictates the temperature to which the steel is to be heated and the time the steel is to remain at this temperature. If the steel has been cast, it should be normalized at a temperature well above the upper critical temperature for a long enough time to insure uniformity of composition. After this, the grain size is refined and the part is quenched if such properties are required.

The size of the piece is of most importance in controlling the time that a steel remains at a particular temperature and the properties that can be obtained with a given rate of cooling. Even though the surface of a piece of steel may be at the desired temperature, the center may be considerably cooler if the steel has been at this temperature for an insufficient length of time to permit the heat to be conducted to the center. Furthermore, time is required for the formation of austenite. Several rules have been devised to control the heating time and the time at temperature. The time of heating is dependent in a large degree upon the characteristics of the furnace being used. A common rule-of-thumb for steel containing between 0.10 and 0.50 per cent carbon is to heat for one hour for each inch of diameter or maximum thickness and to hold at the required temperature for at least one fifth of the heating time.

For hardening, it is common practice to hold the steel at the desired temperature for 15 min per in. of thickness. For reducing the segregation in castings, it is recommended that the steel be held at temperature for a minimum of 1 hr per in. of thickness. In treating forgings, the time at temperature is usually of the order of 30 min per in. of thickness.

When heating is performed in molten salt baths, the element of time-to-heat may be reduced to a matter of minutes, which is possible because of the increased rate of heat transfer in such baths. In the practical sense, prior experience with similar parts and analyses, tied in with production demands, will be determining factors in heating time.

In tempering steel, the time at temperature depends upon the composition of the steel and the tempering temperature. For plain carbon and relatively low-alloy steels, the time is of the order of 20 to 30 min. With alloy steels, the tempering time may be of the order of 1 hr or more.

The effect of size upon the rate of cooling has been discussed in Article 8.17 on hardenability. The shape of the piece being treated is of great importance, since a combination of thick and thin irregular portions may lead to difficulty in quenching operations. When an irregular piece or one made up of thick and thin parts is quenched, a slower cooling rate may be required to prevent cracking. Cracking in the quenching operation is due to the difference in the rate of cooling of a thick and thin part. Transformation of the thicker sections will take place when the thin sections are at a lower temperature. The expansion in the thick sections on transformation will set up very high stress concentrations which may cause warpage or cracking. With irregular pieces, care must be taken that the thin sections are not at temperature for too long a time as undue grain growth and oxidation may occur, or the temperature may become too great at such points, resulting in "burning" of the steel. A similar condition exists in the heating of parts with sharp edges.

The importance of proper part design should be apparent from the above discussion. It may be said that the successful manufacture of a given part is influenced largely by: first, proper design; second, correct heat treatment; and third, material composition. Close cooperation between the design engineer and the metallurgist is essential in the early blueprint stages of part design rather than after the piece is in production and failures are appearing in heat treatment or service. It is a common occurrence to view part failures where the presence of a sharp radius or an undercut tool mark has been the direct cause of excessive scrap resulting from heat treatment or an early fatigue failure in service. Redesign at such a time may involve costly retooling and processing changes with loss of production. In most instances, the incorporation of a generous radius in the original design of a machined part or a fillet in forgings or castings would not interfere with the proper functioning of the part in service. For years, material selection has been based on prior experience or trial-and-error methods. Although a background of experience will remain a valuable asset, the development of sound metallurgical principles in recent years forces the metallurgist first to recognize proper part design and then to specify suitable material and heat treatment for successful manufacture.

8.19 Surface Protection in Heat Treating. When steel is heated in an open furnace in the presence of air and products of combustion, two surface phenomena may be involved: (1) oxidation, and (2) decarburization. Oxidation of steel is caused by oxygen, carbon dioxide, and/or water vapor in accordance with the following general reactions:

$$O_2 + 2Fe \leftrightharpoons 2FeO$$
$$O_2 + 4FeO \leftrightharpoons 2Fe_2O_3$$
$$CO_2 + Fe \leftrightharpoons CO + FeO$$
$$CO_2 + 3FeO \leftrightharpoons Fe_3O_4 + CO$$
$$H_2O + Fe \leftrightharpoons H_2 + FeO$$
$$H_2O + 3FeO \leftrightharpoons H_2 + Fe_3O_4$$

These reactions are reversible, and the equilibrium relationship between iron and iron oxide in contact with carbon monoxide and carbon dioxide, or hydrogen and water vapor, at the heat treating temperature determines in large measure the degree of oxidation or scaling that will be produced. The interrelation of these four gases is known as the *water-gas reaction* and may be expressed as

$$CO + H_2O \leftrightharpoons CO_2 + H_2$$

This reaction is important since furnace atmospheres tend to adjust themselves at operating temperatures in accordance with the following equilibrium equation:

$$\frac{CO \times H_2O}{CO_2 \times H_2} = K$$

This equation indicates why a change in one component of a complex furnace atmosphere will cause a readjustment of the other components until a new stable equilibrium is established, as illustrated by the addition or removal of water vapor. If the furnace atmosphere is too high in carbon monoxide and low in water vapor, a second reaction known as the *producer-gas reaction* may take place as follows:

$$2CO \leftrightharpoons CO_2 + C$$

Steel serves as a catalyst for this reaction at elevated temperatures, resulting in the etching or oxidation of bright surfaces and the deposition of soot.

Oxidation of steel may range from a tight, adherent straw-colored film that forms at a temperature of about 350 F (180 C) to a loose, blue-black oxide scale that forms at temperature above about 800 F (425 C) with resultant loss of metal. Whereas the above reactions illustrate the oxidation of steel, furnace atmospheres may be provided that neutralize the oxidizing effect or actually reduce the iron oxide already formed by the reverse reaction to form metallic sponge iron in the following manner:

$$FeO + H_2 \rightarrow Fe + H_2O$$
$$FeO + CO \rightarrow Fe + CO_2$$

Decarburization or reduction of surface carbon content occurs upon heating steel to temperatures above about 1200 F (650 C) and progresses to greater depths below the surface as a function of time, temperature, and furnace atmosphere in accordance with the following typical reactions:

$$O_2 + C \leftrightharpoons CO_2$$
$$O_2 + Fe_3C \leftrightharpoons 3Fe + CO_2$$
$$CO_2 + C \leftrightharpoons 2CO$$
$$CO_2 + Fe_3C \leftrightharpoons 2CO + 3Fe$$
$$H_2O + Fe_3C \leftrightharpoons CO + H_2 + 3Fe$$

These reactions are also reversible, and the equilibrium relationship is influenced by the ratio of carbon monoxide to carbon dioxide which will be neutral to a given carbon content at a given temperature. Reference should be made to the literature for a detailed study of the subject of chemical equilibrium and equilibrium constants involved in these reactions.

The problem of oxidation and surface decarburization of steel parts with resultant loss of dimensions and lowered surface hardness and strength may be prevented or minimized in several ways:

1. Removal of decarburized surface by machining operations after heat treating.
2. Application of an electroplated coating, usually 0.0005-0.001 in. of copper, prior to heat treating.
3. Application of proprietary stop-off paint or ceramic coating before treating.
4. Heating parts in a sealed steel box or pot packed in charcoal or cast iron chips.
5. Use of molten salt baths as heating media.
6. Use of protective furnace atmosphere which will prevent (a) oxidation or (b) oxidation and decarburization.

Molten salt baths provide a means of rapid heat transfer for austenitizing and tempering semifinished or finished machined parts with adequate protection against oxidation and decarburization, provided the composition of the austenitizing or "neutral" bath is regularly controlled. Certain limitations such as size of available salt baths, possible corrosion of heat-treated parts caused by ineffective salt removal, the difficulty of quenching parts with blind holes, etc., place obvious restrictions on salt bath heating.

The present widespread application of protective atmospheres has resulted from years of basic research on chemical equilibria of steel in combination with various gases at elevated temperatures, together with the development of modern furnaces of gas-tight construction for effective utilization of furnace atmospheres without infiltration of air. Probably the earliest attempt to provide a protective atmosphere in commercial heating may be traced to the operation of gas or oil-fired furnaces by controlling the air-fuel burner mixture to produce products of combustion that reduced heavy scaling, but with inefficient heating. Prior to 1940, the charcoal generator provided the chief source of generated protective atmosphere. Charcoal gas, generated by passing air through a bed of hot charcoal produces a theoretical composition of 34 per cent carbon monoxide and 66 per cent nitrogen. The process has become practically obsolete, having yielded to less-expensive atmospheres capable of consistent control. Certain gases have been used over the years on a laboratory basis or for commercial applications to satisfy special requirements. Hydrogen is used extensively in connection with furnace brazing and for the bright annealing of stainless steels, silicon iron, etc., and is highly reducing to metal oxides. The hydrogen that is purchased in high pressure cylinders contains water vapor and oxygen as objectionable impurities for bright annealing; it therefore requires purification.

Argon and helium are inert gases which, in view of their high cost, find limited commercial application in the heat treatment of titanium, stainless steel, and for inert arc welding. Nitrogen is employed as a purging medium to remove air from small retorts prior to the introduction of argon or helium. It has limited general use as a protective atmosphere.

Modern protective atmospheres produced or generated for the commercial heat treatment of steel may be summarized as follows:

1. Liquid hydrocarbon atmosphere.
2. Dissociated ammonia.
3. Exothermic gas.
4. Nitrogen.
5. Endothermic gas.

Carburizing atmospheres have been produced for a number of years by cracking hydrocarbon fluid, such as benzol, in a separate cracking unit to produce a mixture of hydrogen, carbon monoxide, carbon dioxide, and methane. This mixture is passed through a sealed pit-type furnace retort. Cracking of the fluid in modern equipment is done directly in the furnace chamber to provide an atmosphere that may be controlled by regulating the fluid flow to provide a carbon potential in equilibrium with the carbon in the steel being treated or adjusted for the purpose of intentional carburization. Such equipment finds application where a small volume of atmosphere is needed but is uneconomical for production use when compared to the generated atmospheres.

Dissociated ammonia is produced by passing anhydrous ammonia gas through a catalyst bed at approximately 1700 F (925 C) wherein dissociation takes place in accordance with the reaction

$$2NH_3 \rightarrow 3H_2 + N_2$$

to produce a dry atmosphere of 75 per cent hydrogen and 25 per cent nitrogen. This atmosphere has a dew point of −60 F (−51 C) or contains 0.0055 per cent water. Dissociated ammonia or a hydrogen-nitrogen atmosphere is used extensively as an economical substitute for hydrogen in the bright annealing of stainless steel, electrical sheet, and other applications where pure dry hydrogen is required. It is also used as a diluent atmosphere in the nitriding process to be discussed in a subsequent chapter.

Exothermic atmosphere is the least expensive of the generated atmospheres and is produced by partially burning natural gas, propane, etc., in the presence of air and a nickel oxide catalyst. The chemical reaction that takes place produces heat and cracks the unburned hydrocarbon in two stages as follows:

$$CH_4 + 2O_2 + 7.6N_2 \rightarrow CO_2 + 2H_2 + 7.6N_2$$
$$2CH_4 + O_2 + 3.8N_2 \rightarrow 2CO + 2H_2 + 3.8N_2$$

By appropriate adjustment of the air-gas ratio, the generator may be operated to produce a lean (completely burned) or rich (flammable) exothermic atmosphere with typical compositions as follows:

	Lean	Rich
Air: gas ratio	10:1	6:1
Hydrogen	1%	14%
Carbon monoxide	1%	10%
Carbon dioxide	12%	5%
Methane	—	1%
Nitrogen	Balance	Balance
Dew point 10 F above temperature of cooling water		

Rich exothermic gas is moderately reducing to metal oxides at elevated temperatures and may be employed for bright annealing, normalizing, and tempering of steel where decarburization is not involved. The presence of carbon dioxide and water vapor causes decarburization of medium- and high-carbon steel unless these impurities are removed. Lean exothermic gas is used as a nonflammable purging gas in certain applications prior to the introduction of flammable atmospheres or as a safe atmosphere for tempering operations below a temperature of about 1000 F (540 C), where flammable gases would present a hazard. It is also used extensively for bright annealing of copper.

Nitrogen may be generated for heat-treating applications by (1) purification of exothermic-type atmosphere or (2) by burning anhydrous ammonia in the presence of air. If a hydrocarbon gas is burned with sufficient air to just permit complete combustion, the products of combustion will be water vapor, carbon dioxide, and nitrogen typical of exothermic gas. The water vapor may be removed by cooling, refrigeration, and desiccant driers, whereas the carbon dioxide may be removed by chemical absorption in a solution of monoethanolamine to produce approximately 99.9 per cent nitrogen dried to a dew point of −50 F (−45 C) (0.0112 per cent moisture). Nitrogen produced by the burning of anhydrous ammonia in the presence of air and a catalyst generates a nitrogen atmosphere containing approximately 99.75 per cent nitrogen, with hydrogen as the only impurity and with freedom from any objectionable unburned hydrocarbon. Nitrogen is employed for purging and for the annealing of medium- and high-carbon steel, particularly where slow cooling would create the hazard of possible explosive mixtures if flammable gas was used.

Endothermic atmosphere is used most extensively at the present time as an atmosphere for industrial heat-treating processes. Its popularity stems from the ability to produce a consistent atmosphere which can be maintained to provide a carbon potential in equilibrium with the carbon content of the steel being heated within a furnace, hence preventing unintentional carburization or decarburization. A pressurized air-gas mixture in the ratio of approximately 3:1 is passed through a catalyst bed of nickel oxide contained in a retort, externally heated to 1900-2200 F (1040-1200 C) to produce an atmosphere in accordance with the following over-all endothermic or heat-absorbing reaction:

$$2CH_4 + O_2 + 3.8N_2 \rightarrow 2CO + 4H_2 + 3.8N_2$$

The completely reacted gas is rapidly cooled to below 600 F (315 C) to prevent the formation of soot which forms upon cooling in the range from 1300 to 900 F (700 to 480 C), in accordance with the following reversible reaction:

$$2CO \leftrightharpoons C + CO_2$$

Typical extreme compositions of endothermic gas are as follows:

	Dry	Wet
Air: gas ratio	2.4:1	3.5:1
Hydrogen	38%	28%
Carbon monoxide	20%	17%
Carbon dioxide	0%	2%
Methane	0.5%	0%
Nitrogen	Balance	Balance
Dew point	−10 F (−25 C)	Saturated

Accurate control of the process is possible by variation of the air:gas ratio to produce an atmosphere free of oxygen, carbon dioxide, and with a negligible amount of undesirable methane. Although natural gas (methane) is most generally available, and hence is used in the illustration above, other hydrocarbons such as propane and manufactured gas are also employed.

The composition of endothermic gas is conveniently controlled by determination of the dew point (moisture content) of the completely reacted gas in the furnace chamber at the operating temperature. As a result of ex-

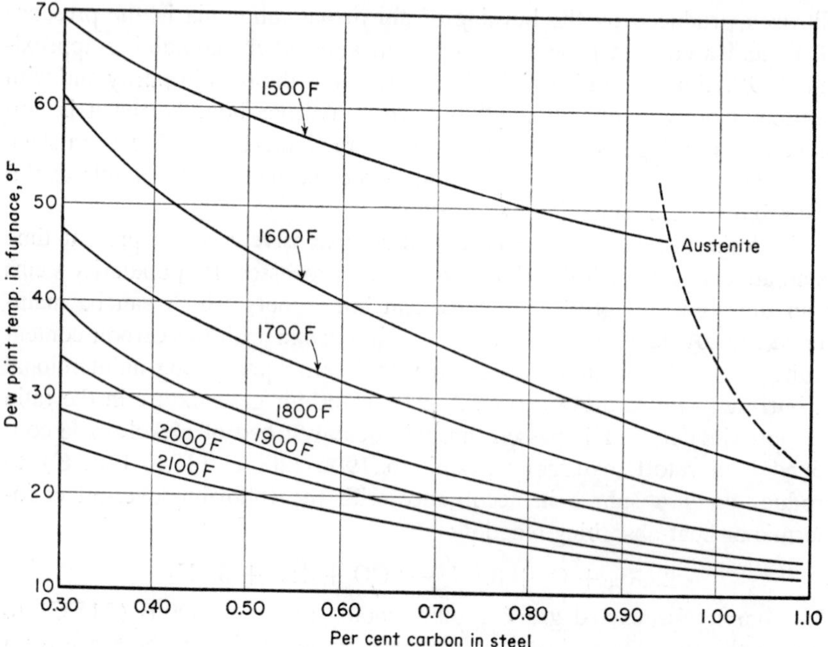

FIG. 8.40 Equilibrium relationship between carbon steels and the dew point of endothermic furnace atmosphere. (Koebel, *Metal Progress* **65,** Feb. 1954, pp. 90-96)

tensive research on samples of steel exposed to furnace atmospheres, equilibrium relationships have been established, of which a typical example is shown in Fig. 8.40. Such relations serve as a means of precision control to provide a carbon potential in equilibrium with carbon content of the steel being heated at any given temperature.

The dew point may be determined with manual instruments or by means of automatic instrumentation. Adjustments of the dew point may be made by (1) variation in the air:gas ratio at the generator or by (2) addition of a small metered quantity of air to the furnace chamber which reacts with the hydrogen present to form water vapor with a resultant increase in dew point and a new condition of equilibrium. Since the dew point curves are based on a completely reacted gas at true furnace equilibrium, it is important that gas samples be taken from the furnace atmosphere after sufficient time has elapsed to establish equilibrium conditions subsequent to any change of the mixture.

Endothermic atmospheres are used extensively for scale-free annealing, normalizing, and hardening of low-, medium-, or high-carbon steel, particularly in the case of finished machined parts where precise control of surface carbon is essential. This type of atmosphere is also used in the carbon restoration of decarburized bar stock or forgings and as a carrier gas for gas carburizing and carbonitriding. In view of the flammable nature of the gas, it is not employed at temperatures below about 1200 F (650 C) to avoid explosive hazards at "black" heat common with the use of hydrogen and dissociated ammonia.

8.20 Flame Hardening. Flame hardening has become a very useful and economical method of hardening the surfaces of various parts. The process consists of heating the surface area that is to be hardened with an oxyacetylene or other type high-temperature flame, until the temperature reaches the proper value above the upper critical temperature. The surface is then flooded with water or suitable coolant in order to quench the heated area. Cylindrical pins, gear teeth, etc., are surface-hardened quite readily by this process. This method depends upon the fact that the heat is applied very rapidly, building up a high thermal gradient and thereby raising the surface temperature to above the upper critical temperature. Immediate application of a coolant causes transformation to take place at a low temperature and prevents heating of the interior. The process thus produces a hard surface with a softer and tougher core. By proper regulation of the operating cycle, the depth of the hardness gradient may be controlled and duplicated in production.

8.21 Induction Hardening. Another method of surface hardening which is applied very satisfactorily to heavy-duty crankshafts, spline shafts, gears, and a number of similar parts is referred to as *induction hardening.* In this process, the part that is to be surface-hardened is surrounded by a copper conductor, very often in the form of a perforated copper block or a tube that is not in contact with the steel to be hardened. A high-frequency current passes through the coil or block, and the surface of the steel is heated by the induced current to above the upper critical temperature. The current is shut off, and water is sprayed out through the perforations in the surrounding block, thereby quenching the surface of the steel. In this manner, a hardness of about Rockwell C 60 may be obtained in certain types of steel to a depth of about $\frac{1}{8}$ in. The heating time varies between 1 and 5 sec, depending on the nature of the equipment and the depth of hardening required. In view of the short time at temperature, there is no tendency for decarburization, grain growth, distortion, or serious oxidation. In the hardening of the bearing surfaces of crankshafts, special equipment has been developed under the trade name *Tocco Process*. In this process, an induction block fits around each of the bearing surfaces, and by proper adjustment of controls, all surfaces are hardened simultaneously. The effect of induction-hardening a spur gear is shown in Fig. 8.41.

FIG. 8.41 Etched section of an induction-hardened spur gear. (*Courtesy of Ohio Crankshaft Co.*)

8.22 Ausforming. When mechanical work is applied to steel in the metastable austenitic condition, a substantial increase in tensile strength and yield strength takes place. This technique of thermal-mechanical processing extends the strength level of high hardenability steels, such as AISI 4340, from former limits of the order of 300,000 lb/in.^{2} tensile strength resulting from conventional quench and temper methods to tensile strengths in excess of 400,000 lb/in.^{2} with satisfactory ductility. An improved technique, known as the *Ausform Process*,[7] consists of mechanical working in

[7] Trademark of Ford Motor Co.

the bainite temperature range of the isothermal transformation diagram, followed immediately by oil quenching to prevent the formation of non-martensitic transformation products. The resultant microstructure consists of fine martensitic plates, the size and dispersion of which is governed by prior austenitic grain size and the magnitude of plastic deformation.

Test results on an alloy steel containing 0.63 per cent carbon indicate a tensile strength of 464,000 lb/in.^{2}, a yield strength of about 320,000 lb/in.^{2} and an elongation of 8 per cent in 2 in. after tempering at 212 F (100 C). Tempering at 600 F (315 C) caused an increase in yield strength to 400,000 lb/in.^{2} with reduction in elongation to about 4 per cent in 2 in.

REFERENCES

Atlas of Isothermal Transformation Diagrams, United States Steel Corp., Pittsburgh, Pa., 1943.

Bullens-Battelle, *Steel and Its Heat Treatment,* John Wiley and Sons, New York, 1948.

Crafts and Lamont, *Hardenability and Steel Selection,* Pitman Publishing Corp., New York, 1949.

End-Quench Test for Hardenability, American Society for Testing Materials, ASTM Standards, Part 3, A255-48T, Philadelphia, Pa., 1949.

Grange and Kiefer, *Transformation of Austenite on Continuous Cooling and Its Relation to Transformation at Constant Temperature, Transactions,* American Society for Metals **29,** Metals Park, Ohio, 1941, p. 85.

Grange and Stewart, *The Temperature Range of Martensite Formation, Trans. AIME* **167,** New York, 1946, p. 467.

Grossmann, *Elements of Hardenability,* American Society for Metals, Metals Park, Ohio, 1952.

Grossmann, *Principles of Heat Treatment,* American Society for Metals, Metals Park, Ohio, 1940.

Hotchkiss and Webber, *Protective Atmospheres,* John Wiley and Sons, New York, 1953.

Induction Heating, American Society for Metals, Metals Park, Ohio, 1946.

Koebel, *Dew Point—A Means of Measuring the Carbon Potential of Prepared Atmospheres, Metal Progress* **65,** Metals Park, Ohio, 1954, p. 90-96.

Metals Handbook, American Society for Metals, Metals Park, Ohio, 1961.

Schmatz, Shyne, and Zackay, *Austenitic Cold Working for Ultra High Strength, Metal Progress* **76,** Metals Park, Ohio, 1959, p. 66-69.

Standard Classification of Austenitic Grain Size in Steels, American Society for Testing Materials, ASTM Standards, Part 1, E19-46, Philadelphia, Pa., 1949.

QUESTIONS

1. Define heat treatment.
2. What is the effect of the rate of heating and the rate of cooling on the critical temperatures of steel?

3. How are the critical temperatures of steel on heating and cooling designated?
4. Draw a graph showing the change of length that occurs during the heating and cooling of steel when it passes through the critical temperatures.
5. What is the mechanism of the transformation of a mixture of ferrite and cementite to austenite on heating?
6. What is the effect of grain size and the presence of alloy elements on the time and the temperature required for the formation of austenite?
7. What is the mechanism of the transformation of austenite to pearlite on cooling?
8. Describe the structure of pearlite.
9. Upon what does the alteration of the mechanical properties of steel by heat treatment depend?
10. Describe the method by which an isothermal transformation diagram is determined.
11. Draw a typical isothermal transformation diagram for a eutectoid steel.
12. What is meant by the M_s and M_f temperatures?
13. Within what general temperature range is pearlite formed on transformation from austenite?
14. What is bainite, and in what temperature range is it generally formed?
15. How does the mechanism of the formation of bainite differ from the mechanism of the formation of pearlite?
16. What is martensite? In approximately what temperature range is it formed?
17. Indicate the mechanism by which martensite is formed from austenite.
18. What is retained austenite?
19. What dimensional change occurs when austenite transforms to martensite?
20. What factors influence the M_s and M_f temperatures?
21. What is the effect of undissolved carbides on the M_s and M_f temperatures?
22. What is the effect of having large amounts of retained austenite present in a structure?
23. What procedure may be used to decrease the amount of retained austenite in a structure?
24. Draw an isothermal transformation diagram characteristic of a hypoeutectoid steel.
25. How does a continuous cooling transformation curve differ from an isothermal transformation diagram?
26. What is a split transformation?
27. Explain how it is possible to obtain a structure composed of fine pearlite, bainite, and martensite.
28. Define the critical cooling rate of a steel.
29. Define annealing.
30. Distinguish between full annealing and process annealing.
31. Define isothermal or cyclic annealing.
32. Define normalizing. What is the purpose of normalizing?
33. Why is it important to know the austenitic grain size of a steel?
34. How is the austenitic grain size of a hypoeutectoid steel determined?

35. How is the austenitic grain size of a eutectoid steel determined?
36. How is the austenitic grain size of a hypereutectoid steel determined?
37. Distinguish between fine-grained and coarse-grained steels.
38. How does the oxidation practice alter the grain-growth tendencies of steels?
39. Compare some of the properties of coarse-grained and fine-grained steel.
40. How is the grain size of a mixture of ferrite and pearlite in a hypoeutectoid steel related to the austenitic grain size of that steel?
41. What is the character of the Widmanstätten structure?
42. What is the effect of the austenitic grain size on the rate of transformation of austenite to pearlite?
43. What three factors influence the transformation of austenite to pearlite?
44. By what process is the hardening of steel accomplished?
45. In what range of carbon content is the greatest percentage of increase of hardness obtained by quenching a steel?
46. How is the cooling rate for hardening of steel controlled?
47. What factors control the quenching characteristics of liquid coolants?
48. By what methods may the conversion of retained austenite to martensite be accomplished?
49. What benefits may be derived by the cold treatment of steel after quenching?
50. What is an interrupted quench and for what purpose is it used?
51. What is martempering? What is the purpose of this treatment?
52. Define tempering. What is the purpose of tempering?
53. Explain why a steel becomes softer upon tempering.
54. What is the process of spheroidizing a steel and what is its purpose?
55. What is the dimensional effect of tempering a steel?
56. What is the effect of time upon the tempering of steel at a given temperature?
57. What is the brittle tempering range in steel? What is its significance?
58. What is temper brittleness, and how might it be avoided?
59. What is austempering, and what are its advantages and limitations?
60. Define hardenability.
61. Explain how isothermal transformation diagrams reflect the hardenability of a steel.
62. How may the hardenability of a steel be determined?
63. Define ideal critical diameter and severity of quench.
64. Describe the end-quench test and its purpose.
65. A steel is quenched in water with a severity of quench (H) of 1.0. A bar of this steel 1.30 in. in diameter quenched in this manner is found to have a structure of 50 per cent martensite and 50 per cent pearlite at the center. (a) What is the ideal critical diameter of this steel? (b) At what distance from the quenched end of an end-quenched bar of this steel would a structure of 50 per cent martensite and 50 per cent pearlite occur?
66. Compare the properties obtained by tempering a fully martensitic structure and tempering an incompletely hardened structure to the same hardness?

67. What factors must always be specified for the proper heat treatment of a part?
68. What factors determine the heat treatment specifications for a steel?
69. Discuss briefly each of the factors in question 68.
70. What factors influence the success of the manufacture of a given part?
71. What is the effect of heating a steel in an open furnace in the presence of air?
72. What methods may be employed to prevent oxidation and surface decarburization of steel?
73. What protective atmospheres are frequently used in the commercial heat treatment of steel?
74. Explain the method of production and the characteristics of exothermic atmospheres.
75. For what purposes is nitrogen gas employed in heat treatment, and how may it be produced?
76. Explain the method of producing an endothermic atmosphere and indicate its use in the heat treatment of steel.
77. What is the process and purpose of flame hardening and induction hardening of steel?
78. What is ausforming, and what significant results are obtained by its use?

IX

FUNCTION OF ALLOYING ELEMENTS IN STEEL

9.1 Classification of Steels. A wide variety of steels are in common use today, and many of them are sold under trade names. As a result, their identification becomes difficult. All steels, however, may be classified according to: (1) kind, (2) class, (3) grade, and (4) quality.

Kind

The method by which the steel is produced determines its kind. The five distinctly different kinds of steel now produced in the United States are listed below in the order of decreasing tonnage:

Basic open-hearth
Electric
Basic oxygen process
Acid Bessemer
Acid open-hearth

Basic open-hearth steels are further classified as *killed* or *rimmed,* depending on the degree of degasification at the time of solidification. Killed steels are those that have been completely deoxidized in the refining process and in which there is practically no gas evolution upon solidification. Their uniformity of composition and freedom from gas pockets, or *blowholes,* make them particularly desirable for forging, carburizing, and heat treating. Rimmed steels, on the other hand, are those in which deoxidation is only partially completed upon pouring into the ingot mold. As solidification occurs, a dense rim of steel solidifies adjacent to the mold wall, accompanied by a rapid evolution of gas. The rimming action may be properly controlled so as to produce an ingot having a sound surface with blowholes located some distance beneath. Subsequent hot-rolling operations will cause welding of these voids and produce a product with a particularly clean surface. Although marked variation in composition will be noted across a rimmed steel section, surface and other characteristics make these steels highly desirable for deep-drawing and forming applications. The carbon content of rimmed steels is normally under 0.15 per cent.

By far the largest proportion of steel is produced by the basic open-hearth process. In recent years, the amount of steel produced by the direct oxygen process has proportionately increased. Some idea of the relative proportions of steel produced by each method can be obtained from Table 9-I.

TABLE 9-I. ANNUAL STEEL CAPACITY IN UNITED STATES (TONS).
(1 January 1960)

Basic open-hearth	125,867,040
Electric	14,395,900
Basic direct oxygen	4,157,400
Acid Bessemer	3,396,000
Acid open-hearth	754,590

Class

Steel may be classified according to form and use. In the broad sense, the form may be either cast or wrought. Steel products may originate in semifinished form either as a steel casting or as a wrought shape, such as a bar, billet, rod, plate, sheet, forging, etc., produced by any hot-working process. Although steels are used for a wide variety of purposes, custom has led to the classification of certain steels according to their use. A steel, however, placed in a so-called class because of its use for a certain purpose, might also be employed in many other applications. Several typical commercial classes of steel are listed below:

Boiler, flange, and fire-box steel: Steels that are particularly suited for the construction of boilers and accessories with particular reference to those steels that may be formed cold without cracking.

Case-hardening steel: Steels that are particularly suited to the carburizing process.

Corrosion- and heat-resistant steel: Steels that are particularly suitable for applications under corrosive or high temperature conditions.

Deep-drawing steel: Steels that are principally used for forming into automobile bodies and fenders, refrigerators, stoves, etc.

Electrical steel: Steels well-suited to the manufacture of electrical equipment, usually of high silicon content.

Forging steel: Any steel that is particularly well adapted for hot-working operations, such as in forging, pressing, etc.

Free-cutting steel (screw stock): Steels that are readily machinable and used for high-rate production of bolts, nuts, screws, etc.

Machinery steel: Steels used for manufacture of automotive and machinery parts.

Pipe, skelp, and welding steel: Very soft steels which are particularly suited for the production of welded pipe, usually of a low carbon content.

Rail steel: Steels employed principally for the production of railroad rails.

Sheet-bar, tin-bar, and sheet steel: Steels that are particularly well suited for manufacture of tin plate and sheet products.

Spring steel: Steels employed in the manufacture of springs of all types.

Structural steel: Steels used in the construction of ships, cars, buildings, bridges, etc.

Tool steel: Steels employed principally for the machining of metals by hand or by power equipment.

Grade

A more specific classification of steels is found under the term *grade,* which particularly refers to the composition of the steel. These classifications are as follows:

Plain carbon steel includes those steels in which the properties are primarily derived from the presence of carbon. Other elements, such as manganese, silicon, phosphorus, and sulfur, may be present in relatively small amounts, but their purpose is not principally that of modifying the mechanical properties of the steel. The carbon content of this grade may vary in the range from a trace to 1.7 per cent, although rarely over 1.3 per cent. The plain carbon steels may be further divided according to the carbon content as follows:

Low-carbon steel, containing from 0.10 to 0.30 per cent carbon.
Medium-carbon steel, containing from 0.30 to 0.85 per cent carbon.
High-carbon steel, containing between 0.85 and 1.3 per cent carbon.

Plain carbon steels containing more than 1.3 per cent carbon are seldom produced or used.

Alloy steel includes those steels that contain other elements added for the purpose of modifying the mechanical properties of the plain carbon steels.

Quality

The quality of steel depends upon the care exercised in processing, close metallurgical control, and rigid mill inspection; it includes consideration of internal soundness, relative uniformity, and freedom from surface defects. The term is used particularly in connection with the suitability of a material for a specific purpose, such as aircraft parts, bearings, rifle barrels, tools, forgings, etc. Special quality standards may be set up to control close composition limits, segregation, grain size, nonmetallic inclusions, response to heat treatment, etc. Quality limits which are more restrictive than those commercially accepted as standard will require closer control and selection of material, with the added costs passed on to the consumer.

9.2 Specifications for Steel Compositions. Specifications covering the carbon and alloy steel grades have been prepared by various sources including the Society of Automotive Engineers, the American Iron and Steel Institute, the American Society for Testing Materials, branches of the Federal government, and individual companies representing industries such as automotive, aircraft, railroad, etc. Many of these specifications cover mechanical properties, quality standards, dimensional tolerances, manufacturing methods, etc., in addition to composition limits. As such, they are employed as purchasing standards for specific products and serve as a basis for rejection of substandard material.

Classification of the various carbon and alloy steel analyses into a system of steel numbers was established by the Society of Automotive Engineers as an SAE standard as early as 1911 in an effort to standardize and limit the large number of existing steel compositions offered by steel producers. The SAE system for the machinery steels specifies analysis limits only, without reference to quality. Code designations indicate the type and approximate carbon content. Chemical analysis of the plain carbon steels is shown in Appendices A and B. The first digit in the classification 1XXX indicates a plain carbon steel. The second digit indicates a modification of the class. The 10XX series are the true plain carbon steels. The 11XX series are those carbon steels which contain larger amounts of sulfur and are commonly referred to as free-cutting steels. The last two digits refer to the average carbon content in points, where 1 point is equal to 0.01 per cent. Alloy steel compositions are shown in Appendix C and will be discussed in detail in Chapter X.

In 1941, the Society of Automotive Engineers, together with the American Iron and Steel Institute, formulated a revision of the basic system to incorporate narrower ranges of analysis and the addition of certain new features. An AISI system of numbers was established corresponding identically with the revised SAE numbers, but with the addition of a letter prefix to indicate melting practice according to the following code:

A: Basic open-hearth—alloy
B: Acid Bessemer—carbon
C: Basic open-hearth—carbon
D: Acid open-hearth—carbon
E: Electric furnace

An SAE 1020 steel contains from 0.18 to 0.23 per cent carbon and corresponds to AISI C1020 which designates a basic open-hearth carbon grade. Absence of a letter prefix in an AISI number implies that such steel is predominantly open-hearth. The use of the prefix TS refers to a tentative standard specification.

There is a growing demand on the part of steel consumers toward specification on the basis of mechanical properties rather than chemical analysis. The design engineer is chiefly interested in the mechanical characteristics he may expect in a given steel. The steel producer, however, finds it more convenient to control steel melting practice on the basis of chemical analysis without guarantee of minimum mechanical properties. Nevertheless, two heats of steel of essentially the same chemistry may differ widely in strength values because of variables in melting practice, mill operation, and many other factors. The mechanical properties of the various SAE steels given in the literature of the steel companies should be considered as typical values to be expected from a given composition under stated conditions of treatment and testing. They represent conservative values based on averages of many test results taken from numerous heats of steel. When the over-all chemical analysis of a given heat falls on the low side of the specification range, it should be recognized that the mechanical properties will, in turn, differ from the normally expected values.

In order to guarantee response to heat treatment, many of the common SAE alloy steels are offered under hardenability band specifications identified by the suffix "H" added to the regular SAE code number. The steel producer is permitted somewhat wider chemical limits within which he may adjust the composition to meet specified hardenability values. The student is referred to Appendix E for currently accepted hardenability bands and modified steel analyses.

9.3 Low-Carbon Sheet and Strip Steel. Steel that is rolled into sheets or strips accounts for a large proportion of the total output of steel. In these forms, it is used in many different grades and in a variety of gauges for automobiles, furniture, refrigerators, stampings of all types, tin plate, porcelain enamelware, galvanized sheets for roofing, ducts, etc.

Most sheet steel contains less than 0.20 per cent carbon, and in a majority of applications, particularly for deep-drawing operations, the steel contains from 0.04 to 0.10 per cent carbon. The low-carbon rimmed steels are particularly well suited for deep drawing because of their great ductility and superior surface finish. Characteristics of deep-drawing steel are considered in the discussion of cold-working operations in Chapter XX. A steel that is used for automobile fenders and bodies containing about 0.05 per cent carbon in the form of hot-rolled strip may have a tensile strength of the order of 55,000 lb/in.^{2} and a yield point of 20,000 lb/in.^{2} with an elongation in 2 in. of 28 per cent. When this strip is reduced 50 per cent by cold rolling, the tensile strength may be as high as 96,000 lb/in.^{2} and the elongation, 2 per cent in 2 in. If this cold-rolled material is annealed, the tensile strength will drop to about 44,000 lb/in.^{2} and the yield point to 23,000 lb/in.^{2}, whereas the elongation will be raised to about 38 per cent in 2 in. This shows the marked influence of cold-rolling and annealing on these low-carbon steels.

The annealed low-carbon steels have a very distinctive yield point. When this steel is employed in this condition for deep drawing, a rippled effect may be produced on the surface called *orange peel, stretcher strains, worms,* etc. This effect can be prevented by a very small amount of cold-rolling prior to the deep-drawing operation. *Temper rolling,* as this process is called, tends to eliminate the yield point in the stress-strain diagram. However, if a very low carbon steel is temper-rolled and then aged, the yield point may reappear with some loss in ductility.

9.4 Structural Steel. Another very extensive use of low-carbon steel is in beams, plates, channels, angles, etc., for construction purposes. The carbon content in these steels varies between 0.15 and 0.25 per cent. Boiler steels usually come within this same range, although the carbon content may be as high as 0.3 per cent. There has been some use of low-alloy steel for this purpose.

9.5 Cold-Heading Steel. Steels that are used for cold-forming bolts, rivets, and the like, require a soft steel which may be easily formed cold in the upsetting machine. A C1020 steel is particularly well adapted for this purpose. Where a somewhat higher strength may be required, a C1035 steel may be used.

9.6 Cold-Finished Bars and Shafting. The terms *cold-rolled* and *cold-drawn* have been used widely for the designation of plain carbon, cold-finished steels produced from properly cleaned, hot-rolled bar stock which has been passed through a set of rolls or drawn through a die. They are particularly well suited for shafting, pins, and other purposes where a good surface finish and close dimensional tolerances are required. The surface finish of these steels can be further improved by machining or grinding. The general term *cold-rolled steel,* when not qualified by other designation, usually refers to a low-carbon steel, such as C1020 steel, which has been cold-finished.

9.7 Free-Cutting Steel. The free-cutting steels have been developed for the purpose of improving machinability and thereby decreasing machining cost, particularly in automatic screw machines. One might expect that a very low-carbon steel would machine very easily, but, because of the strain-hardening characteristics of these low-carbon steels, this is not the case. Any means by which chip formation and breaking can be facilitated will improve machinability. The chip formed with a soft, tough material tears from the work, producing a rough surface with severe heating of the tool. The free-cutting steels owe their advantages to the presence of sulfides which tend to cause chip formation by breaking the continuity of the matrix. In the case of manganese and phosphorus, the improved machinability may be attributed directly to the hardening effect of these alloying elements. The analyses of the free-cutting steels are given in Appendix B.

Increasing the manganese content of a plain low-carbon steel, such as in C1024, will increase the hardness of the steel somewhat. The machinability of these steels, though not as good as some of the others, shows marked improvement over the usual plain carbon steels. The C11XX steels attain their greater machinability by the use of sulfur in amounts up to 0.33 per cent for some grades. Furthermore, the manganese content is also increased above the normal amount. Such steels will contain manganese sulfide and iron sulfide inclusions which render the turnings more brittle and serve as effective chip-breakers, tending to produce smaller chips. The B11XX steels are commonly referred to as Bessemer screw stock. The amount of phosphorus in these steels may be as high as 0.12 per cent, and the manganese and sulfur content is maintained at high values. These steels have about the best machinability of any of the free-machining steels.

Another type of free-machining steel which has been developed is the leaded steel, containing lead up to 0.25 per cent. Lead in steel exists in the structure in the form of very small submicroscopic globules. This distribution breaks up the structure sufficiently to produce a somewhat embrittled chip when the steel is machined. Lead has the particular advantage over some of the other elements which are used for improving machinability in that the ductility and toughness of the steel are not reduced and there is very slight effect on any of the other mechanical properties of the steel at normal temperatures.

A comparison of the machining characteristics of several steels is given in Table 9-II.

TABLE 9-II. RELATIVE MACHINABILITY OF SOME COLD-DRAWN STEELS.
(Based on AISI B-1112 as 100. As-rolled except *Annealed. After *Metals Handbook.*)

Steel AISI	Rating	Steel AISI	Rating	Steel AISI	Rating	Steel AISI	Rating
C-1010	55	A-2330	45	A-4032	70*	E-6150	48*
C-1020	64	A-2330	61*	A-4130	51	A-8625	51
C-1045	51	A-2515	48	A-4130	67*	A-8640	55*
C-1045	57*	A-3115	67	A-4140	61*	A-8650	45*
B-1112	100	A-3140	36	A-4340	51*	A-8740	55*
C-1120	82	A-3140	57*	E-4620	55	A-8750	42*
C-1144	79	E-3310	48*	A-4820	45*	A-9442	51*
A-1320	55	A-4023	73	A-5140	61*	A-9747	42*
A-2317	64	A-4032	55	A-5150	51*	A-9840	57*

9.8 Carburizing Steel. A low-carbon steel is tough, but it is not very resistant to wear, whereas a high-carbon steel when properly heat-treated is highly resistant to wear. A combination of these two characteristics can be obtained by the carburizing process in which the carbon content of the surface is increased. The C1015 and C1020 steels are used very extensively for this purpose. These materials can be formed while cold with great facility, and then by the application of carburizing the surface can be made extremely hard. This process will be discussed in detail under Chapter XI.

9.9 Medium-Carbon Steel. Steels containing from 0.30 to 0.83 per cent carbon are probably the most widely used steels for the construction of equipment. In the chapter on the heat treatment of steel, it was stated that those steels which contained less than approximately 0.35 per cent carbon did not harden appreciably by quenching. The properties of those steels containing 0.35 per cent carbon or more are greatly improved by heat treating and, therefore, are best adapted for machine parts. For these purposes, it is customary to use steels containing between 0.35 and 0.55 per cent carbon. It has already been shown that increased carbon content lowers the ductility of steel. Therefore, where greater toughness is required, a lower carbon steel may be necessary, provided the desired strength can be obtained through suitable heat treatment.

It is advisable to use a steel with as low a carbon content as possible which will give the strength required. The C1040 steel is used extensively in the automotive industry for tubing, torque tubes, axles, bolts, crankshafts, connecting rods, etc. Where somewhat greater hardness and wear resistance are required, the C1050 steel is quite suitable for gears and heavy steel forgings. This steel is usually quenched in oil, although, in large uniform sections, water may be employed in order to obtain greater hardening. Steels containing between 0.55 and 0.83 per cent carbon are used for springs and some woodworking tools. Railroad rails containing 0.65 to 0.75 per cent carbon are within the medium-carbon classification.

9.10 High-Carbon Steel. When maximum response to heat treatment in the plain carbon grade of steels is required, it is preferable to use a high-carbon variety. The C1090 and C1095 steels are employed for many springs, either leaf or coil type, and also for punches, dies, chisels, saws, hammers, wire and cable, cutting tools, etc. In tools that require greater hardness and in which ductility is not such an important factor, the carbon content may vary between 1 and 1.2 per cent. Saws, files, razors, jewelers' files, balls and races for ball bearings are usually made from steels containing between 1.2 and 1.5 per cent carbon.

9.11 Limitations of Plain Carbon Steel. Plain carbon steels are used successfully where strength and other requirements are not too severe. At ordinary temperatures and in atmospheres that are not of a severely corroding nature, plain carbon steels will be highly satisfactory. However, the relatively low hardenability of the plain carbon steels limits the strength that can be attained except in relatively small cross sections.

One of the purposes of tempering a hardened steel is to decrease the *internal stresses* that may have been produced in the quenching operation. Plain carbon steels exhibit continued softening with increasing tempering temperature. The higher the tempering temperature, the more complete is the relief of the internal stress. However, the hardness or strength attained at the stress-relieving temperature may be lower than required. One would desire a steel that could be tempered to secure stress relief without sacrifice of hardness.

The low resistance of plain carbon steels to corrosion and oxidation, and their loss of strength at elevated temperature, limits their broad usefulness.

The engineer must bear these limitations in mind when considering the selection of steels. In most instances they can be surmounted. In summary the most common limitations are as follows:

1. Low hardenability.
2. Major loss of hardness on tempering (stress-relieving).
3. Low corrosion and oxidation resistance.
4. Low strength at elevated temperature.

The most common and practical method of overcoming the deficiencies of the plain carbon steels is to employ alloy steels. Alloys are added to steel as a means of enhancing the already outstanding characteristics of the plain carbon steels.

Alloying elements may not only help to overcome the limitations of plain carbon steels that have been cited, but may also effect an improvement in some other properties. While fully hardened plain carbon steels are usually quite resistant to abrasion, improvements may be secured by the presence of alloying elements. There are some instances in which a steel cannot be heat treated because of the physical limitations of the part or structure in which it is employed. Some improvement in strength can be secured by the use of alloying elements without a hardening heat treatment.

The purposes of using alloying elements may be summarized as follows:

1. To increase hardenability.
2. To increase resistance to softening on tempering.
3. To increase resistance to corrosion and oxidation.
4. To improve high temperature properties.
5. To increase resistance to abrasion.
6. To strengthen steels that cannot be subjected to quenching.

The engineer must bear in mind that he should make every attempt to use plain carbon steels in preference to alloy steels, since the latter are more expensive and may require more elaborate handling or treatment. One must remember that alloying elements in steel in many instances do not improve the inherent properties of the plain carbon steels, but they do make it possible to attain the equivalent properties uniformly throughout larger sections and certain other special properties.

Sometimes *stiffness* of a part is an important consideration. For this requirement, alloy steels do not offer any advantages over the plain carbon steels. Stiffness is determined by the modulus of elasticity or the relationship between the stress and the strain within the elastic limit. An alloy steel may have a higher elastic limit than a plain carbon steel, but the modulus of elasticity will be the same, regardless of heat treatment. The solution to stiffness problems, when steel is demanded, is one of design and not metallurgy.

9.12 General Effect of Alloy Elements. The selection of the proper steel to be used for a particular job is one of the most difficult tasks of the designer. Plain carbon steels have advantages, such as simplicity of heat treatment and low cost. Nevertheless, these steels have limitations, as has been indicated in the previous article. On the other hand, by using the proper alloy steel, desired properties throughout heavy sections may be obtained which are unattainable with a plain carbon steel. Moreover, certain economies may result from the use of alloy steels for certain purposes. Yet the engineer should not use alloy steels blindly, with the thought that they are the "cure-all" for the deficiencies of plain carbon steel. There are instances where the use of an alloy steel may intensify rather than diminish the difficulties present. Consequently, the engineer must understand the underlying principles involved in the alteration of characteristics by the addition of alloy elements. He should also be familiar with the general effects of alloying elements, and he should recognize those situations in which alloy steels will be of material assistance to him in his design.

The addition of various elements to the iron-iron carbide alloys changes the equilibrium diagram; however, usually no new structural constituents appear, although the composition and proportion of the phases in the structure may be altered considerably. Alloy elements effect a modification of the transformation temperatures and rates, thus modifying the properties of the plain carbon steels. With certain alloy additions, new structural constituents may appear, as when copper in amounts greater than approximately one per cent is employed. In this case, the solubility limit of copper in either alpha or gamma iron is about 0.8 per cent. Hence, particles of elemental copper will be found in the structure. The principles that have been established in the preceding chapters are not altered in any respect by the addition of alloying elements. The entire effect of the addition of alloying elements is to modify or to enhance the excellent properties of the plain carbon steels.

Carbon is the most important element present in any steel whether plain carbon or alloy. The maximum hardness obtainable in a steel by quenching to form a structure consisting of 100 per cent martensite is governed by carbon rather than alloy content. This relationship is shown in Fig. 9.1,

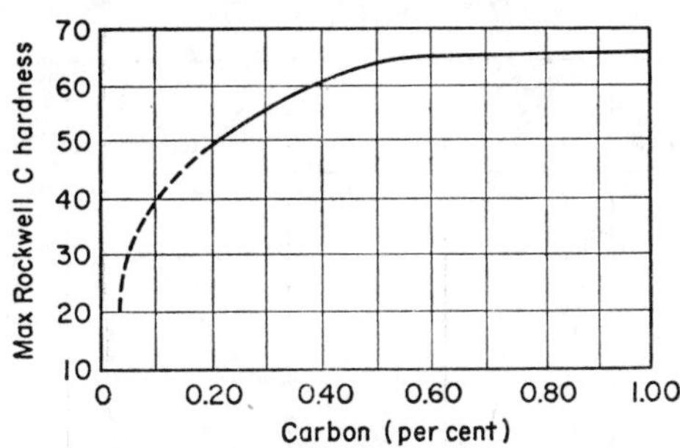

FIG. 9.1 Maximum hardness vs. carbon content for alloy and carbon steels. (After Burns, Moore, and Archer, *Trans. ASM* **26,** 1938, p. 1)

where it should be noted that the maximum hardness value of Rockwell C 65 occurs at a carbon content of approximately 0.55 per cent. By the addition of alloying elements, the effect of carbon may be intensified, diminished, or neutralized.

The possible effects that may result from the addition of alloying elements to steel may be summarized as follows:

1. The alloying element may form solid solutions or intermetallic compounds.
2. The element may alter the temperature at which phase transformations occur.
3. The element may alter the solubility of carbon in gamma and in alpha iron.
4. The element may alter the rate of reaction of the transformation of austenite to its decomposition products and, likewise, the rate of solution of cementite into austenite upon heating.
5. The presence of elements may decrease the softening on tempering.

9.13 Mode of Combination of Alloying Elements in Annealed State. Practically all of the alloying elements which are added to steel, including nickel, silicon, aluminum, zirconium, manganese, chromium, tungsten, molybdenum, vanadium, titanium, phosphorus, sulfur, and copper are soluble in ferrite to a varying degree. Some of the elements will form carbides when sufficient carbon is present. Nonmetallic compounds may be formed, locating themselves in the structure as inclusions. In some cases, intermetallic compounds are formed. The combining tendencies of the elements in annealed steel are presented in Table 9-III. The first seven elements, namely, nickel, silicon, aluminum, zirconium, phosphorus, sulfur, and copper, have no tendency to form carbides. Silicon, aluminum, and zirconium, however,

TABLE 9-III. COMBINING TENDENCIES OF ELEMENTS IN ANNEALED STEEL.

Element	Dissolve in Ferrite		Dissolve in Austenite	Form Carbide	In Nonmetallic Inclusions	Form Special Intermetallic Compounds	In Elemental State
Nickel	Ni		Ni			NiSi (?)	
Silicon	Si		Si		$SiO_2 \cdot M_xO_y$		
Aluminum	Al	with	Al		Al_2O_3 etc.	Al_xN_y	
Zirconium	Zr	→	Zr		ZrO	Zr_xN_y	
Phosphorus	P	ease	P				
Sulfur	S		S		{MnFeS, ZrS}		
Copper	Cu		Cu				Cu when $> \pm 0.8\%$
Manganese	Mn		Mn ↔	Mn	{MnS, MnFeO, $MnO \cdot SiO_2$}		
Chromium	Cr	time	Cr ↔	Cr	Cr_xO_y		
Tungsten	W	→	W ↔	W			
Molybdenum	Mo	and	Mo ↔	Mo			
Vanadium	V	temp.	V ↔	V	V_xO_y	V_xN_y	
Titanium	Ti		Ti ↔	Ti	Ti_xO_y	{$Ti_xN_yC_z$, Ti_xN_y}	
Lead							Pb (?)
Principal effect:	Strengthen		Increase hardenability	Reduce hardenability Fine-grain Toughness	Deoxidizers and grain-growth control	Increase hardness	Increase machinability

tend to oxidize and to form oxides. Silicon and aluminum are particularly useful as deoxidizers in steel. The manner in which a steel is deoxidized is influential in the control of austenitic grain size. When aluminum is the deoxidizing agent, the steel will have fine-grain tendencies. This is attributed to the very fine distribution of aluminum oxide which acts as nuclei to promote a finer grain structure. The elements manganese, chromium, tungsten, molybdenum, vanadium, and titanium have stronger tendencies to form carbides in the order named, titanium having the greatest tendency. In the absence of carbon, these elements dissolve to a certain degree in ferrite; but those lower in this series have a greater tendency to form a carbide than those higher in the group.

Phosphorus in the amounts commonly found in steel dissolves in the ferrite with considerable ease. In larger amounts, it forms an iron phosphide, promoting brittleness. Sulfur forms an iron sulfide which locates at the grain boundaries of ferrite and pearlite and imparts poor ductility at forging temperatures (hot-shortness). In the presence of manganese, an iron-manganese sulfide is formed which is uniformly distributed throughout the structure and is less harmful than the iron sulfide. Copper does not form a carbide, and only about 0.8 per cent is soluble in ferrite. The limited solubility of copper is employed in certain cast copper-bearing steels to improve strength properties through precipitation hardening. Lead is known to be very insoluble in steel and will not form a carbide.

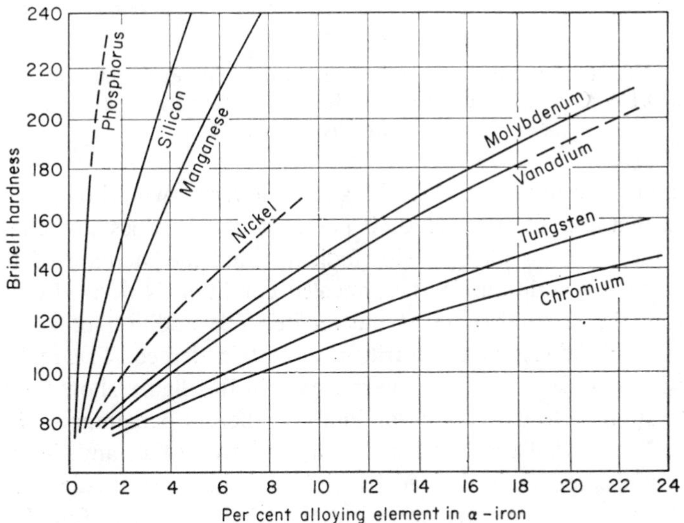

FIG. 9.2 Probable hardening effects of various elements as dissolved in pure iron. (After Bain, *Alloying Elements in Steel*)

9.14 Hardening Effect of Elements in Iron. In Fig. 9.2 is shown the probable hardening effect of several of the elements when dissolved in alpha iron. Those elements, described in the preceding article, that have the greatest solubility in ferrite have the greatest effect on the hardness and strength of alpha iron. Those elements that are normally associated with increased hardness and strength in steel have a small effect upon the hardness of iron. However, in the presence of carbon, these elements, such as chromium, tungsten, vanadium, and molybdenum, have a marked influence on the response to heat treatment. Fig. 9.3 shows the effect of manganese and

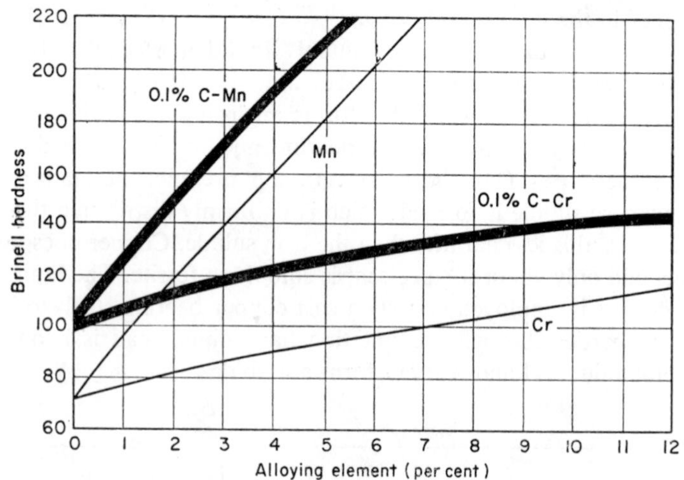

FIG. 9.3 Effect of manganese and chromium in increasing hardness of pure iron and 0.1% C steel as dissolved in ferrite. (After Bain, *Alloying Elements in Steel*)

chromium on the hardness of ferrite and on a very low-carbon steel. It will be noted that the addition of 0.10 per cent carbon does not change the slope of the hardness curves. This indicates that the increase of hardness derived by alloy additions in the annealed condition is derived almost entirely from the strengthening of the ferrite. The addition of carbon also increases the hardness of the ferrite to a certain degree. The real story is told, however, in Fig. 9.4. The lower curves in this figure show the relation of tensile strength to per cent chromium for different percentages of carbon. It is to be noted that the lower curves are relatively flat, and, in this case, the increased strength is derived by virtue of the strengthening of ferrite. If, however, these steels are cooled at the same rate, it is found that the effects of alloying elements become much greater with increasing amounts under rapidly cooled conditions. For example, in the plain iron-carbon alloys, the increase in strength due to the increase in carbon content from 0.1 per cent to 0.3 per cent is of the order of 60 per cent; whereas, by the addition of 3 per cent chromium, the improvement gained with this same range of carbon content present is more than 100 per cent. It can be seen, then, that the addition of a small amount of chromium has made the strengthening effect of carbon considerably greater. The strengthening effect of alloys dissolved in ferrite does not usually warrant the extensive use of

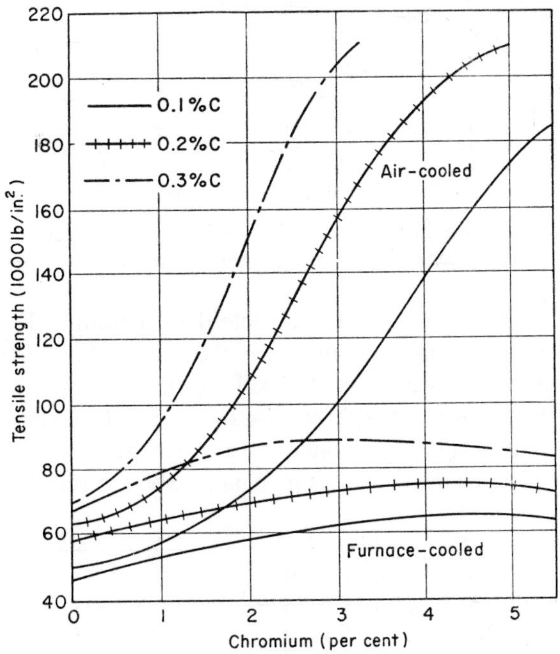

FIG. 9.4 Effect of chromium on tensile strength of 0.10, 0.20, 0.30% C steels, furnace-cooled and air-cooled. (After Bain, *Alloying Elements in Steel*)

costly alloy steels. It is only after suitable heat treatment that the best combination of superior properties is brought out.

9.15 Effect of Alloy Elements on Transformation Temperature. The addition of alloy elements alters the temperature at which gamma iron transforms to alpha iron; likewise, it changes the temperature of the eutectoid transformation. The effect may be either to raise or to lower the transformation temperatures. Furthermore, the critical temperature on heating and cooling may not be affected in exactly the same manner, since the critical temperature on heating may be raised, while that upon cooling may be lowered. A further significant factor is an alteration in the composition of the eutectoid by the addition of alloying elements. Fig. 9.5 shows the effect

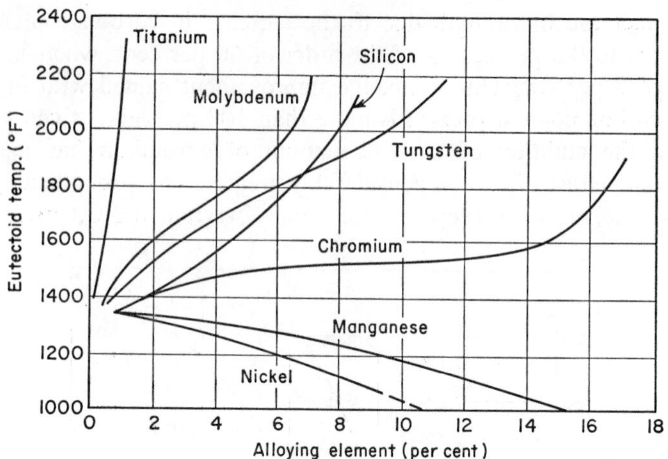

FIG. 9.5 Effect of alloying elements in steel on the eutectoid temperature. (After Bain, *Alloying Elements in Steel*)

of several alloying elements on the eutectoid temperature. Fig. 9.6 shows the effect of the same alloying elements on the composition of the eutectoid. It will be noted that manganese and nickel are the only elements that lower the eutectoid temperature, the others raise it, and all elements shift the composition of the eutectoid to lower amounts of carbon.

The effect of an element such as manganese, on the transformation tem-

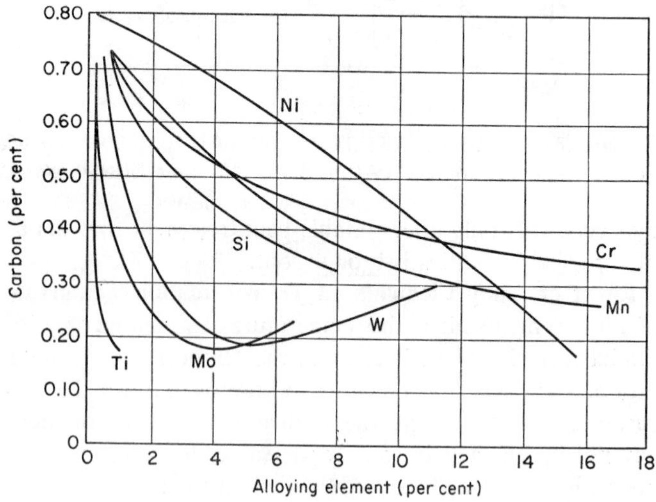

FIG. 9.6 Effect of elements on the carbon content of the eutectoid. (After Bain, *Alloying Elements in Steel*)

perature and eutectoid composition is illustrated in Fig. 9.7. In this case (typical of manganese and nickel) the normal position of the critical temperature is shown by a dashed line. The critical temperature is lowered, and the eutectoid occurs with less than the normal carbon content as indicated by the solid lines.

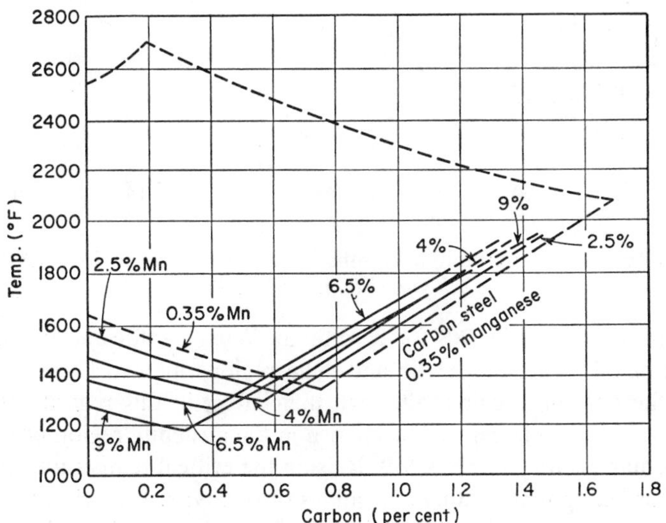

FIG. 9.7 Effect of manganese on the austenite phase region. (After Bain, *Alloying Elements in Steel*)

The critical temperatures are further decreased by larger amounts of alloying elements of this type. The critical temperature may be lowered sufficiently to prevent the transformation of austenite on slow cooling; thus, it is retained at room temperature. Both manganese and nickel increase the difference between the critical temperature obtained on heating and on cooling and are known as *austenite-stabilizing elements*.

Since the eutectoid occurs in alloys of this type with less carbon than in the plain carbon steels, the properties obtained in the eutectoid plain carbon steels can be obtained with a lower carbon content in these alloys. This fact usually leads to somewhat greater ductility in the alloy steels. Since the eutectoid temperature is lower in this type of alloy, quenching temperatures can be lower than with the plain carbon steels, thus tending to lessen the strain usually experienced in hardening.

The effect of chromium, tungsten, vanadium, molybdenum, silicon, and titanium is to shift the critical temperature to higher values. A shift of this type tends to bring the delta solid solution region and the ferrite region to-

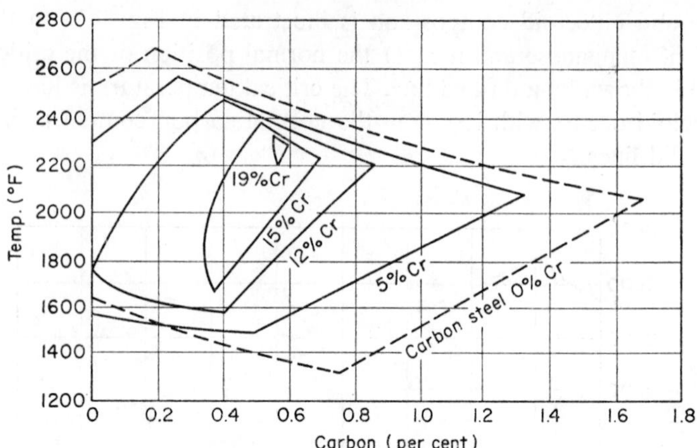

FIG. 9.8 Effect of chromium on the austenite phase region. (After Bain, *Alloying Elements in Steel*)

gether as shown in Fig. 9.8 for chromium. With a sufficient amount of an element of this type, the ferrite and delta regions merge. These alloys also reduce the austenite region, although their effect in this regard varies considerably with the elements. The change in the austenite region produced by molybdenum is shown in Fig. 9.9. Raising the critical temperatures requires higher hardening temperatures for alloys of this type. Likewise, the tempering temperatures may be higher than those used for plain carbon steels to obtain the same hardness. The effect of an alloying element upon the austenite region is important in hardening. In order to harden a steel fully, it must be heated to a temperature at which the structure is composed entirely

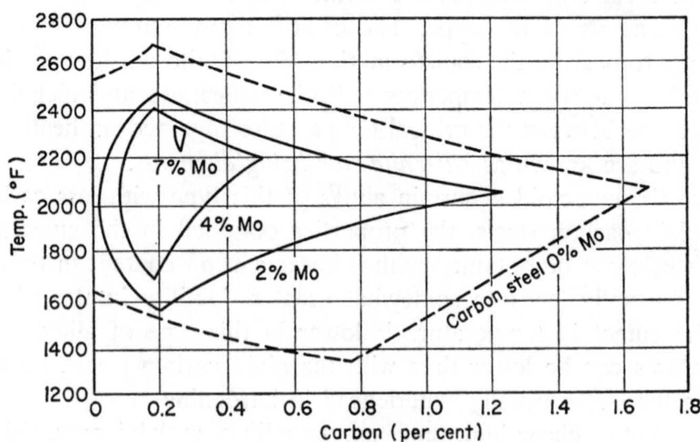

FIG. 9.9 Effect of molybdenum on the austenite phase region. (After Bain, *Alloying Elements in Steel*)

of austenite. If the effect of the element is as shown for chromium or the other elements that contract the austenite region, hardening can be secured only within a limited range of carbon content.

9.16 Effect of Alloying Elements on Critical Cooling Rate. A significant characteristic of alloying elements is to alter the isothermal transformation diagram. As might be expected, the alloying elements not only alter the temperature at which austenite transforms to pearlite under equilibrium conditions but also the temperatures at which other transformation products are formed. Of more importance is the change effected in the rate at which these products are formed. Fig. 9.10 illustrates schematically the tendency

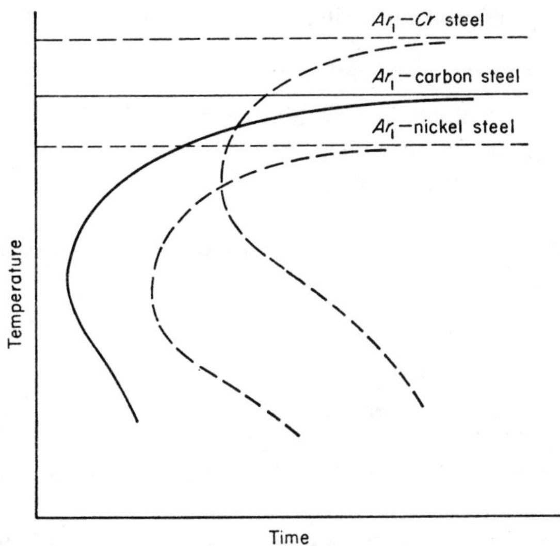

FIG. 9.10 Schematic effect of nickel and chromium on the isothermal transformation diagram.

in the change in the isothermal transformation diagram. The general effect of most alloying elements is to make it possible to obtain full hardening with lower rates of cooling than can be employed in the plain carbon steels, provided the element is dissolved in the austenite. This is because the isothermal transformation diagram is shifted to the right, thus requiring more time for the beginning and completion of the austenite transformation. Although the equilibrium transformation temperature is lowered by nickel and raised by chromium, the critical rate of cooling is reduced by both elements.

9.17 Effect of Alloying Elements on Hardenability. In discussing the heat treatment of steel in Chapter VIII, considerable space was devoted to hardenability. In that discussion, some reference was made to the advantages derived from the use of alloy elements by obtaining greater depth of hardening in larger sections.

Hardenability is dependent largely upon the mode of distribution of the alloy elements. This factor, as well as others, is shown clearly in the following tabulation after Bain:

Factors that decrease hardenability:

1. Fine grains of austenite.
2. Undissolved inclusions.
   a. Carbide or nitride
   b. Nonmetallic inclusions
3. Inhomogeneity of austenite.

Factors that increase hardenability:

1. Dissolved elements in austenite (except cobalt).
2. Coarse grains of austenite.
3. Homogeneity of austenite.

It is apparent then that there are five fundamental factors that influence hardenability. They are:

1. Mean composition of the austenite.
2. Homogeneity of the austenite.
3. Grain size of the austenite.
4. Nonmetallic inclusions in the austenite.
5. Undissolved carbides and nitrides in the austenite.

The effect of dissolved elements on the hardenability of steel is shown in Fig. 9.11. In this case, steels of different chromium content were heated to dissolve all carbides and were then quenched in oil. It has been found that those elements that combine with carbon in preference to dissolving in ferrite have the greatest influence in increasing hardenability if they are dissolved in the austenite before quenching. A carbide-forming element that is not dissolved in austenite has no effect on hardenability except that, as a carbide, it may restrict grain growth, thus reducing the hardenability of the steel. Undissolved carbides reduce both the alloy and carbon content of the austenite. This concept may give the reader a clue as to the reason why, in some instances, quenching from a higher temperature promotes deeper hardening. A higher temperature may increase the austenitic grain size and also cause solution of additional carbide. A material in which vanadium has been used will tend to be fine-grained; that is, coarsening will not take place until a very much higher temperature is reached than with a plain carbon

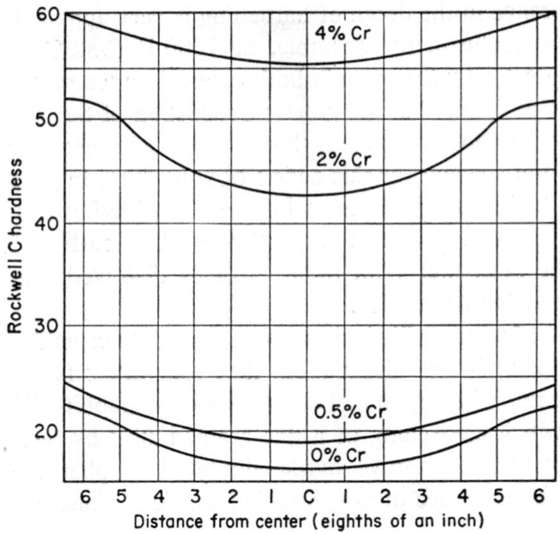

FIG. 9.11 Hardness distribution across $1\frac{5}{8}$-in. round in 0.35% C steel with different amounts of chromium as quenched in oil, originally free of undissolved carbide. (After Bain, *Alloying Elements in Steel*)

steel. The effect of vanadium in this respect is shown in Fig. 9.12. It is to be noted here that the coarser grains of the plain carbon steel tend to harden deeper than the vanadium steel when heated to approximately the same temperature. When quenched from 1650 F (899 C), for example, the grain size of the plain carbon steel is ASTM 3-4, whereas the vanadium is ASTM

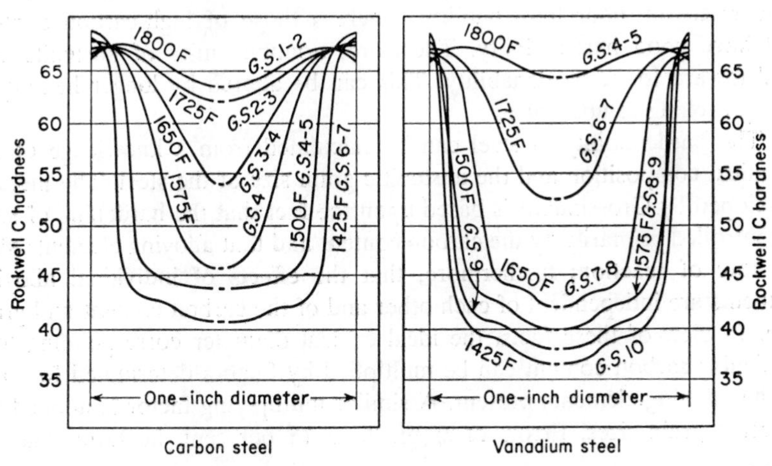

FIG. 9.12 Hardness distribution across 1-in. rounds of 0.9% C steel and 0.9% C, 0.27% V steel. (After Bain, *Alloying Elements in Steel*)

7-8. The difference in the depth of hardening is very apparent. With higher quenching temperatures, for example, at 1800 F (982 C), the depth of hardening is practically the same in both cases. At a temperature of 1800 F (982 C), it is probable that the greater proportion of the vanadium is dissolved. It may be concluded, then, that the marked grain growth and solution of vanadium account for this action of vanadium. In general, it can be said that the higher temperature stability of carbide-forming elements restrains grain growth, provided a portion of fine carbide is distributed throughout the austenite. With these carbides dissolved, the element has very deep-hardening tendencies.

The action of nonmetallic inclusions is to restrict austenitic grain growth; and, because of their fine distribution, they may intensify nucleation, thus increasing the rate of reaction and decreasing hardenability. There are some instances in which coarsening of a fine-grained steel by heating to a high temperature may not improve hardenability. This phenomenon may be attributed to the existence of nonmetallic inclusions which maintain a high rate of nucleation.

Iron carbide dissolves in gamma iron very rapidly. Some of the more complex carbides formed by chromium, molybdenum, etc., have a considerably lower rate of diffusion than pure iron carbide. Therefore, to obtain complete diffusion of the complex carbide into austenite, it may be necessary to maintain the steel at temperature for a longer time or at a higher temperature than is common for the plain carbon steels. In the event that the steel is quenched before the carbide is completely dissolved and uniformly distributed in the austenite, there will be a lack of homogeneity in the composition of the austenite. The regions of low-carbon content will, of course, have relatively poor hardenability, whereas those of high-carbon content will have good hardenability. Therefore, nonuniformity of austenite will tend toward lower hardenability. This can be altered by longer heating at the quenching temperature.

The hardenability of steel can be calculated from a knowledge of the chemical composition and the austenitic grain size of the steel. The method developed by Grossmann[1] is based upon the fact that the hardening of steel is controlled primarily by the carbon content and that alloying elements alter the rate of reaction; furthermore, that the effects of individual alloying elements are independent of each other and of the carbon content and grain size. In view of these facts, the ideal critical diameter corresponding to a particular carbon content can be multiplied by factors determined for each of the alloying elements present. A similar multiplying factor is applied for different grain sizes. Errors of as much as 15 per cent, however, may be

[1] Grossmann, *Hardenability Calculated from Chemical Composition, Trans. AIME* **150,** 1942, pp. 227-259.

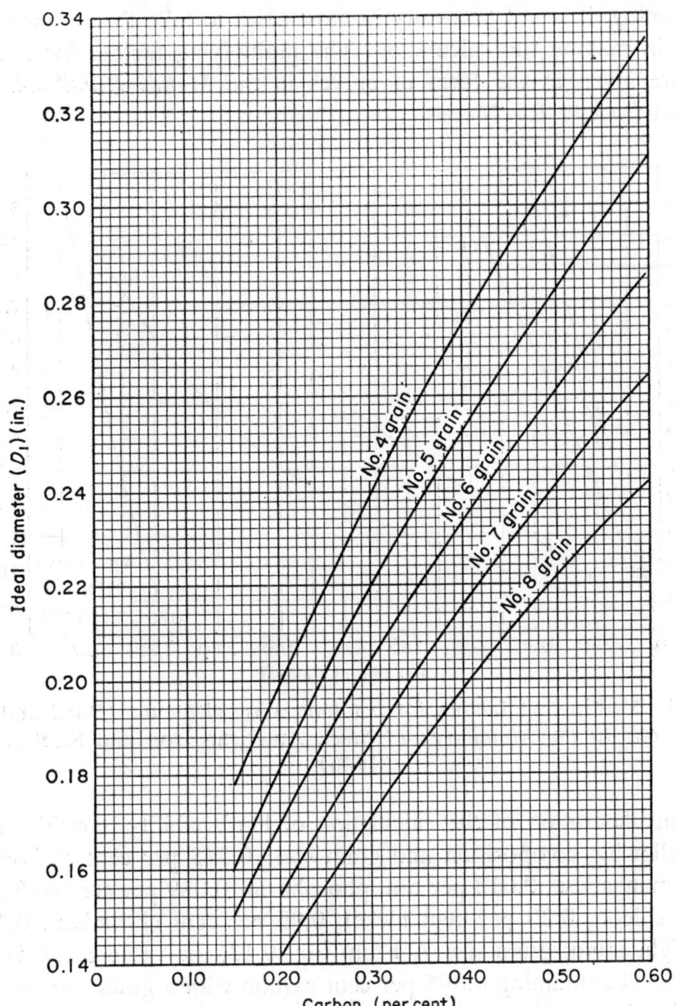

FIG. 9.13 Relation between ideal diameter (D_I), carbon content, and grain size. (Boyd and Field, *Contributions to the Metallurgy of Steel,* American Iron and Steel Institute, 1946)

expected in the calculation of hardenability by this method. The effect of grain size upon the ideal critical diameter is shown in Fig. 9.13. The multiplying factors for the common alloying elements are presented in Fig. 9.14.[2] Because the multiplying factors for sulfur and phosphorus are very nearly

[2] Boyd and Field, *Calculation of the Standard End-Quench Hardenability Curve From Chemical Composition and Grain Size, Contributions to the Metallurgy of Steel,* No. 12, American Iron and Steel Institute, New York, 1946, 25 pp.

compensating, it is not necessary in most cases to take these elements into account in making the calculation. The multiplying factors for other elements are given in the form of curves in the *Metals Handbook* of the American Society for Metals.

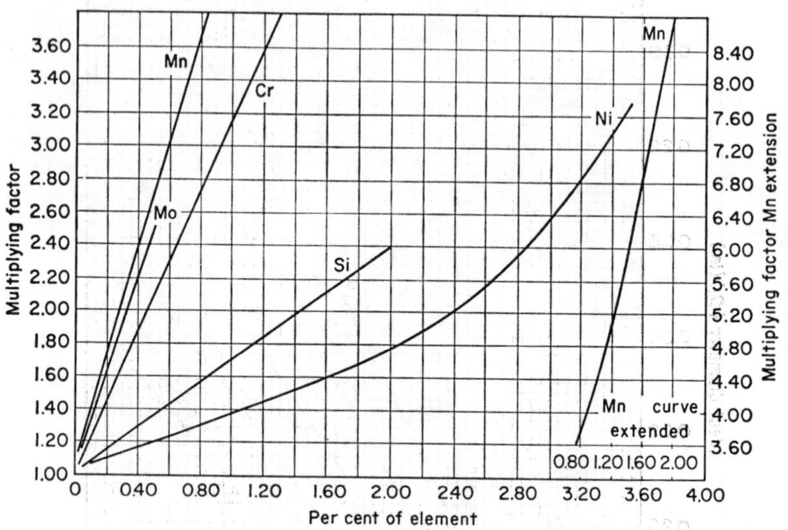

FIG. 9.14 Multiplying factors for common alloy elements. (Boyd and Field, *Contributions to the Metallurgy of Steel,* American Iron and Steel Institute, 1946)

As an illustration of the calculation of hardenability, consider a steel of the following composition and grain size: 0.395 per cent carbon, 0.83 per cent manganese, 0.018 per cent phosphorus, 0.029 per cent sulfur, 0.31 per cent silicon, 0.20 per cent nickel, 0.99 per cent chromium, 0.17 per cent molybdenum; grain size 7. From Fig. 9.13, the value of D_I is 0.214 in. for a steel containing 0.395 per cent carbon with a grain size of 7. The multiplying factors obtained from Fig. 9.14 are as follows:

0.83%	Mn: 3.80
0.31%	Si : 1.20
0.20%	Ni : 1.07
0.99%	Cr : 3.15
0.17%	Mo: 1.52

Hence, $D_I = 0.214 \times 3.80 \times 1.20 \times 1.07 \times 3.15 \times 1.52 = 5.0$ in.

If the ideal critical diameter is known, an end-quench curve can be computed by a method developed by Field.[3] This method can be used for purposes of estimation and is not a substitute for the actual end-quench test. Three assumptions form the basis for this method of calculation: (1) the hardness of the first sixteenth inch of the end-quench bar, designated initial hardness (*IH*), is solely a function of the carbon content of the steel; (2) the hardness at any distance from the end of the end-quench bar, designated as distance hardness (*DH*), is a function of the ideal critical diameter of the steel; (3) the ratio of the initial hardness (*IH*) to the distance hardness (*DH*) is a constant function of ideal critical diameter. The initial hardness (*IH*) is determined by means of Fig. 9.15. Since the ideal critical diameter

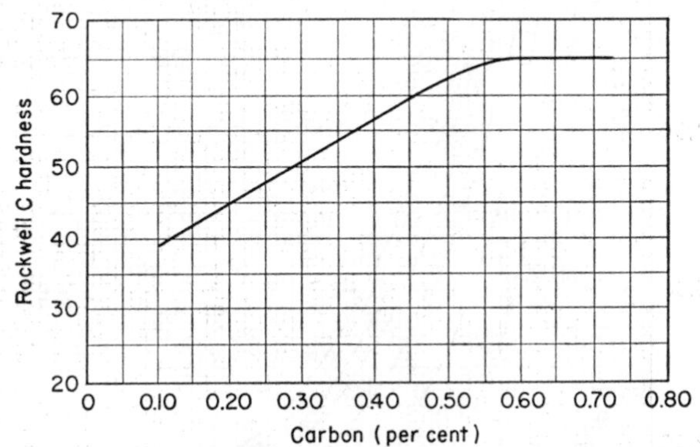

FIG. 9.15 Relation between carbon content and initial hardness (*IH*) at one-sixteenth inch from quenched end of end-quench bar. (Boyd and Field, *Calculation of the Standard End-Quench Hardenability Curve from Chemical Composition and Grain Size, Contributions to the Metallurgy of Steel, No. 12,* American Iron and Steel Institute, New York, 1946, 22 pp.)

(D_I) is known, the ratios of initial hardness to distance hardness can be determined for different distances from the quenched end of the end-quench bar by means of Fig. 9.16. With the value of IH/DH known and the value of *IH* known, it is then possible to calculate the hardness for the different positions on the end-quench bar. Other methods are available for calculating the end-quench curves.[4]

The method of calculating an end-quench curve from a known value of D_I is illustrated by application to the steel considered for the calculation of D_I. The values of IH/DH for different positions on the end-quench bar for

[3] Field, *Calculation of Jominy End-Quench Curve from Analysis, Metal Progress* **43,** 1943, pp. 402-405.

[4] Crafts and Lamont, *Addition Method for Calculating Rockwell C Hardness of the Jominy Hardenability Test, Trans. AIME* **167,** 1946, pp. 698-718.

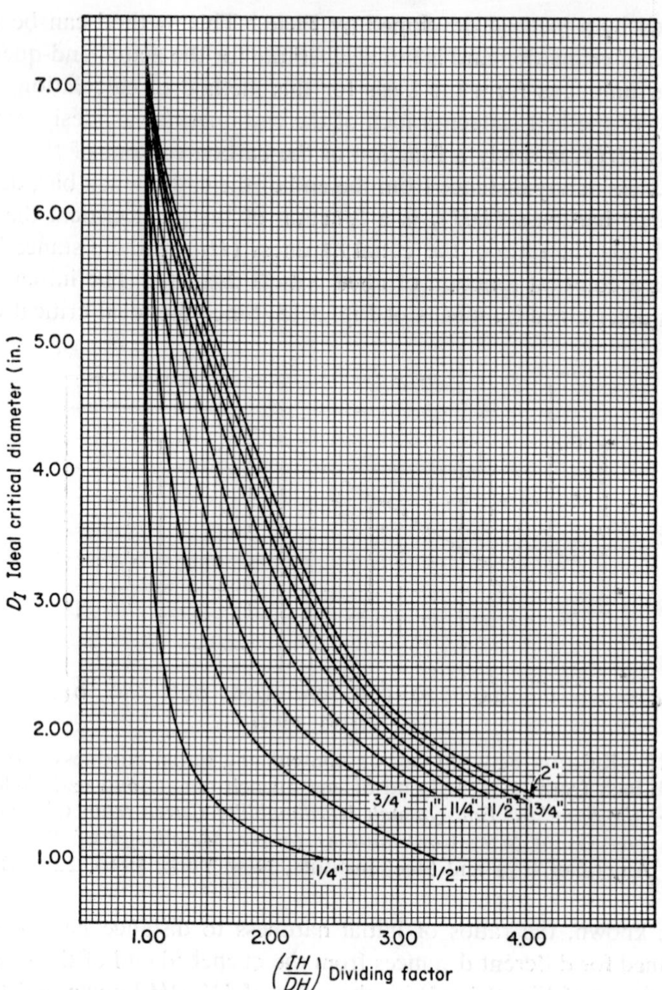

FIG. 9.16 Relation between ideal critical diameter (D_I) and ratio of initial hardness to distance hardness $\left(\frac{IH}{DH}\right)$. (Boyd and Field, *Contributions to the Metallurgy of Steel,* American Iron and Steel Institute, 1946)

$D_I = 5.0$ can be obtained from Fig. 9.16. These are given in the following table. The value of IH for a carbon content of 0.395 per cent is determined from Fig. 9.15 to be 56 RC. The hardness values corresponding to the different positions on the end-quench bar can be computed by dividing the value of IH by the IH/DH ratio. The results are indicated in the table and plotted in Fig. 9.17, which also presents the experimental end-quench curve for this steel.

Position $\frac{1}{16}$		$\frac{IH}{DH}$	RC
1	....	1	56
4	....	1	56
8	....	1.08	51.8
12	....	1.18	47.5
16	....	1.30	43.0
20	....	1.40	40.0
24	....	1.46	38.4
28	....	1.51	37.1
32	....	1.55	36.1

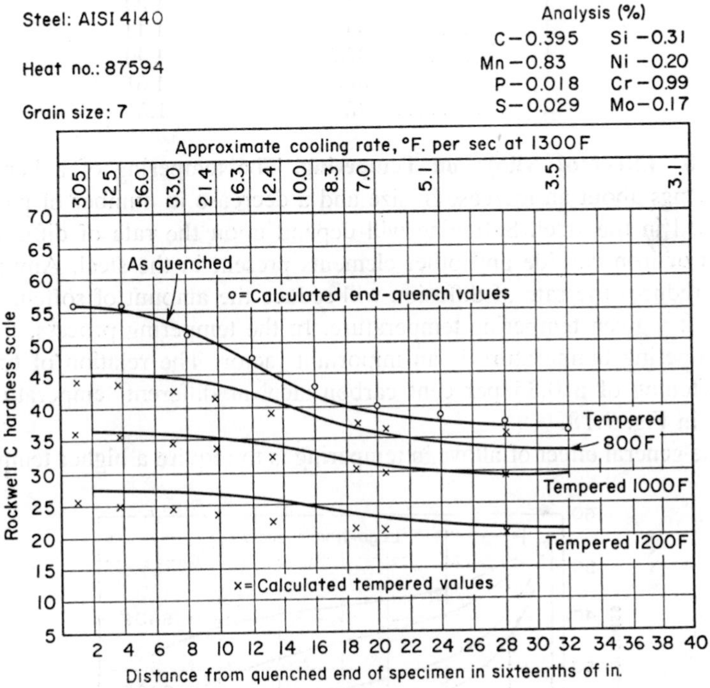

FIG. 9.17 End-quench and tempered end-quench curves for AISI 4140 steel and calculated values.

The ideal critical diameter can be determined from the end-quench curve by reversing the procedure used in the preceding example. The hardness values corresponding to positions on the end-quench curve are determined from the experimental curve as in Fig. 9.17. The value of *IH* is 56 RC for the steel represented in Fig. 9.17. The ratios of *IH*/*DH* are then computed as indicated in the following table. These values are plotted on the edge of a piece of paper to the scale given on the abscissa of Fig. 9.16.

The paper is then placed on the figure to obtain the best fit of these points with the curves. The value of D_I is then indicated at the intersection of the paper with the ordinate. In this example, the value of D_I is determined to be 5.0 in.

Distance	RC	$\frac{IH}{DH}$
$\frac{1}{16}$	56	1
$\frac{1}{4}$	55	1.02
$\frac{1}{2}$	52	1.08
$\frac{3}{4}$	47	1.19
1	42	1.33
$1\frac{1}{4}$	39	1.44
$1\frac{1}{2}$	37.7	1.49
$1\frac{3}{4}$	37	1.51
2	37	1.51

9.18 Effect of Alloys on Tempering. The tempering of a hardened steel brings about an increase in size and a decrease in number of particles dispersed in the steel. Softening will depend upon the rate of diffusion of carbon or iron carbide and other elements present in the steel. Any factor which reduces the rate of diffusion will reduce the amount of softening obtained at a given tempering temperature. In the tempering process, time at the tempering temperature is an important factor. The relation of time to the softening of a 0.35 per cent carbon steel at different temperatures is shown in Fig. 9.18.

The general effect of alloys in tempering is to require a higher tempering

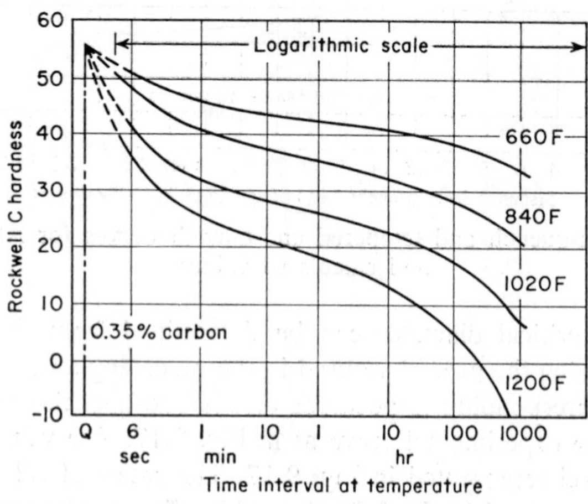

FIG. 9.18 Effect of time at four tempering temperatures on hardness of quenched 0.35% C steel. (After Bain, *Alloying Elements in Steel*)

temperature or a longer holding time to secure a given hardness, thus permitting greater stress relief when compared to carbon steel. A comparison between (1) plain carbon, (2) non-carbide-forming alloy, and (3) carbide-forming alloy steels is given in Fig. 9.19.

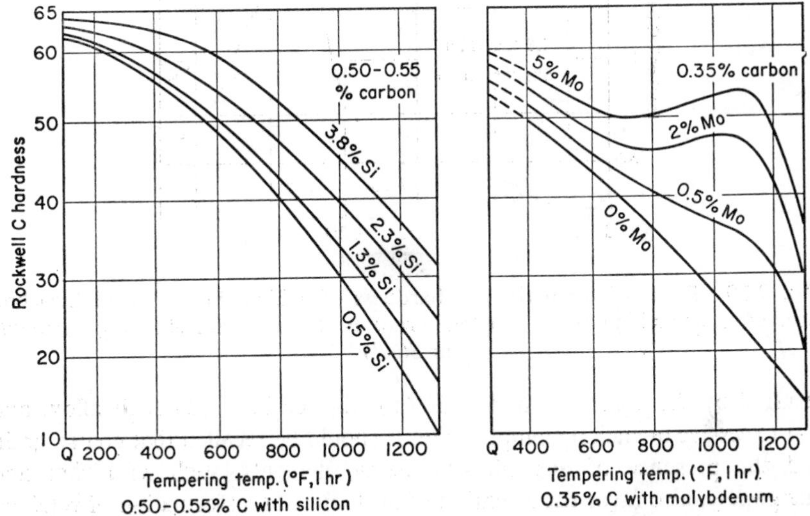

FIG. 9.19 Comparison of softening characteristics in tempering of a steel containing silicon (non-carbide-forming) with a steel containing molybdenum (carbide-forming). (After Bain, *Alloying Elements in Steel*)

The presence of appreciable amounts of strong carbide-forming elements, such as chromium, molybdenum, tungsten, vanadium, and titanium, may cause softening to be retarded or may result in an actual increase in hardness when tempered over a certain range of temperature. This phenomenon is known as "secondary hardening" and is indicated by the tempering curve for the 5 per cent molybdenum steel in Fig. 9.19. Greater amounts of molybdenum would also show a hardness increase. A probable explanation for secondary hardness may be in the formation of complex alloy carbides at elevated tempering temperatures. When first formed, these carbides probably exist as minute particles which act as effective hardening agents. At somewhat higher temperatures, diffusion and coalescence take place with subsequent softening. This property of certain high alloy steels is particularly useful in the heat treatment of alloy tool steels to be discussed in Chapter XII.

Another example of hardness increase upon tempering is noted in the transformation of retained austenite as shown in Fig. 9.20. This illustration is an extreme case in that an abnormal quantity of retained austenite has

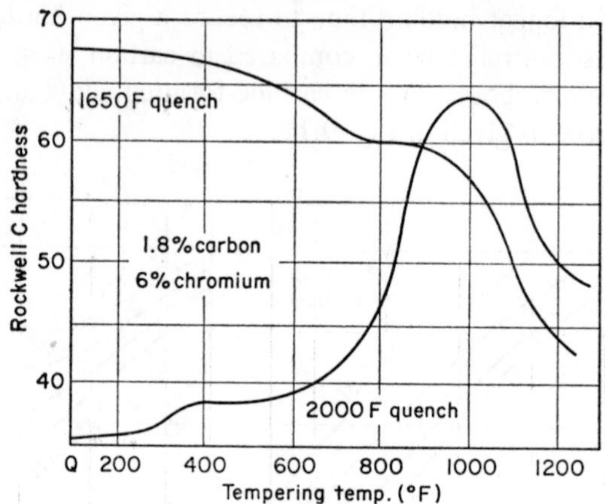

FIG. 9.20 Effect of tempering on hardness of a high-chromium, high-carbon steel after quenching from two temperatures. (After Bain, *Alloying Elements in Steel*)

resulted by the combined action of (1) high carbon, (2) high alloy, and (3) excessive quenching temperature. It should be recalled that coarse grain and the presence of austenite-stabilizing elements, such as nickel and manganese, also favor the retention of austenite. Transformation of retained austenite upon tempering at a given temperature will be in accordance with

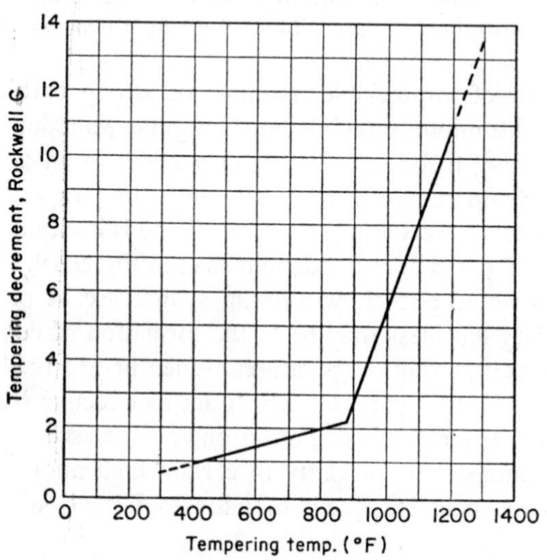

FIG. 9.21 Hardness decrement (*D*) for "unhardened" steel. (Crafts and Lamont, *Trans. AIME* **172,** 1947)

the direct isothermal transformation at that temperature level, possibly accelerated by the presence of carbide nuclei already formed at lower temperatures.

A method has been developed by Crafts and Lamont[5] by which the hardness of a quenched steel after tempering for 2 hr at a particular temperature can be estimated. This method resulted from an extensive study of the effect of the alloying elements on the tempering of steel. The relation between the tempering temperature and the decrease of hardness in plain

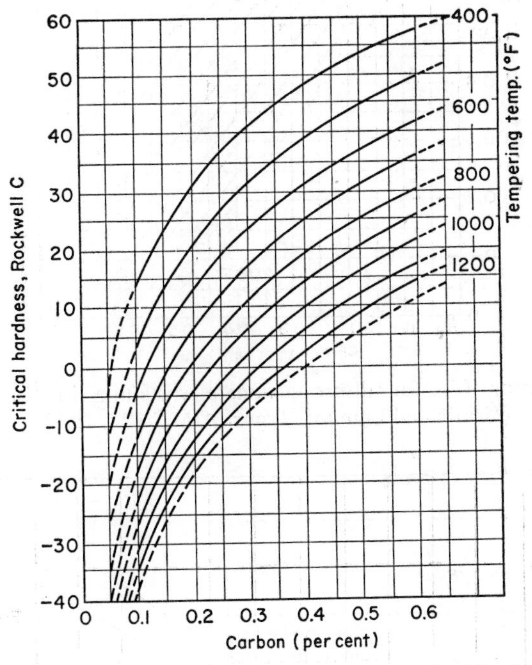

FIG. 9.22 Critical hardness (*B*) for alloy-free steel. (Crafts and Lamont, *Trans. AIME* **172,** 1947)

carbon steels was determined and is shown in Fig. 9.21. This decrease of hardness is called the *hardness decrement* (*D*). For steels that produce a high as-quenched hardness, there is a further decrease of hardness in tempering, as shown in Fig. 9.22, which is called the *critical hardness* (*B*) and which is a function of tempering temperature and carbon content. There will be a disproportionate loss of hardness in tempering, depending upon the carbon and alloy content. This *disproportionate softening* (*f*) as influ-

[5] Crafts and Lamont, *Effect of Alloys in Steel on Resistance to Tempering. Trans. AIME* **172,** 1947, pp. 222-243.

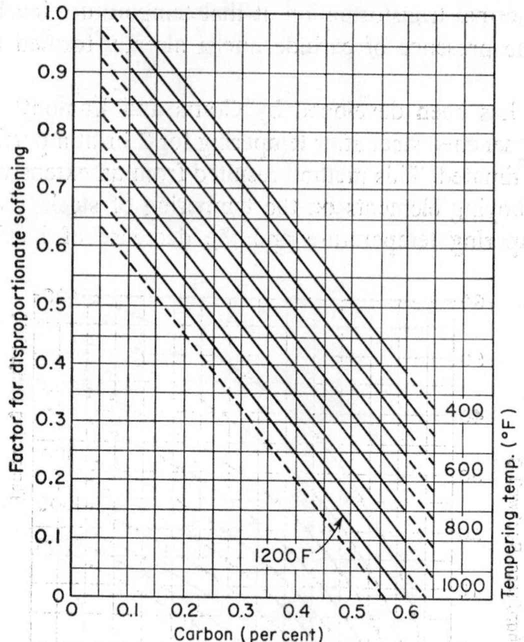

FIG. 9 23 Factor (f) for disproportionate softening in "hardened" steel. (Crafts and Lamont, *Trans. AIME* **172,** 1947)

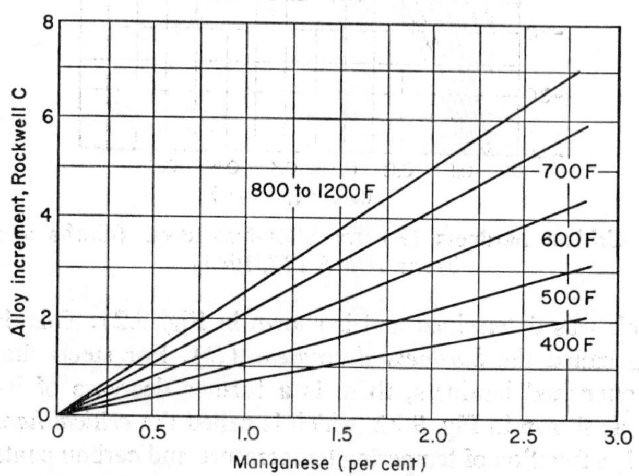

FIG. 9.24 Effect of manganese on resistance to softening. (Crafts and Lamont, *Trans. AIME* **172,** 1947)

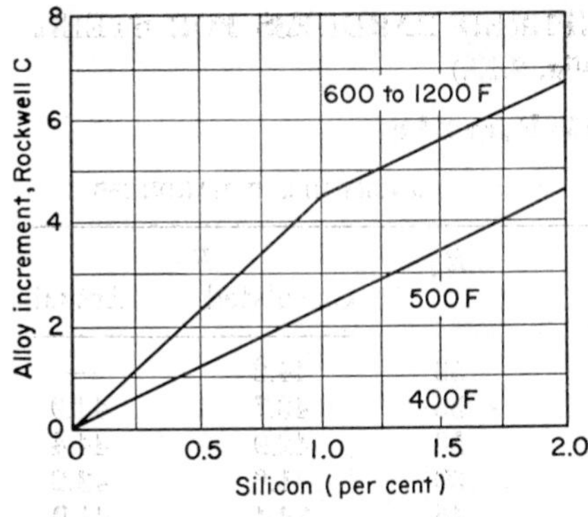

FIG. 9.25 Effect of silicon on resistance to softening. (Crafts and Lamont, *Trans. AIME* **172,** 1947)

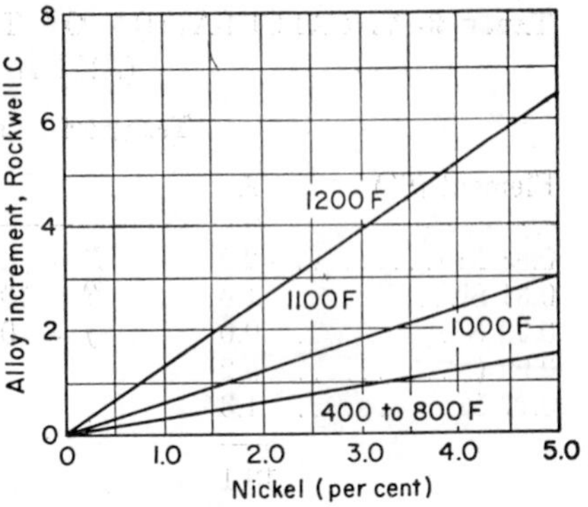

FIG. 9.26 Effect of nickel on resistance to softening. (Crafts and Lamont, *Trans. AIME* **172,** 1947)

enced by carbon content is shown in Fig. 9.23. The alloy elements provide a greater hardness for a given tempering temperature than would be obtained in a plain carbon steel. The factors representing these alloy increments are shown in Figs. 9.24, 9.25, 9.26, 9.27, and 9.28 for manganese, silicon, nickel, chromium, and molybdenum, respectively. The relation between these factors that has been developed for estimating the hardness after tempering a steel for two hours is:

$$R_T = (R_Q - D - B)f + B + A$$

where R_T = tempered hardness,
R_Q = as-quenched hardness,
D = tempering decrement (Fig. 9.21),
B = critical hardness (Fig. 9.22),
f = tempering factor (Fig. 9.23),
A = sum of alloy increments (Figs. 9.24, 9.25, 9.26, 9.27, 9.28).

This method for calculating tempered hardness may be illustrated by considering the end-quench data for an AISI 4140 steel, presented in Fig. 9.17. The calculations and the results for tempering this steel to temperatures of 800, 1000, and 1200 F are presented in Table 9-IV and are plotted in Fig. 9.17. As one might expect, the correlation between the experimental and calculated results is not perfect, but the agreement is sufficient to warrant this method of estimation.

Table 9-IV. CALCULATION OF TEMPERED HARDNESS FOR STEEL. (AISI 4140, Fig. 9.17.)

Tempering at 800 F for 2 hr

Element (%)	A		Rockwell C hardness R_Q	R_T Calculated	R_T Actual
0.83 Mn	2.2	$D = 2$			
0.31 Si	1.4	$B = 19$			
0.20 Ni	0.0	$f = 0.43$			
0.99 Cr	5.2		56	44.2	—
0.17 Mo	1.3		55	43.7	44.9
	—		53	42.9	44.4
	10.1		50	41.6	43.2
			45	39.4	41.0
			40	37.3	36.2
			39	36.8	35.5
			37	36.0	33.2

Tempering at 1000 F for 2 hr

Element (%)	A		Rockwell C hardness R_Q	R_T Calculated	R_T Actual
0.83 Mn	2.2	$D = 5.3$			
0.31 Si	1.4	$B = 10$			
0.20 Ni	0.1	$f = 0.34$			
0.99 Cr	5.2		56	36.0	—
0.17 Mo	3.2		55	35.6	35.9
	—		53	34.9	35.6
	12.1		50	33.9	35.0
			45	32.2	33.5
			40	30.5	31.0
			39	30.2	30.7
			37	29.5	29.3

Tempering at 1200 F for 2 hr

Element (%)	A		Rockwell C hardness R_Q	R_T Calculated	R_T Actual
0.83 Mn	2.2	$D = 10.8$			
0.31 Si	1.4	$B = 2.5$			
0.20 Ni	0.3	$f = 0.24$			
0.99 Cr	5.2		56	25.1	—
0.17 Mo	3.2		55	24.8	27.0
	—		53	24.4	26.9
	12.3		50	23.6	26.5
			45	22.4	25.5
			40	21.2	23.0
			39	21.0	22.5
			37	20.5	21.2

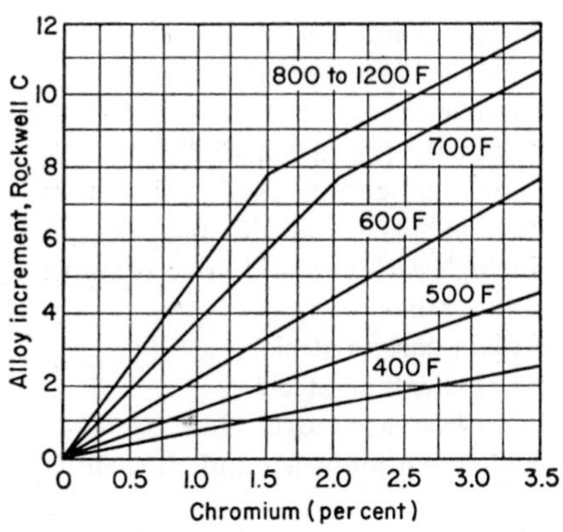

FIG. 9.27 Effect of chromium on resistance to softening. (Crafts and Lamont, *Trans. AIME* **172,** 1947)

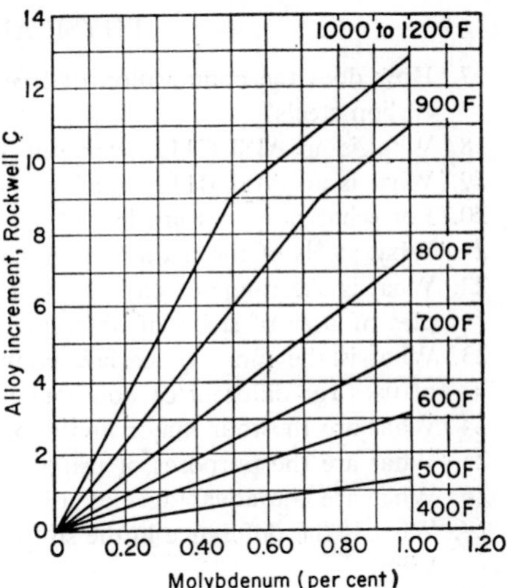

FIG. 9.28 Effect of molybdenum on resistance to softening. (Crafts and Lamont, *Trans. AIME* **172,** 1947)

REFERENCES

Bain, *Functions of the Alloying Elements in Steel,* American Society for Metals, Metals Park, Ohio, 1939.

Machining—Theory and Practice, American Society for Metals, Metals Park, Ohio, 1950.

Metals Handbook, American Society for Metals, Metals Park, Ohio, 1961.

S.A.E. Handbook, Society of Automotive Engineers, New York, 1961.

QUESTIONS

1. In what four ways may steels be classified?
2. Distinguish between killed and rimmed steel.
3. For what particular application is rimmed steel well suited?
4. Name the five principal kinds of steel in order of decreasing production capacity.
5. Name at least five different classes of steel, indicating their principal uses.
6. What grades of steel are recognized in practice?
7. What factors are included in the classification of steel by quality?
8. Identify the following steels: AISI C1015, AISI E1045, AISI D1030.
9. For what purposes is low-carbon sheet and strip steel used?
10. What is orange peel or stretcher strains? What is the cause and how may it be prevented?
11. What is temper-rolling, and what is its purpose?
12. What steel is commonly used for structural uses?
13. What steel is commonly used for boiler plate?
14. What steels are commonly used for the cold-forming of bolts and rivets?
15. What is cold-rolled steel and for what is it commonly used?
16. Why is a very low carbon steel not easily machined?

17. How does the composition of free-cutting steels differ from ordinary plain carbon steels?
18. What is an AISI C1120 steel and for what is it usually employed?
19. What is an AISI B1120 steel and why does it have superior machinability?
20. For what purposes are leaded steels employed?
21. What steels of the plain carbon variety are used for carburizing?
22. What is the range of carbon content of medium-carbon steels? Give examples of uses of steels of different carbon contents within this range.
23. What is the range of carbon content in high-carbon steel? Give examples of uses for different carbon contents in this range.
24. What are the four major limitations of plain carbon steels?
25. What are the purposes of using alloying elements in steel?
26. What are the advantages of using plain carbon steels over alloy steels?
27. What is the difference in the stiffness of plain carbon steels and alloy steels? Why?
28. In general, what is the effect of the addition of alloying elements in steel compared to plain carbon steels?
29. What is the most important element present in any steel, whether it be plain carbon or alloy?
30. What possible effects may result from the addition of alloying elements to steel?
31. In what form may the alloying elements added to steel be found when the steel is in the annealed condition?
32. What alloying elements form carbides as well as a solid solution with iron?
33. What is the effect of relatively large amounts of both phosphorus and sulfur on the properties of steel?
34. What is the effect upon strength when any alloy is dissolved in ferrite?
35. What is the effect of alloying elements upon the eutectoid transformation temperature of steel?
36. What is the effect of alloying elements upon the composition of the eutectoid in steel?
37. What alloying elements are known as austenite-stabilizing elements? Why?
38. In general, what is the effect of alloying elements upon the critical cooling rate of steel?
39. What five factors influence the hardenability of steel?
40. What is the effect of the presence of a carbide-forming element that is not dissolved in austenite prior to a hardening operation?
41. In what way does the austenitic grain size affect hardenability?
42. What is the effect of the presence of nonmetallic inclusions on hardenability? Why?
43. Calculate the ideal critical diameter for a bar of AISI 4340 steel using the minimum chemical analysis given in the appendix and a grain size of No. 8. What would the critical diameter be if quenching was in still water? In still oil? What severity of quench would be required to fully harden a bar $3\frac{3}{4}$ inches in diameter?
44. Calculate the end-quench hardenability for an AISI 4130 steel having a

grain size of No. 5 using the average chemical analysis given in the appendix. Plot the calculated values obtained and compare with the actual end-quench band for this steel.

45. What factor controls the softening of a steel by tempering as a function of temperature? Show, by means of typical curves of hardness versus temperature, the effect of the presence of non-carbide-forming elements and carbide-forming elements in comparison with plain carbon steel in tempering.
46. What is secondary hardening and to what may it be attributed?
47. What type of alloying elements exhibit secondary hardening?
48. Calculate the tempered hardness values for the heat of AISI 6150 shown in Appendix F for a tempering temperature of 1000 F employing the method of Crafts and Lamont. Plot the calculated values obtained and compare with the actual curve as shown.

X

LOW-ALLOY STEELS

10.1 General Consideration. Much has already been said about the effect of alloy elements on steels. In this chapter, the individual alloying elements will be discussed with respect to their effects on the mechanical properties of plain carbon steel. Although reference is made to standard AISI-SAE steels wherever possible, similar advantages will be found in other compositions containing the respective alloying elements in both cast and wrought form.

10.2 Classification of Alloy Steels. Alloy steels may be classified on the basis of total alloy content in the following manner:

Low alloy—less than 10% alloy.
High alloy—more than 10% alloy.

The low-alloy steels are often referred to as pearlitic steels since their microstructures are similar to the plain carbon grades. The high-alloy group includes the alloy tool steels together with the corrosion-, scale-, and wear-resistant steels, the structures of which may consist largely of austenite or ferrite rendered stable at room temperature by the high alloy content.

The AISI-SAE designations for steels were introduced in Chapter IX. Analysis specifications established for some of the alloy machinery steels are given in Appendix C. The first digit of the AISI number designates the type of alloy; the second digit refers to the series within that alloy group; and the last two digits indicate the average carbon content. The general classification of the AISI-SAE steels is given in Table 10-I.

There are many other alloy steels produced that are not included within the AISI-SAE specification given in Table 10-I. Classification of the high-alloy steels will be covered in subsequent chapters. In many cases, trade names obscure detailed chemical analyses, although recommended heat treatments, resultant mechanical properties, and hardenability values are furnished by the steel producer. Such steels are sold on the basis of satis-

Table 10-I. CLASSIFICATION OF AISI-SAE STEELS—1958.

Type	AISI-SAE
Carbon steels	1XXX
Plain carbon	10XX
Free-machining	11XX
Manganese (intermediate)	13XX
Nickel	2XXX (Series deleted 1958)
3.5% Ni	23XX
5.0% Ni	25XX
Nickel-chromium	3XXX
1.25% Ni, 0.65% Cr	31XX
3.50% Ni, 1.60% Cr	33XX
Molybdenum	4XXX
C-Mo (0.25% Mo)	40XX
Cr-Mo	41XX
Ni-Cr-Mo (1.80% Ni)	43XX
C-Mo (0.40% Mo)	44XX
C-Mo (0.55% Mo)	45XX
Ni-Mo (1.80% Ni)	46XX
Ni-Cr-Mo (1.05% Ni)	47XX
Ni-Mo (3.50% Ni)	48XX
Chromium	5XXX
Low Cr (0.35% Cr)	50XX
Med. Cr. (0.90% Cr)	51XX
Low Cr. (0.50% Cr) bearing steel	50XXX
Med. Cr. (1.0% Cr) bearing steel	51XXX
High Cr. (1.45% Cr) bearing steel	52XXX
Chromium-vanadium	6XXX
0.75% Cr	61XX
Triple-alloy steels	
0.30% Ni, 0.40% Cr, 0.12% Mo	81XX
0.55% Ni, 0.50% Cr, 0.20% Mo	86XX
0.55% Ni, 0.50% Cr, 0.25% Mo	87XX
0.55% Ni, 0.50% Cr, 0.35% Mo	88XX
3.25% Ni, 1.20% Cr, 0.12% Mo	93XX
1.00% Ni, 0.80% Cr, 0.25% Mo	98XX
Silicon-manganese	
2% Si	92XX
Boron (0.0005% B minimum)	
C	TS 14BXX
0.50% Cr	50BXX
0.80% Cr	51BXX
0.30% Ni, 0.45% Cr, 0.12% Mo	81BXX
0.55% Ni, 0.50% Cr, 0.20% Mo	86BXX
0.45% Ni, 0.40% Cr, 0.12% Mo	94BXX

factory performance in specific applications rather than strict conformity to chemical limits. This is common practice among tool-steel manufacturers and has become popular among consumers of other steel products. In the practical sense, the design engineer is interested in meeting certain requirements directly, as indicated in the previous chapter. Adherence to chemical specifications as an indirect method of selection assumes that similar analyses produced under like conditions will provide the average mechanical properties typical of each grade. In view of the large number of test samples used to prepare such data, this assumption is reasonable. Experience has shown, however, that many heats of steel which fall outside chemical specifications for the selected grade are suitable for the intended application on the basis of hardenability and mechanical properties. The reader is referred to available handbooks or manufacturers' literature covering the mechanical properties, treatment, and type analyses of other alloy steels.

10.3 Manganese Steel. The plain carbon and free-machining steels, as previously discussed under the 10XX and 11XX series, contain normal amounts of residual manganese ranging from 0.25 to about 1.65 per cent. The use of 1.6 to 1.9 per cent manganese further increases the strength of steel in the as-rolled condition and contributes to improved strength and ductility in the heat-treated condition. The group of so-called intermediate manganese steels, making up the 1300 series, develop high-strength ductility and are superior to the plain carbon steels of the same carbon content at very low alloy cost. To obtain maximum advantage from the use of manganese, such steels should be heat-treated. Since they possess good as-normalized mechanical properties, they find application in many forged parts. Temper brittleness is noted in steels containing more than about 0.60 per cent manganese.

Manganese also has played an important part in tool-steel metallurgy; some of the nondeforming varieties contain 1 to 1.7 per cent manganese with about 0.8 per cent carbon. In some rail steels, the manganese content is increased to about 1.5 per cent for greater strength and good ductility.

Manganese in amounts between 2 and 10 per cent imparts brittleness to steel. If the manganese content is increased to between 11 and 14 per cent in steels containing 1 to 1.2 per cent carbon, a tough, wear-resistant product may be obtained after suitable treatment. This material, known as Hadfield's austenitic manganese steel, was one of the first alloy steels produced commercially. With the large amount of manganese present, the critical temperature is sufficiently lowered and the critical cooling rate is so reduced that a martensitic structure will be obtained on slow cooling. If the same steel is quenched rapidly, it will be austenitic and possess toughness with ductility. A material of this composition is very resistant to impact abrasion, wherein the surface is continually work-hardened in service but wears rapidly in applications involving only pure abrasion. In view of its severe work-hardening characteristics, it cannot be economically machined but must be cast and ground to shape. This material is used extensively for frogs and switches in railroad trackwork, power-shovel teeth, rock-crusher jaws, dredge buckets, etc. It is noteworthy that, over the years, no satisfactory substitute has been developed for this steel.

10.4 Nickel Steel. One of the first elements alloyed with iron and carbon in steel was nickel. The principal advantage in the use of this element lies in the higher tensile strength that can be obtained without appreciable decrease of elongation and reduction of area. Since nickel lowers the critical temperatures, lower heat-treating temperatures can be used. In many respects, nickel resembles manganese in its effect on the mechanical properties of steel. The nickel steels possess good as-normalized mechanical properties. Nickel decreases the critical cooling rate; therefore, a less rapid quench is required to obtain hardness equal to that of the plain carbon steel. Although the 2XXX series of nickel steels is shown deleted from the AISI-SAE classification of alloy steels, they are discussed in this chapter in view of their historical significance in the development of the multiple-alloy steels.

The properties of a 2340 steel and a C1045 steel are compared in Fig. 10.1. When these steels are treated to have the same tensile strength, the ductility, as indicated by reduction of area, is very much higher in the nickel steel than it is in the plain carbon steel. A considerable improvement in

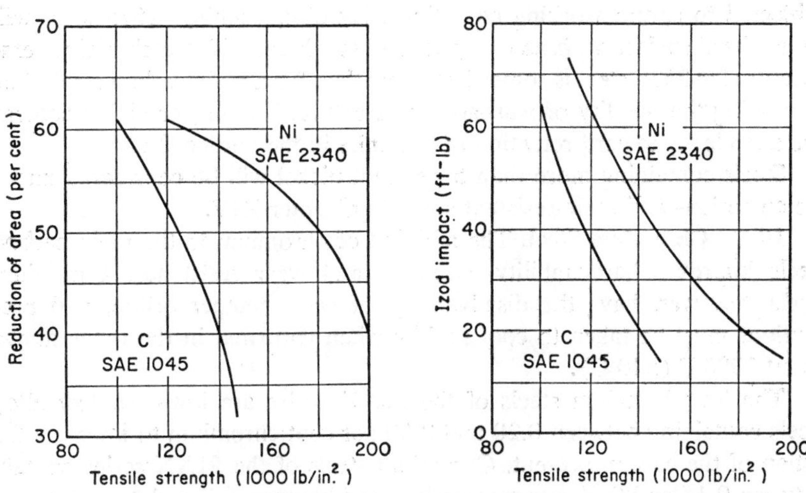

FIG. 10.1 Comparison of reduction of area and impact values with tensile strength of SAE 1045 and SAE 2340 steels. (*Nickel Alloy Steels*, 2nd ed., International Nickel Co., New York, Section 5, Data Sheet E)

notch sensitivity, as indicated by the Izod impact value, is also to be found in the nickel steels and is particularly noticeable at extremely low temperatures.

The first group in the nickel series is 23XX. Because of the particularly satisfactory properties that can be attained, this group was probably the most important of the nickel steels. These steels were used extensively, since less carbon is required to obtain the same results as in plain, higher carbon steels without the necessity of extremely rapid cooling. The low-carbon, 3.5 per cent nickel steels, such as 2317, were used for carburizing because they developed hard surfaces and very tough cores. There has been some use of the 3.5 per cent nickel steels in parts where additional strength is necessary and in which heat treatment is impossible, such as in steels for boilers and bridge structures. The higher carbon, 3.5 per cent nickel steels have been widely used for highly stressed bolts, gears, axles, nuts, castings, and various machine parts. For machine parts, the carbon content ranges between about 0.35 and 0.50 per cent; with larger amounts of carbon, the nickel steel develops greater hardness and strength when carefully heat-treated.

The 2515 steel which contains approximately 5 per cent nickel has been employed for carburizing where extremely severe conditions are encountered. It will give a case having excellent wear resistance and fatigue strength with a very high-strength core, but is rather costly, owing to the high nickel content. It has been used principally for wrist pins, king pins, transmission gears, ring gears, pinions, engine cams, and other machinery parts that are subjected to severe working conditions. By oil-quenching, 2515 steel will have a hard surface with an exceptionally tough core. The carburizing temperature for 2515 steel is somewhat lower than for a plain carbon steel, thus decreasing the liability of warpage and fracture. The higher nickel content has a tendency toward retention of austenite in the hardened case.

Steels containing more than 5 per cent nickel will be considered under the corrosion- and scale-resistant steels in Chapter XIV.

10.5 Chromium Steel. The addition of chromium to the plain carbon steels improves hardenability, strength, and wear resistance. Chromium steels, however, have the disadvantage of being temper brittle, and precautions must be taken to cool rapidly when tempered in the range above about 1000 F (540 C).

The low-chromium steels of the 50XX series are low-cost, low alloy steels containing between 0.20 and 0.50 per cent chromium to intensify the action of the carbon present. Chromium steels of the 51XX series contain between 0.15 and 0.65 per cent carbon and between 0.7 and 1.15 per cent chromium. The very low-carbon group of the chromium steels are suitable for carburizing and produce very hard, wear-resistant surfaces but do not possess the tough, fibrous core that characterizes the nickel steels. The steels of this series that contain larger amounts of carbon are used for springs, gears, bolts, etc. Plain chromium-carbon steels are very useful for keen cutting-edge tools, but they lack the toughness that is sometimes required of die blocks used in drop forging. The decrease in the critical cooling rate produced by the addition of chromium leads to air-hardening characteristics in the high-chromium steels. Therefore, such steels must be carefully cooled after rolling and forging to prevent cracking.

A steel that is particularly well suited for the races and balls or rollers of antifriction bearings is the 52100 steel which contains from 0.95 to 1.1 per cent carbon and from 1.3 to 1.6 per cent chromium. This steel has very excellent wear resistance and high strength. The higher-chromium steels will be considered under the corrosion- and scale-resistant steels in Chapter XIV.

10.6 Nickel-Chromium Steel. A proper combination of nickel and chromium will improve the general serviceability of a carbon steel. The improvement in ductility and toughness conferred upon the carbon steels by the addition of nickel can be combined with the increased strength, surface hardness, and depth of hardening produced by the addition of chromium. The inherent temper brittleness of the chromium steels also prevails. Experience has shown that there is an optimum ratio between the amount of nickel and chromium that can be added to steel. If the chromium content exceeds a certain proportion with respect to the nickel content, the steel will be more difficult to heat treat successfully, since the temperature limits in heat treatment will be narrower, and the possibility of poor results will be greatly increased. The ratio of nickel to chromium is about 2.5 to 1. In general, it may be said that these steels find uses similar to those which were cited for the nickel series, especially in cases where service conditions are more severe and the added cost of a quaternary steel is warranted. The low-carbon steels in this series are also used for carburizing.

The 31XX series containing 1.25 per cent nickel and about 0.65 per cent chromium gives excellent properties at a relatively low cost. These steels are used for drive and axle shafts, transmission gears, connecting rods, highly stressed pins, etc.

The 33XX series contains 3.5 per cent nickel and 1.5 per cent chromium. These steels are deeper-hardening than the preceding series and are used to meet severe requirements in carburized parts, such as heavy-duty rear axles and transmission gears.

Those steels that contain more than about 3.5 per cent nickel and 1.5 per cent chromium are considered under the corrosion- and scale-resistant steels in Chapter XIV.

10.7 Molybdenum Steel. The pronounced effect of molybdenum when added to steel in relatively small amounts in combination with other elements has led to the popular use of these steels. Molybdenum is used as an alloy element in steel primarily to enhance the properties imparted by other alloying elements, such as manganese, nickel, and chromium. Molybdenum, however, will also in itself improve the properties of the plain carbon steels to a certain degree. The amount of molybdenum usually employed in any of the molybdenum steels varies between 0.15 and 0.60 per cent.

The resultant effects of molybdenum in steels may be summarized as follows:

1. Higher tempering temperatures are required to develop mechanical properties equivalent to the plain carbon steels.
2. Greater ductility and toughness may be obtained for a given elastic limit.
3. Greater machinability at higher hardness is obtained.
4. Greater hardenability is obtained, particularly when chromium is present.
5. Temper brittleness is greatly reduced or eliminated.
6. Longer holding time is required for heating at quenching or normalizing temperatures to permit complete solution of the complex molybdenum-iron carbide.
7. Creep resistance at high temperatures is improved because of the stability of the carbide.

The 40XX, 44XX, and 45XX series of carbon-molybdenum steels may be used at low alloy cost whenever better mechanical properties are desired than can be obtained with the plain carbon steels. The 4023-4028 steels are carburizing grades employed for spline shafts, transmission gears, pinions, and similar applications for passenger automobiles where service is not severe. Since these steels are fine-grained and shallow-hardening, a higher carbon content is used compared to normal carburizing grades. The higher-carbon, 4063 steels have been used for some time in coil and leaf springs of light section for passenger automobiles.

The 41XX series of chromium-molybdenum steels includes those containing about 1.0 per cent chromium, with from 0.15 to 0.25 per cent molybdenum. This series enjoys great popularity because it is relatively cheap and possesses good deep-hardening characteristics, ductility, and weldability. The 4130 steel has been used extensively for pressure vessels, aircraft structural parts, automobile axles, and steering knuckles and arms. The higher carbon steels are used for parts of heavier section.

The 43XX and 47XX series of nickel-chromium-molybdenum steels have the characteristics of the nickel-chromium series with the added advantages of the properties imparted by molybdenum. Heavy sections of these steels may be conveniently heat-treated with excellent improvement in properties because of their deep-hardening characteristics. The steels of the 46XX and 48XX series include those steels containing nickel and molybdenum. They have the advantage of high strength and ductility conferred by nickel, combined with the added deep-hardening property and improved machinability imparted by molybdenum. These steels have good toughness together with high fatigue and wear resistance. They are used for transmission gears, roller bearings, chain pins, shafts, and other parts where high fatigue and tensile properties are important.

The low-carbon steels in the molybdenum series find application as carburizing grades where excellent properties are required. It was previously indicated that the chromium and nickel-chromium steels have the disadvantage of temper brittleness which can be remedied only by quenching from the tempering temperature. The addition of molybdenum to these steels almost completely eliminates temper brittleness.

Other combinations of molybdenum are employed, such as with the manganese, silicon, and vanadium steels. Chromium-silicon-molybdenum steel finds application in automotive exhaust valves. Molybdenum-vanadium and chromium-molybdenum-vanadium steels find use in springs and bolts to resist creep in high-temperature service. The use of molybdenum in amounts up to about 4 per cent is found in certain grades of corrosion-resistant steels and in some tool steels up to approximately 8 per cent.

10.8 Vanadium Steel. In many ways, vanadium acts in a manner similar to that of molybdenum. It tends to accentuate or to enhance the properties developed by other alloying elements. The principal alloying element that is employed with vanadium is chromium. The 61XX series contains about 0.75 per cent chromium and up to 0.18 per cent vanadium.

The advantages of vanadium are derived from its ability to decrease the tendency toward grain growth and to improve effectively the fatigue resistance. The 6150 steel is used in parts subjected to severe conditions requiring high strength and fatigue resistance, such as leaf and coil springs, heavy-duty axles, shafts, driving parts, gears, pinions, valves, etc. Resistance to creep makes this steel suitable for valves and springs operating at relatively high temperatures. It is well adapted for use in springs because of a very high elastic limit and fatigue strength in the heat-treated condition. The 6150 steel, however, is being replaced for automotive springs by steels in the 40XX and 86XX series. Low-carbon grades, such as 6118 and 6120, are ideal for carburizing in view of the grain-growth-inhibiting properties of vanadium.

Vanadium is more readily oxidized in the production of steel than is molybdenum, and, therefore, contributes to deoxidation. Since the vanadium carbides are somewhat more difficult to dissolve in austenite than are the molybdenum carbides, it may be necessary either to use somewhat higher heating temperatures for proper quenching or to maintain the temperature for a longer time.

10.9 Tungsten Steel. The greatest use of tungsten is in tool and die steels. In this discussion, only those steels that contain less than about 6 per cent tungsten will be considered.

Tungsten dissolves in gamma iron to a certain extent and remains in solution when gamma changes to alpha. It also forms a hard, stable carbide which imparts wear and abrasive resistance to the steel. In the dissolved form, tungsten increases hardenability. Certain combinations of tungsten, chromium, and vanadium serve to decrease the tendency of steel to crack and distort in heat treatment. Steels containing from 1 to 2 per cent tungsten and 1 to 1.3 per cent carbon with approximately 0.5 per cent chromium are water-hardened. Steels of this type are used in keen-edged tools, dies, shredding dies, broaches, reamers, taps, drills, hack-saw blades, and other tools for which requirements are beyond those of ordinary plain carbon steels.

Tungsten in the range of 4.5 to 6 per cent, with from 1 to 1.5 per cent chromium, is used with success for finishing tools where light cuts are to be taken at high speeds.

Tungsten also has the characteristic of chromium and molybdenum in decreasing the rate of softening in tempering operations. Much higher tempering temperatures may be employed with less loss in hardness with material reduction in internal strain than can be obtained in the plain carbon steels. A more detailed discussion of the tungsten steels will be found in Chapter XII with specific reference to the high-tungsten, high-speed steel.

10.10 Silicon Steel. Alloys of iron containing between 0.5 and 4.5 per cent silicon are used in the electrical industry as magnetically soft materials for transformer and generator laminations. The carbon content is maintained below 0.01 per cent to minimize interstitial and precipitation effects on the crystal lattice, which tend to increase hysteresis loss and reduced magnetic permeability. Silicon has a strong influence on increasing the grain size and decreasing the solubility of carbon in iron. Since silicon eliminates the allotropic transformation in iron, it is possible to produce a coarse-grained structure by annealing that is not refined by subsequent cooling. Magnetic properties are enhanced by the minimized surface effects found in coarse-grained structures. The electrical resistivity of iron is also increased by the addition of silicon, thus reducing eddy current effects with alternating current.

Wherever high-density permeability is required, grain direction is as important as grain size since iron-silicon alloys are more easily magnetized in a direction parallel to their crystallographic cubic edge. By a combination of rolling and annealing it is possible to produce grain oriented iron-silicon sheet with most of the grains aligned in the direction of rolling.

Since silicon increases brittleness, about 4 per cent of this element is the maximum that is used commercially for parts which are subject to vibration. High-grade transformer sheets may be composed of 4.5 to 5 per cent silicon, 0.10 per cent manganese, 0.02 per cent phosphorus, 0.02 per cent sulfur, and 0.05 per cent carbon.

The silicon-manganese steels, such as 9255 and 9260, containing between 1.8 and 2.2 per cent silicon and 0.7 and 1.0 per cent manganese, have been used extensively for leaf springs because of their high elastic ratio. This series of alloys requires very careful handling in heat treatment to avoid decarburization and grain growth. In general, they are not as shock-resistant as some of the other spring steels, such as carbon-molybdenum, nickel-chromium-molybdenum, and chromium-vanadium.

The silicon content of some of the structural steels has been increased somewhat in order to obtain greater strength than can be obtained with the plain carbon steels. These steels may contain between 0.20 and 0.40 per cent carbon, 0.25 and 1.25 per cent silicon, and 0.60 and 0.90 per cent manganese.

10.11 Triple-Alloy Steels. The triple-alloy, or low-alloy, nickel-chromium-molybdenum steels, are an outgrowth of studies on the multiplying effects of alloy elements on the hardenability of steel. The effect on the 43XX series of combining several alloying elements has already been noted. The total alloy content is generally greater than in the triple-alloy group. The triple-alloy steels offer excellent hardenability at lower cost in many applications where it was formerly considered necessary to specify steels in the 4XXX series.

10.12 Boron Steel. The continued need for conservation of strategic elements such as manganese, nickel, chromium, and molybdenum has directed attention toward boron as a substitute element in the production of hardenable alloy steels. A few thousandths of one per cent of boron causes a marked increase in the hardenability of a steel under certain conditions and may replace several hundred times its weight of critical elements. The hardenability increases linearly as a function of the boron content with less than 0.001 per cent boron. However, with larger amounts of boron up to 0.003 per cent, the effect is diminished. Further additions of boron cause a condition of both hot- and cold-shortness. The hardenability response to additions of boron is directly related to the carbon content of the steel, the greatest increase being noted in the range of 0.15-0.45 per cent carbon, whereas the eutectoid and hypereutectoid steels show no improvement. The influence of boron on the end-quench curves of a plain carbon and an alloy steel is shown in Fig. 10.2.

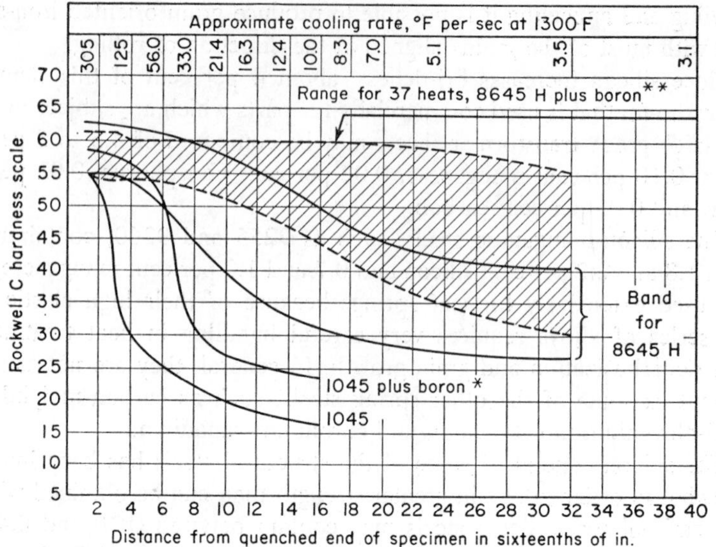

FIG. 10.2 Influence of boron on end-quench curves of SAE 1045 and SAE 8645H steels. * Robbins and Lawless, *Metal Progress* **57,** January, 1950, p. 81. ** *Boron Steels, Metal Progress* **60,** August, 1951, p. 81.

The mechanism by which boron increases the hardenability of steel is not definitely known. Boron is believed to act either directly by dissolving in the austenite or indirectly by changing the condition of the steel with respect to the presence of some other element. A "boron constituent" may be observed in the grain boundaries of boron-treated steel in the form of a row of dark-etching dots. It has been noted that prolonged heating at hot-working temperatures above about 2200 F (1200 C) causes this constituent to be absent from the resultant microstructure at room temperature. When this structural change occurs, any potential increase in hardenability is permanently lost and cannot be recovered by subsequent heat treatment, even though the boron content remains unchanged in the analysis.

When boron is added to a given composition, the effect on the isothermal transformation diagram is to delay the beginning of transformation, with negligible change in the time required for completion or the temperature level of the pearlite nose. This unique property permits cooling to a temperature at which ferrite and pearlite form isothermally in a shorter time than would be required for a full annealing treatment of a steel of the basic composition. Austenitic grain growth takes place readily in the boron steels unless suitable inhibitors, such as aluminum oxide, are present in sufficient quantity to restrict excessive growth. When austenitized at temperatures higher than normal, the boron steels show a considerable decrease in hardenability, which is contrary to the behavior of other steels. Special precautions are required for the boron carburizing grades in providing a low quenching temperature and control of maximum surface carbon if optimum hardenability is to be realized. Lower tempering temperatures are usually required in the tempering of hardened boron steels to produce a given hardness as compared to the other alloy steels.

The boron steels satisfy a demand for substitute steels from the standpoint of hardenability and related properties during periods of alloy shortage. In addition to improved hardenability, these steels generally possess better hot- and cold-working characteristics, permit more economical annealing, and provide better machinability than other alloy steels having equal or higher hardenability. The chemical analysis specifications for boron steels currently recommended as alternates for some of the standard alloy grades are given in Appendix C. The 14BXX and 50BXX series are modifications of plain carbon and low-chromium steels, respectively.

10.13 Low-Alloy Structural Steels. The requirements of the transportation industry for steels with higher yield strength without materially increased cost has brought about the development of steels of low-alloy content. A weight saving of about 25 per cent is possible with certain of these steels, with resultant increase in pay load. The higher-carbon steels in many cases would meet some of these requirements, but usually the lack of ductility of these steels is a disadvantage if they are to be used in structural work. By decreasing the carbon content and adding alloying elements, an excellent combination of ductility, yield strength, and weldability can be obtained. These alloys are sometimes referred to as the high-strength, low-alloy steels and are supplied as-rolled or normalized. Typical examples of these steels are given in Table 10-II.

TABLE 10-II. COMPOSITION (%) AND STRENGTH PROPERTIES OF LOW-ALLOY STRUCTURAL STEELS.
(Based on *Metals & Alloys Data Book*, S. L. Hoyt, Reinhold Pub. Corp., New York, 1943.)

NOMINAL COMPOSITION AND PROPERTIES

Name	C	Mn	Si	P	Ni	Cr	Cu	Mo	Zr	Y.S. 1000 lb/in.²	T.S. 1000 lb/in.²	% Elong.
Cor-Ten	0.10	0.25	0.75	0.15	—	0.75	0.40	—	—	55	70	23 in 8 in.
Hi-Steel	0.10	0.60	0.15	0.12	0.50	—	1.10	0.10	—	55	70*	20 in 8 in.
Mayari R	0.10	0.75	0.30	0.10	0.50	0.75	0.50	—	—	50	70	25 in 8 in.
N-A-X	0.13	0.70	0.80	0.04	0.20	0.60	0.20	—	0.12	50	70	30 in 2 in.
Otiscoloy	0.10	1.10	0.05	0.12	—	—	0.35	—	—	50	70	22 in 2 in.
Yoloy	0.08	0.50	0.25	0.07	2.0	—	1.0	—	—	55	70*	28 in 8 in.

* Additional increase in strength of about 20,000 lb/in.² can be produced in the 1% copper steels by heating to 900-925 F.

Since many of these steels are used in thinner sections than are ordinarily employed for transportation purposes, it was found necessary to improve their atmospheric corrosion resistance. Such improvement can be obtained by the addition of less than one per cent copper usually associated with an increased phosphorus content. Although the copper-bearing steels are not corrosion-resistant in the strict sense, they are superior to the plain carbon steels. Usually, the carbon content of these steels is less than 0.12 per cent. In view of the low carbon content, welding is generally possible without hardening effects in the welded area which otherwise would require annealing of the structure.

10.14 Ultra High-Strength Steels. A group of steels, known as ultra high-strength steels, has been developed to meet the needs of the aircraft industry for high-strength airframe structural components. Originally, it was considered essential that heat-treated airframe parts be tempered at temperatures above about 700 F (370 C) to avoid the brittleness associated with the brittle-tempering range between the temperatures of 450 and 650 F (230 and 290 C). This limitation established a tensile strength ceiling of 200,000 lb/in.^{2} for heat-treated steel parts. In view of the high hardenability of 4340 steel and its ability to harden to a fully martensitic structure throughout heavy sections, this steel had been widely adopted for landing gear and other airframe structural applications within the 200,000 lb/in.^{2} tensile strength limitation. This steel was found to possess satisfactory ductility, toughness, and fatigue strength when tempered at a temperature of 450 F (230 C) to a tensile strength of 270,000 lb/in.^{2} provided certain heat-treating and processing controls were observed. Since temper brittleness is believed to be directly associated with transformation of retained austenite, process specifications for the heat treatment of 4340 steel at high-strength levels generally include such precautionary measures as austempering or double tempering and cold treatment. Process controls that have contributed to the success of 4340 steel at ultra high strengths include: (1) control of decarburization by the use of protective atmospheres or removal by finish grinding, (2) baking after grinding to remove grinding stresses, (3) baking after plating to eliminate hydrogen embrittlement, and (4) use of shot-peening after grinding or straightening of critical portions.

To satisfy the need for steels having tensile strengths between 200,000 and 270,000 lb/in.^{2}, steels were developed which are capable of being tempered with good ductility and toughness within the temper-brittle range encountered with 4340 steel. The principal steels in this group are 4330 Modified, Hy-Tuf, and HS-220, as shown in Table 10-III compared with AISI 4340 steel. The lower carbon content of these steels improves the ductility at low tempering temperatures. An increased silicon content in Hy-Tuf and HS-220 has the added effect of shifting the brittle-tempering range to higher temperatures, thus causing the range of 450 to 650 F (230 to 345 C) to be the most desirable from the standpoint of toughness.

Table 10-III. COMPOSITION (%) AND STRENGTH PROPERTIES OF ULTRA HIGH-STRENGTH STEELS. NOMINAL COMPOSITION AND PROPERTIES.

Designation	C	Mn	Si	Ni	Cr	Mo	V	Tempered °F	Tempered °C	Tensile Strength lb/in.²	Yield Strength lb/in.²	Elongation in 2 in. %	Reduction of Area %	Charpy Impact ft-lb
AISI 4340	0.40	0.75	0.30	1.83	0.80	0.25	—	850	455	200,000	180,000	15	50	25
								450	230	270,000	215,000	10	35	19
AISI 4330 modified	0.30	0.90	0.30	1.83	0.85	0.43	0.08	500	260	250,000	210,000	10	42	17
HS-220	0.30	0.70	0.55	2.05	1.20	0.45	—	600	315	235,000	195,000	11	42	16
Hy-Tuf	0.25	1.35	1.50	1.83	0.30	0.40	—	550	285	230,000	190,000	13	49	30
Tricent	0.43	0.80	1.60	1.83	0.85	0.38	0.08	500	260	300,000	240,000	8	23	18
Super Tricent	0.55	0.80	2.10	3.60	0.90	0.50	—	400	205	340,000	—	—	—	12
HiC Super Hy-Tuf	0.47	1.28	2.42	—	1.11	0.42	0.25	500	260	325,000	—	—	24	10
98BV40	0.40	0.75	0.30	0.85	0.80	0.20	Boron	450	230	290,000	235,000	7	28	13
USS Strux	0.43	0.90	0.55	0.75	0.90	0.55	Boron	450	230	290,000	—	—	—	—
Type H11 tool steel	0.40	—	—	—	5.00	1.50	0.50	1000	540	300,000	240,000	6	—	16

In addition to the above four steels, which played a prominent role in establishing the initial ultra high-strength steels, a second group has been developed to extend the ceiling of tensile strength beyond the 300,000 lb/in.^{2} level. A comparison of the compositions and strength properties of some typical steels in this group is presented in Table 10-III. These steels incorporate a higher carbon and silicon content in general than the previous group.

A third group of ultra high-strength steels has evolved from the hot-work die steels with particular reference to the 5 per cent chromium, Type H11 tool steels. These steels develop high strength with ductility by means of secondary hardening which results from double tempering in the range of 950 to 1200 F (510 to 650 C). In addition to the high strength at room temperature, as shown in Table 10-III, these steels maintain satisfactory strength at elevated temperatures up to about 1100 F (595 C).

10.15 Correlation between Tempered Low-Alloy Steels. A remarkable correlation exists in the constant relationship between hardness, tensile strength, yield strength, elongation, and reduction of area of quenched and tempered low-alloy steels regardless of composition. This similarity prevails, provided that the steel is quenched to a fully martensitic structure prior to tempering and the tensile strength does not exceed 200,000 lb/in.^{2} A plot

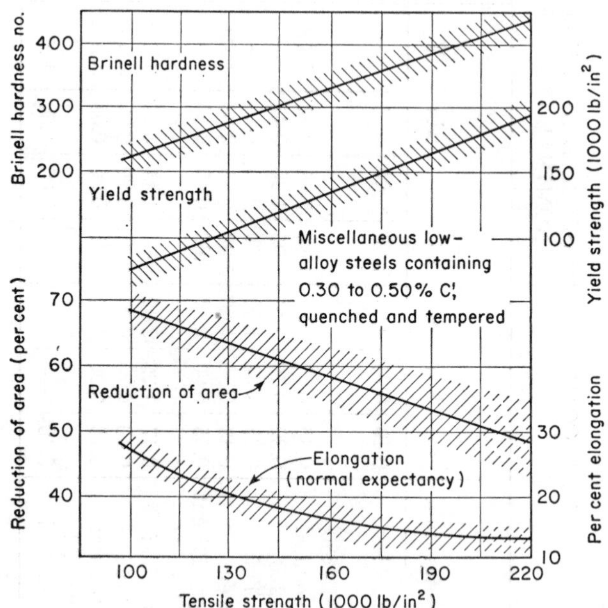

FIG. 10.3 Probable mechanical properties of tempered martensite. (After *Metals Handbook*)

of this relationship is shown in Fig. 10.3. It should be recognized that other properties and distinctive characteristics may differ greatly between the respective alloy groups.

10.16 Relative Cost of Steel. The price of steel changes from time to time, but, in general, the relative cost remains approximately the same, except at times when there is a shortage or a high demand on certain types. A comparison of the prices of plain carbon and alloy steels can be obtained from Table 10-IV.

TABLE 10-IV. COMPARATIVE PRICES OF CARBON AND ALLOY STEEL.
(November, 1960)

Hot-Rolled Bars			Cents per lb
Plain carbon (basic open-hearth) C1020, C1040, C1060, C1095			6.175
Alloy (AISI No.)	Cents per lb	AISI No.	Cents per lb
1340	7.625	E52100	10.925
3140	9.325	6150	9.375
4042	7.975	8640	8.675
4140	8.525	8740	8.825
4340	10.675	9255	7.975
5140	7.775	9840	10.025

Cold-Finished Bars			
AISI No.	Cents per lb	AISI No.	Cents per lb
C1020	7.80	E3310	17.125
C1040	7.80	4320	12.775
C1060	7.80	4620	12.475
C1095	7.80	4820	14.725
C1213	9.00	5120	10.325

Subject to extras for sizes, restricted quality, specifications, quantity, etc.

10.17 Steel Selection and Heat Treatment. In selecting a steel for a particular application, the engineer must consider whether a steel satisfies the service requirements of the part, and at the same time is available and economical to process. A comparison of the prices of steel given in Table 10-IV will indicate the saving in the cost of material that can be effected if a plain carbon steel will provide the required properties of a design. In many applications, however, the size or shape of the part will not permit the use of a plain carbon steel since the necessary properties cannot be attained with the lower hardenability of these steels. It should be emphasized once again that the hardness obtainable by quenching is a function of carbon content, whereas the hardenability is governed largely by alloying elements and austenitic grain size.

Several alloy steels may satisfy the strength requirements of a given design by providing sufficient hardenability for the section size. The exact choice may be limited by certain special requirements, such as low-temperature toughness, creep resistance, corrosion or wear resistance, freedom from temper brittleness, etc. Over-alloying adds to the cost of steel and increases the tendency toward quenching cracks and the retention of austenite. In some applications, it may be desirable to employ a shallow-hardening steel with a softer core to provide greater impact strength. Steels that can be tempered at higher temperatures for a given hardness provide an opportunity for greater stress relief after quenching.

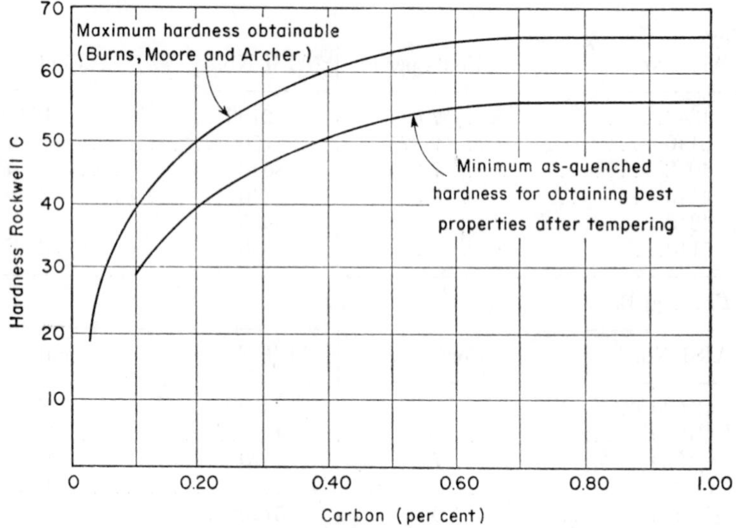

FIG. 10.4 As-quenched hardness for best properties after tempering. (Boegehold, *Trans. SAE* **52,** 1944, pp. 472-485)

The importance of quenching a steel to secure complete transformation to martensite before tempering has been indicated in Chapter VIII. A. L. Boegehold [1] has suggested a curve of as-quenched hardness vs. per cent carbon content that will be acceptable for producing suitable properties in steel after quenching. This is shown in Fig. 10.4, together with the Burns, Moore, and Archer curve of maximum hardness obtainable.

[1] A. L. Boegehold, *Selection of Automotive Steel on the Basis of Hardenability, Trans. SAE* **52,** 1944, pp. 472-485.

The hardenability of available steels must be considered in relation to the thickness of a part, and an appropriate tempering temperature must be selected to produce the desired properties. Steel selection and heat treatment recommendations may be made with the aid of end-quench curves, tempered end-quench curves, and the known relations between hardness and other properties. Typical end-quench curves are available for many steels, but each curve applies to a specific heat of steel. In some instances, the results of tempering end-quenched specimens at different temperatures are available for a specific lot of steel. These are useful in specifying the heat treatment of parts to be made from a particular heat of steel. This type of curve is illustrated in Appendix F for several lots of steel. In most cases, however, the tempered hardness curves are not available. If the end-quench curve is available for a particular heat of steel, a close estimate of the tempered hardness can be obtained by means of the method of calculation formulated by Crafts and Lamont, as described in Chapter IX. If the end-quench curve is not available, the method of calculating hardenability, also explained in Chapter IX, may be employed for the actual analysis of the steel, or the end-quench test may be conducted.

A designer is frequently called upon to specify a steel and its heat treatment to meet a strength requirement. As an illustration, consider a 2-in.-diameter shaft which is required to have a yield strength of 125,000 lb/in.^{2} throughout. According to Fig. 10.3, this corresponds to a Brinell hardness of approximately 310 which, in turn, may be converted to a hardness of Rockwell C 33 by the table in Appendix G.1. Assuming that the shaft will be quenched in oil, the center of the 2-in. section will be equivalent to the 12/16-in. position on an end-quench bar, according to Fig. 8.35.

If the steel selection is to be based only on hardenability, the minimum as-quenched hardness given by Fig. 10.4 should be maintained at the 12/16-in. position. The hardenability bands for the "H" steels, as given in Appendix E, serve as a basis for determining what steel will most likely meet this requirement. A survey of the minimum end-quench curves will show that three steels satisfy this specification as follows:

Carbon content (per cent)...	0.30	0.35	0.40	0.45	0.50
Required hardness, as-quenched (*RC*)........	45	48	50	52	53
Hardenability specification..	$J_{45} = 12$	$J_{48} = 12$	$J_{50} = 12$	$J_{52} = 12$	$J_{53} = 12$
Suitable "H" steel..........	None	4337H	4340H	None	4150H
Band range at $\frac{12}{16}$ in. (*RC*)...	—	49-57	51-59	—	53-63

Individual heats of other grades of steel might qualify for selection, depending on their analysis and hardenability; however, only the above steels are certain of meeting the specification within the allowable analysis variations for the grade. The actual choice of alternate steels will be governed by availability and cost.

The recommended heat treatment for the selected steel may be determined from Appendix D with the aid of tempered end-quench curves. Heating times were discussed in Chapter VIII. For example, if a steel of the composition shown for AISI 4340H, Heat MBE, in Appendix F, was selected for the above shaft, then the heat treatment may be specified as follows:

Normalize at 1650 F (900 C) for 2 hr;
Heat at 1500 F (815 C) for 30 min;
Quench in oil;
Temper at 1100 F (595 C) for 2 hr;
Cool as desired.

If the minimum hardness of Rockwell C 33 cannot be attained as a result of this initial treatment, then the tempering temperature may be reduced in subsequent treatments.

Another illustration of the application of the principles of hardenability is in the selection of a suitable steel for a part of irregular shape. The relationship between cooling rate in a round bar and positions on the end-quench bar will not hold in this instance. The actual cooling rates in an irregular part could be determined by thermocouple measurements while a sample part is quenched in a suitable medium. A simpler method consists of making a sample part of a low-hardenability steel which will provide a wide range of hardness over a cross section after quenching from the proper temperature. An end-quench test bar is then prepared from a bar of the identical steel and end-quenched from the same temperature. Variations in hardness found on the cross section of the sample part represent relative cooling rates which may be evaluated from corresponding hardness positions on the end-quench bar.

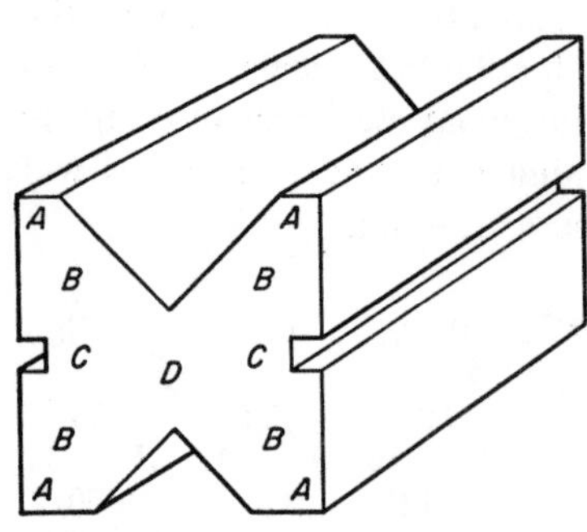

FIG. 10.5 V-block.

To illustrate this method, a sample steel V-block, shown in Fig. 10.5, was quenched from a known temperature, sectioned, and hardness measurements were made at the indicated positions. A hardenability curve was prepared from an end-quench test bar machined from the same steel and end-quenched at the same temperature. A comparison of the hardness values at the various locations on the cross section of the V-block with corresponding distances on the end-quench bar was made as shown in Table 10-V. After a cooling rate relationship had been established between the irregular V-block and the standard end-quench bar, a steel was selected to satisfy a given design specification. For example, if the as-quenched hardness of the V-block were specified as shown in Table 10-V, a survey of the hardenability bands in Appendix E would indicate that, although five possible steels satisfied the minimum hardness of Rockwell C 48 at the center, only AISI 4147H could meet the specifications at locations A, B, and C.

TABLE 10-V. DATA FOR SELECTION OF STEEL FOR V-BLOCK.

Location on V-block		A	B	C	D
Sample block hardness (*RC*)		56	45	37	27
Sample end-quench bar corresponding distance ($\frac{1}{16}$ in.)		1.0	4.0	4.8	12.1
Equivalent cooling rate (deg/sec at 1300 F)		490	125	85	16
SPECIFIED HARDNESS IN V-BLOCK					
As-Quenched (*RC*)		55 min	55 min	63 max	48 min
Hardenability bands for possible steels (*RC*)	4145H	55-63	**54-62**	54-62	48-59
	4147H	57-64	56-64	56-63	51-62
	4150H	58-65	58-65	**58-65**	53-63
	4337H	**52-59**	**51-58**	51-58	49-57
	4340H	**53-60**	**53-60**	53-60	51-59

10.18 Advantages and Disadvantages of Alloy Steel. The important advantages and disadvantages in the choice of alloy steel from the general point of view in relation to plain carbon steel are listed in the following tabulation:

Advantages That May be Attained

1. Greater hardenability.
2. Less distortion and cracking.
3. Greater stress relief at given hardness.
4. Less grain growth.
5. Higher elastic ratio and endurance strength.
6. Greater high temperature strength.
7. Better machinability at high hardness.
8. Greater ductility at high strength.

Disadvantages That May be Encountered

1. Cost.
2. Special handling.
3. Tendency toward austenite retention.
4. Temper brittleness in certain grades.

REFERENCES

Boron Steels, Metal Progress **60,** No. 2, August 1951, p. 81.

Crafts and Lamont, *Hardenability and Steel Selection,* Pitman Publishing Corp., New York, 1949.

Grange and Garvey, *Factors Affecting the Hardenability of Boron-treated Steels, Trans. ASM* **37,** Metals Park, Ohio, 1946, p. 136.

Hollomon and Jaffe, *Ferrous Metallurgical Design,* John Wiley & Sons, New York, 1947.

Hoyt, *Metal Data,* Reinhold Publishing Corp., New York, 1952.

Metals Handbook, American Society for Metals, Metals Park, Ohio, 1961.

Modern Steels and Their Properties, Bethlehem Steel Co., Bethlehem, Pa., 1958.

Molybdenum Steels, Irons, Alloys, Climax Molybdenum Co., New York, 1948.

Nickel Alloy Steels, International Nickel Co., New York, 1949.

Republic Alloy Steels, Republic Steel Co., Cleveland, Ohio, 1949.

Ryerson Alloy Steel Reference Book, Jas. T. Ryerson and Son, Inc., Chicago, Illinois, 1947.

S.A.E. Handbook, Society of Automotive Engineers, New York, 1961.

Samans, *Engineering Metals and Their Alloys,* Macmillan Co., New York, 1949.

Vanadium Steels and Irons, Vanadium Corporation of America, New York.

Woldman, *Engineering Alloys,* American Society for Metals, Metals Park, Ohio, 1954.

QUESTIONS

1. How may alloy steels be classified?
2. Identify the following: AISI A2340, AISI A3140, AISI A4140, AISI A5140, AISI A6150, AISI A8640, AISI A9840.
3. What is the effect of manganese as an alloying element in steel?
4. What structure is produced in the slow cooling of a steel containing between 11 and 14 per cent manganese and 1 to 1.2 per cent carbon? For what purposes is such a steel employed?
5. What advantages are derived from the use of nickel as an alloying element in steel?
6. What is the effect of additions of chromium on the characteristics of steel?
7. What disadvantage may accrue through the use of chromium as an alloying element in steel?
8. What advantages are derived by using a combination of nickel and chromium in steel?
9. What is the effect of the addition of molybdenum on the characteristics of steel?
10. What is the effect of the addition of molybdenum to chromium-bearing steels which exhibit temper brittleness?
11. What is the primary advantage of the use of vanadium as an alloying element in steel?
12. For what purposes is tungsten used in steel in amounts of less than 6 per cent?

13. Why is silicon useful as an alloying element in very low-carbon steels?
14. What difficulties may be encountered in using steels containing up to approximately 2 per cent silicon?
15. What are the principal elements present in the triple-alloy steels? Why are they of particular value?
16. What is the effect of the addition of boron to steel? Approximately what proportions of boron are employed?
17. What are the so-called low-alloy structural steels? Why are they of importance?
18. What ultra high strengths have become possible in steels?
19. What process controls have contributed to the success of 4340 steel at ultra high strengths?
20. Why are the low-alloy structural steels more weldable than the higher-carbon alloy machinery steels?
21. State both the advantages and disadvantages of alloy steels.
22. An AISI 4150 steel has been used successfully in the manufacture of a given part $3\frac{1}{2}$ in. in diameter in its heaviest section. It has been possible to consistently harden the piece to Rockwell C 49 throughout. Because of material shortages it is necessary to make a substitution for this steel from the following list of AISI steels currently in stock: 3140, 4140, 4340, 5150, 6150 and 9260. Which steel is best suited for substitution?
23. Given the end-quench and tempered end-quench curves for the heat of AISI 8642 steel shown in Appendix F. What tempering temperature should be recommended to produce a tensile strength of 125,000 lb/in.^{2} at the center of a cylindrical part 2 in. in diameter after an agitated water quench? What is the yield strength, elongation, reduction of area, and Brinell hardness at the center of the bar when so treated?

XI

CASEHARDENING AND SURFACE TREATMENT

11.1 General. The parts of some equipment must have a very hard, wear-resistant surface and also possess great toughness. Such a combination is usually not possible in a piece of heat-treated steel. If it is treated to give maximum surface hardness, it is too brittle. If it is treated for maximum toughness, it will not be hard enough. Consequently, several processes, known as casehardening, have been developed by which this combination can be attained commercially. These include carburizing, cyaniding, carbonitriding, nitriding, and localized surface hardening. The surface hardness obtained by the first three processes (carburizing, carbonitriding, and cyaniding) depends upon heat treatment after the composition of the case has been altered. The fourth process (nitriding) alters the composition of the case in such a way that the compounds formed are inherently hard. The last process of casehardening (surface hardening) depends entirely upon heat-treating the surface of a hardenable steel. The mechanism involved in the hardening of steel was considered in Article 8.13 of Chapter VIII.

11.2 Carburizing. Carburizing is a process by which the carbon content of the surface of a steel is increased. Since the object of casehardening is to obtain a hard, wear-resistant surface with a tough interior, the first consideration is the selection of a low-carbon steel which will be tough. The relation between carbon content, strength, and hardenability has already been discussed. The low-carbon steels are tougher than the medium- or high-carbon steels, but their hardenability is quite low. Therefore, any means by which the carbon content of a low-carbon (0.15 per cent carbon) steel can be raised to eutectoid proportions will produce the desired results, provided the steel is properly heat-treated. The use of alloy steels may prove to be beneficial by directly improving toughness, hardenability, or wear-resistance.

There are three general methods of carburizing, depending upon the form of the carburizing medium: first, solid or pack carburizing, employing solid carburizing material; second, gas carburizing, employing suitable hydrocarbon gases; and third, liquid carburizing, employing fused baths of carburizing salts.

The choice of the method that is to be used in carburizing depends mostly upon the characteristics of the case which are desired, the equipment available, and the quantity of parts which is to be handled.

11.3 Pack Carburizing. In pack carburizing, the parts that are to be carburized are packed in a suitable container with the carburizing medium. The essential part of the carburizing medium or compound is some form of carbon, usually hardwood charcoal or charcoal produced from cocoanut shells, peach pits, etc. Other substances have been used from time to time, but the charcoal base of the carburizing compound seems to be the most successful. The boxes containing the parts to be treated, which have been packed with carburizing compound, are placed in the furnace and heated to the desired temperature for carburization. This temperature must be high enough to form austenite. The parts are maintained at this temperature until the desired degree of penetration has been obtained.

Although charcoal is the basic substance of any carburizing compound, experiments have shown that no carbon is absorbed by the steel if all the air or oxygen is removed from the carburizing box, an indication that there is no direct transfer of carbon to the steel. It is probable that the carbon absorption at the surface occurs by a reaction between iron and carbon monoxide, the latter serving as an active carrier of carbon. The carbon monoxide is provided by the reaction of oxygen with carbon or between carbon dioxide and carbon, in accordance with the following reaction:

$$CO_2 + C \leftrightharpoons 2CO$$

At the carburizing temperature, this reaction always goes to the right. The carbon monoxide is free to combine with iron in the following manner:

$$2CO + 3Fe \leftrightharpoons Fe_3C + CO_2$$

Although the reaction may be represented in this manner, the carbon is not in the form of iron carbide, but instead is in the form of elemental carbon dissolved in gamma iron, forming austenite. As the carbon content at the interface between solid and gas is increased, the carbon, by diffusion at these temperatures, begins to migrate in toward the region of lower carbon concentration, i.e., the center of the piece being carburized. The carbon dioxide produced by this reaction is liberated at the surface and reacts with the carburizing compound to form more carbon monoxide at the expense of the carbon of the charcoal.

The penetration of carbon into the steel will depend upon the temperature, the time at temperature, and the carburizing agent. Since the solubility of carbon in steel is greatest above the Ac_3 temperature, carburization takes place most readily above this temperature. Furthermore, the higher the temperature, the greater the rate of carbon penetration, since the rate of diffusion is greater. It is customary to select a temperature about 100 F (40 C) above the Ac_3 point. The time at the carburizing temperature is the most influential factor in the control of the depth of carbon penetration. Fig. 11.1 shows the relation between the depth of penetration and time.

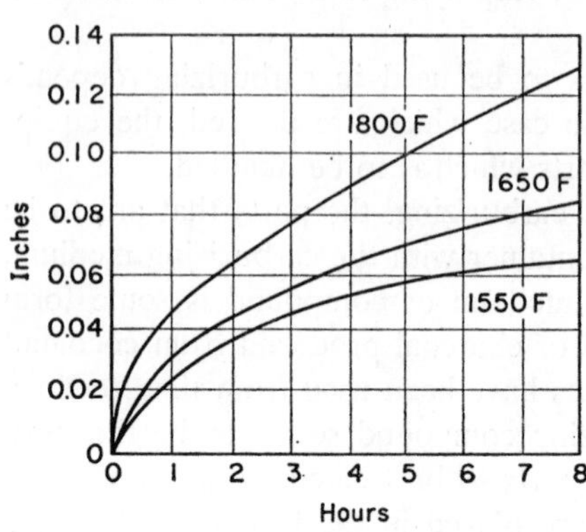

FIG. 11.1 Case depth vs. time and temperature for carburizing in ordinary solid compound. (Bullens-Battelle, *Steel and Its Heat Treatment,* 5th ed., John Wiley and Sons, New York, 1948)

The variation of carbon content from the surface inward for different carburizing times at a temperature of 1700 F (930 C) is shown in Fig. 11.2.

In some applications, it is desirable to carburize only certain portions of the part. This is known as selective carburizing and can be done by plating the entire piece with copper and machining off those regions that are to be carburized, or by masking the part prior to plating so that only

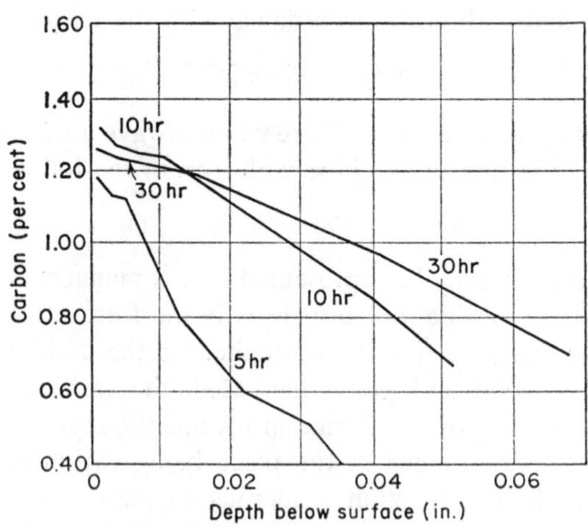

FIG. 11.2 Carbon content vs. depth below surface of a 0.20% C steel, carburized at 1700 F (930 C) for times indicated. (After Manning, *Trans. ASM* **31,** 1943, p. 8)

those portions are plated where it is desired to prevent carburizing. The copper plate serves as a barrier to the absorption of carbon. An alternate method commonly employed is to carburize the entire piece and then machine off the undesired case prior to hardening. Proprietary paints consisting of copper powder in a silicate vehicle may also be applied and baked onto the part. However, these paints do not provide the protection against carbon absorption that is possible with dense copper plating.

The process of carburizing depends upon the formation of carbon monoxide. Since the carburizing box, in many instances, is sealed rather tightly with clay, there is liable to be an insufficient source of carbon dioxide; hence, carburization may be very slow. Energizers, which are usually carbonates, such as barium or sodium carbonate, have been employed to insure a proper supply of carbon monoxide. By the application of heat to barium carbonate, the following reaction occurs:

$$BaCO_3 \leftrightharpoons BaO + CO_2$$

thus providing a source of carbon dioxide which will react with the carbon to produce carbon monoxide for the carburizing process.

A typical carburizing compound may consist of 53 to 55 per cent hardwood charcoal, 30 to 32 per cent solvay coke, 2 to 3 per cent sodium carbonate, 10 to 12 per cent barium carbonate, and 3 to 4 per cent calcium carbonate. The function of the coke in this carburizing compound is to act as a filler; also it may increase the thermal conductivity of the compound.

The boxes, in which parts are carburized, vary widely in size, shape, and composition. Since they are heated for long periods to high temperatures, materials that are easily oxidized are not practical for continued use. Plain carbon steel, either cast or welded, oxidizes rapidly and distorts badly. Therefore, it is common practice to employ heat-resistant materials, such as an 18 per cent chromium-8 per cent nickel steel or a 65 per cent nickel-16 per cent chromium alloy.

The pack carburizing process is high in labor cost yet low in equipment cost, since special furnaces are not required to heat the carburizing boxes. Heat transfer efficiency is poor because the carburizing compounds act as ideal insulating agents requiring long heating periods to bring the pack up to temperature. Conversely, this is an advantage when slow cooling is required to produce a soft case and permit additional machining prior to hardening. Uniform case depths are difficult to control since parts near the outer wall of the box reach the carburizing temperature much sooner than those at the center. For this reason, the process is seldom used for case depths less than 0.025 in. and is being superseded by the more efficient and controllable gas carburizing method for modern production carburizing.

11.4 Gas Carburizing. The principal source of carbon in gas carburizing is methane, CH_4, which is present in manufactured and natural gas. The reaction may be represented as follows:

$$CH_4 + 3Fe \leftrightharpoons Fe_3C + 2H_2$$

In using hydrocarbons for the carburizing medium, there is a tendency to form a deposit of soot on the work which inhibits carburization. To prevent this, methane is diluted with a carrier gas, such as generated endothermic gas, although other diluents such as flue gas have been used.

Gas carburizing requires specially designed atmosphere-tight furnaces capable of maintaining a positive pressure of atmosphere to prevent the infiltration of air. These furnaces are generally equipped with a high-alloy fan to provide the circulation of atmosphere throughout the heating chamber or retort. The furnace is initially purged with endothermic gas, while operating at the carburizing temperature of 1700 or 1750 F (930 or 955 C), and adjusted to a furnace dew point of between 20 and 25 F (−6 and −4 C). Reference to the dew point equilibrium curves in Chapter VIII will indicate that this atmosphere will have a carbon potential of 0.80-0.90 per cent carbon and will be carburizing to low-carbon steels. To increase the carburizing potential of the atmosphere, a metered addition of from 5 to 10 per cent of natural gas (methane) is added after the work is charged and the furnace has recovered to the carburizing temperature. This addition of natural gas is discontinued after the predetermined time cycle is between 50 and 75 per cent completed. This action permits diffusion of the high-carbon case and re-establishment of surface equilibrium with the carbon potential of the carrier atmosphere. By means of this diffusion technique, it is possible to maintain the surface carbon content at or about a desired eutectoid composition. A typical example of the effect of temperature and time upon the depth of carbon penetration in gas carburizing is shown in Fig. 11.3.

Although the limitations of alloy furnace parts restrict the economical operation of gas carburizing to a temperature of about 1750 F (955 C), special furnaces employing refractory substitutes have extended the process to a temperature of 2000 F (1095 C). High-temperature gas carburizing requires close control of temperature and composition of the atmosphere. Excessive grain growth at elevated temperatures may require a reheating operation to permit grain refinement. A three-hour carburizing cycle at a temperature of 2000 F (1095 C) will produce a total case depth of about 0.100 in., compared to 0.040 in. at 1700 F (925 C).

Gas carburizing is particularly adapted to large volume production and provides for accurate control of case depth and surface carbon content.

Although labor costs are lower than in pack carburizing, higher skilled personnel are required to maintain the necessary controls.

11.5 Liquid Carburizing. The application of liquid carburizing is an outgrowth of the older process of cyaniding. It is employed principally for relatively shallow cases which can be produced at lower cost with existing equipment than with pack or gas carburizing. Deep-case baths are also available. The fused bath is usually composed of sodium cyanide and alkaline earth salts, although barium, calcium, or strontium salts, such as chlo-

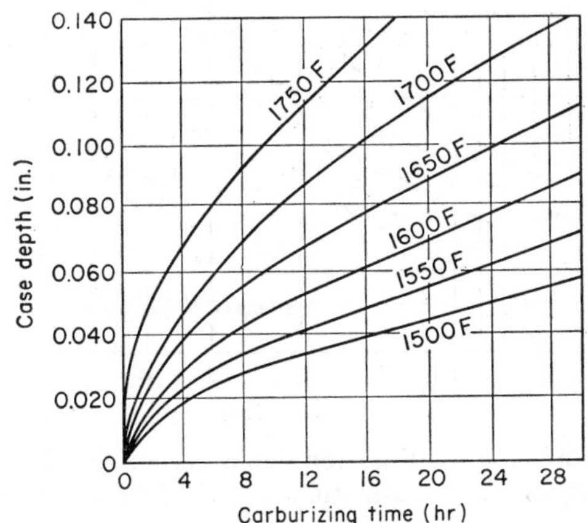

FIG. 11.3 Depth of carbon penetration for different times and different temperatures in gas carburizing an SAE 1020 steel. (After Williams, *Symposium on Carburizing*)

ride, may be used. The salt reacts with the cyanide to form a cyanide of the alkaline earth metal, which then reacts with iron to form Fe_3C. The reaction may be written:

$$Ba(CN)_2 + 3Fe \rightarrow Fe_3C + BaCN_2$$

In this process, the carbon is liberated to combine with the iron. A small amount of nitrogen is liberated and also absorbed. The parts are placed in the molten salt for the desired time and then quenched.

The relation of time, depth of case, and temperature for activated baths is shown in Fig. 11.4. In this type of liquid carburizing, a certain amount of nitrogen is present, although the amount is much less than when straight cyanide is used. The characteristics of the case produced by liquid carburizing are somewhat similar to the case produced by pack or gas carburizing. Salt-bath carburizing offers advantages of rapid heat transfer, low distortion, negligible surface oxidation or decarburization, and rapid absorption of carbon and nitrogen.

Another method of liquid carburizing is by activating the cyanide bath with ammonia gas. This process, commercially known as chapmanizing, increases the nitrogen content of the surface more than the other liquid carburizing processes.

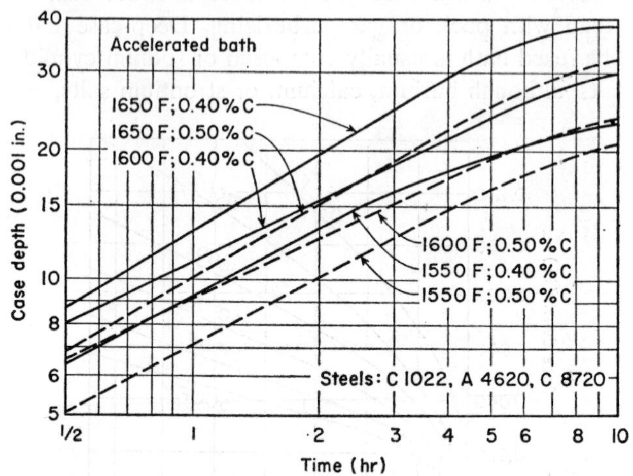

FIG. 11.4 Depth of case vs. time for liquid carburizing in an activated bath at three temperatures. (After Komarnitsky, *Metal Progress* **50,** 1946, p. 665)

11.6 Heat Treatment. If a steel that has been carburized is slowly cooled from the carburizing temperature, the surface hardness will be low enough to permit machining, since it is effectively an annealed high-carbon steel. If a cross section of the steel is observed under the microscope, it may have the appearance of that shown in Fig. 11.5. The outer surface is a hypereutectoid steel, and, with increasing depth, the carbon content decreases until the hypoeutectoid structure of the core is encountered. Hence, a carburized steel is effectively a combination of steels containing different amounts of carbon. The heat treatment of a carburized steel is somewhat complex because of the variation of carbon content that occurs in one piece of steel. What is done with the surface must be done with the interior. In spite of this complication, treatments can be given that will produce the desired results without introducing any new metallurgical principles.

The heat treatment to be given a carburized steel will depend upon the carburizing temperature employed, the composition of the core and the case, and the purpose for which the part is intended or the properties that must be obtained to meet specifications.

Heat-treating processes for carburized steel may be divided into two broad classifications, depending upon the temperature of carburization and upon the carbon content of the case. The first class of treatments (I) in-

FIG. 11.5 Carburized steel (100×)—high-carbon case at top; low-carbon core at bottom.

cludes those that are employed for steels carburized at temperatures only slightly above the Ac_3 temperature of the core. The second class of treatments (II) includes those in which the steel has been carburized at a temperature considerably above the Ac_3 temperature of the core. The specific treatment given in each classification may be divided into three groups, depending upon the carbon content of the case. Treatments in the first group (a) are applied to those steels in which the case is of hypoeutectoid composition. The second group (b) includes those treatments for steels in which the case is of hypereutectoid composition, but for which the temperature of carburization has been greater than the A_{cm} temperature. The third group (c) includes those treatments for steels in which the case is of hypereutectoid composition and for which the temperature of carburization was lower than the A_{cm} temperature. These conditions are shown graphically in Fig. 11.6.

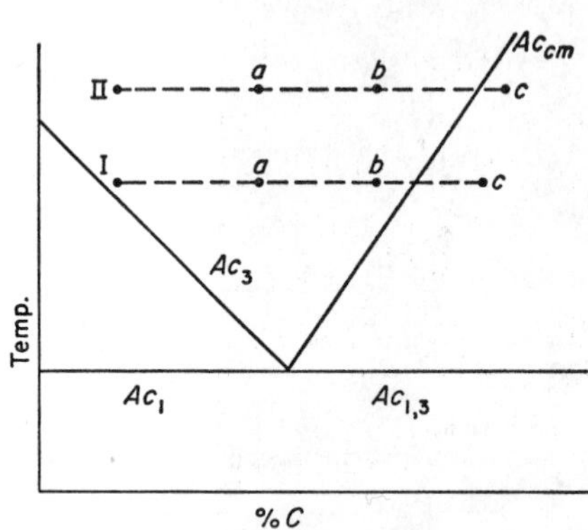

FIG. 11.6 Carburizing conditions controlling selection of heat treatment of carburized steels.

Although the discussion of heat-treating procedures to follow is referred specifically to parts that have been pack-carburized, similar procedures may be employed for parts that are processed by other methods.

Class I (a). When the carburizing temperature has been just above the Ac_3 temperature and the carbon content of the case does not exceed eutectoid proportions, treatment may be relatively simple. Two treatments may be used. The first consists of quenching the steel directly from the carburizing box. This is by far the cheapest method of heat treatment, since no heating of the steel is necessary other than tempering to reduce the hardness of the case slightly when required. If the carbon content of the case corresponds to the eutectoid composition, the case may not be as fine-grained as desired. In this event, the steel may be treated by the second method, in which the steel is allowed to cool in the carburizing box. After the steel is removed from the box, it is reheated to slightly above the $Ac_{3,1}$ temperature, quenched, and tempered. This treatment is superior to the preceding one, particularly if the steel has a tendency to coarsen at the carburizing temperature. It is not often that the carbon content of the case is below the eutectoid composition; in fact, the case is usually hypereutectoid.

Class I (b). When the case is of hypereutectoid composition and the steel has been carburized at a temperature above the A_{cm} temperature, the case will have very high hardness and wear resistance when properly heat-treated. If a steel of these characteristics is slowly cooled in the carburizing box from the carburizing temperature, the proeutectoid cementite will precipitate at the grain boundaries. If, on the other hand, the steel is quenched from the box, the case will be hardened and proeutectoid cementite will be absent. However, if the steel is a coarse-grained type, some coarsening may occur, resulting in less toughness. This again is the most inexpensive treatment, but it is not to be recommended where best properties are required.

If the steel is slowly cooled from the carburizing temperature, then reheated to above the A_{cm} temperature, quenched and tempered, the core will have excellent properties of grain refinement. The case will be hardened by this treatment but will not be any tougher than when treated by the direct-quench method.

Although more expensive, highly satisfactory results are obtained with a double heat treatment. The steel is slowly cooled in the box, and is then removed and heated to above the A_{cm} temperature, followed by a quench. This treatment refines the core. The steel is then heated to above the $Ac_{3,1}$ temperature, quenched, and tempered. This portion of the treatment refines the grain of the case and completes the hardening.

Double heat treatment will give the best combination of a hard, wear-resistant, fine-grained case and a tough, refined core. Unless the parts are relatively small and can be handled easily, danger may be encountered in quenching direct from the box, because it is often difficult to remove the parts without the surface temperature decreasing to a value very close to or below the A_{cm} temperature before the part reaches the quenching bath.

Class I (c). If, at the carburizing temperature, the carbon content of the case is high enough to produce a structure of austenite and cementite, the A_{cm} temperature of the case will not be exceeded. As a result a carbide network is present at the austenite grain boundaries. If the steel is quenched from the carburizing temperature, the network will be preserved, thus producing a poor case. This condition can be eliminated by heating the steel to above the A_{cm} temperature, followed by quenching and tempering. However, treating the steel in this manner is likely to coarsen the core as well as the case; hence, the steel will not be as tough as required for some applications. These conditions can be improved after the first quench by heating the steel to the $Ac_{3,1}$ temperature, quenching, and tempering. Even with this treatment, the core will remain in the coarsened condition.

Class II. When the carburizing temperature is considerably above the Ac_3 temperature of the core, the steel is likely to be coarsened. Therefore, the double heat treatment is to be preferred. However, if it is necessary to refine only the case, the steel, if hypoeutectoid, can be slowly cooled and then heated to the Ac_3 temperature, quenched, and tempered. It is very

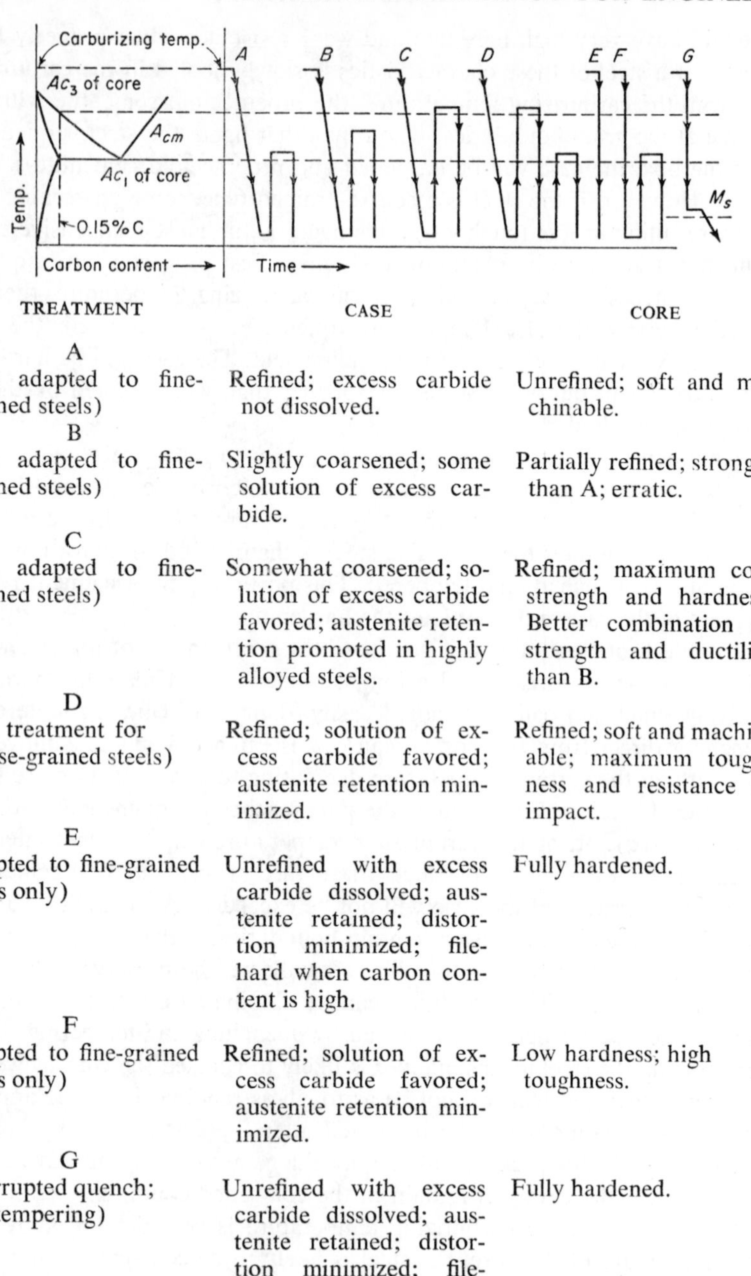

TREATMENT	CASE	CORE
A (Best adapted to fine-grained steels)	Refined; excess carbide not dissolved.	Unrefined; soft and machinable.
B (Best adapted to fine-grained steels)	Slightly coarsened; some solution of excess carbide.	Partially refined; stronger than A; erratic.
C (Best adapted to fine-grained steels)	Somewhat coarsened; solution of excess carbide favored; austenite retention promoted in highly alloyed steels.	Refined; maximum core strength and hardness. Better combination of strength and ductility than B.
D (Best treatment for coarse-grained steels)	Refined; solution of excess carbide favored; austenite retention minimized.	Refined; soft and machinable; maximum toughness and resistance to impact.
E (Adapted to fine-grained steels only)	Unrefined with excess carbide dissolved; austenite retained; distortion minimized; file-hard when carbon content is high.	Fully hardened.
F (Adapted to fine-grained steels only)	Refined; solution of excess carbide favored; austenite retention minimized.	Low hardness; high toughness.
G (Interrupted quench; martempering)	Unrefined with excess carbide dissolved; austenite retained; distortion minimized; file-hard when carbon content is high.	Fully hardened.

FIG. 11.7 Heat-treating cycles for carburized steels. (After *Metals Handbook*)

difficult to obtain satisfactory results if carburization has been under the conditions of Class II (c), since any heating to just above the critical temperature of the core for the purpose of refinement will permit the carbide to form at the grain boundaries, which will impair the toughness of the case.

These methods of heat treatment are schematically presented in Fig. 11.7. The heat treatments recommended for several typical steels are presented in Table 11-I.

TABLE 11-I. HEAT TREATMENTS FOR CARBURIZED STEELS.
(From *Metals Handbook*)

STEEL[1] SAE NO.	APPROXIMATE TRANS. TEMP. (°F) Case Ac_1	Core Ac_3[2]	PREFERRED TREATMENTS (See Fig. 11.7)	QUENCHING MEDIUM	RETENTION OF AUSTENITE
1015	1355	1585	A,D,F	H_2O or 3% NaOH sol.	Very slight
1019	1350	1550	A,D,F	H_2O or 3% NaOH sol.	Very slight
1117	1345	1520	A,D,F	H_2O or 3% NaOH sol.	Slight
1320	1325	1500	E,C	Oil	Slight
3115	1355	1500	E,C	Oil	Moderate
3310	1330	1435	C,D,F	Oil	Strong[3]
4119	1395	1500	E,C	Oil	Moderate
4320	1350	1475	C,E	Oil	Strong[3]
4615	1335	1485	E,C	Oil	Moderate
4815	1300	1440	C,E	Oil	Strong[3]
8620 / 8720	1350	1540	E,C	Oil	Moderate

[1] All steels preferably of ASTM grain size 5 or finer.
[2] Ac_3 of core will decrease as carbon content is increased.
[3] Spheroidization of excess carbide by subcritical treatment before hardening may be used to decrease the amount of austenite retained.

Cold treatment is often required in connection with the hardening of carburized parts. This is particularly true of those alloy steels that tend toward austenite retention such as AISI-SAE 3310, 4320, 4620, 4820, and 9310. Table 11-II illustrates the effect of multiple cold treatment—retempering cycles on a carburized gear that was intentionally carburized with a high carbon content on the surface. The greatest improvement in hardness takes place after the first cold treatment and is most pronounced near the surface where the carbon content is a maximum.

11.7 Grain Size. The relation of austenitic grain size to the hardenability of steel has already been discussed. However, the austenitic grain size also has considerable influence on the carburizing characteristics of the steel and the response to hardening.

Some steels when carburized and heat-treated do not harden uniformly. McQuaid and Ehn, in working on the problem of soft spots, discovered that those steels in which soft spots were present usually had a structure of the type shown in Fig. 11.8, as carburized. Steels that harden uniformly usually have a structure as shown in Fig. 11.9, as carburized. These photomicrographs were taken of two steels after being slowly cooled from the carburizing temperature. Steels having these characteristics have been called abnormal and normal steels, respectively. These structures are distinguished

TABLE 11-II. EFFECT OF COLD TREATMENT ON CASE HARDNESS OF AISI-SAE 3310 CARBURIZED GEARS.

DEPTH BELOW SURFACE (INCH)	HARDNESS ROCKWELL C CONVERTED FROM 500 GRAM KNOOP		
	As-quenched from 1700 F (925 C)	1st Cycle −100 F (−75 C)	2nd Cycle −100 F (−75 C)
0.002	48	64	65
0.004	47	64	65
0.006	44	64	65
0.008	48	64	65
0.010	52	63	65
0.012	53	63	65
0.014	53	63	64
0.016	53	64	64
0.018	59	64	64
0.020	59	64	64
0.030	57	59	59
0.040	49	51	51
0.050	43	44	44

by the character of the proeutectoid cementite and the pearlite. In the normal steel, the proeutectoid cementite which forms the network around the austenite grains, and hence is observed around the pearlite grains, is immediately adjacent to the lamellae of pearlite grains; in the abnormal steel, there is a film of ferrite between the cementite network and the pearlite grains.

McQuaid and Ehn also observed that the abnormal steels tended to have a finer austenitic grain than the normal steels. As a result of their investigation, they devised a test known as the McQuaid-Ehn test, which consists of packing samples of the steel in a carburizing compound, heating to a temperature of 1700 F (925 C) for 8 hours, and then slowly cooling. This treatment produces a hypereutectoid case, and the slow cooling forms a cementite network at the austenitic grain boundary, thereby permitting the determination of the austenitic grain size and the normality of the steel. It must be remembered that this test is applied under specific temperature conditions. In discussing grain size, it was stated that the temperature to which a steel is heated has a marked influence upon the austenitic grain size. It is necessary, therefore, when determining the austenitic grain size of

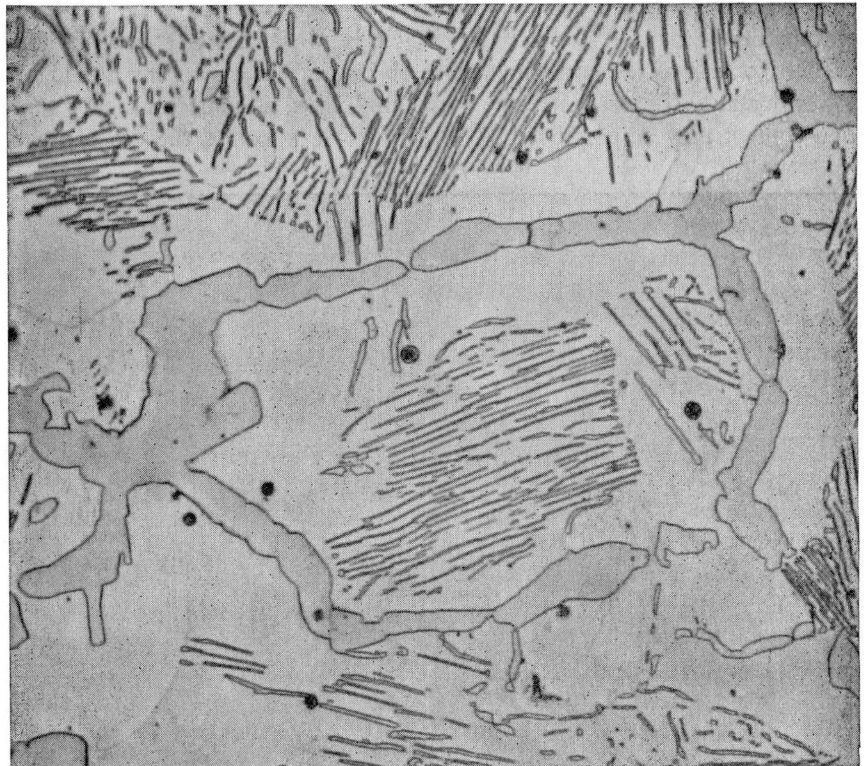

FIG. 11.8 "Abnormal" steel (Vilella) (1000×). (*Courtesy U. S. Steel Co.*)

carburized steels, to make the determination under conditions that are similar to the heat-treating procedure.

The presence of soft spots is related to the grain size of the steel. It has already been shown that fine-grained steels tend to be shallow-hardening, whereas coarse-grained steels are deeper-hardening. Hence, a coarse-grained steel will tend to harden more uniformly than a fine-grained steel. If, however, other alloying elements are present which tend to alter the rate of transformation in favor of deeper hardening, grain size may have a somewhat less effect upon the uniformity of hardening of the carburized steel.

11.8 Steels for Carburizing. A wide variety of steels are used for carburizing. The most common practice is to use steels containing between 0.15 and 0.25 per cent carbon. This low carbon content is important for superior toughness. Many benefits, such as considerably greater toughness and improved hardness of the case, can be derived by use of the low-carbon alloys steels. If shallow cases are employed and highly concentrated loads are involved, indentation of the surface, referred to as Brinelling, may result. When this occurs, it may be necessary either to increase the depth of carburizing or to use a stronger steel. There is no particular reason why the carbon content of the core should not be higher, except for the attendant lower ductility. If greater strength is required, then, of course, a higher carbon content may be employed. A present trend is toward the use of light

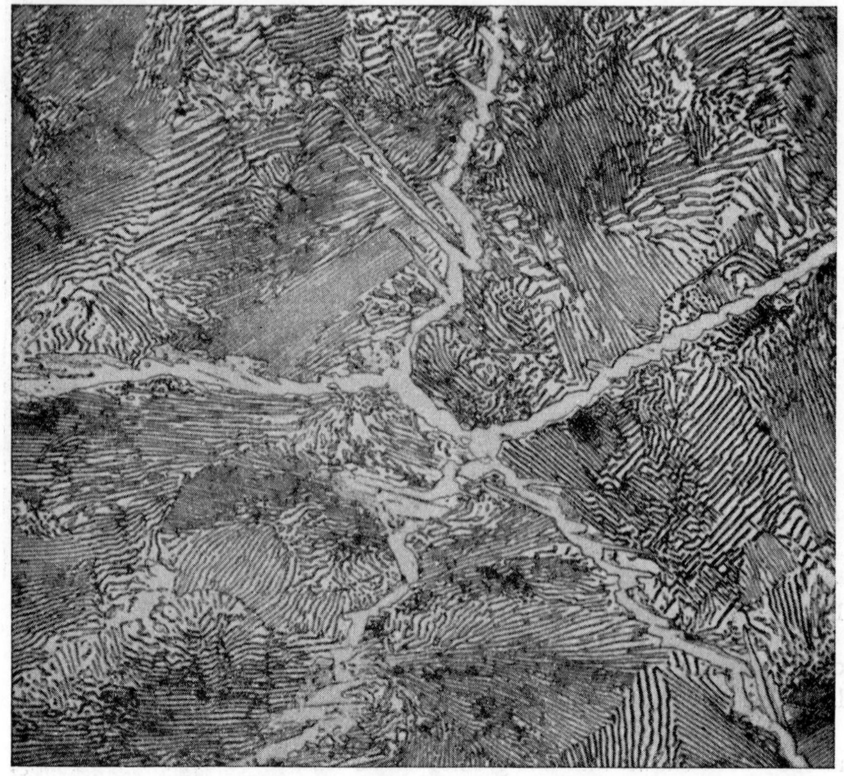

FIG. 11.9 "Normal" steel (Vilella) (1000×). (*Courtesy U. S. Steel Co.*)

cases on higher carbon cores containing up to about 0.40 per cent carbon.

11.9 Case Depth. The measurement of total case depth generally includes both the hypereutectoid and eutectoid regions. Classification on the basis of total case depth has been made as follows:

Thin: less than 0.020 in.
Medium: 0.020-0.040 in.
Medium heavy: 0.040-0.060 in.
Heavy: more than 0.060 in.

There is a trend toward specifying *effective case depth* measured as the depth at which the hardness of the hardened case reaches a level of 50 Rockwell C converted from microhardness survey data.

Thin cases are cheaper to produce and are satisfactory for applications involving light wear; when properly supported, they can resist crushing under load. The cost of heavier cases is justified in certain situations to withstand maximum wear, to prevent collapse of the case under heavy loads, to allow grinding to size and finish after hardening, and to improve the mechanical properties of the surface layer to resist fatigue. In the latter case, it should be noted that, conversely, a hard, brittle surface is susceptible to early fatigue failure if stress raisers are present in the form of tool marks, etc.

11.10 Application of Carburizing. Casehardening by carburizing is used extensively for all sorts of machine parts, including cams, piston pins, gears, pump shafts, etc. In some applications, only certain portions of the part require a hard surface. This can be done either by slowly cooling the part after carburizing, machining away the surface of those portions that are not to be hardened, and then heat-treating; or by copper plating those portions that are not to be carburized.

11.11 Cyaniding. Closely related to carburizing is the process known as *cyaniding* in which an extremely thin case (usually less than 0.010 in.) is produced. This process is similar to liquid carburizing except that the bath used is generally fused sodium cyanide. The steel is immersed in a molten bath of sodium cyanide at a temperature between 1475 and 1600 F (800 and 870 C) for a period of between 30 min and 3 hr, depending upon the depth of case required. The depth of penetration obtained by cyaniding is shown in Fig. 11.10.

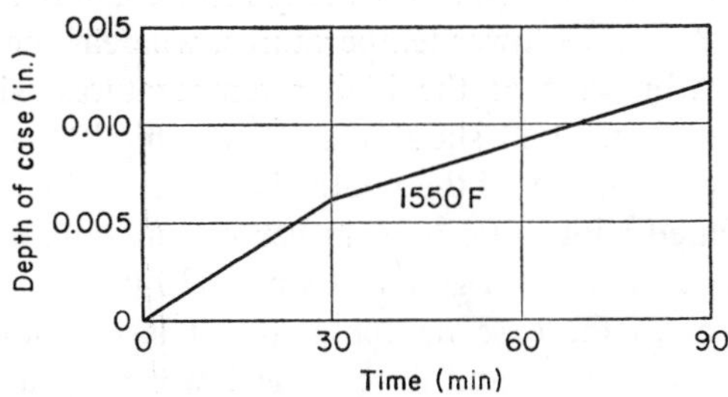

FIG. 11.10 Depth of penetration vs. time in cyaniding. (After Beckwith, *Carburizing Symposium*)

The sodium cyanide decomposes in two possible ways on heating in contact with air:

(1) $2NaCN + 2O_2 \rightarrow Na_2CO_3 + CO + 2N; \quad 2CO \leftrightarrows CO_2 + C$

(2) $2NaCN + O_2 \rightarrow 2NaCNO$

also $NaCN + CO_2 \rightarrow NaCNO + CO;$

$$3NaCNO \rightarrow NaCN + Na_2CO_3 + C + 2N$$

The sodium cyanide is oxidized to form sodium cyanate which further dissociates to liberate carbon and nitrogen. The last two elements, in turn, are absorbed by the iron.

In order to obtain hardness after cyaniding, it is necessary to quench directly into oil or water from the cyaniding bath. In cyaniding, both carbon and nitrogen are absorbed at the surface of the steel. Nitrogen imparts inherent hardness, whereas the increased carbon content makes the surface of the steel respond to a quenching treatment. Probably the greatest use of cyaniding is for parts that are to be subjected to relatively light loads and that require improvement in the surface-wear resistance. The containers for the molten cyanide are usually pressed steel, and the area above the container should be well ventilated to allow adequate removal of fumes.

11.12 Carbonitriding. *Carbonitriding* is a modification of the gas carburizing process with the addition of anhydrous ammonia gas to the furnace atmosphere to cause both carbon and nitrogen to be absorbed by the surface of the steel at the carbonitriding temperature. The process is also known as *dry cyaniding, gas cyaniding, nitrocarburizing,* and *ni-carbing.* Although a wide variety of gas mixtures are possible for this process, depending on the work load and furnace construction, a common atmosphere consists of 15 per cent anhydrous ammonia gas with 5 per cent natural gas in a carrier atmosphere of endothermic gas having a furnace dew point of 20 to 25 F (−6 to −4 C). Carbonitriding is conducted at lower temperatures than gas carburizing. The temperatures range from 1200 to 1625 F (650 to 885 C) with 1550 F (845 C) being most common. The absorption of nitrogen is greater at the lower temperatures, whereas the process closely approaches gas carburizing at the higher temperatures with a minimum transfer of nitrogen. Fig. 11.11 shows the relationship between temperature, time, and case depth. In view of the relatively low temperatures used in carbonitriding compared with those used in gas carburizing, it is seldom economical to specify a case depth greater than 0.020 in.

Nitrogen increases the case hardenability of steel, increases the solubility of carbon in austenite, and lowers the lower critical temperature. It is possible to harden small carbonitrided parts by cooling from a temperature as low as 1200 F (650 C) by transferring the part to an atmosphere cooling chamber without the necessity of oil quenching, thus minimizing distortion. The increased hardenability made possible by the alloying effect of nitrogen permits the oil quenching of carbonitrided plain carbon steels that would otherwise require drastic water quenching to develop effective

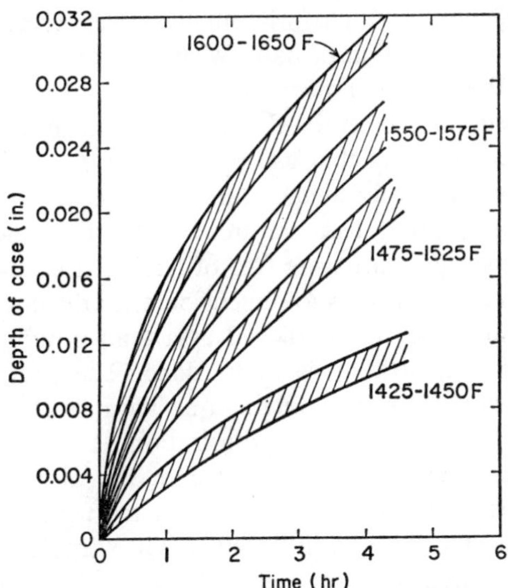

FIG. 11.11 Depth of penetration vs. time and temperature in carbonitriding (After *Metals Handbook,* 1954 Supplement)

hardening. A comparison of the hardenability of carbonitrided and carburized plain carbon steel is shown in Fig. 11.12. In view of the high carbon and nitrogen content of carbonitrided cases, it is often necessary to employ a cold treatment at −100 F (−75 C) to convert retained austenite and to increase the surface hardness.

11.13 Nitriding. *Nitriding* is a casehardening process by which the surface hardness of certain alloy steels may be increased by heating in con-

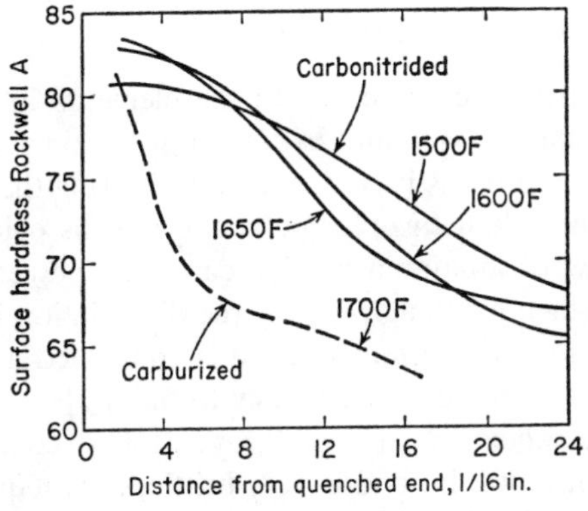

FIG. 11.12 End-quench hardenability of SAE 1020 steel carbonitrided vs. carburized. (After *Metals Handbook,* 1954 Supplement)

tact with a nitrogeneous medium, without the necessity of quenching. The process involves the formation of hard, wear-resistant nitrogen compounds on the surface of the steel by absorption of nascent nitrogen.

Prior to nitriding, it is essential that the required core properties be developed by heat treatment of the steel in its rough machined state. In such heat treatment the tempering temperature must be at least 100 F (40 C) above the intended nitriding temperature. Furthermore, the structure must consist of fully tempered martensite free of ferrite. If considerable machining is required to bring the parts to dimension after the core has been heat-treated, the part must be stress-relieved at a temperature above the temperature at which the nitriding is to be done. The presence of residual stresses leads to nonuniform nitriding and an unsatisfactory case. The surface of the steel must be free of decarburization or there will be excessive growth, and the case will be very brittle and susceptible to cracking and spalling. If the surface of the steel is contaminated with such substances as oil, the absorption of nitrogen will be impeded. Nitrogen absorption is restricted by highly buffed or burnished surfaces.

The gas nitriding process consists of placing properly racked and supported steel parts in an alloy retort or specially constructed furnace heated to a temperature between 930 and 1050 F (500 and 565 C), while anhydrous ammonia gas is circulated around the parts to be nitrided. Retorts or boxes for nitriding are usually constructed of Inconel or high-nickel alloys and generally incorporate a fan to assure adequate circulation of ammonia in intimate contact with the work. Anhydrous ammonia gas partially dissociates or cracks on the surface of the workpiece at the nitriding temperature in accordance with the following reaction:

$$\begin{aligned} NH_3 &\leftrightarrows N + 3H \\ 2N &\leftrightarrows N_2 \\ 2H &\leftrightarrows H_2 \end{aligned}$$

The nascent nitrogen combines with the elements in the steel to form nitrides. Both nascent nitrogen and hydrogen are rapidly converted to their molecular forms. The process is controlled by metering the flow of ammonia gas and checking the exhaust gas for percentage of dissociation or cracking. Decreasing the flow of ammonia gas increases the percentage of dissociation. In the single-stage nitriding process, the dissociation is generally maintained at 30 per cent and the temperature is controlled at 975 F (525 C). In long nitriding cycles there is a tendency to build up a high nitrogen content on the surface, known as a "white layer." This compound layer may extend to a depth of 0.002 in. and is very brittle, thus requiring removal by finish grinding or lapping if chipping or spalling is to be avoided.

A two-stage nitriding process has been developed, known as the Floe

Process,[1] which makes it possible to minimize the white layer to a depth of from about 0.0002 to 0.0005 in. The first stage is generally accomplished at a temperature of 975 F (525 C) for a period of from 5 to 10 hr. After this time, the temperature is raised in the second stage to 1025 F (550 C) and the dissociation is increased to 80 per cent for the remainder of the cycle. Dissociated ammonia from an external cracking unit must be admitted to produce the high percentage of dissociation in the second stage. The second stage serves as a diffusion cycle permitting the high nitrogen surface layer to diffuse inward in a manner similar to the diffusion in gas-carburized cases.

Although gas nitriding is the most widely used production process of nitriding, salt baths have been developed that are useful for small lots of parts or for shallow cases. These baths are essentially alkali cyanides, which partially oxidize to form cyanates and then liberate active nitrogen. Anhydrous ammonia is also bubbled through salt baths in a modification of the process in which case nitriding probably takes place directly from the active nitrogen which is derived from the ammonia.

The variation of hardness produced by nitriding is shown in Fig. 11.13 for various time cycles. During the process of nitriding, the dimensions of a part will be found to increase somewhat. Therefore, allowance must be made for this growth in the original dimensions. The growth is of the order of 0.002 in. in diameter and 0.0005 in./in. in length for a solid round bar. Once the growth pattern has been established for a given part and the nitriding cycle, the growth will remain constant for subsequent lots, provided all processing variables are constant. Parts may be placed in service in the "as-nitrided" condition or may be finished to close dimensions by grinding, lapping, or honing. Since the nitriding temperature is relatively low, there is less chance for warpage to occur than in other hardening processes. However, distortion may be serious if the part has not been properly stress-relieved before nitriding. This can be very serious if the surface area of one side of a part is greater than the other and internal stresses are present at the time of nitriding. This condition will cause undue warping.

Areas on which nitriding is not desired may be plated with copper, bronze, or nickel. Proprietary paints containing tin oxide and a vehicle may also be applied. The tin oxide is reduced in the nitriding atmosphere to metallic tin which effectively prevents nitriding in selected areas.

11.14 Steels for Nitriding. The addition of such elements as aluminum, chromium, vanadium, titanium, tungsten, molybdenum, and manganese to steel greatly improves the nitriding characteristics. Most of the steels that are commonly used for nitriding contain combinations of aluminum, chromium, molybdenum, vanadium, and, in some instances, nickel. The carbon

[1] U. S. Patent No. 2,437,249.

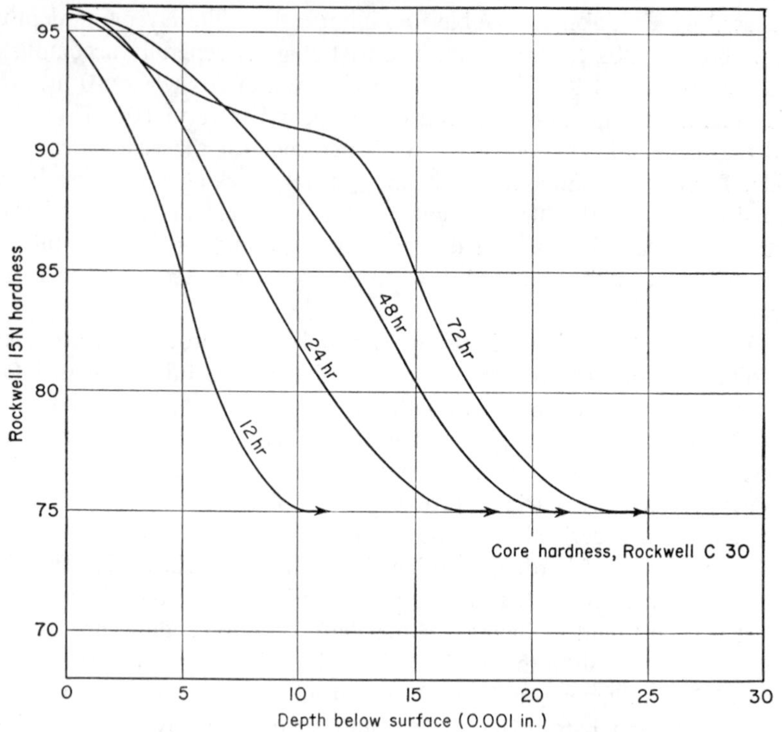

Fig. 11.13 Hardness vs. depth below surface of nitrided nitralloy 135 modified for various time cycles at 975 F and 30% dissociation.

content is somewhat higher than is employed for carburizing grades to provide support for the hard, brittle case. Steels in the SAE 4000 series respond well to nitriding, but do not develop the high surface hardness typical of special steels containing aluminum.

Steels of special composition which are used principally for the purpose of nitriding are referred to as nitralloys. The compositions of four of the

Table 11-III. COMPOSITION (%) OF NITRIDING STEELS.
(After *Metals Handbook*)

Steel	C	Cr	Mo
Nitralloy 135 (Type G)	0.30-0.40	0.90-1.40	0.15-0.25
Nitralloy 135 modified (Aircraft Spec.)	0.38-0.45	1.40-1.80	0.30-0.45
Nitralloy N ($3\frac{1}{2}$% Ni)	0.20-0.27	1.00-1.30	0.20-0.30
Nitralloy EZ (Type G with 0.15-0.25% Se)	0.30-0.40	1.00-1.50	0.15-0.25

Note: All these steels contain 0.85 to 1.20% Al; 0.20 to 0.40% Si and 0.40 to 0.70% Mn, except Nitralloy EZ, which contains 0.50 to 1.10% Mn.

nitriding steels are given in Table 11-III. Characteristic properties of the N135, N135 modified, and EZ steels are given in Table 11-IV. These steels have excellent properties and, therefore, are used not only for nitriding, but also for machine parts where high strength is required. The presence of nickel in nitriding steels tends to reduce the hardness of the case somewhat, but improves toughness and strength. Nickel also tends to impart greater hardness and strength to the core by the mechanism of precipitation hardening taking place during the nitriding cycle to produce a core hardness of Rockwell C 40-44. For improved machinability, the sulfur- or selenium-bearing steels may be employed. Hardenable stainless steels may be nitrided. The austenitic stainless steels nitride with difficulty to shallow case depths and low hardness, because of the presence of the impervious oxide film on the steel. Corrosion resistance of the stainless steels is greatly reduced by nitriding.

TABLE 11-IV. PROPERTIES OF HEAT-TREATED NITRALLOY 135 (G), 135 (MODIFIED), AND EZ.
(From *Metals Handbook*)

Steel	Tempering Temp. (°F)	Y.P. (1000 lb/in.²)	T.S. (1000 lb/in.²)	Elong. (%) in 2 in.	Red. of Area (%)	Brinell Hard. No.	Izod Impact (ft-lb)
Nitralloy 135[1]	1100	137.5	155	15.0	52.0	310	52
	1200	120	138	20.0	58.0	280	65
	1300	103	121	23.0	62.0	230	80
Nitralloy 135 (modified)[2]	1000	181.5	206	13.2	45.8	415	—
	1100	165	181	15.5	54.3	368	—
	1200	141	158.75	17.5	55.8	320	—
	1300	125	145	20.5	64.5	285	—
Nitralloy EZ[3]	1100	67.5[4]	139	16.0	43.6	293	—
	1200	90.0[4]	125.5	17.0	44.6	255	—

[1] Test specimens quenched in oil from 1750 F and then tempered at temperature indicated.

[2] Test specimens quenched in oil after ½ hr at 1700 F and then tempered 1 hr at temperatures indicated.

[3] Quenched in oil from 1700 F and then tempered at temperatures indicated.

[4] Proportional limit.

11.15 Application of Nitriding. In addition to high hardness and wear resistance, nitrided steels are more corrosion-resistant and have a higher endurance limit than most of the plain carbon, carburized steels. Such parts as gauges, bushings, forming dies, bearing parts, cylinder liners, dies, aircraft crankshafts, wrist pins, valve slides and sleeves, gears, thrust washers, etc., are surface-hardened by nitriding. The hardness and wear resistance of valve parts that are operated at elevated temperatures can be maintained by using nitrided steel.

The cost of the nitriding process is a limitation to its use. When extreme hardness is required which cannot be obtained by carburizing, then nitriding becomes feasible. In general, the time required to complete the process is much longer than with carburizing, requiring from 50 to 90 hr to produce a maximum case depth of from 0.020 to 0.030 in. Furthermore, nitriding requires close control by regulating ammonia flow and dissociation.

11.16 Surface Treatment of Steel. Coatings may be applied to the surface of iron and steel to improve corrosion, heat, or wear resistance; to rebuild worn or undersized parts; or merely to serve as an ornamental finish. Metallic coatings may be classified as (1) hot-dipped; (2) electroplated; (3) impregnated; (4) sprayed; (5) faced; or (6) clad. The principal nonmetallic coatings include: (1) oxide, (2) phosphate, and (3) vitreous enamel.

The application of a coating may serve economically to lengthen the useful life of a part manufactured from a low-cost material having surface characteristics unsuited for a given installation. It would be impossible within the scope of this text to discuss all of the surface treatments that have been applied to steel. The commercially accepted processes of greatest interest to the engineer will be covered.

11.17 Hot-Dipped Coatings. Iron and steel products may be dipped in molten baths of the relatively low melting point metals such as zinc, tin, lead, and aluminum. The metal surface must be thoroughly cleaned by pickling in an acid bath prior to dipping. A suitable flux is usually employed to assist in the formation of a good bond between the coating and the base metal.

Zinc coating, or *galvanizing,* provides the best atmospheric corrosion resistance of the commercial hot-dipped coating processes. Galvanizing serves in two ways as a popular method of protecting steel sheet, wire, and formed products. Protection is secured, first, by the formation of a thin oxide layer on the surface of the zinc which prevents rapid oxidation; second, the zinc is attacked in preference to the iron at exposed surfaces, such as sheared edges, scratches, seams, blemishes, etc. The subject of galvanic protection will be covered in detail in a subsequent chapter.

In the hot-dipping process, irregular objects are dipped by hand, whereas sheet and wire are coated in galvanizing machines with exit wipers to control the smoothness and thickness of the coating. The pattern of the zinc crystal "spangles" on the surface of sheet products may be controlled by air blast, sulfur dioxide, ammonium chloride, etc., while the coating is in the process of solidification.

The outer layer of a galvanized coating consists of commercially pure zinc with a secondary alloy layer of an iron-zinc intermetallic compound believed to be $FeZn_7$. A third region may sometimes exist, consisting of an iron-zinc compound of the type $FeZn_3$. The time and temperature of dipping control the character and thickness of the iron-zinc compound layer formed as a bonding agent. A long time in the molten zinc increases the amount of the hard, brittle compounds and causes subsequent difficulty in bending and forming of the final product. Special galvanizing treatments have been developed to promote diffusion and more effective bonding.

Galvanized sheet is specified on the basis of total surface area covering both sides of the sheet, the corrosion resistance being directly proportional to the thickness and tightness of the coating. Commercial coatings vary between 0.25 and 3.00 oz of zinc per ft^2 of surface, depending on the section size and degree of formability or protection required. Galvanizing is used for outdoor hardware, pipe, fence wire, nails, roofing, and a wide variety of sheet metal products.

Tin-coated sheet, or *tin plate,* fills the need for a nontoxic coating that is resistant to oxidation by food acids. The high cost of tin makes it uneconomical in competition with galvanizing for applications other than the packaging of food. Tin plate provides good corrosion resistance as long as the coating is intact; however, once the coating is ruptured, the exposed iron base is readily attacked by corroding agents.

Tin plate is usually sold by the *base box,* a term that is unique with the tin plate industry. A base box contains a unit of sheets consisting of 31,360 $in.^2$ of any gauge; for example, 112 sheets of a size 14 in. by 20 in. The thickness of tin coating is expressed in terms of pounds of tin per base box and ranges between one and 2.5 lb for standard "coke" plate. Heavier coatings, known as "charcoal" plates, are produced for special purposes with from 2 to 7 lb per base box.

In the manufacture of tin plate, cold-rolled low carbon steel sheets are passed through rollers in a bath of molten tin and wiped for a smooth surface finish. A thin layer of a tin-iron intermetallic compound, $FeSn_2$, forms the bonding agent, producing a ductile, adherent coating. The manufacture of tin plate requires rigid quality control of the steel through all stages of processing, together with critical surface inspection of the final coating.

Lead-tin coated sheet, known as *terneplate,* is produced in a manner similar to tin plate but with a less expensive alloy of 25 per cent tin, with the remainder lead. It is sold in units of a double base box coated with from 10 to 40 lb of alloy per double base box. Terneplate is ductile, is readily soldered, and provides an excellent base for paint. It remains bright and free from products of corrosion in contrast to galvanized sheet. The lubricating value of the high lead coating aids in metal-forming operations. Applications include roofing sheets, furniture, caskets, oil drums, gasoline tanks, and containers other than for food.

Lead-coated sheet and wire are used to a limited extent, the bond in this case being mechanical since lead does not alloy with iron. Small amounts of tin and antimony are generally added to the lead bath to improve bonding. In some applications, sheet containing a coating of about 20 lb of lead per base box is used as a substitute for terneplate and galvanized sheet.

Aluminum-coated sheet is prepared by a process known as *aluminizing,* which consists of hot-dipping properly prepared sheet in a molten bath of aluminum alloy. Aluminized sheet incorporates the distinctive advantages of aluminum combined with a higher strength steel base. These advantages include appearance, reflectivity, atmospheric corrosion resistance, galvanic protection, and heat resistance. Aluminized steel is a relatively new commercial product which has found use in the aircraft and automotive fields as an economical corrosion- and heat-resistant material for mufflers, tail pipes, heater jackets, etc.

11.18 Electroplated Coatings. Metallic coatings may be deposited on steel by electroplating for ornamentation, to provide corrosion resistance, to protect against wear, to rebuild worn parts, or for special purposes such as in selective carburizing. The elements that are commercially plated include copper, nickel, chromium, zinc, cadmium, and tin, as well as the precious metals, silver, gold, and platinum.

Copper is chiefly employed as an undercoat which can be economically buffed to a high degree of surface finish for subsequent nickel or chromium plating. Copper tarnishes readily in the atmosphere, particularly in the presence of sulfur-containing gases; thus, it requires a transparent lacquer coating to preserve any polished luster. Nickel possesses excellent corrosion resistance and ductility but is also tarnished by the atmosphere, although to a lesser degree than copper. It may be used to rebuild worn parts where maximum hardness is not needed. Bright chromium plating of the order of 0.00002 in. thick is used extensively for decorative purposes, because of its relative freedom from tarnishing. Corrosion protection is derived largely from impervious undercoats of copper or nickel. Hard chromium plating of the order of 0.001 to 0.010 in. thick, deposited with higher current density, is used for industrial applications where wear is involved in gauges, taps, drills, reamers, saws, shafts, rolls, molds, etc. Worn or otherwise undersized parts may be restored to size by depositing hard chromium up to about 0.050 in. where hardness is required. The application of porous chromium coatings for cylinders and piston rings is a specialized development for the retention of lubricating oil at wearing surfaces. All applications of hard chromium plate should be followed immediately by heat treatment for about one hour at 300 F (150 C) to prevent hydrogen embrittlement resulting from the electrolytic process.

Zinc-plated sheet and wire are produced on a continuous basis with freedom from the formation of intermetallic compounds and resultant brittleness found in the hot-dip galvanized product. Adherent coatings as thin as 0.0005 in. are possible by this method, although heavier coatings are recommended for maximum corrosion resistance.

Cadmium-plated steel resists atmospheric corrosion without the formation of the white deposits present with zinc coatings, particularly in marine atmospheres. Applications include nuts, bolts, screws, and many hardware items for household, aviation, and marine use.

Tin-plated sheet is produced electrolytically on a continuous basis with coatings as thin as 0.5 lb per base box as a means of conserving the strategic supply of tin.

11.19 Electroless Plated Coatings. The process of electrodeposition of metals has certain inherent disadvantages such as a tendency toward nonuniform coating thickness, excessive "build-up" on edges, gas porosity, and the difficulty of plating blind holes, internal recesses, etc., without conforming anodes which are often impractical. The development of electroless nickel coatings solves these shortcomings and is the answer to many special plating problems.

Electroless nickel consists of a deposition of an alloy of approximately 90 per cent nickel and 10 per cent phosphorus from a suitable chemical solution without the use of an external electropotential. The coating is an amorphous deposit of laminated layers parallel to the base metal with a microstructure typical of an undercooled liquid. Coatings as thin as 0.0002 in. are free of porosity, as determined by the salt spray test. The coating as deposited has an initial hardness of approximately 500 Vickers (DPH) or 50 Rockwell C which may be increased by a precipitation hardening treatment at a temperature of 750 F (400 C) for one hour to develop a hardness of 1000 Vickers (DPH) or 70 Rockwell C. This treatment brings about the precipitation of nickel phosphide (Ni_3P) and establishes a crystalline structure.

Dense coatings of uniform thickness are obtained by tank immersion or by pumping the plating solution through interior and remote recesses of large parts, such as tanks and piping. Coatings may be applied to steel, copper alloys, aluminum, beryllium, titanium, and many nonmetallic materials. In view of the dimensional uniformity of the coating, the process finds application on finish-machined parts and for rebuilding worn components where corrosion and wear resistance is involved. Applications include tank cars, piping, valves, pumps, heat exchangers, hydraulic components, piston rings, etc.

11.20 Impregnated Coatings. The art of impregnation, or penetration of elements into a metal by atomic diffusion processes, has been practiced for hundreds of years. Damascus steel was produced by the cementation process of diffusing carbon into iron. To be successful, an alloy system must possess a wide range of intersolubility or must produce intermetallic compounds of the two elements. The depth of penetration is a function of temperature and time at temperature, as in the case of carburizing and nitriding.

Direct-contact impregnation may be accomplished by heating steel to a high temperature with the cementation medium in the form of a solid, liquid, or gas. A common method consists of packing the steel part with a metal powder and carrier, such as chlorine or a volatile chloride. Complex alloy coatings may be produced by impregnation with more than one metal.

Indirect methods of impregnation involve the diffusion by heat treatment of coatings previously applied by hot dipping, electroplating, spraying, cladding, etc.

Many binary systems of iron have been studied to determine which of the elements might be applied as a coating for steel by the impregnation process; however, relatively few of the elements have gained commercial acceptance. Those systems having present or potential value are shown in Table 11-V.

TABLE 11-V. DIFFUSION COATINGS ON IRON AND STEEL.

COMMERCIAL PROCESSES		PROMISING SYSTEMS	
Element	Process	Element	Characteristics of Coating
Al	Calorizing	B	Hard; resists HCl
C	Carburizing	Be	Very hard; resists atmospheric corrosion
Cr	Chromizing		
N	Nitriding	Cd	Similar to galvanizing
Si	Siliconizing	Co	Improved corrosion resistance
Sn	Stannizing	Mn	Fair corrosion resistance
Zn	Sherardizing	Mo	Very hard; fair corrosion resistance
		Ni	Good corrosion resistance
		Sb	Resists HCl and dilute H_2SO_4
		Ta	Hard; resists dilute HCl and H_2SO_4
		Ti	Hard
		V	Very hard; resists HNO_3
		W	Very hard; resists H_2SO_4
		Zr	No information

One process of aluminum impregnation, or *calorizing,* consists of heating steel parts in an airtight box with a mixture of powdered aluminum, aluminum oxide, and aluminum chloride for a period of from 6 to 24 hr at a temperature between 1500 and 1800 F (815 and 980 C). Impregnation is designed to take place in the first 6 hr, followed by a homogenizing anneal to promote diffusion of the coating. The diffused layer usually extends from 0.025 to 0.040 in. in depth, with the surface containing about 25 per cent aluminum. Calorized steel resists the attack of flue gases containing sulfur that are found at high temperatures in oil refineries. Other uses include bolts, salt pots, furnace parts, etc., for service up to about 1500 F (815 C).

Chromium impregnation, or *chromizing,* may be accomplished by heating iron or low-carbon steel parts in a container with a mixture of powdered chromium and alumina in a hydrogen atmosphere for about 4 hr at a temperature in the range of 2375 to 2550 F (1300 to 1400 C). Gaseous methods have also been developed by the reaction of hydrogen and hydrogen chloride with chromium to form $CrCl_2$ as the active carrier. The process is relatively expensive and produces hard, brittle coatings varying from 0.0005 to 0.005 in. in depth with the surface containing about 10 or 30 per cent chromium. Chromizing has been employed to protect turbine buckets from corrosion or erosion and to provide protection for steel parts from atmospheric corrosion and attack by nitric acid.

Silicon impregnation, or *siliconizing,* consists of heating iron and low carbon steel parts in a closed container with a mixture of silicon carbide and chlorine for about 2 hr at a temperature within the range of 1700 to 1800 F (925 to 980 C) to produce a case depth of about 0.025 in., with the surface containing about 14 per cent silicon. Depths of between 0.005 and 0.100 in. may be obtained by the process. The surface layer is brittle, has good wear resistance, and cannot be machined by ordinary methods. Silicon-impregnated steels resist nitric, sulfuric, and hydrochloric acids and provide good scale resistance at temperatures up to about 1800 F (980 C). Typical uses include valve parts for the chemical industry, water-pump shafts, cylinder liners, valve guides, etc.

Zinc impregnation, or *sherardizing,* involves the heating of iron and steel parts in a rotating sealed drum in contact with zinc powder for a period of from 3 to 12 hr at a temperature of from 660 to 700 F (350 to 370 C). A coating about 0.002 in. thick is obtained in 3 hr at 700 F (370 C). Sherardized coatings are uniform in thickness and particularly adapted to the protection of bolts, nuts, washers, and miscellaneous small parts.

11.21 Sprayed, Faced, and Clad Coatings. The coating methods discussed in the preceding articles require parts that can be placed in a processing pot, tank, or retort. Large parts and heavy coatings are more economically prepared by special techniques such as metal spraying, facing, or cladding.

Metal spraying consists of melting metal wire or powder with an oxyhydrogen or oxyacetylene flame in a special metallizing gun which sprays the metal particles by an air blast. Molten particles impinging on the previously prepared surface to be coated are flattened and interlocked to provide mechanical bonding rather than alloying with the base metal. The process requires roughening of the surface by sand blasting, rough turning, or other techniques. A subsequent diffusion heat treatment may be required to obtain the best bonding conditions. In general, coatings of from 0.001 to 0.015 in. thick are applied for the corrosion protection of steel. Somewhat heavier coatings are used to rebuild worn or undersize parts. Almost any metal that is obtainable in wire or powder form may be employed, provided it can be melted in the oxyacetylene flame. Metal spraying is adapted to the coating of large structures, such as bridges, storage tanks, tank cars, etc., and the building up of worn pistons, shafts, bearings, etc.

Flame plating is a process wherein tungsten carbide particles are propelled from a special gun at supersonic velocity which is produced by the detonation of oxygen and acetylene. This action heats the particles so that they become plastic prior to becoming imbedded in the surface of the work. Coatings 0.002 to 0.010 in. thick may be applied, and the temperature of the workpiece seldom exceeds 400 F (210 C). The process finds application in increasing the wear resistance of such items as gauges.

Metal facing, or hard surfacing, is the operation of welding metal to the surface of a part in order to improve the abrasion, corrosion, or heat resistance of the surface, or to rebuild worn or eroded parts. Facing materials include the high-carbon and high-alloy steels, cobalt- and nickel-base alloys, and the carbides of tungsten, tantalum, and boron mounted in suitable alloy binders to increase their toughness. Hard facing is applied to valve seats, metal-working dies, oil-well drilling tools, dredge-bucket teeth, etc.

When a surface coating exceeds about 3 per cent of the total mass of the steel base, the resultant product is known as a *composite steel* and the process is called *cladding*. Cladding may be accomplished in several ways, including casting the clad metal around or adjacent to a previously formed steel ingot, rolling composite plates to produce bonding by pressure welding, or building up a thick surface layer by fusion welding in a manner similar to hard facing.

Examples of composite or clad steels include high-carbon, alloy, or high-speed steel bonded to mild steel for use in hacksaw blades, circular knives, broaches, blanking dies, jailbar, vault plate, etc. Stainless clad steels are produced with a 10 to 20 per cent thickness of stainless steel cladding on one or both sides of a mild steel backing for use in cooking vessels for the varnish, canning, and soap industries. The backing serves to increase the thermal conductivity of the stainless cladding and avoids localized hot spots. Nickel-clad and nickel-alloy clad steel plates are manufactured with coatings of from 5 to 25 per cent of the thickness for use in pressure vessels incorporating excellent corrosion resistance with high strength. Copper-clad steel wire is produced with high electrical conductivity on the surface, combined with a high-strength steel core.

11.22 Nonmetallic Coatings. Artificial black or blue-black oxide coatings may be produced on steel by heat alone or by contact with various solid, liquid, or gaseous oxidizing agents. Room-temperature oxidation with chemical oxidizers or by electrolytic treatment has the advantage of avoiding possible tempering of hardened steel parts, but it does not, in general, produce the lasting finish that may be obtained at higher temperatures. A prerequisite for successful blueing and blackening of steel is a "chemically clean" surface, free of oil and other contaminating agents. Oxide finishes are employed chiefly for appearance, with very limited protection against corrosion. Typical applications include gun parts, screws, nails, bicycle chain, blued sheet products, etc.

Phosphate coatings are applied to steel by dipping or spraying with a solution of phosphoric acid and manganese oxide to produce an insoluble crystalline coating of iron phosphate. One commercial process, known as *Parkerizing,* produces coatings of from 0.00015 to 0.003 in. thick. Phosphate coatings readily absorb oil, grease, wax, and dyes. They resist corrosion, aid in deep-drawing of sheet products, and reduce wear in such applications as piston rings, valve tappets, gears, camshafts, gun parts, etc.

Bonderizing is a process of applying a thin phosphate coating to serve as a base for paint on sheet products, such as refrigerator cabinets, typewriter frames, automobile bodies, etc.

Vitreous or porcelain enameling involves the application of fused silicate glasses to the surface of steel. A "frit" is prepared by melting refractories, fluxes, and opacifiers in a furnace and then pouring the charge into a tank of water. The resulting glass particles are ground in a ball mill with special chemicals, clay, pigments, water, etc., to form "slip." This material is usually applied to the metal surface by dipping, drying, and "firing" in an enameling furnace at a temperature of from 1500 to 1600 F (815 to 870 C) to produce a *ground coat.* This coating adheres to the base metal and prevents reaction between the metal and one or more *cover coats.* The cover coats are applied by spraying and are fired in a similar manner to provide service protection and decorative qualities. Cover coats may be white or may contain suitable inorganic pigments.

In contrast to the organic enamels, paints, varnishes, etc., vitreous enamels possess durability, attractiveness, hardness, and resistance to heat, abrasion, and chemical attack. In general, heat resistance is satisfactory up to 600 to 1000 F (315 to 540 C) with some special coatings formulated for short-period service up to about 1700 F (925 C). Thermal shock resistance is influenced by the difference in thermal expansion between the coating and the base metal. Since the coating is actually a glassy phase, it has very poor resistance to mechanical shock. Sheet-metal coatings range between 0.007 and 0.015 in. with heavier coatings being used on castings. To minimize chipping, thin coatings are preferred. Chipping and cracking of the coating cannot be economically repaired.

Vitreous enamel resists the action of nearly all chemicals except hydrofluoric acid and hot caustic solutions. In fact, its resistance to chemical attack makes it uneconomical to strip defective coatings for the salvage of steel parts for subsequent recoating.

Iron and steel enameling stock requires close control of composition. Low-carbon rimmed steel and ingot iron are generally used with manganese, phosphorus, and sulfur closely controlled. Gas-forming elements, such as carbon, tend to react with the enameling frit during firing and to produce brown spots, known as "copperheads," on the surface. Sheet-enameling stock also requires good surface finish, forming qualities, weldability, and resistance to sagging at the firing temperature.

The successful application of vitreous coatings to cast iron is complicated by the high carbon content and variations in foundry practice. Cast irons containing low combined carbon and high silicon are considered best for enameling.

Vitreous enameled products include bathtubs, refrigerators, dishwashers, washing machines, water heaters, stove bodies, cooking utensils, sign panels, etc.

REFERENCES

Bullens, Battelle, *Steel and Its Heat Treatment,* John Wiley & Sons, New York, 1948. (Vol. 2.)

Burns and Schuh, *Protective Coatings for Metals,* Reinhold Publishing Corp., New York, 1939.

Carburizing Symposium, American Society for Metals, Metals Park, Ohio, 1937.

Metals Handbook, American Society for Metals, Metals Park, Ohio, 1961.

Nitriding, American Society for Metals, Metals Park, Ohio, 1929.

Surface Treatment of Metals, American Society for Metals, Metals Park, Ohio, 1941.

Ziehlke, Dritt, and Mahoney, *Heat Treating Electroless Nickel Coatings,* Metal Progress, Metals Park, Ohio, 1960, pp. 84-87.

QUESTIONS

1. Define carburizing.
2. What three methods of carburizing are commonly employed?
3. Describe the process of pack carburizing.

4. Discuss briefly the mechanism by which the carbon content of the surface of a steel is increased in the carburizing process.
5. What substances are commonly used as solid carburizers? What is the function of an energizer in solid carburizing?
6. What is the mechanism involved in gas carburizing?
7. What materials are commonly used in liquid carburizing?
8. Upon what factors does the heat treatment of a carburized steel depend?
9. Describe in general terms the heat treatment to be given to a steel that has been carburized at a temperature just above the Ac_3 temperature, and in which the carbon content of the case does not exceed the eutectoid proportions.
10. Specify in general terms the heat treatment of a part that has been carburized, in which the case is of hypereutectoid composition and the steel has been carburized at a temperature above the A_{cm} temperature.
11. Indicate the heat treatment, in general terms, given to a steel in which the carbon content, while the steel is at carburizing temperature, is sufficient to produce a structure of austenite and cementite.
12. What is the advantage of the double quench over the single quench of carburized parts?
13. Under what circumstances is the cold treatment of carburized parts desirable?
14. Distinguish between normal and abnormal steels.
15. Describe the McQuaid-Ehn test. For what purpose is it employed?
16. To what may soft spots in a carburized steel be attributed?
17. What grades of steel are commonly employed for carburizing?
18. Give some examples of parts that are carburized.
19. Define cyaniding.
20. Explain the mechanism involved in the cyaniding of steel.
21. For what purposes is cyaniding employed?
22. What is the process of carbonitriding?
23. What advantage does carbonitriding have over carburizing?
24. What is the process of nitriding?
25. What are some of the advantages of nitriding in comparison with other casehardening processes?
26. How is the "white layer" formed in nitriding and what is its effect?
27. By what process may the "white layer" be minimized?
28. How is nitriding accomplished with salt baths?
29. What problems exist in nitriding with respect to dimensional stability?
30. What steels are commonly used for nitriding? In what way does the composition of these steels differ from others?
31. What is galvanizing?
32. Explain how galvanizing provides protection against corrosion.
33. In what way is the use of galvanized steel superior to tin-coated steel?
34. What is terneplate?
35. For what purposes is aluminized steel used?

36. What elements are commonly electroplated onto steel?
37. What is electroless plating?
38. What are the advantages of electroless plating?
39. What are some of the applications of electroless plating?
40. Give some examples of coatings that are produced by impregnation.
41. What is calorizing? For what purposes are parts that are calorized principally employed?
42. What is chromizing? For what purposes is it employed?
43. What is sherardizing? For what purpose is it employed?
44. What is siliconizing? What are the characteristics of a siliconized steel?
45. What is the process of metal spraying, and for what purposes may it be employed?
46. What is flame plating, and what are some of its applications?
47. How is metal facing accomplished, and for what purposes?
48. Give some examples of clad steel as to composition and uses.
49. What is the purpose of applying phosphate coatings to steel? What are two commercial names for processes of phosphate coating?
50. What are the advantages of vitreous enamels?

XII

TOOL STEELS

12.1 Classification. There is a very large group of steels whose application is intended especially for working and shaping metals. The uses for these materials include hand tools, such as chisels, punches, hammers, etc.; tools intended for cutting metals with machine tools; shears for cutting metals and other materials; dies for deep drawing, extrusion, forging, die casting, hot drawing, etc. From this partial list, it will be evident that the requirements of each steel in this group may be quite different.

The tool steels, as this group is called, are refined by the basic electric process. Every effort is made in their processing to produce what is known as "tool-steel quality." Although sold under trade names on the basis of performance rather than analysis, close metallurgical control has been employed to keep porosity, segregation, nonmetallic inclusions, and impurities, such as phosphorus and sulfur, at a minimum.

The large variety of alloys included in tool steels makes their classification difficult. Although several systems of classification have been employed, the most commonly accepted system in use is that established by the American Iron and Steel Institute. The AISI system of identification and type classification of tool steels is based on quenching method, application, special characteristics, and steels for particular industries. By this system, the commonly used tool steels are grouped into six major classifications with subgroups assigned a letter symbol. The classifications are as follows:

1. Water hardening . . . . W
2. Shock resisting . . . . S
3. Cold work:
   - Oil hardening . . . . O
   - Medium-alloy air hardening . . . . A
   - High-carbon, high-chromium . . . D

4. Hot work:
   - Chromium base . . . . H1 to H19 Incl.
   - Tungsten base . . . . H20 to H39 Incl.
   - Molybdenum base . . . . H40 to H59 Incl.
5. High-speed:
   - Tungsten base . . . . T
   - Molybdenum base . . . . M
6. Special-purpose:
   - Low-alloy L
   - Carbon-tungsten . . . . F
   - Mold steels
     - Low-carbon . . . . P1 to P19 Incl.
     - Other types . . . . P20 to P39 Incl.

The analyses of tool steels classified within this system are given in Table 12-I with typical applications for each class. Since tool steels are not sold on the basis of specified analysis, deviations from these typical compositions may exist in tool steels produced by different manufacturers under the same AISI-type designation.

Tools may be divided into three main classes: cutting tools, shearing tools, and forming tools. *Cutting tools* include those implements that remove chips from a material, as in machining operations. Steels for this service require resistance to wear, retention of hardness at excessive temperatures, and toughness to prevent breakage of the tool. *Shearing tools* are those implements that are used principally to part materials, either hot or cold, with a shearing action. Such tools include shears, punches, blanking dies, etc. They require resistance to wear, resistance to chipping or breakage, and resistance to loss of hardness by exposure to high temperature or by heat generated in the operation. In some instances, tools of this type must withstand high compressive stresses and, in others, must have a high endurance limit. Since, in many applications, shearing tools are of complex shape, the steels employed must be readily machinable. In complicated shearing tools, the steel must be hardenable without undue warpage. *Forming tools* are those implements that transmit their shape to the work. These tools may be used at high temperature, as in hot drawing or hot forging; or at ordinary temperatures, as in cold drawing or in cold forging. The impression may be transmitted to the work by sudden blows, as in hammer forging; or by slow pressure, as in press forging, extrusion forging, etc. In addition to the properties of the previous groups, forming tools must resist distortion in heat treatment and cracking caused by thermal shock and temperature gradients in service.

It is immediately recognized that the requirements for any tool of any

Table 12-I. CLASSIFICATION OF TOOL STEELS.
(After AISI—1955)

Type	Group	Typical Composition (%) C	Mn	Si	Cr	Ni	V	W	Mo	Co	% Total Alloy Content	Typical Application
W	Water hardening	0.60-1.40	—	—	0.25-0.50*	—	0.20-0.50*	—	—	—	0.45-1.00	General purpose.
S	Shock-resisting	0.50-0.55	0.80	1.00-2.00*	0.75-1.50*	—	—	1.00-2.50*	0.40-0.50*	—	3.95-7.30	Chisels, punches, shears, heading dies.
O	Cold work, oil hardening	0.90-1.20	1.00-1.60*	—	0.50-0.75*	—	—	0.50-1.75*	—	—	2.00-4.10	General purpose tools, dies, broaches.
A	Cold work, air hardening	0.70-1.00	2.00-3.00*	—	1.00-5.00	—	—	—	1.00	—	4.00-9.00	Less distortion than Type O.
D	Cold work, high-C, high Cr	1.00-2.35	—	1.00*	12.00	—	4.00*	1.00*	1.00*	3.00*	22.00	Rolls, mandrels, punches, dies, shear, high wear resistance.
H	Hot work, Cr base	0.35-0.55	—	—	5.00-7.00	—	0.40-1.00*	1.50-7.00*	1.50-5.00*	—	8.40-20.00	Gripper, bending, heading dies—light work to 600 F.
	W base	0.25-0.50	—	—	2.00-12.00	—	1.00*	9.00-18.00	—	—	12.00-30.00	Blanking, forming, extrusion, casting dies to 1100 F.
	Mo base	0.55-0.65	—	—	4.00	—	1.00-2.00	1.50-6.00*	5.00-8.00	—	11.50-16.00	Less expensive than above.
T	High-speed, W base	0.70-1.50	—	—	4.00-4.50	—	1.00-5.00	12.00-20.00	—	5.00-12.00*	22.00-41.50	Cutting tools all types, severe hot work.
M	High-speed, Mo base	0.80-1.50	—	—	4.00	—	1.00-5.00	1.50-6.50	3.50-8.75	5.00-12.00* (1.25 Cb*)	16.25-37.50	Less expensive than above.
L	Special-purpose, low alloy	0.50-1.10	0.35-1.00*	—	0.75-1.50	1.50*	0.20-0.25*	—	0.25-0.40*	—	3.05-4.65	Similar to Type O, higher Cr content.
F	Special-purpose, C-W	1.00-1.25	—	—	0.75*	—	—	1.25-3.50	—	—	2.00-4.25	Tungsten finishing steels.
P	Special-purpose, mold steels											
	Low-C types	0.07-0.10	—	—	0.60-5.00*	0.50-3.50*	—	—	0.20*	—	1.30-8.70	Carburized hubbed molds.
	Other types	0.30	—	—	0.75	—	—	—	0.25	—	1.00	Carburized cut molds.

* May not contain any.

class are not unique with that particular class. In general, the following factors must be considered in the selection of a steel for a tool: hardness, wear resistance, toughness, distortion or warping in hardening, depth of hardening, resistance to softening at elevated temperatures, machinability, and cost. Some of the characteristics of the steels listed in Table 12-I are listed in Table 12-II. While no information is presented concerning the

TABLE 12-II. CHARACTERISTICS OF TOOL STEELS.
(After AISI-1955)

Type	Group	Quench Medium	Wear Resist.	Toughness	Warpage	Hard. Depth	Resistance to Softening at High Temp.
W	Water hardening	Water or brine	Fair to good	Good	Poor	Shallow	Poor
S	Shock resisting	Oil or water	Fair	Best	O: Fair W: Poor	Medium	Fair
O	Cold work oil hardening	Oil	Good	Fair	Very good	Medium	Poor
A	Cold work air hardening	Air	Good	Fair	Best	Deep	Fair
D	Cold work high C-high Cr	Air or oil	Best	Poor	A: Best O: Very good	Deep	Good
H	Hot work, Cr base	Air or oil	Fair	Good	A: Good O: Fair	Deep	Good
	W base	Air or oil	Fair to good	Good	A: Good O: Fair	Deep	Very good
	Mo base	Oil or air or salt	Good	Poor	O: Fair A: Good S: Good	Deep	Very good
T	High-speed, W base	Oil or air or salt	Very good	Poor	Good	Deep	Best
M	High-speed, Mo base	Oil or air or salt	Very good	Poor	Good	Deep	Very good
L	Special-purpose low alloy	Oil or water	Fair to good	Fair to good	O: Fair W: Poor	Medium	Poor
F	Special-purpose, C-W	Water or brine	Very good	Poor	Poor	Shallow	Poor
P	Special-purpose, mold	Air or oil or water	Good*	Good	Good	Shallow	Poor

* After carburizing.

relative cost of the different types of tool steels, the cost is usually closely related to the total alloy content.

In this chapter, each class of tool steel will be discussed in detail with reference to individual characteristics and applications. The heat treatment of tool steel does not involve any new principle, the principles given in Chapter VIII being the same for all types of steel. Although the principles are the same, the practice, however, is slightly modified with respect to the heat-treating temperature and time at temperature. This is to be expected, since most of the tool steels contain an appreciable amount of alloying elements.

12.2 Water Hardening Tool Steel. The water hardening tool steels are usually of the plain carbon type. However, a small amount of alloy may be incorporated, as indicated in Table 12-I. Since plain carbon steel is the cheapest material for tools, it has been used for many purposes. It is true that these steels cannot withstand the severe conditions to which alloy tool steels can be subjected, but they are quite suitable for certain applications. Whenever possible, therefore, they should be used because of their low cost, good machinability, and high hardness.

The carbon content of a tool steel depends mostly upon the use to which the tool is to be put. For shear blades, rock drills, striking dies, hammers, etc., the carbon content varies between about 0.60 and 0.85 per cent. Chisels, dies, etc., are made from carbon steels containing between 0.80 and 0.95 per cent carbon. Steels used for trimming dies, trimming and planer tools, cutters, drills, milling cutters, large taps, small shear blades, and small cold chisels contain between 0.95 and 1.1 per cent carbon. For wood-cutting tools, small taps, drills, some turning tools and razors, the steel may contain between 1.1 and 1.40 per cent carbon.

In Chapter IX, it was shown that by increasing the carbon content above about 0.5 per cent, the as-quenched hardness of small sections does not increase. The advantage of a higher carbon content lies in greater abrasion resistance because of the presence of excess cementite. The hypereutectoid steels are not quite as tough as those of lower carbon; therefore, the latter are preferred if wear resistance is not of major concern. The tensile properties of tool steels do not necessarily reflect their performance in tools. The properties of a 0.9 per cent carbon steel and a 1.10 per cent carbon steel, quenched and tempered to different temperatures, are shown in Figs.

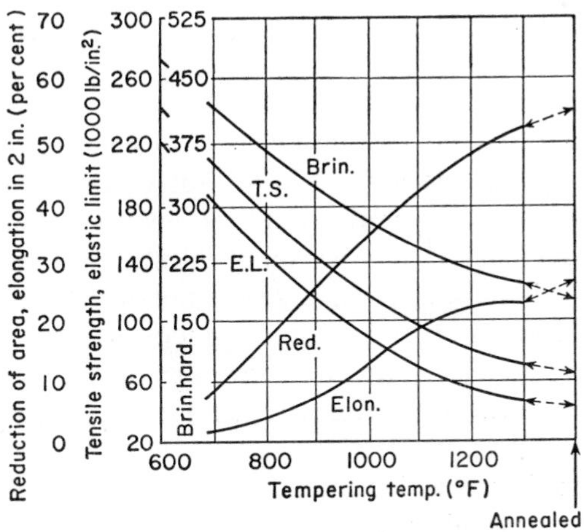

FIG. 12.1 Mechanical properties of carbon tool steel (approx. 0.90% C) vs. tempering temperatures. (Gill, *Tool Steels,* 1934)

12.1 and 12.2, respectively. The most suitable test of performance of any tool steel is its behavior under operating conditions.

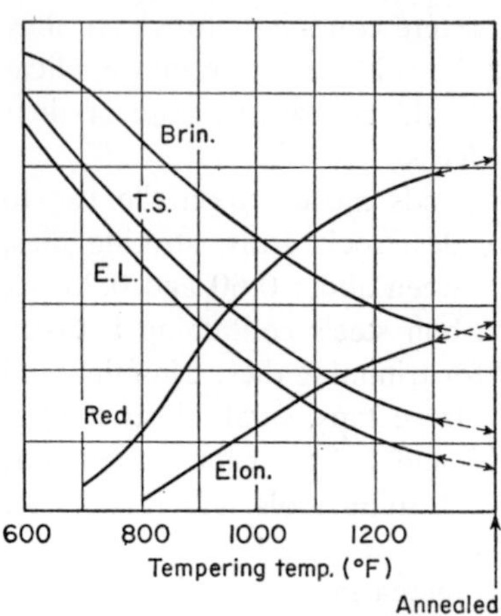

Fig. 12.2 Mechanical properties of carbon tool steel (approx. 1.10% C) vs. tempering temperature. (Gill, *Tool Steels,* 1934)

In order to obtain the hardness required in tools, plain carbon steels must be rapidly quenched in water or brine. Considerable distortion of the tool may occur in hardening because of the change of dimensions resulting from the change in structure and change of temperature.

The plain carbon tool steels can be machined readily if the cementite is in the form of small spheroids. Spheroidization may be attained by cooling the steel rapidly and heating to just below the critical range for an appreciable length of time.

The hardening of plain carbon tool steel is accomplished by slowly heating the piece to above the upper critical temperature, followed by quenching. Steels containing between 0.85 and 0.95 per cent carbon are heated to a temperature in the range 1450-1500 F (790-815 C), whereas steels containing more than 1.10 per cent carbon are heated to a temperature between 1400 and 1475 F (760 and 800 C). These temperatures are approximations and depend upon the analysis of the steel at hand. Because high-carbon steels are susceptible to serious decarburization of their surface when heated under oxidizing conditions in an open furnace, finished tools should be heated in a neutral protective atmosphere or else packed in a box with cast iron chips. Salt baths provide an excellent means of rapidly heating such work with little decarburization. The piece is held at the quenching temperature until it is heated uniformly throughout, with about an additional 15 min per in. of thickness for soaking. The steel is usually quenched in brine or water. The tempering range of 300 to 650 F (150 to 345 C) depends upon the quenching temperature employed and the use for which the tool is intended. Tempering of the carbon tool steels provides relief of quenching stresses rather than appreciable reduction in hardness.

Maximum toughness, as indicated by the torsion impact test, is shown by Fig. 12.3 to occur within a narrow tempering range which shifts to higher tempering temperatures with correspondingly higher quenching temperatures. The toughness decreases noticeably when the steel is tempered at a temperature above this narrow range, but, by tempering at higher temperatures, the toughness again increases. This variation of toughness is probably associated with two factors, namely, internal strain and retained austenite. When the steel is quenched, high internal strains are produced which tend to be relieved when tempering is applied, thus giving rise to increased toughness. However, with an increase of the tempering temperature, retained austenite transforms to a harder, less tough structure which may also introduce further internal strain; hence, the toughness decreases. Further increase of the tempering temperature relieves this condition, and the toughness improves. If very high hardness is required with a maximum toughness, one

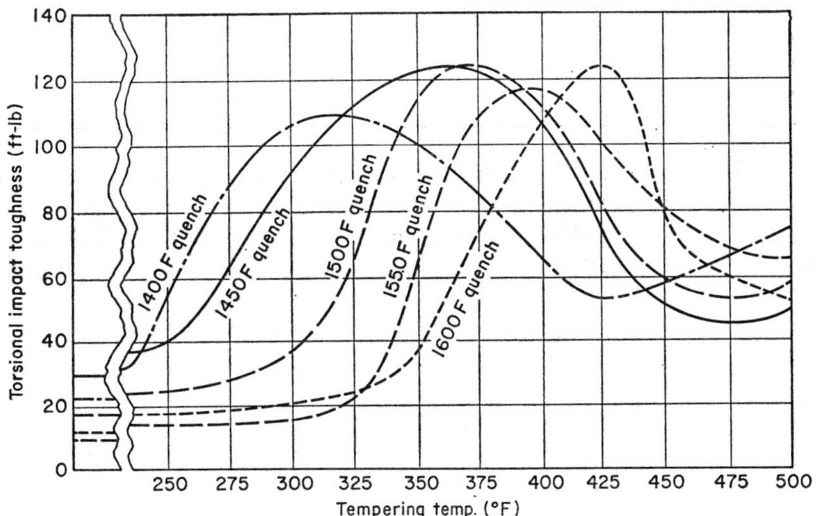

FIG. 12.3 Toughness (impact) of carbon tool steel (approx. 1.05% C) vs. tempering temperature for different quenching temperatures. (Palmer and Luerssen, *Tool Steel Simplified,* Carpenter Steel Co., 1948)

must be careful to temper in the proper temperature range for the quenching temperature employed for the particular steel.

Vanadium is used on occasion to obtain better control of grain growth. With the addition of vanadium, the tendency for grain growth is decreased, and also the steel is shallower-hardening. However, the depth of hardening can be controlled by carefully selecting the quenching temperature.

Some tools require greater hardness and depth of hardening than can be obtained with plain carbon tool steel. These properties can be obtained by using chromium alloy steels. Although the class under consideration contains relatively small amounts of alloy elements, the improved properties justify their use in place of plain carbon tool steels when the requirements are more severe. The increased depth of hardening and the elimination of soft spots obtained with chromium is particularly valuable in tool steels. The analyses of Type-W steels are given in Table 12-III.

TABLE 12-III. ANALYSIS (%) OF WATER HARDENING TOOL STEELS. (After AISI-1955)

AISI Type	C	Cr	V	Remarks
W1	0.60-1.40	—	—	Plain carbon tool steel.
W2	0.60-1.40	—	0.25	V controls grain growth.
W3	1.00	—	0.50	
W4	0.60-1.40	0.25	—	Cr corrects tendency toward soft spots.
W5	1.10	0.50	—	
W6	1.00	0.25	0.25	Combination of above.
W7	1.00	0.50	0.20	

12.3 Shock-Resisting Tool Steel. The steels listed in Table 12-IV have been used successfully to resist shock in tools such as chisels and punches. Types S1 and S3, known as tungsten chisel and punch steels, combine toughness resulting from their relatively low carbon content with the high hardness and wear resistance obtained by the use of tungsten and chromium. Type S1 is oil-quenched from a temperature of between 1650 and 1800 F (900 and 980 C) and tempered at a temperature in the range of 400 to 1200 F (205 to 650 C) to produce a hardness of Rockwell C 58-40. Type S3 may be quenched in brine or water from a temperature between 1500 and 1600 F (815 and 870 C) and tempered at a temperature in the range of 300 to 600 F (150 to 315 C) for a hardness of Rockwell C 59-50.

TABLE 12-IV. ANALYSIS (%) OF SHOCK-RESISTING TOOL STEELS. (After AISI-1955)

AISI Type	C	Mn	Si	Cr	W	Mo	Remarks
S1	0.50	—	—	1.50	2.50	—	W chisel and punch steel.
S2	0.50	—	1.00	—	—	0.50	Low Si for increased toughness.
S3	0.50	—	—	0.75	1.00	—	Less wear resistance than S1.
S4	0.55	0.80	2.00	—	—	—	Si-Mn punch and chisel steel, susceptible to decarburization.
S5	0.55	0.80	2.00	—	—	0.40	Mo increases hardenability.

Steels of the silicon-manganese group, Types S2, S4, and S5, are relatively inexpensive and owe their wide use for chisels, punches, and shear blades to high hardness, toughness, and satisfactory wear resistance. Originally, the silicon-manganese steels were used for springs, but were found to possess properties that made them more suitable for tools. Typical of steels containing high silicon content, they possess very poor resistance to decarburization. With lower silicon and manganese, the toughness is somewhat greater. The addition of chromium or molybdenum provides deeper-hardening characteristics. Type S2 is brine- or water-quenched from a temperature in the range of 1550-1650 F (845-900 C). Type S4 is oil- or water-quenched from a temperature in the range of 1600-1750 F (870-955 C). Type S5 is oil-quenched from a temperature in the range of 1600-1700 F (870-925 C). These grades of tool steel develop hardnesses of Rockwell C 60-50 when tempered at temperatures between 350 and 800 F (175 and 425 C).

12.4 Cold Work Tool Steels. Parts that must be finish-machined to close tolerance before heat treatment and those tolerances maintained after heat-treating, require the use of the "nondeforming" variety of steels. The term "nondeforming" is misleading since all steels exhibit a change of dimensions in varying amounts during the heat-treating operations. The hazards of distortion and cracking common to the water hardening steels are minimized in the steels classified as Cold Work Tool Steels. Although these steels are primarily intended for cold work applications, certain of the higher-alloyed types possessing good resistance to softening at elevated temperatures are employed where shock is not a factor.

Cold work tool steels are of three principal types: (1) oil hardening, Type O, (2) medium-alloy air hardening, Type A, and (3) high-carbon, high-chromium, Type D. The Type O manganese and tungsten oil hardening steels are indicated in Table 12-V. Types O1 and O2 are oil-quenched from

TABLE 12-V. ANALYSIS (%) OF COLD WORK TOOL STEELS OIL HARDENING TYPES.
(After AISI-1955)

AISI Type	C	Mn	Cr	W	Remarks
O1	0.90	1.00	0.50	0.50	Cr and W controls growth.
O2	0.90	1.60	—	—	Subject to grain growth.
O7	1.20	—	0.75	1.75	Best wear resistance.

a temperature in the range 1400-1500 F (760-815 C) and tempered in the range of 300-500 F (150-260 C) to develop a hardness of Rockwell C 62-57. Type O7 may be oil-quenched from a temperature in the range 1560-1625 F (845-885 C) or water-quenched from a temperature between 1450 and 1525 F (790 and 830 C) and tempered at a temperature between

325 and 550 F (165 and 290 C) to a hardness of Rockwell C 64-58. The increased hardness gained by water quenching is offset by the greater warpage. Steels of this type are used for a wide variety of purposes, including dies for molding plastics which usually cannot be ground after forming because of complicated designs, stamping and trimming dies, punches, thread-rolling dies, some taps and broaches.

The Type A chromium and manganese air hardening steels are listed in Table 12-VI. Type A2 is air-hardened from a temperature between 1700

TABLE 12-VI. ANALYSIS (%) OF COLD WORK TOOL STEELS MEDIUM-ALLOY, AIR-HARDENING TYPES.
(After AISI-1955)

AISI Type	C	Mn	Cr	Mo	Remarks
A2	1.00	—	5.00	1.00	5% Cr die steel.
A4	1.00	2.00	1.00	1.00	Less distortion with addition of Mn.
A5	1.00	3.00	1.00	1.00	
A6	0.70	2.00	1.00	1.00	Good shock resistance.

and 1800 F (925 and 980 C) and tempered in the range of 350-100 F (175-540 C) to produce a hardness in the range of Rockwell C 62-57. Types A4, A5, and A6 are air-hardened from a temperature in the range 1450-1600 F (790-870 C) and tempered at a temperature between 300 and 800 F (150 and 425 C) to obtain a hardness in the range of Rockwell C 60-54. These steels have wear resistance that is intermediate between the Type O and Type D cold work steels being used in similar applications where improved wear resistance and nondeforming properties are required.

Type D high-carbon, high-chromium tool steels may be divided into two groups, as indicated in Table 12-VII, namely, the oil-hardening and the air-hardening types. These steels show very little change in dimensions with heat treatment and harden quite deeply. They are very difficult to machine and have extreme wear resistance. When steels of this type are tempered, the hardness decreases very little until a temperature of approximately 800 F (425 C) is reached. The hardness of certain of these steels may be much lower immediately after quenching from a high temperature, but, when tempered at a temperature of 1000 F (450 C), the hardness is increased markedly as a result of elimination of retained austenite and the occurrence of secondary hardening, as previously discussed in Chapter IX. Grain coarsening is usually not noticeable at temperatures below about 1900 F (1040 C).

Types D1, D2, D4, and D5 are air-hardened from a temperature between 1775 and 1875 F (970 and 1025 C) and tempered in the range 400-

TABLE 12-VII. ANALYSIS (%) OF COLD WORK TOOL STEELS, HIGH CARBON-HIGH CHROMIUM TYPES.
(After AISI-1955)

AISI Type	C	Si	Cr	V	W	Mo	Co	Remarks
D1	1.00	—	12.00	—	—	1.00	—	Air hardening. Toughness and machinability.
D2	1.50	—	12.00	—	—	1.00	—	Air hardening.
D3	2.25	—	12.00	—	—	—	—	Oil hardening. Wear resistance.
D4	2.25	—	12.00	—	—	1.00	—	Air hardening. Wear resistance.
D5	1.50	—	12.00	—	—	1.00	3.00	Air hardening.
D6	2.25	1.00	12.00	—	1.00	—	—	Oil hardening. Wear resistance.
D7	2.35	—	12.00	4.00	—	1.00	—	Air hardening. Highest wear resistance.

1000 F (205-450 C), with a resultant hardness of Rockwell C 61-54. Type D7 is air-hardened from a temperature in the range 1850-1950 F (1010-1065 C) and tempered in the range 300-1000 F (150-450 C) to produce hardness in the range of Rockwell C 65-58. Types D3 and D6 are oil-quenched from a temperature in the range 1700-1800 F (925-980 C) and tempered in the range 400-1000 F (205-450 C), yielding a hardness range of Rockwell C 61-54.

The range of applications of this type of steel is extremely wide. It has been found satisfactory for use as drawing dies, blanking dies, forming dies, bushings, coining dies, thread-rolling dies, trimming dies, shear blades, punches, lathe centers, cold-forming rolls, cutting tools for brass and bronze, and hand-threading dies.

12.5 Hot Work Tool Steels. Steels that are used for the forming of materials while hot must be resistant to deformation at the processing temperatures. Furthermore, the dies for hot working must be resistant to erosion, wear, cracking, and heat-checking under conditions of severe thermal shock.

Hot work tool steels are of three principal types: (1) chromium base, Type H11-H16, (2) tungsten base, Type H20-H26, and molybdenum base, Type H41-H43, as indicated in Table 12-VIII. The chromium base, hot work die steels find considerable use in application where hot-forming is to be done in the absence of shock and under conditions in which the temperature of the die does not exceed approximately 600 F (315 C). Uses that come within these requirements include stripper dies, hot-heading dies, riveting dies, and bending dies. Because the major softening of these steels

TABLE 12-VIII. ANALYSIS (%) OF HOT WORK TOOL STEELS.
(After AISI-1955)

AISI Type	C	Cr	V	W	Mo	Remarks
			Chromium-Base Types			
H11	0.35	5.00	0.40	—	1.50	Serviceable to 600 F (315 C).
H12	0.35	5.00	0.40	1.50	1.50	Air hardening.
H13	0.35	5.00	1.00	—	1.50	
H14	0.40	5.00	—	5.00	—	Air hardening.
H15	0.40	5.00	—	—	5.00	Air or oil hardening.
H16	0.55	7.00	—	7.00	—	
			Tungsten-Base Types			
H20	0.35	2.00	—	9.00	—	Serviceable up to 1000 F. (540 C).
H21	0.35	3.50	—	9.00	—	Cr lowered to increase toughness.
H22	0.35	2.00	—	11.00	—	
H23	0.30	12.00	—	12.00	—	Increased W raises serviceability to
H24	0.45	3.00	—	15.00	—	1100 F (595 C); hardenability and
H25	0.25	4.00	—	15.00	—	brittleness rises with C content.
H26	0.50	4.00	1.00	18.00	—	Low-carbon, high-speed steel.
			Molybdenum-Base Types			
H41	0.65	4.00	1.00	1.50	8.00	
H42	0.60	4.00	2.00	6.00	5.00	Low-carbon, high-speed steels.
H43	0.55	4.00	2.00	—	8.00	

on tempering occurs at a temperature of about 600 F (315 C), there is a limitation on sustained operation at higher temperatures. The steels containing higher tungsten and molybdenum possess improved elevated temperature properties and red-hardness.

The tungsten base, hot work die steels find application for hot-forming operations at temperatures of about 1000 F (540 C), made possible by the red-hardness characteristics imparted by tungsten. Increased carbon content tends toward deeper hardening whereas high chromium content promotes brittleness. Modifications of the basic tungsten-chromium compositions containing up to 1.00 per cent silicon have been extensively used for dies in die-casting aluminum-base alloys. The addition of silicon tends to improve resistance to erosion and to the flow of molten metal. The tungsten base hot work steels are used for dies in forging brass, die-casting dies, permanent molds for casting brass and bronze, dies for extruding brass, and hot-forming and blanking dies.

Low-carbon variations of high-speed tool steels extend the red-hardness range of the hot work die steels, as in the case of type H26 and the less expensive molybdenum base types, H41-H43. The heat treatment of the hot work tool steels is summarized in Table 12-IX.

TABLE 12-IX. HEAT TREATMENT OF HOT WORK TOOL STEELS.

Type	Preheat Temp. F	Preheat Temp. C	Quenching Temp. F	Quenching Temp. C	Quenching Medium	Tempering Range F	Tempering Range C	Hardness Rockwell C
H11, H12	1500	815	1825-1875	995-1025	Air	1000-1200	540-650	54-38
H13	1500	815	1825-1900	995-1040	Air	1000-1200	540-650	53-38
H14	1500	815	1850-1950	1010-1065	Air	1100-1200	595-650	47-40
H15	1500	815	2100-2300	1150-1260	Air or oil	1100-1200	595-650	49-36
H16	1500	815	2050-2150	1120-1175	Air or oil	1050-1250	565-675	60-45
H20	1500	815	2000-2200	1095-1260	Air or oil	1100-1250	595-675	54-36
H21	1500	815	2000-2200	1095-1260	Air or oil	1100-1250	595-675	54-36
H22	1500	815	2000-2200	1095-1260	Air or oil	1100-1250	595-675	52-39
H23	1550	845	2200-2325	1205-1275	Air or oil	1200-1500	650-815	47-30
H24	1500	815	2000-2250	1095-1230	Air or oil	1050-1200	565-650	55-45
H25	1500	815	2100-2300	1150-1260	Air or oil	1050-1250	565-675	44-35
H26	1600	870	2150-2300	1175-1260	Molten salt or air or oil	1050-1250	565-675	58-43
H41	1450	785	2000-2175	1095-1190	Molten salt or air or oil	1050-1200	565-650	60-50
H42	1450	785	2050-2225	1120-1220	Molten salt or air or oil	1050-1200	565-650	60-50
H43	1450	785	2000-2175	1095-1190	Molten salt or air or oil	1050-1200	565-650	58-45

12.6 High-Speed Steel. Steels that maintain their hardness at elevated temperatures, that are deep-hardening, and that are quite resistant to wear contain relatively large amounts of tungsten or molybdenum, together with chromium, cobalt, or vanadium. These steels are referred to as the high-speed steels.

The high-speed steels can be divided into two types: tungsten, Type T, and molybdenum, Type M, high-speed steels. Typical analyses of each of these groups are given in Tables 12-X and 12-XI, respectively.

TABLE 12-X. ANALYSIS (%) OF HIGH-SPEED TOOL STEELS, TUNGSTEN-BASE TYPES.
(After AISI-1955)

AISI Type	C	Cr	V	W	Co	Remarks
T1	0.70	4.00	1.00	18.00	—	Popular, general-purpose grade.
T2	0.80	4.00	2.00	18.00	—	Cutting properties increase with carbon content. (T2, T3)
T3	1.05	4.00	3.00	18.00	—	
T4	0.75	4.00	1.00	18.00	5.00	For cutting hard, gritty, or tough materials. Cutting ability varies as total alloy content. (T4, T5, T6)
T5	0.80	4.00	2.00	18.00	8.00	
T6	0.80	4.50	1.50	20.00	12.00	
T7	0.75	4.00	2.00	14.00	—	Roughing cuts; somewhat erratic in hardening.
T8	0.75	4.00	2.00	14.00	5.00	
T9	1.20	4.00	4.00	18.00	—	
T15	1.50	4.00	5.00	12.00	5.00	

TABLE 12-XI. ANALYSIS (%) OF HIGH-SPEED TOOL STEELS, MOLYBDENUM BASE TYPES.
(After AISI-1955)

AISI Type	C	Cr	V	W	Mo	Co	Cb	Remarks
M1	0.80	4.00	1.00	1.50	8.00	—	—	Popular grade.
M2	0.80	4.00	2.00	6.00	5.00	—	—	Most popular grade
M3	1.00	4.00	2.70	6.00	5.00	—	—	
M4	1.30	4.00	4.00	5.50	4.50	—	—	
M6	0.80	4.00	1.50	4.00	5.00	12.00	—	
M7	1.00	4.00	2.00	1.75	8.75	—	—	
M8	0.80	4.00	1.50	5.00	5.00	—	1.25	
M10	0.85	4.00	2.00	—	8.00	—	—	Popular grade.
M15	1.50	4.00	5.00	6.50	3.50	5.00	—	
M30	0.80	4.00	1.25	2.00	8.00	5.00	—	
M34	0.90	4.00	2.00	2.00	8.00	8.00	—	
M35	0.80	4.00	2.00	6.00	5.00	5.00	—	
M36	0.80	4.00	2.00	6.00	5.00	8.00	—	

These steels are not as tough as some others, but the retention of hardness at elevated temperatures makes them particularly useful for high-speed cutting tools of all types, blanking dies, hot-forming dies, lathe centers, wearing plates, etc. The hardness and toughness of these steels are controlled principally by the carbon content. Those steels containing the lower range of carbon are tougher but less hard, while those with higher carbon content are less tough but considerably harder and more wear-resistant. The best combination of toughness, wear resistance, and cutting ability is to be found in the tungsten high-speed steel with 18 per cent tungsten. With larger amounts of tungsten, the toughness decreases.

Chromium is one of the principal elements that controls hardenability in these steels. The optimum amount is about 4 per cent chromium. With less than 4 per cent, the hardenability, wear resistance, and cutting ability are decreased, and the steel is more brittle.

Vanadium seems to increase the cutting ability of high-speed steel. A typical steel of this type is the 18-4-2 (18 per cent tungsten, 4 per cent chromium, 2 per cent vanadium).

The tungsten high-speed steels are hardened by heating to a temperature between 2250 and 2400 F (1230 and 1315 C) followed by cooling in oil, air, or molten salt. The steel is heated slowly and preheated at temperatures of the order of 1500 and 1600 F (815 and 870 C) to prevent undue oxidation and then rapidly raised to the quenching temperature where it is held for only a matter of minutes. The greatest hardness is developed in these alloys by tempering to a temperature of about 1050 F (565 C) at which secondary hardening occurs. The steel is held at the tempering temperature for from one to two hours after the piece has reached the tempering temperature. This is usually followed by a second and sometimes a third tempering at the same temperature to obtain hardnesses of Rockwell C 60-67.

The replacement of tungsten with molybdenum has produced a high-speed steel at much lower cost. The molybdenum high-speed steels contain 1.5 to 6 per cent tungsten, 4 to 9 per cent molybdenum, and between 3.5 and 4.5 per cent chromium. These steels perform as well as the conventional 18-4-1 type tungsten steels; however, they are somewhat more difficult to heat-treat. They are more susceptible to oxidation at the quenching temperatures and, therefore, must be coated with a substance, such as borax; otherwise heating must be in a fused salt bath or a protective atmosphere. These steels are also more susceptible to grain growth than the tungsten type.

The hardening of the molybdenum high-speed steels is accomplished by heating to temperatures of the order of 2150 to 2300 F (1175 to 1260 C), which are somewhat lower than the temperatures prescribed for the tungsten high-speed steels. The tendency toward decarburization is marked in the molybdenum high-speed steels.

The addition of cobalt to the high-speed steels permits cutting hard, gritty, or scaly materials, since cutting ability increases with the cobalt content. However, cobalt also reduces hardenability, and consequently the carbon content must be increased above that normally used in the tungsten high-speed steels. High-speed steels containing cobalt are somewhat more susceptible to cracking in quenching than the tungsten high-speed steels. The hardening temperatures for these high-speed steels are about 50 F (10 C) higher than those used in the tungsten type, namely, between 2325 and 2400 F (1275 and 1310 C). These steels are also subject to decarburization if held for too long at the preheating or quenching temperature.

12.7 Special-Purpose Tool Steels. The special-purpose tool steels included in the AISI system are: (1) low-alloy, Type L, (2) carbon-tungsten, Type F, and (3) mold steels, Type P, as indicated in Table 12-XII. The Type L steels are chromium cutting tool steels used interchangeably with the Type O, oil-hardening cold work steels, where increased wear resistance is required. The heat treatment of these steels is given in Table 12-XIII.

Type F, or tungsten finishing-steels, are particularly useful for finishing-tools because of their very high hardness and retention of a keen cutting edge. Type F1 is somewhat erratic in response to heat treatment, whereas Types F2 and F3 possess low toughness and a tendency to warp and crack. These steels are water- or brine-quenched from a temperature between 1450 and 1600 F (790 and 870 C) and tempered in the range 300-500 F (150-260 C) to develop a hardness of Rockwell C 66-60.

The Type P, mold steels or irons, are used for molds in molding plastics,

TABLE 12-XII. ANALYSIS (%) OF SPECIAL-PURPOSE TOOL STEELS. (After AISI-1955)

AISI Type	C	Mn	Cr	Ni	V	W	Mo	Remarks
				Low-Alloy Types				
L1	1.00	—	1.25	—	—	—	—	Oil-hardening cutting tool steels.
L2	0.50-1.10	—	1.00	—	0.20	—	—	
L3	1.00	—	1.50	—	0.20	—	—	
L4	1.00	0.60	1.50	—	0.25	—	—	
L5	1.00	1.00	1.00	—	—	—	0.25	Mo increases hardenability.
L6	0.70	—	0.75	1.50	—	—	0.25	High toughness and shock resistance.
L7	1.00	0.35	1.40	—	—	—	0.40	
				Carbon-Tungsten Types				
F1	1.00	—	—	—	—	1.25	—	Hard, keen edge for cutting hard materials.
F2	1.25	—	—	—	—	3.50	—	Deeper hardening.
F3	1.25	—	0.75	—	—	3.50	—	Deepest hardening.
				Mold Steels—Low-Carbon Types				
P1	0.10	—	—	—	—	—	—	Hubbed molds for carburizing.
P2	0.07	—	2.00	0.50	—	—	0.20	
P3	0.10	—	0.60	1.25	—	—	—	
P4	0.07	—	5.00	—	—	—	—	
P5	0.10	—	2.25	—	—	—	—	
P6	0.10	—	1.50	3.50	—	—	—	
				Mold Steels—Other Types				
P20	0.30	—	0.75	—	—	—	0.25	Cut molds for carburizing.

TABLE 12-XIII. HEAT TREATMENT OF TYPE L SPECIAL-PURPOSE TOOL STEELS.

Type	Quenching Temperature F	C	Quenching Medium	Tempering Range F	C	Hardness Rockwell C
L1	1450-1550	790-845	Oil or water	300- 600	150-315	64-56
L2	1550-1700	845-925	Oil	300-1000	150-540	63-45
	1450-1550	790-845	Water			
L3	1500-1600	815-870	Oil	300- 600	150-315	64-56
	1425-1500	775-815	Water			
L4	1475-1600	800-870	Oil or water	300- 600	150-315	64-56
L5	1450-1600	790-870	Oil	300- 600	150-315	64-56
L6	1450-1550	790-845	Oil	300-1000	150-540	62-45
L7	1500-1600	815-870	Oil	300- 600	150-315	64-56

wherein the impression is either forced (hubbed) into the soft mold blank in the case of Types P1 to P6 or cut by machining methods in the case of the higher-carbon Type P20. After forming the cavity, the mold is carburized and hardened by conventional practice to develop a hardness of Rockwell C 58-64 on the surface.

12.8 Summary. It should be clear that the development of tool steel has been the result of experimentation. The evidence of satisfactory performance is actual shop trial on a fairly large scale and under controlled conditions.

Most of the tool steels are sold under trade names, which have been listed conveniently in the *Metals Handbook* or other references given at the end of this chapter. Each of these brands may be identified by basic composition and accordingly correlated with the discussion that preceded this article. It is to be remembered that the higher-alloy steels are far more expensive than the lower-alloy or plain carbon steels; therefore, their performance must warrant the additional cost involved. Whether the steel is to be used for a tool or a machine part, it is preferable to give first consideration to the plain carbon steels. If these will not suffice, then it may be necessary to use the more expensive alloy steels which in the end may prove to be more economical. The plain carbon steels are simple to use and to heat-treat. With alloy steels, difficulties may be encountered in heat treatment which may give rise to erratic results and lead to complicated production schedules.

In conclusion, it may be stated that alloy tool steels are justified in applications requiring (1) abrasive hardness, to provide the superior wear resistance of complex alloy carbides; (2) toughness, as required by cold-forming tools and punches; (3) red-hardness, as required in hot-working tools and high-speed cutting tools to resist tempering and retain hardness; and (4) nondeforming properties, to resist distortion and cracking during the heat treatment of complicated forming dies.

REFERENCES

Gill, Roberts, Johnstin, and George, *Tool Steels,* American Society for Metals, Metals Park, Ohio, 1944.

Hoyt, *Metal Data,* Reinhold Publishing Corp., New York, 1952.

Lement, *Distortion in Tool Steels,* American Society for Metals, Metals Park, Ohio, 1959.

Metals Handbook, American Society for Metals, Metals Park, Ohio, 1961.

Palmer and Luerssen, *Tool Steel Simplified,* Carpenter Steel Co., Reading, Pa., 1948.

Steel Products Manual—Tool Steels, American Iron and Steel Institute, New York, 1955.

Woldman, *Engineering Alloys,* American Society for Metals, Metals Park, Ohio, 1954.
Manufacturers' Literature.

QUESTIONS

1. By what system are tool steels classified?
2. What is the basis of the AISI system of classification of tool steels?
3. How may tools be classified?
4. What requirements are placed upon steels to be used for cutting tools?
5. What are the requirements for steels to be used for shearing tools?
6. What is the application of forming tools? What are the requirements for steels to be used for forming tools?
7. What factors must be considered in the selection of a steel to be used for a tool?
8. What are the advantages of the use of plain carbon steel for tools?
9. What are some of the limitations in the use of plain carbon steel for tools?
10. What types of tools are frequently made from plain carbon steels containing between about 0.6 and 0.85 per cent carbon?
11. What order of magnitude of carbon content should plain carbon steels that are to be used for shear blades contain?
12. For what types of tools are plain carbon steels that contain between approximately 0.80 and 0.95 per cent carbon employed?
13. For what types of tools are plain carbon steels that contain between approximately 0.95 and 1.1 per cent carbon employed?
14. For what purposes are plain carbon steels that contain between about 1.10 and 1.40 per cent carbon employed?
15. Indicate the advantages and disadvantages of using very high carbon content in steels for tools.
16. What precautions must be taken in the heat treatment of plain, high-carbon steels used in tools?
17. The toughness of plain carbon steels decreases somewhat within a narrow range of tempering temperature. To what two factors may this decrease of toughness on tempering be ascribed?
18. What is the effect of adding small amounts of vanadium to the plain carbon steels?
19. What advantage is derived from the use of low-chromium, or a chromium-vanadium, steel for tools in the place of the standard carbon steels?
20. What types of tools are included in the shock-resisting classification, and what elements are frequently included in steels for this purpose?
21. What is meant by "nondeforming" tool steels?
22. What are the three principal types of cold work tool steels?
23. What types of elements are used for alloying in the cold work tool steels?
24. What requirements are placed upon hot work tool steels?
25. What are the three principal types of hot work tool steels?
26. What elements are primarily utilized for the hot work tool steels?

27. What elements are used to impart retention of hardness at high temperatures?
28. What is high-speed steel? Name the two types of high-speed steel.
29. What heat treatment will produce the greatest hardness in high-speed steels?
30. What are the advantages and disadvantages of adding cobalt in high-speed steels?
31. What types of steels are included in the classification "Special-Purpose Tool Steels"?

XIII

PRINCIPLES OF CORROSION

13.1 The Corrosion Problem. Corrosion is one of the most serious destructive agents that the engineer is called upon to combat. It is a subject extremely complex and difficult to analyze because of the large number of variables involved. Yet familiarity with the principles of corrosion is mandatory for the engineer. In his design problem, he must be able to attain not only the strength required of a structure or a machine, but also to prevent the serious deterioration of metal as a result of the environment to which it will be exposed. Although it is impossible in this text to give a complete dissertation on the subject of corrosion, the important facts that should be useful to the engineer will be presented from the fundamental point of view.[1]

Corrosion is the destruction of a material by chemical or electrochemical reaction with its environment. This includes the destruction of metals in all types of atmospheres and liquids, and at any temperature. The study of the basic aspects of corrosion may be divided into four parts: first, the fundamental mechanism involved in corrosion; second, the nature of corroding agents; third, those factors that control the rate of corrosion; and fourth, the manifestations of corrosion.

After considering the principles of corrosion, one may examine the methods by which metals may be protected and the tests that can be employed to determine the relative corrosion resistance of different materials under different conditions.

13.2 Direct Chemical Attack. Corrosion occurs by direct chemical action when the metal enters into a chemical reaction with other elements, such as oxygen, chlorine, etc., to form a nonmetallic compound. The direct chemical attack of zinc by hydrochloric acid is an example of this type of corrosion although it may be considered as galvanic action, which is discussed in the next section. The surface of the metal is usually smoothly

[1] For a more detailed discussion of the subject, the reader is referred to the bibliography at the end of this chapter.

corroded, with an etched appearance, and may look clean as though it were ground, or it may be darkened by the appearance of the nonmetallic compounds which are formed. The pickling of parts to remove scale is a corrosion process by direct chemical attack. In this instance, the metal is attacked slightly under controlled conditions.

13.3 Galvanic Action (Two Metals). Probably, galvanic action is the most common mechanism by which corrosion occurs. When two dissimilar metals, connected by an electrical conductor, are placed in a conducting solution (an electrolyte), a potential between the two metals is established. This potential is related to the relative tendency of each of the metals to go into solution. Although the electromotive-force series of pure metals indicates the solution potentials, it must be remembered that these potentials are determined on the basis of an open circuit, that is, with no current flowing. Sometimes, these potentials are referred to in the electrochemical series of metals which gives an indication of the relative ease with which each of the elements goes into solution. In the usual electromotive-force series, the elements are arranged in order of decreasing potential with respect to a hydrogen electrode. For practical purposes, the elements and alloys may be arranged in a galvanic series for a particular environment. A series of several metals is given in Table 13-I for sea water. Considerable care must

TABLE 13-I. GALVANIC SERIES IN SEA WATER.
(From Uhlig, *Corrosion Handbook*)

Anodic (corroded) end
Magnesium
Magnesium alloys
Zinc
Galvanized steel or galvanized wrought iron
Aluminum 5052-H38
Aluminum 3004-0
Aluminum 3003-0
Aluminum 6053-T6
Alclad
Cadmium
Aluminum 2117-T4
Aluminum 2017-T4
Aluminum 2024-T4
Mild steel
Wrought iron
Cast iron
Ni-Resist
13% Cr stainless steel (410) (active)
50-50 Lead-tin solder
18-8 Stainless steel (304) (active)
18-8, 3% Mo stainless steel (316) (active)
Lead
Tin
Muntz metal
Manganese bronze
Naval brass
Nickel (active)
78% Ni, 13.5% Cr, 6% Fe (Inconel) (active)
Yellow brass
Admiralty brass
Aluminum bronze
Red brass
Copper
Silicon bronze
5% Zn, 20% Ni, Bal. Cu (Ambrac)
70% Cu, 30% Ni
88% Cu, 2% Zn, 10% Sn (comp. G bronze)
88% Cu, 3% Zn, 6.5% Sn, 1.5% Pb (comp. M bronze)
Nickel (passive)
70% Ni, 30% Cu (Monel)
18-8 Stainless steel (304) (passive)
18-8, 3% Mo stainless steel (316) (passive)
Cathodic (protected) end

be exercised in using the galvanic series, since the electrode reactions dissipate energy and alter the electrode potentials, thus changing the relative solution tendencies of the electrodes.

If two pieces of metal of different composition are placed in a solution, such as salt water, and no electrical connection is made between the two metals, the metals may be slightly attacked by direct chemical action. If, however, the two pieces of metal are connected by means of a wire, a current will flow in the circuit. If the metals are iron and brass, the iron, being anodic to brass, will corrode rather rapidly whereas the brass, being cathodic to iron, will not be attacked. Under such a condition, brass is cathodic to iron.

During any galvanic action, polarization may be expected. *Polarization* may be defined as the change of potential of an electrode that results from the flow of current. This flow of current changes the potential of the anode toward the potential of the cathode; the potential of the cathode, in turn, is changed in the direction of the potential of the anode. Therefore, it should be appreciated that complete change of potential can occur during galvanic action, in which a metal which would normally be expected to be cathodic would be changed to anodic. Situations of this character have accounted for the unexpected behavior of metals and alloys in corrosion tests.

Even corrosion by direct chemical attack may be considered to be the result of galvanic action. When zinc is placed in hydrochloric acid, the zinc goes into solution at the anodes while hydrogen forms at the cathodes. This means that there are many very small anode and cathode areas over the entire surface of the metal. The anodic and cathodic areas exchange, probably as a result of polarization. The existence of these anodic and cathodic areas may be related to lack of homogeneity, grain orientation, nonuniform stresses, inclusions, surface imperfections, etc.

There are two requirements for galvanic corrosion: first, that a good electrical connection be made between the two metals; second, that the solution be an electrical conductor. Distilled water and pure, dry air are very seldom involved in galvanic corrosion because of their poor electrical conductivity. Acids, alkalies, and salt solutions are very good conductors; therefore, corrosion by galvanic action is common in the presence of these substances.

If two different metals are joined, as by riveting, and the joint is exposed to a moist atmosphere, such as would be found near the ocean, corrosion will probably take place rapidly by galvanic action. In this case, moisture deposits on the metal joint, supplying the electrolyte. The relative areas of the cathodic and anodic regions of riveted or bolted joints are of considerable importance. If the anodic area is relatively small with respect to the cathodic area, very intense corrosion may be expected. In practice, it is preferable to make the bolts or rivets of a material which is cathodic to the material being joined. By this practice, the anodic area becomes large in comparison with the cathodic area.

Two-metal galvanic action depends not only upon the character of the two metals that are in contact, but also upon the characteristics of the electrolyte. These characteristics are as follows:

1. Conductivity.
2. Composition.
3. Temperature.
4. Formation of hydrogen gas or other products.
5. Resistance of the circuit which controls the current flow and, hence, the degree of polarization.
6. Velocity of flow of the electrolyte past the electrode.

The relative areas of anode and cathode in many instances may be of considerable importance. These are factors that control the rate of corrosion and, therefore, will be considered in a later article.

Under normal conditions, the greater the difference in potential between the metals in the galvanic series, the greater will be the tendency for galvanic corrosion, modified, however, by a consideration of the current flow in the galvanic circuit which may alter the effective anode and cathode potentials.

13.4 Concentration Cells. Corrosion by galvanic action may take place in a somewhat different manner from that described in the previous article. Corrosion may also occur when two pieces of the same metal are placed in two different electrolytes, or the same electrolyte of different concentration, and separated by some porous substance. In these two cases, corrosion will be related to the current flow resulting from the potential differences established by differences in the concentration of the electrolyte. A very common

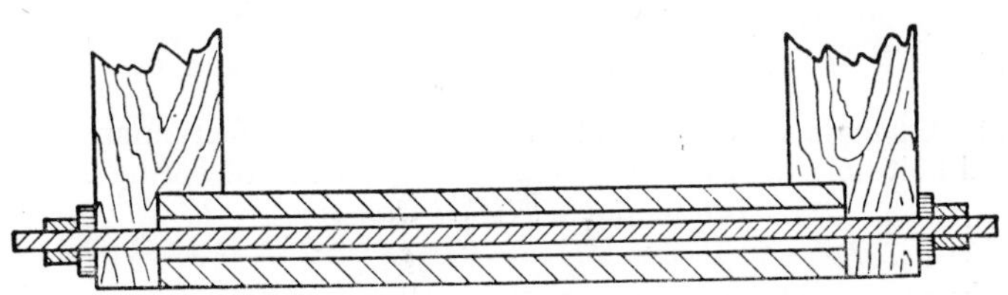

FIG. 13.1 Section view of bottom of pickling tank.

example is that in which a metal is exposed to an electrolyte varying in concentration. Consider a stay bolt employed to tie together the sides of a wooden pickling tank. A section view is shown in Fig. 13.1. The tank is filled with a pickling solution such as hydrochloric acid or sulfuric acid. If the joints where the wooden bottom joins the sides are not tight or there is some leakage of the fluid through the bottom into the space provided for the stay bolt, the concentration of the solution will vary along the length of the bolt. This concentration gradient forms an electrolytic cell, and deterioration of the stay bolt will occur at a greater rate in certain regions than in others.

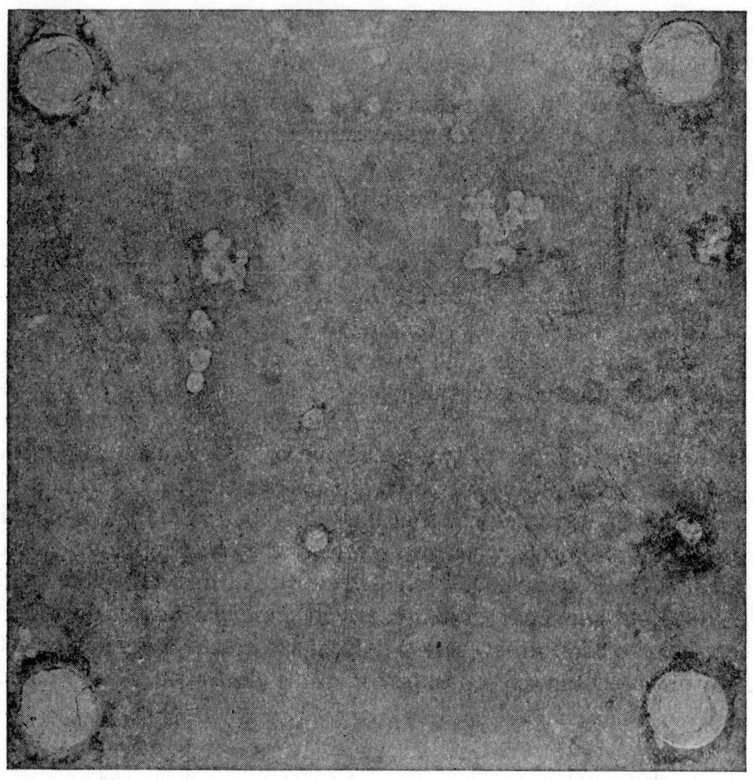

FIG. 13.2 Example of concentration cell corrosion under nonmetallic washers around bolt holes and attack produced by presence of barnacles. (*Courtesy International Nickel Co., Inc.*)

Another example of a concentration cell is found in situations in which rust or scale accumulates at the bottom of a steel tank. This material permits the formation of a stagnation region, thus permitting the establishment of a concentration gradient and corrosion of the tank. This is further illustrated in Fig. 13.2 which shows concentration cell corrosion under nonmetallic washers around bolt holes and also shows the corrosive attack produced by the presence of barnacles.

13.5 The Rate of Corrosion. The factors that influence the rate of corrosion may be divided into those dependent upon the metal and those dependent upon the environment.

The metal-dependent factors are:

1. The effective electropotential of a metal in solution.
2. The overvoltage of hydrogen on the metal.
3. The chemical and physical homogeneity of the metal surface.
4. The inherent ability of the metal to form an insoluble protective film.

The environment-dependent factors are:

1. The hydrogen-ion concentration (pH) of the solution.
2. The influence of oxygen in solution adjacent to the metal.
3. The specific nature and concentration of other ions in solution.
4. The ability of the environment to form a protective deposit on the metal.
5. The rate of flow of the solution in contact with the metal.
6. The temperature.
7. The application of cyclic stresses (corrosion fatigue).

The subject of the electropotential of a metal in a solution has been discussed briefly. With two metals in contact in an electrolyte, the rate of corrosion will depend largely upon the difference in solution tendency of the metals. Large differences in solution tendency will tend to increase the rate of corrosion. Also, a greater conductivity of the solution tends to increase the rate of corrosion.

The formation of hydrogen on or near the electrode has a tendency to alter greatly the rate of corrosion. It does so principally by decreasing the electrical conductivity of the solution in the region of the electrode. This tendency is referred to as the hydrogen overvoltage. The rate of corrosion may be greatly retarded if the overvoltage is high.

The uniformity of composition of a metal is important in the presence of an electrolyte since nonuniformity will permit the formation of a two-metal galvanic cell. Similarly, a difference in structure may stimulate corrosion in the presence of an electrolyte. A material in which there is a nonuniformity of internal stress will produce potential differences which may lead to serious corrosion.

The character of the film or the product of corrosion that forms on a metal is of great importance in corrosion. The production of an insoluble and tenacious protective film may greatly reduce conductivity between an electrode and the solution, thereby tending to decrease the rate of corrosion. However, porosity or inadequate bonding of the film may permit the establishment of concentration cells, resulting in an increased rate of corrosion. Insoluble and adherent products of corrosion tend to protect the material by preventing the corroding agent from coming in contact with the metal.

The film formed on certain metals is highly protective. Such surface films are said to be cathodic or passive. They are usually of the oxide type, are very thin, and produce an effective protective coating on the material. Aluminum, for example, may be passivated by the formation of an aluminum oxide film on the surface which is impervious to many corroding media. Similarly, those steels that are referred to as the corrosion-resistant type owe their corrosion resistance to the formation of a protective film, principally a chromium oxide. Such materials are said to be spontaneously passive under ordinary atmospheric oxidation. The rupture of such films is self-healing under atmospheric conditions; however, the presence of certain acids, chlorides, etc., may destroy the film to the point where the material becomes active. With constant removal of the protective film, even this supposedly corrosion-resistant material is corroded.

The concentration of hydrogen ions in the solution is of considerable importance in determining the general reaction which will take place. In an aqueous solution, a product of the concentration of the hydrogen and hydroxide ions at room temperature is equal to a constant 10^{-14}. The acidity of a solution is due entirely to the hydrogen-ion concentration, whereas alkaline solution properties are due to the concentration of hydroxide ions. In the case of distilled water, these ions are present in equal amounts, each having a concentration of 10^{-7} moles per liter. Acid solutions are those in which the concentration of hydrogen ions is greater than 10^{-7} moles per liter, and alkaline solutions are those in which the concentration of the hydroxide ions is greater than 10^{-7} moles per liter.

A more convenient method of expressing the acidity or alkalinity of a solution is to use the logarithm of the reciprocal of the gram ionic hydrogen equivalents per liter. This is indicated as the pH value of the solution. Thus $\mathrm{pH} = \log \frac{1}{(\mathrm{H}^+)}$ per liter. Pure water with a concentration of H^+ ion of 10^{-7} moles per liter will have a pH value of 7. Acids have pH values less than 7, while alkalies have pH values greater than 7. The pH value indicates the extent of ionization, which is a very important factor in considering the rate of corrosion.

Oxygen is very influential in the control of the rate of corrosion. The general term *oxidation,* as it is generally considered, is the increase in valency of an element. A metal is converted into a compound when in contact with an oxidizing agent, and there is a valency change from zero to some positive value. The effect of oxidation is similar to the acid reaction in direct chemical attack. Oxygen may be in atmospheric form or it may be dissolved in other corrosive agents, thereby increasing their potency.

The presence of other ions in a corroding solution may be very influential in the control of the rate. These ions may alter the character of the film produced at the surface, thereby either increasing or decreasing the rate of corrosion, depending upon whether they form a soluble or insoluble type of film.

The tendency to form a protective film on a metal is dependent not only upon the character of the metal, but also upon the character of the corroding medium or environment. Certain metals in the presence of concentrated nitric acid form a very protective oxide which will prevent further rapid corrosion. This method of anodic passivation has been applied to steel to provide temporary protection. In a dilute solution, however, this same material may be corroded appreciably owing to the difference in the character of the oxidation and the soluble nature of the film produced.

The rate of flow of a solution in contact with a metal is an important factor in the control of the rate of corrosion. If fresh solution is continually in contact with the metal by virtue of agitation or motion of the solution, corrosion may take place more readily, either by the actual application of fresh solution or by the washing away of protective films. The silicon and tin bronzes, for example, are equally resistant to salt water corrosion; yet only the latter will satisfactorily withstand impingement attack by salt water because of the persistent nature of its protective film.

The rate of corrosion is also greatly influenced by temperature. There is a greater tendency for oxidation to occur at elevated temperature because of the greater rate of reaction. Increasing the temperature tends to increase the rate of diffusion and the rate of ionization. An increase in temperature may also increase the solubility of protective films. A material that is normally resistant to corrosion at room temperature may be susceptible to corrosion at some higher temperature.

The effect of stress condition and of repeated application of such stresses will be considered in a later section of this chapter. It may be said at this point that the existence of a stress condition will tend to increase the rate of corrosion.

13.6 Corrosive Agents. Under the subject of corrosive agents, one may consider all of those substances that are the cause of corrosion. These may be of the type that produces direct chemical attack or forms electrolytes for galvanic action.

There is a large number of corrosive agents of which air is possibly the most active and the most prevalent. Air is composed principally of nitrogen, oxygen, and gases of a corrosive nature, such as sulfur compounds, chlorides, oxides, etc., resulting from industrial fumes. Some of these gases in themselves are not necessarily corrosive, but, in the presence of moisture, reaction will occur to produce a solution of corrosive character. An example of this may be found in atmospheres carrying sulfur dioxide or sulfur trioxide; the latter, upon reacting with water, forms sulfuric acid. Certain chlorides may tend to form hydrochloric acid. Air carries variable amounts of moisture, depending upon the general atmospheric conditions for the locality. In regions where there are large bodies of water, the humidity is usually relatively high; therefore, an electrolyte may be easily deposited with subsequent corrosion. Air-corrosive conditions are usually encountered in localities in which heavy fogs and dews are experienced. In some localities, a certain amount of solid matter, such as silica, carbon, etc., may be carried in the air which may further accentuate corrosive conditions by serving as particles on which moisture may condense.

The corrosive action of soil is frequently a serious problem for those who maintain pipe lines. The composition of soils varies extensively: many of them are acid; others, basic. They may differ in texture, in moisture content, and may contain oxidizing agents derived from organic substances. As a result of these wide variations in character and composition, the action of soil in corrosion may be expected to vary sharply from locality to locality. Very little corrosion is encountered in dry soil. In marshy land, where the acid concentration is usually high, severe corrosion of pipe lines may be experienced. When a pipe line runs through many different types of soil, concentration cell action as well as galvanic action may take place in some instances. The proximity of a pipe line to high-tension power lines may introduce a serious corrosion problem. Power lines running parallel to pipe lines may cause considerable current induction, leading to an accelerated form of corrosion of the galvanic type. At times, the mistake is made of grounding certain electrical apparatus to water lines. If there is appreciable leakage of current to the ground, there may be undue corrosion where the connection is made or at some point on the metal in the ground where an electrolyte is present.

Corrosion by acid is usually considered as a direct chemical action. Acids, such as sulfuric, hydrochloric, phosphoric, and acetic, are extremely severe in their corrosive action. They tend to form sulfate, chloride, phosphate, and acetate salts during the reaction. The weaker acids as exemplified by the fruit and organic acids are, in general, less active; but under certain conditions of concentration and electrolysis, corrosion by these acids may be quite rapid.

Highly oxidizing acids, such as nitric, perchloric, and nitrous, are extremely corrosive. Products of their reaction are, in practically all instances, oxides. The highly oxidizing corrosive agents, however, such as concentrated nitric acid, may form a protective oxide film, thus passivating the material and preventing further corrosion. Salt solutions or brines are particularly corrosive, since most chlorides are soluble and form excellent electrolytes for galvanic corrosion.

The alkalies are usually not as corrosive as the acids. Both iron and steel are relatively corrosive-resistant to strong alkalies in dilute solutions, although higher concentrations may tend to produce localized corrosion, resulting in pitting. Alkaline solutions, in certain concentrations in contact with iron or steel, may liberate hydrogen in the nascent, or highly active, state. This nascent hydrogen will usually pass along the grain boundaries reacting with oxides and sulfides in the steel, thus causing embrittlement. This phenomenon, sometimes referred to as *caustic* or *hydrogen embrittlement,* is common in boilers where alkaline water is employed or in applications where hard water is not properly treated. Sulfur and many of its compounds are highly corrosive, since they combine directly with the metal. Difficulties with sulfur are frequently encountered in petroleum-refining equipment. At ordinary temperatures, the sulfur compounds are not usually troublesome, but the high temperatures employed in refining may be favorable for reaction with the metal.

13.7 Manifestations of Corrosion. After corrosion has progressed to an advanced stage, it is often extremely difficult to determine the manner by which it has taken place. It may not be certain whether corrosion has been by direct chemical action or by one of the cell actions. Frequently, however, if the damage has not progressed too far, the cause and the mechanism by which corrosion occurs can be detected. Corrosion may appear as a general eating-away of the original material, leaving it free of all corrosion products. Such an appearance is usually an indication of direct chemical attack, although there are instances in which galvanic action may accomplish the same result. Usually the source of corrosion can be traced by carefully examining the entire system that is involved in the corrosive action. When a film or heavy deposit is formed, it may be difficult to establish the specific cause of corrosion because many different things can produce the same appearance. The several types of corrosion that may occur are as follows:

When corrosion occurs in localized areas, it is referred to as *pit corrosion* (Fig. 13.3).

There are some cases in which localized areas of an alloy may be attacked in such a way as to dissolve one or more constituents, thus leaving that area devoid of those elements. This type of corrosion is called *dezincification* and is illustrated in Fig. 13.4. The term was originally applied to the removal of zinc in brasses by a corrosive environment.

Intergranular corrosion is frequently observed in which attack occurs at the grain boundaries and penetrates into the material as shown in Fig. 13.5.

Corrosion may also occur when a material is in a state of stress with attendant cracking of the material. This is called *stress corrosion cracking.*

The presence of alternating stresses may accelerate corrosion. By the combined presence of a corroding medium and alternating stresses, fatigue failure may occur more rapidly than under ordinary noncorrosive conditions. This combination is called *corrosion fatigue.*

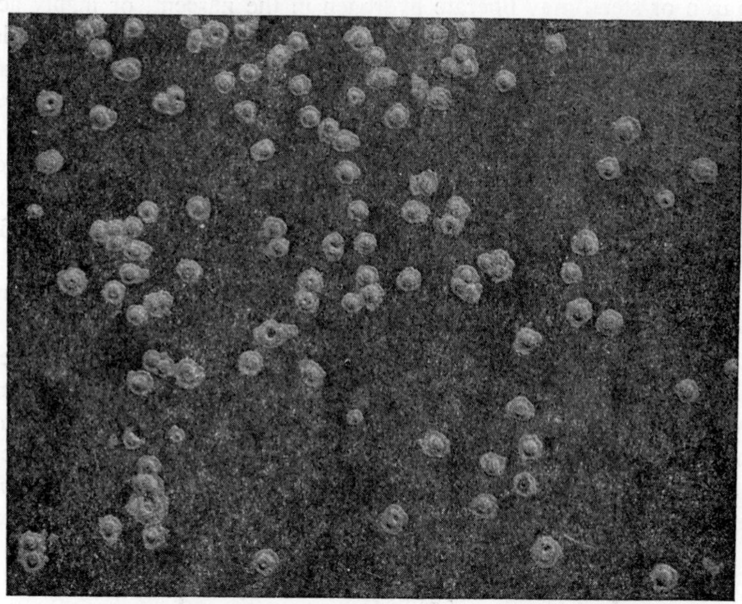

FIG. 13.3 Example of pit corrosion. (*Courtesy International Nickel Co., Inc.*)

Each of these types of corrosion will be discussed in detail in the following section of this chapter.

13.8 Pit Corrosion. Pit corrosion is quite readily recognized because small localized areas of the metal are acted upon to produce pits of varying size. Pitting may be the result of a nonuniform distribution of a protective film formed during the process of corrosion. Pits may also be produced by the localized destruction of a normally protective film, which leaves some areas of the surface exposed to the continued action of the corroding agent. Sometimes, the nonadherent products of corrosion collect in localized areas as a result of convection or eddy currents. These accumulations may form an electrolytic cell with the metal or create a concentration cell, resulting in localized action and pitting.

Pipe lines or tanks carrying or containing corrosive substances may experience less corrosive action if the substance is in continuous motion than if at rest. When the substance is not in motion, concentration cells may be formed. Sometimes pit corrosion occurs in a tank in which a corroding or semi-corroding medium is at rest for a long period of time, whereas, if in continuous circulation, no damage is experienced. Pit corrosion resulting from impingement attack in copper-alloy condenser tubes is another example of this destructive type of corrosion. Inhomogeneity of the metal is one of the most common causes of pitting.

Nonmetallic inclusions also may lead to pit corrosion. The corrosive agent may attack the inclusion directly; or, if the inclusion is more cathodic than the metal, localized attack by galvanic action will occur. The difference in potential existing between unstrained and strained metal may lead to pit corrosion.

FIG. 13.4 Example of dezincification type of attack. (*Courtesy International Nickel Co., Inc.*)

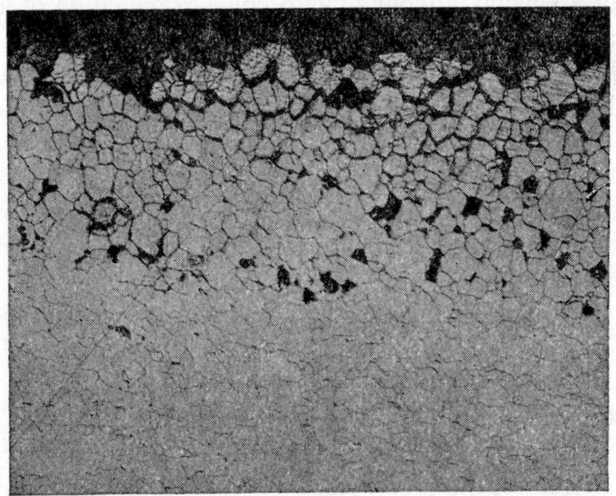

FIG. 13.5 Example of intergranular attack (section of structure—original surface at top). (*Courtesy International Nickel Co., Inc.*)

13.9 Dezincification. The term *dezincification* may be somewhat misleading, because it implies that only zinc can be removed from an alloy. It was first used to refer to the action of a corrosive agent on brass in which the zinc was removed. In a general sense, however, the term is applied to any condition of corrosion in which a specific element is removed from an alloy. This phenomenon is well known in the case of brass and is recognized by transformation of the brass from its typical yellow appearance to a distinct copper-red. The strength of brass is greatly reduced by dezincification.

Dezincification is usually localized in character and is associated with galvanic action. It occurs most commonly in solutions of high conductivity, such as salt water and acid solutions, and is often found in brass condenser tubes handling salt water.

13.10 Intergranular Corrosion. The phases that exist at the grain boundaries of some alloys may be less resistant to corrosion than the grain itself. When this situation exists and the alloy is placed in contact with a corrosive agent, attack begins at the surface in the region of this grain boundary material and then penetrates into the body of the alloy, following the boundaries. This action is seriously detrimental to the strength of the alloy. Certain compositions of the 18 per cent chromium-8 per cent nickel type of stainless steel, under certain conditions, are quite susceptible to this type of corrosion. Penetration of the corroding media follows the grain boundaries, thereby destroying the bonds between grains. Hydrogen or caustic embrittlement, which has been referred to in a previous article, takes place in much the same way.

Intergranular corrosion cannot always be detected easily by examining the surface of the metal, although microscopic examination of a section will clearly reveal the penetration of corrosion along the grain boundary.

13.11 Stress Corrosion Cracking. Prolonged static stresses or internal strains resulting from cold work may cause some materials to crack and corrode by intergranular penetration in the presence of a corroding medium. Certain brasses, which normally withstand corrosive conditions in the annealed or stress-relieved state, fail in the cold-worked condition. The most common occurrence is in the "season cracking" of brass cartridge cases (see Article 15.5).

13.12 Corrosion Fatigue. Fatigue failure is the progression of a fine crack through a material as a result of a large number of cycles of stress-reversal, usually originating in a surface notch or other stress-raiser. High stresses existing at the bottom of the crack make attack by a corroding medium more severe than the corrosion might be under ordinary unstressed conditions. Therefore, the presence of a corroding medium which may initiate notches in the form of corrosion pits, and the presence of alternating or cyclic stresses which tend to produce fine cracks, go hand in hand to produce early failure of parts subjected to such conditions.

The fatigue strength or endurance limit of materials is greatly decreased by the presence of a corroding medium. The general phenomenon is referred to as corrosion fatigue and the fracture is usually of a transcrystalline nature. Some typical endurance curves for a material tested in air and the same material tested in the presence of water are shown in Fig. 13.6.

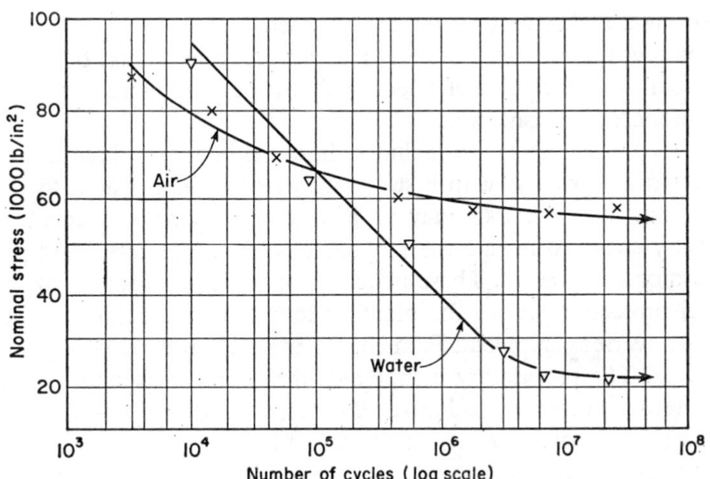

FIG. 13.6 Comparison of endurance curves for a 0.50% C, 5% Ni steel, annealed, 122,000 lb/in.² tensile strength tested in air and water. (Battelle, *Prevention of Fatigue of Metals,* John Wiley and Sons)

The existence of alternating stresses may tend to cause considerable damage to any protective film formed in the normal course of the action of a corrosive medium. Continual damage to this film may be a source of the further development of cracks which accelerate fatigue failure.

13.13 Methods of Protection. The means by which corrosion can be prevented or reduced may be classified into four groups as follows: first, by the use of corrosion-resistant metals and alloys; second, by the covering of the metal with a corrosion-resistant substance by painting, electroplating, etc.; third, by the treatment of the surface of the metal so that the surface composition is changed without the application of a coating; fourth, by cathodic protection through the use of a metal higher in the electropotential series in contact with the metal to be protected.

The first classification includes such metals as the corrosion-resistant steels, special brasses, bronzes, nickel, Monel, Inconel, etc., which are used because of their inherent ability to resist corrosion under specified conditions. Great care, however, must be exercised in selecting materials to resist corrosion. Each of the metals or alloys has its advantages and its limitations. There is no universal metal that will resist all types of corrosion. Before selecting an alloy for a particular application, all available data on its corrosion resistance and the conditions under which it may be used must be examined.

On first glance, one thinks of the use of corrosion-resistant materials as involving a higher cost of finished product. This is usually true, but one must consider the increased life that may accrue from the use of such materials. From the standpoint of effective life, it may be more economical to use these more expensive materials.

The second classification includes those methods by which the surface is protected by covering with a substance that will be resistant to the corrosive agent. Such methods may be of a temporary or permanent nature. Temporary coatings include the familiar oils, greases, waxes, and proprietary slushing compounds. These materials may provide short-term interoperational protection within the plant or long-term protection for outdoor storage or overseas shipment. Film characteristics may be dry, oily, or waxy as required. The proprietary compounds may contain chemical inhibitors, wetting agents, and special additives. For application to finished parts, such products should possess effective water-displacing power, inertness to the metal, the ability to neutralize fingerprint residues, and ease of application and removal in addition to required corrosion-protective properties.

The more-or-less permanent coatings include painting, electroplating, galvanizing, spraying, anodizing, coronizing, blueing, Parkerizing, and vitreous enameling. Paint serves as an excellent method of preventing atmospheric corrosion and, in some instances, corrosion by certain substances in liquid form. Paint is composed of a vehicle and a pigment. The vehicle must be of such a constitution that it will bond the pigment particles together and to the metal. The pigment imparts color and hardness to the paint. There are many different types of vehicles and pigments. The paints include the oil paints, varnishes, enamels, lacquers, synthetic resins, etc.

Anodizing is a protective treatment particularly well adapted to aluminum alloys. In the discussion of the fundamentals of corrosion, reference has been made to the production of thin films by an oxidizing solution for the purpose of securing protection against corrosion. Anodizing is a method of producing such a protective film. Parts that are to be treated are made the anode in an electrolytic tank. The tank, usually of steel, is made the cathode. The electrolyte may consist of a 5 to 10 per cent solution of chromic acid in water. Usually about 30 minutes is required for treatment.

The third classification of methods to prevent corrosion includes those processes by which the surface metal is changed in composition in such a way that its resistance to corrosion increases. Processes of this type include siliconizing, chromizing, and calorizing. These are specialized diffusion processes wherein the metal surface is impregnated with silicon, chromium, or aluminum, respectively.

The fourth classification, known as *cathodic protection,* is in many instances an economical and effective way of decreasing or even eliminating the deterioration of metal parts by corrosion. This is accomplished by placing a metal that is higher in the electropotential series in contact with the metal to be protected. In this manner, the inserted metal, the anode, which should not form a functional part of the structure, will establish a potential with the metal to be protected and will prevent corrosion. From time to time, it may be necessary to replace the anode. Care must be exercised in cathodic protection that polarization does not occur, which would counteract the expected benefits. Furthermore, the relative areas of cathode and anode must be properly balanced. If the anode area is large compared with the cathode area, the expected protection may not be secured. The possibilities of reversal of behavior were discussed in Article 13.3.

13.14 Corrosion Testing. Probably one of the most difficult aspects of corrosion studies is the development of a means by which the relative resistance of different metals to certain corroding media and conditions can be evaluated. Since many variables are involved in the process of corrosion, it becomes practically impossible to devise a single test that will give results which are commensurable with service conditions. Often a test is set up in the laboratory to show relative corrosion resistance of material as measured by loss of weight. When these materials are placed in service under apparently identical conditions, the behavior may be quite different. It therefore becomes necessary in a great many instances, such as in weathering experiments, to test the materials out in the open in various parts of the country on a long-term basis involving a period of years. These samples are inspected at regular intervals, and the results are noted. The weight loss, after removal of the corrosion products, may be expressed in milligrams per square decimeter per day (mdd) and converted to inch penetration per year (ipy). This is a long and tedious process but, in general, is far more informative than many of the laboratory tests. A comparison of several common materials exposed to total immersion in sea water is shown in Table 13-II.

Salt spray tests have been used to a considerable extent for the purpose of indicating the relative merits of different materials under accelerated and standardized test procedures. In such tests, the samples are placed in a salt spray contained in a box of particular design, and a salt solution of specific concentration is employed. Other laboratory tests include partial, alternate,

TABLE 13-II. RESULTS OF EXPOSURE OF SEVERAL MATERIALS TO CORROSION BY SEA WATER UNDER NATURAL CONDITIONS (CONTINUOUS IMMERSION).

(Compiled from Uhlig, *Corrosion Handbook*)

Material	Original Surface Condition	Exposure (years)	Wt. Loss (mdd*)	Geographical Location
Steel, 0.20% C, 0.02% Cu	Sandblasted	1.2	27	Eastport, Me.
Wrought iron	Pickled	15	25	Halifax, N.S.
Cast iron, 3.41% C	As-cast	15	32	Halifax, N.S.
Cast iron, 2.8% C, 20% Ni, 2.5% Cr	As-cast	6	8	Kure Beach, N.C.
Copper, 99.9%	Machined	3	2.4	Eastport, Me.
Copper, 99.9%	Cold-rolled strip	1	9.7	Kure Beach, N.C.
Copper 96%, Zn 1%, Si 3%	Pickled plate	1	2.3	Kure Beach, N.C.
Brass 80-20	Cold-rolled strip	1	11.0	Kure Beach, N.C.
Brass 65-35	Cold-rolled strip	1	12.1	Kure Beach, N.C.
Brass, 76% Cu, 22% Zn, 2% Al	Cold-rolled strip	1	4.5	Kure Beach, N.C.
Brass, 70% Cu, 29% Zn, 1% Sn	Cold-rolled strip	1	10.0	Kure Beach, N.C.
Brass 60-40	Cold-rolled strip	1	12.7	Kure Beach, N.C.
Brass, 62% Cu, 37% Zn, 0.7% Sn	Cold-rolled strip	1	10.0	Kure Beach, N.C.
Bronze, 95% Cu, 5% Sn, 0.1% P	Cold-rolled sheet	1	7.6	Kure Beach, N.C.
Bronze, 91% Cu, 7% Al, 2% Ni	Cold-rolled strip	1	6.7	Kure Beach, N.C.
Copper 70%, nickel 30%	Cold-rolled sheet	1	1.7	Kure Beach, N.C.
Copper 85%, Sn 5%, Zn 5%, Pb 5%	As-cast	1	2.6	Kure Beach, N.C.
Copper 88%, Sn 9%, Zn 3%	As-cast	1	1.1	Kure Beach, N.C.
Nickel, 99%+	Hot-rolled plate	3	6.1	Kure Beach, N.C.
Nickel 68%, Cu 29%, Fe 1%	Hot-rolled plate	3	2	Kure Beach, N.C.
Nickel 79.5%, Cr 13.6%, Fe 6.4%	Cold-rolled sheet	2.7	4.5	Kure Beach, N.C.
Nickel 56%, Mo 22%, Fe 22%	Hot-rolled plate	6	7.1	Kure Beach, N.C.
Nickel 36%, Fe Bal.	Hot-rolled plate	1	9.0	Kure Beach, N.C.
Cadmium plate on steel	Plated panel	1	3.7	Kure Beach, N.C.
Galvanized steel 2.5 oz/ft²	Plate	1	5.7	Kure Beach, N.C.
Lead, 99.96%	Bar	4	4.0	Bristol Channel, Weston-super-Mare
Solder, 60% Pb, 40% Sn on Cu	Plate	2.1	2.1	Kure Beach, N.C.
Stellite 6	Plate	2	0.7	Kure Beach, N.C.
Tin, 99.2%	Cast bar	4	0.15	Bristol Channel, Weston-super-Mare
Zinc	Sheet	0.5	9.5	Kure Beach, N.C.

* mdd = milligrams per square decimeter per day.

and total immersion in various corroding media. The results are usually determined by the time required for appearance of initial corrosion or on the basis of weight loss per unit time. For a more complete discussion of corrosion-testing procedure, the reader is referred to the bibliography.

REFERENCES

Borgmann *et al.*, *Corrosion of Metals,* American Society for Metals, Metals Park, Ohio, 1946.

Burns and Bradley, *Protective Coatings for Metals,* Reinhold Publishing Corp., New York, 1955.

Corrosion Tests for Corrosion-Resisting Steels, A.S.T.M. Standards, Part 3, pp. 239-353, American Society for Testing Materials, Philadelphia, Pa., 1958.

McKay and Worthington, *Corrosion Resistance of Metals and Alloys,* Reinhold Publishing Corp., New York, 1936.

Symposium on Corrosion-Testing Procedures, American Society for Testing Materials, Philadelphia, Pa., 1937.

Uhlig, *Corrosion Handbook,* John Wiley & Sons, New York, 1948.

QUESTIONS

1. Define corrosion.
2. Into what four parts may the basic aspects of corrosion be divided?
3. Explain how corrosion may occur by direct chemical attack.
4. What is the most common mechanism by which corrosion occurs?
5. In what way is the electromotive-force series related to corrosion problems?
6. Explain the mechanism of the process of corrosion by electrolysis.
7. What is polarization, and what is its effect upon galvanic action?
8. What are the two requirements for corrosion by galvanic action?
9. How may corrosion be altered by the relative areas of cathode and anode regions?
10. What characteristics of the electrolyte control corrosion by galvanic action?
11. Explain the mechanism of corrosion by the action of concentration cells.
12. What factors, associated with the metal, influence the rate of corrosion?
13. What factors, associated with the environment, influence the rate of corrosion?
14. What is hydrogen overvoltage and what is its effect upon the rate of corrosion?
15. What is the effect of the character of film formed on metals in relation to corrosion?
16. How is the acidity or alkalinity of a solution expressed? What is the order of magnitude of values for acids and alkalies?
17. What is the effect of the rate of flow of a corrosive solution upon the rate of corrosion?
18. Name some of the more common corrosive agents.
19. What is the cause of caustic embrittlement?
20. Describe each of the following: pit corrosion, dezincification, intergranular corrosion, stress corrosion cracking, corrosion fatigue.
21. What may be the causes of pit corrosion?
22. Explain the mechanism of dezincification.
23. Explain the mechanism by which fatigue properties may be decreased when a material is in the presence of a corroding medium.
24. Indicate the four general methods that may be used to prevent or reduce corrosion.
25. Name some typical corrosion-resistant metals.
26. Give some examples of coatings used for the prevention or reduction of corrosion.
27. What is anodizing? By what process does it tend to reduce corrosion?
28. What processes are sometimes used to improve resistance to corrosion by altering surface composition of the material?
29. What types of tests are sometimes used for the purpose of comparing the corrosive resistance of metals?

XIV

CORROSION- AND SCALE-RESISTANT ALLOYS

14.1 General. Steels containing 5 per cent or more of chromium are of great interest to the engineer because they are more resistant to corrosion and stronger at elevated temperatures than the low-alloy steels. The combination of chromium with iron and other ferrous alloys produces a type of iron or steel which has been referred to by some as stainless iron or steel. Since only a very few metals and alloys are stainless in the strict sense of the word, these alloys should be called the corrosion- and scale-resistant alloys. Resistance to corrosion and the effects of elevated temperature in alloys are the result of accidental discovery and intensive research. Some of those who have been closely associated with these developments include Marsh, Haynes, Brearley, Armstrong, Johnson, Becket, Strauss, and Maurer.

The engineer constantly faces new and stringent corrosion and thermal problems arising particularly in the aircraft, missile, and nuclear energy fields. Supersonic aircraft require materials to withstand aerodynamic and engine heating. Liquid-propelled rockets require materials to handle highly corrosive liquids such as nitrous oxide, whereas nuclear reactors require materials to combat the effects of liquid metals used as heat-transfer media. Many of the alloys introduced in this chapter as corrosion- and scale-resistant materials also find application as high-strength, heat-resistant alloys, as discussed in Chapter XV.

Chromium is the principal alloying element in many of these alloys, but other elements, such as nickel and silicon, enhance the properties imparted by chromium. Chromium strengthens ferrite slightly compared to its strong effect on increasing hardenability in those alloys which are hardenable by heat treatment. Resistance to tempering at temperatures up to about 1000 F (540 C) allows desirable elevated strength properties in certain alloys. The corrosion resistance obtained with the use of chromium is derived both from its inherent ability to resist corrosion and from the ability to form a protective film of chromium oxide.

The fundamental aspects of the corrosion- and scale-resistant alloys may be secured most readily by studying the effect of the various elements on the iron-iron carbide equilibrium diagram. The determination of equilibrium diagrams for the chromium alloys is extremely difficult because of the low rate of diffusion. Nevertheless, the careful work of many individuals has made it possible to approximate the diagrams which are of value in studying these alloys. Diagrams that present the phase relations under non-equilibrium conditions may be called constitution diagrams in distinction to equilibrium diagrams.

There is a large number of corrosion- and scale-resistant alloys available on the market. Many of them are sold under trade names which are sometimes confusing. When the analyses of these alloys are known, their characteristics can be evaluated by any engineer who is familiar with the effects of the elements employed.

14.2 Iron-Chromium Alloys. The iron-chromium constitution diagram is shown in Fig. 14.1, and the iron-rich end of this diagram is shown in Fig. 14.2.

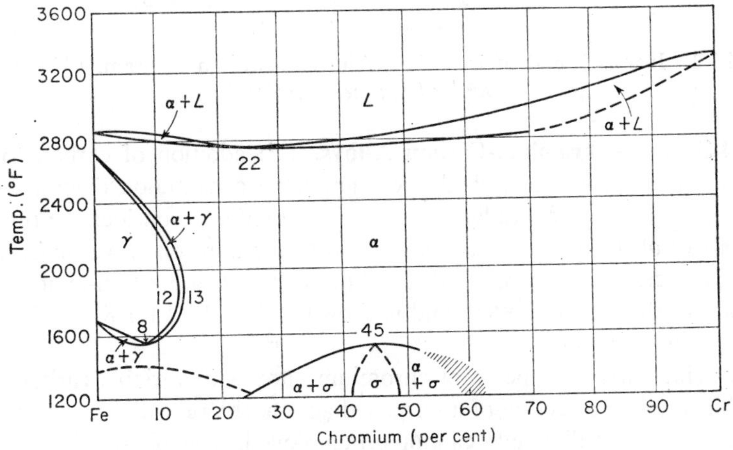

FIG. 14.1 Iron-chromium constitution diagram.

The addition of chromium to iron lowers the A_4 transformation appreciably and slightly decreases the A_3 transformation. With about 12.5 per cent chromium, the gamma solid solution region is nonexistent. Alloys containing more than about 12.5 per cent chromium consist of a solid solution of chromium dissolved in a body-centered cubic iron.

Long exposure of alloys of iron and large amounts of chromium to temperatures of the order of 900 F (480 C) may allow the formation of the sigma phase, which will greatly reduce the ductility at room temperature. The sigma phase is a hard, brittle, intermetallic compound, and its formation in engineering alloys has been a source of difficulty. This same situation exists in systems such as iron-molybdenum and others.

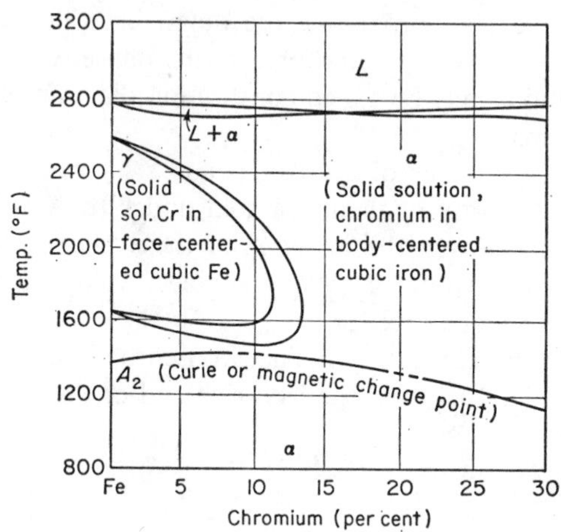

FIG. 14.2 Iron-rich end of iron-chromium constitution diagram. (Thum, *The Book of Stainless Steels*)

14.3 Iron-Chromium-Carbon Alloys. The addition of carbon to the chromium alloys has a marked effect upon the constitution diagram. One of the simplest ways of studying a ternary system is to consider the proportion of two of the components constant, with the third one a variable. This, then, is a cross section of a three-dimensional, composition-temperature diagram of the ternary system and is shown in Fig. 14.3 for 8, 12, 15, and 20 per cent chromium.

Consider first an 8 per cent chromium alloy with variable carbon content. Since this is a section of a ternary diagram, the rules that are normally applied to binary diagrams cannot be employed. The phases that coexist are indicated in each region of the diagram. For the ordinate corresponding to zero per cent carbon, the critical points will be found to coincide with those of the 8 per cent chromium alloy shown in Fig. 14.2. The general effect of chromium follows the discussion presented in the chapter on low-alloy steels, in which it was noted that the addition of chromium shifts the concentration of carbon in the eutectoid to lower values.

The effect of carbon on a 12 per cent chromium-iron alloy shows that the trend established in the preceding diagram is continued in the alloys containing larger amounts of chromium. With the 12 per cent chromium-iron alloy, the delta solid solution-plus-austenite and ferrite-plus-austenite regions are almost joined. In these alloys, the terms *ferrite, austenite,* and *delta solid solution* refer to solid solutions of chromium and carbon in alpha iron, in gamma iron, and in delta iron, respectively.

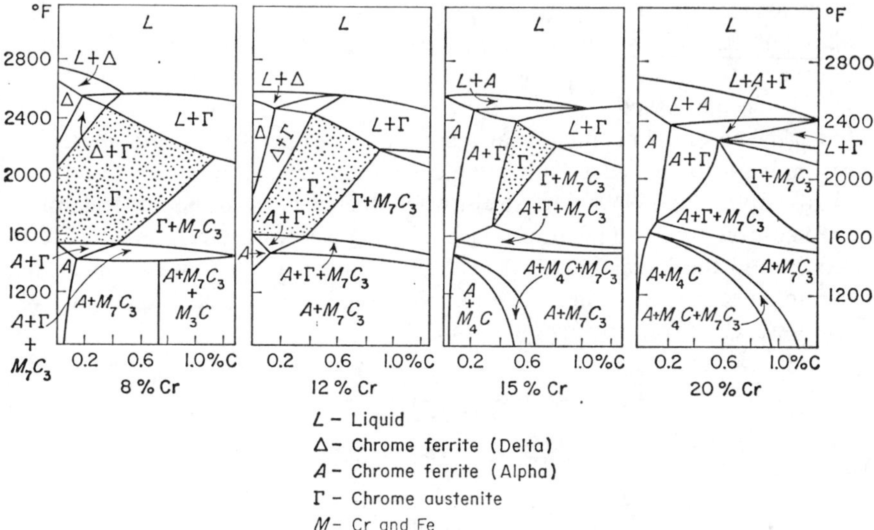

FIG. 14.3 Sections of iron-chromium-carbon ternary constitution diagram for 8, 12, 15, and 20% chromium.

Those alloys that at elevated temperature exist in the form of austenite and that upon cooling transform to a mixture of chrome ferrite and chrome carbide will respond to heat treatment. The response to hardening will be more intense with higher percentages of carbon. With more than about 0.35 per cent carbon, slow cooling from the austenite region will give rise to the precipitation of excess chrome-iron carbide. Since many of the 12 per cent chromium-iron alloys containing more than a small percentage of carbon pass through a phase change on cooling from the austenite region, they will be affected by heat treatment.

The constitution diagram of iron-chromium-carbon alloys containing 15 per cent chromium also shows that modification of the constitution diagram continues in the manner already indicated. Referring to the binary diagram of iron and chromium, Fig. 14.2, one observes that a 15 per cent chromium-iron alloy is outside of the gamma loop. In fact, it consists of a solid solution of chromium in body-centered cubic iron at all temperatures below the liquid region. The austenite region in iron-carbon alloys that contain 15 per cent chromium is considerably restricted. Those alloys that contain very small amounts of carbon are the ones that cannot be hardened by heat treatment. With the carbon content sufficiently high so that austenite may exist at elevated temperature, hardening to a certain degree may be expected. Hardenability in these alloys will not, in general, be equal to those of lower chromium content because of the presence of a relatively large amount of chrome ferrite. If the carbon content exceeds about 0.40 per cent, then the alloy will be entirely composed of chrome austenite when heated to about 2000 F (1095 C). If such alloys are rapidly cooled, satisfactory hardening may be accomplished.

If the chromium content is increased to 20 per cent, the ferrite-austenite region is greatly widened with complete elimination of the austenite region. High-carbon, high-chromium steels, when quenched from the austenite plus chrome-iron carbide region, produce the high wear resistance and hardness required in the better grades of cutlery. In medium- or low-carbon, high-chromium steels, it is impossible to produce effective hardening because of the presence of soft chrome ferrite.

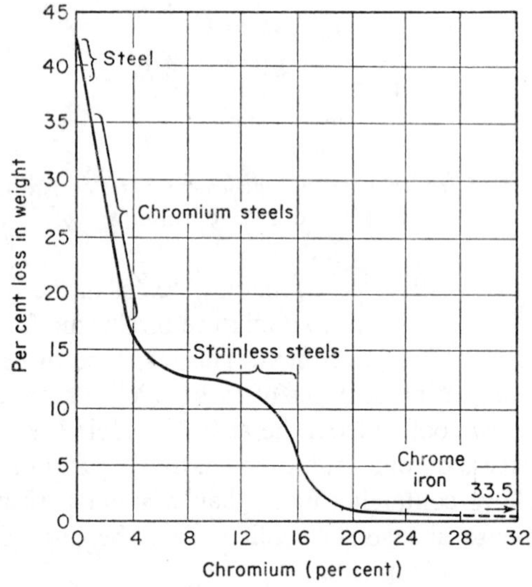

FIG. 14.4 Oxidation of steel at 1830 F vs. chromium content. (Thum, *The Book of Stainless Steels*)

The relation between the chromium content and the oxidation of steels at 1830 F (1000 C) is shown in Fig. 14.4. The resistance to oxidation is expressed in per cent weight loss and is dependent on the ability of the steel to form an adhering, refractory oxide film. As the chromium content increases, the resistance to oxidation increases very rapidly at first, and then less rapidly as the solid solution of chromium in gamma iron becomes more saturated with chromium. With about 13 or 15 per cent chromium, corresponding to the end of the gamma loop, the corrosion resistance improves markedly, reaching a maximum resistance with about 22 per cent chromium.

14.4 Iron-Nickel Alloys. The constitution diagram of the iron-nickel system is shown in Fig. 14.5. Nickel decreases the temperature at which

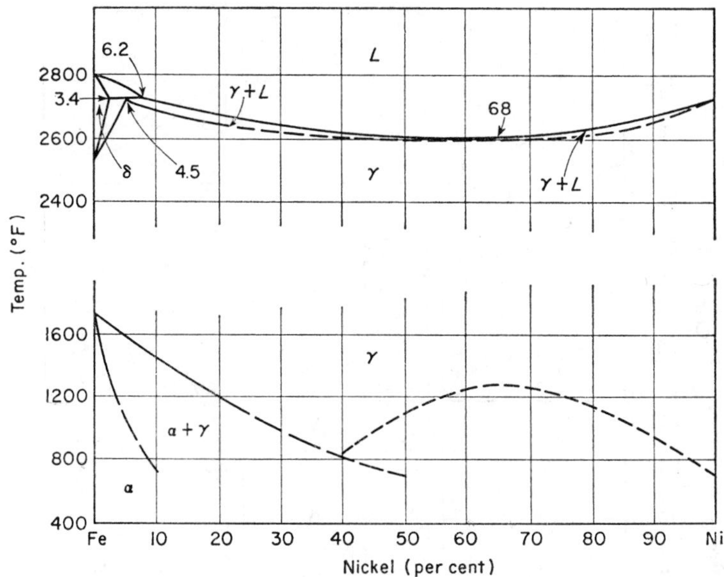

FIG. 14.5 Iron-nickel constitution diagram.

gamma iron transforms to alpha iron upon cooling. Upon heating, the transformation occurs at a much higher temperature for all iron-nickel alloys that contain less than about 35 per cent nickel. This sluggishness of transformation causes such alloys to be known as "irreversible."

The commercially useful alloys of the system contain more than about 35 per cent nickel and exist as a magnetic gamma solid solution at room temperature. An important iron-nickel alloy, known as *Invar,* contains 36 per cent nickel and has a very low coefficient of expansion. A similar alloy, containing additives of chromium and tungsten, known as *Elinvar,* has the additional property of retaining a constant modulus of elasticity over a certain temperature range. Alloys containing greater amounts of nickel are noted for their high magnetic permeability. Since these alloys contain negligible amounts of carbon, however, they cannot be properly classed as steels.

14.5 Iron-Chromium-Nickel Alloys. Although alloys of chromium, nickel, and iron are not of great commercial importance, they form the basis of many corrosion- and scale-resistant steels. The effect of the addition of chromium to iron-nickel alloys in a restricted range is shown in Fig. 14.6. Chromium widens the alpha and alpha-plus-gamma regions and restricts the gamma region. Increasing amounts of nickel tend to extend the limits of the gamma solid solution. The probable distribution of phases at about

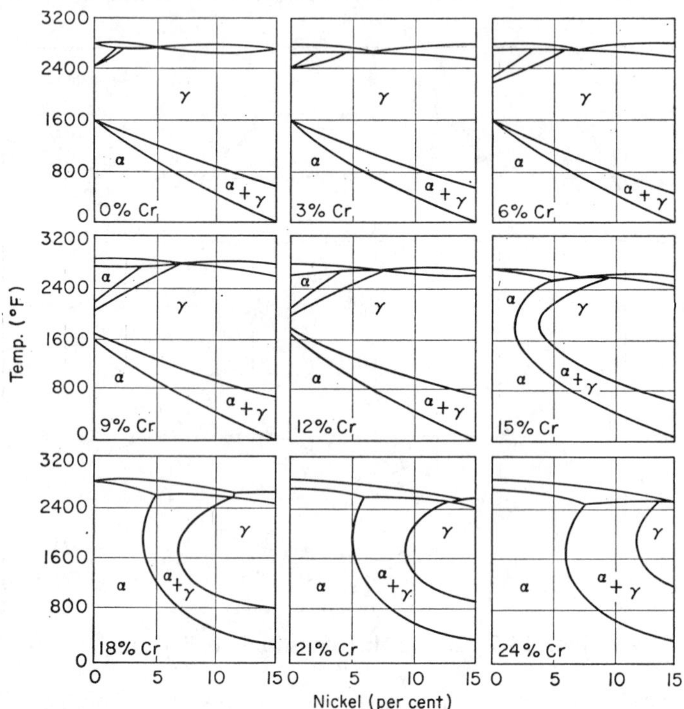

FIG. 14.6 Sections of iron-chromium-nickel ternary constitution diagram for 0, 3, 6, 9, 12, 15, 18, 21, and 24% chromium. (Bain and Aborn, *Metals Handbook,* 1939)

570 F (300 C) under near-equilibrium conditions is shown in Fig. 14.7. Since the diffusion rate of nickel and chromium in iron is very low, even air cooling is probably sufficiently rapid to produce the phases shown. It is possible, therefore, with a normal rate of cooling to obtain gamma solid solution in an alloy containing 20 per cent nickel, 20 per cent chromium, and 60 per cent iron, whereas under true equilibrium conditions, the structure might consist of alpha solid solution and gamma solid solution.

14.6 Iron-Chromium-Nickel-Carbon Alloys. At the present time, it is costly to produce commercial iron-chromium-nickel alloys that are essentially free of carbon. The probable phase relations in alloys containing 18 per cent chromium, 8 per cent nickel, with different amounts of carbon, are shown in Fig. 14.8. Under equilibrium conditions, an alloy containing 0.06 per cent carbon, 18 per cent chromium, and 8 per cent nickel will be composed of chrome-nickel austenite and ferrite with iron-chrome carbide at room temperature. With ordinary rates of cooling, the chrome-nickel austenite will not transform; therefore, the structure will be composed of austenite at room temperature. These alloys are sometimes called *austenitic steels*. Heating this alloy into the two- or three-phase regions for a prolonged

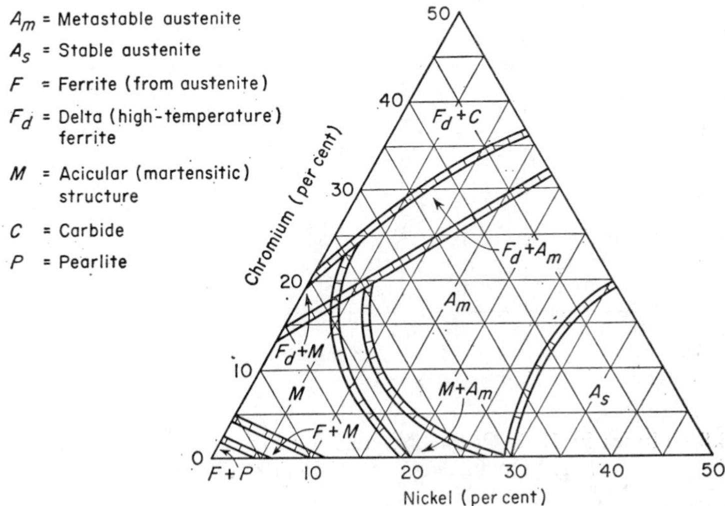

FIG. 14.7 Distribution of phases in iron-chromium-nickel alloys, quenched from temperatures of maximum austenite (0.1% C). (*Metals Handbook*)

time may bring about carbide precipitation, as discussed in Article 14.16.

The retention of austenite at ordinary temperatures produces an alloy that is ductile and quite resistant to corrosion. Since there is no transformation in these alloys of low carbon content, no major alterations in structure can be attained by heat treatment. They can be strengthened only by cold working. These characteristics will be described in a later article.

14.7 Classification. Many of the wrought corrosion- and scale-resistant steels (stainless steels) have been classified and coded by the American Iron and Steel Institute. These are shown in Table 14-I. Some of these steels are also available as castings with some modification of their compositions; they have been coded by the Alloy Casting Institute (ACI). The corrosion- and scale-resistant steels may be divided into three groups. Group A includes the plain chromium steels of the AISI 400 series in which a proper balance is maintained between the carbon and chromium content so that they are hardenable by the austenite-martensite transformation. These alloys are usually magnetic and are referred to as *martensitic stainless steels,* since they respond to hardening by heat treatment.

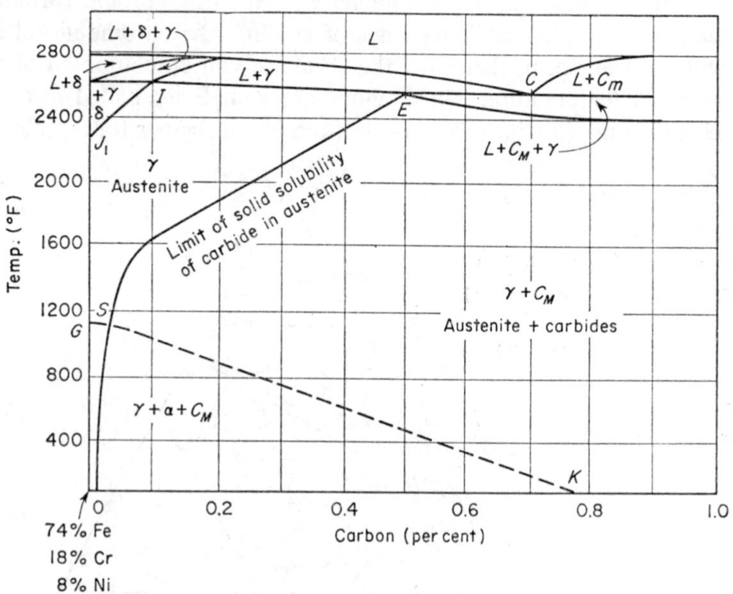

FIG. 14.8 Phases in 18% Cr-8% Ni steel for carbon content between 0 and 1%. (Thum, *The Book of Stainless Steels*)

In the discussion of the phase relations of iron, chromium, and carbon, it was noted that the gamma loop was closed with about 13 per cent chromium. Carbon combines with about 17 times its own weight of chromium. Chromium will combine with all the carbon present, but if there is more than is required by the carbon content, it will enter into solution with the iron. If more than 12.5 or 13 per cent chromium is dissolved in the iron, the steel will be ferritic. Therefore, Group A includes those alloys in which the difference between per cent chromium and 17 times the per cent carbon is less than 12.5 per cent:

$$\% \text{ Cr} - (17 \times \% \text{ C}) < 12.5\%$$

When the difference between the per cent chromium and 17 times the per cent carbon is greater than 12.5 per cent, the alloys are classed in Group B of the AISI 400 series and are known as *ferritic stainless steels:*

$$\% \text{ Cr} - (17 \times \% \text{ C}) > 12.5\%$$

TABLE 14-I. COMPOSITION (%) OF AISI STAINLESS STEELS.
(After *Metals Handbook*)

AISI No.	C	Cr	Ni	Other Elements	Properties and Uses	ACI No. (c)
A. Martensitic						
403	0.15 max	11.5-13.0	—	Mn 1.00 max; Si 0.50 max	Forged turbine blades.	—
410	0.15 max	11.5-13.5	—	(a)	General-purpose, low cost.	CA-15
416	0.15 max	12.0-14.0	—	P, S, Se 0.07 min (a); Zr, Mo 0.60 max	Free-machining 410.	—
420	>0.15	12.0-14.0	—	(a)	High-carbon 410; valves, ball bearings.	CA-40
431	0.20 max	15.0-17.0	1.25-2.50	(a)	High mechanical properties.	CB-30
440A	0.60-0.75	16.0-18.0	—	Mo 0.75 max (a)	Higher hardness than 420.	—
440B	0.75-0.95	16.0-18.0	—	Mo 0.75 max (a)	Cutlery grade.	—
440C	0.95-1.20	16.0-18.0	—	Mo 0.75 max (a)	Bearing races and balls.	—
501	>0.10	4.00-6.00	—	(a)	Mild corrosion resistance; oil-refining still tubes.	—
502	0.10 max	4.00-6.00	—	(a)	Better weldability than 501.	—
B. Ferritic						
405	0.08 max	11.5-13.5	—	Al 0.10-0.30 (a)	Al inhibits air hardening 410 during welding, etc.	—
430	0.12 max	14.0-18.0	—	(a)	Easily formed, auto. trim, chemical equip.	—
430F	0.12 max	14.0-18.0	—	P, S, Se 0.07 min; Zr, Mo 0.60 max (a)	Free-machining 430.	—
446	0.35 max	23.0-27.0	—	N 0.25 max; Mn 1.50 max, Si 1.00 max	High corrosion and scale resistance to 2150 F.	CC-50
C. Austenitic						
201	0.15 max	16.0-18.0	3.50-5.50	Mn 5.50-7.50	Low Ni equivalent of 301.	—
202	0.15 max	17.0-19.0	4.00-6.00	Mn 7.50-10.0	Low Ni equivalent of 302.	—
301	0.08-0.20	16.0-18.0	6.00-8.00	(b)	High-strength, ductile, structural use, trim.	—
302	0.08-0.20	17.0-19.0	8.00-10.0	(b)	General purpose, decorative uses.	CF-20
302B	0.08-0.20	17.0-19.0	8.00-10.0	(b)	More scale resistance than 302.	—
303	0.15 max	17.0-19.0	8.00-10.0	P, S, Se 0.07 min; Zr, Mo 0.60 max (b)	Free-machining 302.	CF-16F
304	0.08 max	18.0-20.0	8.00-11.0	(b)	Low C 302 restricts ppt. of carbides in welding.	CF-8
304L	0.03 max	18.0-20.0	8.00-11.0	(b)	Lower C 304 immune to carbide ppt.	—
305	0.12 max	17.0-19.0	10.0-13.0	(b)	Low work-hardening rate for severe forming.	—
308	0.08 max	19.0-21.0	10.0-12.0	(b)	Welding rod alloy for 301, 302 and 304.	—
309	0.20 max	22.0-24.0	12.0-15.0	(b)	High-temperature strength and scale resistance.	CH-20
309S	0.08 max	22.0-24.0	12.0-15.0	(b)	Low C 309 for welding.	—
310	0.25 max	24.0-26.0	19.0-22.0	Mn 2.00 max; Si 1.50 max	Higher strength and scale resistance than 309.	CK-20
310S	0.08 max	24.0-26.0	19.0-22.0	Mn 2.00 max; Si 1.50 max	Low C 310 for welding.	—
314	0.25 max	23.0-26.0	19.0-22.0	Mn 2.00 max; Si 1.50-3.00	High Si improves scale resistance.	—
316	0.10 max	16.0-18.0	10.0-14.0	Mo 2.00-3.00 (b)	Superior resistance to pit corrosion.	CF-8M; CF-12M
316L	0.03 max	16.0-18.0	10.0-14.0	Mo 1.75-2.50 (b)	Low C 316 for welding.	—
317	0.10 max	18.0-20.0	11.0-14.0	Mo 3.00-4.00 (b)	Higher corrosion and creep resistance.	—
321	0.08 max	17.0-19.0	8.00-11.0	Ti 5 × C min (b)	Stabilized 18-8.	—
347	0.08 max	17.0-19.0	9.00-12.0	Cb 10 × C min (b)	Stabilized 18-8.	CF-8C

(a) Also contains Mn, Si, 1.00 max.
(b) Also contains Mn 2.00 max, Si 1.00 max.
(c) ACI casting designations shown for nominal comparison only; cast alloy composition ranges vary from wrought grades.

Since these steels are ferritic, they will not respond to heat treatment. The structural and hardening characteristics of the plain iron-chromium-carbon alloys have been shown roughly by Bain in Fig. 14.9.

The third group, or Group C, includes those alloys that contain at least 24 per cent combined chromium, nickel, and manganese, and in which the amount of chromium is at least 18 per cent. These alloys make up the AISI 200 and 300 series and are referred to as *austenitic stainless steels*. By examining the phase diagrams of the iron-chromium-nickel alloys, it is evident that alloys of this group are normally austenitic. Under true equilibrium

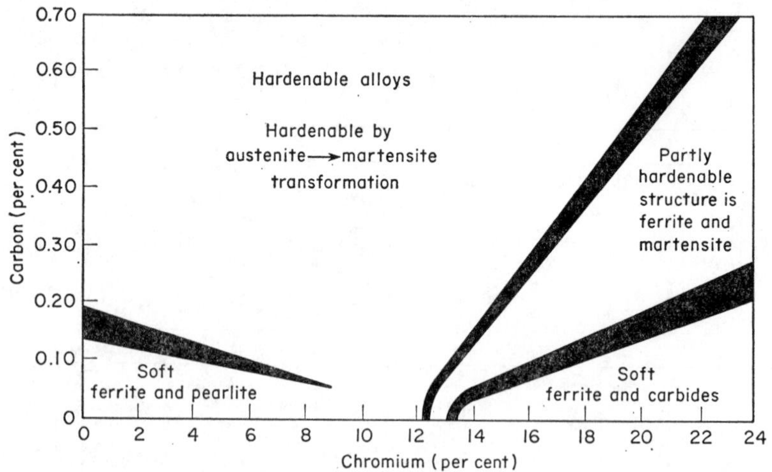

FIG. 14.9 Characteristics of iron-chromium-carbon alloys for different amounts of chromium. (Thum, *The Book of Stainless Steels*)

conditions, ferrite and austenite may be coexistent, but, in view of the sluggishness of these elements in diffusing, the alloys are austenitic.

Alloys in all three of these groups may contain small percentages of copper, silicon, molybdenum, titanium, columbium, selenium, sulfur, etc., which are added for special purposes.

14.8 Low-Alloy, Corrosion-Resistant Steel. A group of steels containing between 4 and 6 per cent chromium is used extensively in places where conditions are not too severe. These steels are designated as Types 501 and 502 in Table 14-I and occupy an alloy position midway between the AISI 5XXX low-alloy series and the AISI 400 martensitic group. These steels are about four to ten times more resistant to sulfide corrosion than ordinary steel under laboratory conditions. Their resistance to oxidation at 1000 F (538 C) is approximately three times as great as ordinary steel.

Steels of this composition are used with success for petroleum-refining equipment in heat exchangers, cracking-still tubes, etc.

The 4 to 6 per cent chromium steels containing more than about 0.10 per cent carbon harden when cooled in air. The properties that may be attained in alloys of this type with different treatments are shown in Table 14-II. Since these alloys are subject to air hardening, care must be taken in welding and in hot forming. For softening, the steel is annealed at a tem-

TABLE 14-II. TENSILE AND IMPACT PROPERTIES OF 5% CHROMIUM STEEL, AIR-HARDENED FROM 1505 F (538 C).
(After Thum, *The Book of Stainless Steels*)

Tempering Temperature	Yield Point (lb/in.²)	Tensile Strength (lb/in.²)	Elong. in 2 in. (%)	Red. of Area (%)	Izod Impact (ft-lb)	Charpy Impact (ft-lb)
5.21% Cr, 0.10% C						
Not Tempered	108,860	181,420	15.5	53.2	23.3	18.5
1022 F (550 C)	128,850	169,480	16.5	57.4	8.1	19.1
1125 F (607 C)	102,180	114,100	20.0	71.0	76.4	44.9
1202 F (650 C)	86,780	100,100	23.0	73.2	96.4	51.2
1292 F (700 C)	76,810	91,810	25.0	76.2	102.6	55.6
1382 F (750 C)	67,290	84,810	28.0	77.9	100.6	60.4
Furnace-Cooled	27,300	66,030	37.5	75 6	76.8	47.0
5.19% Cr, 0.21% C						
Not Tempered	113,950	212,310	9.0	18.5	23.4	15.6
1022 F (550 C)	137,970	194,840	14.5	46.6	16.6	12.5
1125 F (607 C)	119,870	137,560	18.5	58.7	35.7	39.0
1213 F (656 C)	95,090	115,440	22.5	69.4	81.8	45.7
1292 F (700 C)	83,420	104,720	23.0	68.0	85.9	54.4
1382 F (750 C)	72,040	95,640	27.5	71.5	80.2	57.4
1472 F (800 C)	60,730	94,640	29.0	70.2	88.3	62.4
Furnace-Cooled	32,250	75,670	32.0	75.0	84.1	66.6
5.27% Cr, 0.30% C						
Not Tempered	117,260	221,830	13.5	31.7	16.9	15.4
1022 F (550 C)	144,300	195,950	14.5	47.2	13.3	14.5
1125 F (607 C)	116,840	137,400	18.0	60.5	59.1	35.6
1213 F (656 C)	97,040	120,220	21.0	66.7	76.5	41.4
1292 F (700 C)	83,080	108,080	24.0	69.6	85.1	46.9
1382 F (750 C)	73,820	99,700	28.0	72.0	77.0	60.6
1472 F (800 C)	59,610	96,750	30.0	74.1	72.5	55.6
Furnace-Cooled	33,420	78,990	33.5	75.6	79.9	79.4

perature of about 1575 F (857 C) and reheated to approximately 1400 F (760 C) for several hours. The cold-working properties of this alloy in the fully annealed condition are almost as good as those of the low-carbon steels. In the annealed state, the 5 per cent chromium steel will not usually exhibit the normal pearlitic structure. The carbides are commonly globular or spheroidal in form dispersed throughout the ferrite. It is difficult to make a sweeping statement concerning the conditions for which the 4 to 6 per cent chromium steels are best adapted. However, it may be said that, for mild corrosive conditions or at temperatures below 1000 F (538 C), the 4 to 6 per cent chromium steels have given excellent service at relatively low cost as compared to the higher-chromium steels.

The addition of other elements, such as silicon, tungsten, molybdenum, and titanium, to the 4 to 6 per cent chromium steels has proved to be beneficial. When 1 to 3 per cent silicon is present, the scale resistance is increased, making the alloy useful in automotive exhaust valves (silchrome). The addition of about 0.5 per cent molybdenum or 1.0 per cent tungsten increases the strength at elevated temperatures. The addition of about 0.75 per cent titanium tends to reduce the air-hardening characteristics of these steels, which makes it possible to work the steel in the as-rolled condition without annealing. Since titanium has a very high affinity for carbon, it combines with the carbon, leaving the chromium in solution with the iron, thus improving the corrosion resistance. In order to obtain these results, the titanium content must be from five to eight times the carbon content.

The 4 to 6 per cent chromium steels are also used for castings in the petroleum industry for valve bodies, pump diffuser rings, and other fittings that are subject to elevated temperatures and sour-hot crude oil. Castings usually contain about 0.20 per cent carbon, although some alloys that contain up to 0.30 per cent carbon have been used. These steels usually contain either molybdenum or tungsten in amounts of about 0.50 and 1.0 per cent, respectively. Since this material is air-hardenable, it must be very slowly cooled if the parts are to be machined. Where flame cutting is used for removing gates and risers, fine cracks may develop because of the air-hardening characteristics. These can be prevented by heating the whole casting to approximately 900 F (482 C) before cutting. By normalizing this material at a temperature of about 1750 F (955 C), followed by a second normalizing at about 1650 F (900 C), and then tempering at about 1250 F (675 C), the following properties may be developed:

Tensile strength	110,000 to 120,000 lb/in.^{2}
Yield point	75,000 to 85,000 lb/in.^{2}
Elongation in 2 in.	16 to 20%
Reduction of area	30 to 55%
Brinell hardness	210 to 250

14.9 Chromium Steel—Low Carbon. Certain applications require a steel that can be heat-treated for medium-high strength and that will be resistant to corrosion. To meet these requirements, a steel containing between 11 and 14 per cent chromium and less than 0.15 per cent carbon may be used. Types 403 to 416 inclusive respond to heat treatment and are employed for turbine blades, corrosion-resistant castings, and machine parts that must be corrosion-resistant and must possess high strength. The properties that may be obtained with a steel of this group are shown in Fig.

TABLE 14-III. TYPICAL MECHANICAL PROPERTIES OF WROUGHT MARTENSITIC STAINLESS STEELS.*
(From *Metals Handbook*, 1954 Supplement)

AISI No.	Tempering Temp.	T.S. (1000 lb/in.²)	Y.S. (1000 lb/in.²)	Elong. in 2 in. (%)	Red. of Area (%)	Izod Impact (ft-lb)	Rockwell Hardness
403	Annealed	75	40	35	70	90	B82
410	Annealed	70	35	30	—	70	—
403 and 410	As-quenched (1800 F) (980 C)	—	—	—	—	—	C43
	400 F (205 C)	190	145	15	55	35	C41
	600 F (315 C)	180	140	15	55	35	C39
	800 F** (425 C)	195	150	17	55	**	C41
	1000 F** (540 C)	145	115	20	65	**	C31
	1200 F (650 C)	110	85	23	65	75	B97
	1400 F (760 C)	90	60	30	70	100	B89
416	Annealed	75	40	30	60	30	B82
	As-quenched (1800 F) (980 C)	—	—	—	—	—	C43
	400 F (205 C)	190	145	12	45	20	C41
	600 F (315 C)	180	140	13	45	20	C39
	800 F** (425 C)	195	150	13	50	**	C41
	1000 F** (540 C)	145	115	15	50	**	C31
	1200 F (650 C)	110	85	18	55	30	B97
	1400 F (760 C)	90	60	25	60	60	B89
431	Annealed	125	95	20	55	50	C24
	As-quenched (1900 F) (1040 C)	—	—	—	—	—	C45
	400 F (205 C)	205	155	15	55	30	C43
	600 F (315 C)	195	150	15	55	45	C41
	800 F** (425 C)	205	155	15	60	**	C43
	1000 F** (540 C)	150	130	18	60	**	C34
	1200 F (650 C)	125	95	20	60	50	C24
420	Annealed	95	50	25	55	—	B92
	As-quenched (1900 F) (1040 C)	—	—	—	—	—	C54
	600 F (315 C)	230	195	8	25	10	C50
440A	Annealed	105	60	20	45	2	B95
	As-quenched (1900 F) (1040 C)	—	—	—	—	—	C56
	600 F (315 C)	260	240	5	20	4	C51
440B	Annealed	107	62	18	35	2	B96
	As-quenched (1900 F) (1040 C)	—	—	—	—	—	C58
	600 F (315 C)	280	270	3	15	3	C55
440C	Annealed	110	65	14	25	2	B97
	As-quenched (1900 F) (1040 C)	—	—	—	—	—	C60
	600 F (315 C)	285	275	2	10	2	C57

* Bars 1 in. in diameter.

** Tempering in range 750-1050 F (400-565 C) should be avoided because of low and erratic impact values.

14.10 and Table 14-III. As indicated in the diagram, by quenching from 1750 F (954 C) into oil, an ultimate strength of 200,000 lb/in.^{2} may be obtained. By tempering at 1100 F (593 C), the ultimate strength will be approximately 120,000 lb/in.^{2} with an elongation of a little over 20 per cent in 2 in. Steels in this group should not be tempered in the range of 750-1050 F (400-565 C) in view of the low and erratic impact properties shown in Fig. 14.10. By tempering these steels at higher temperatures, the precipitation of carbides reduces the corrosion resistance somewhat. With the carbon content on the low side of the range, lowering of corrosion resistance is not severe.

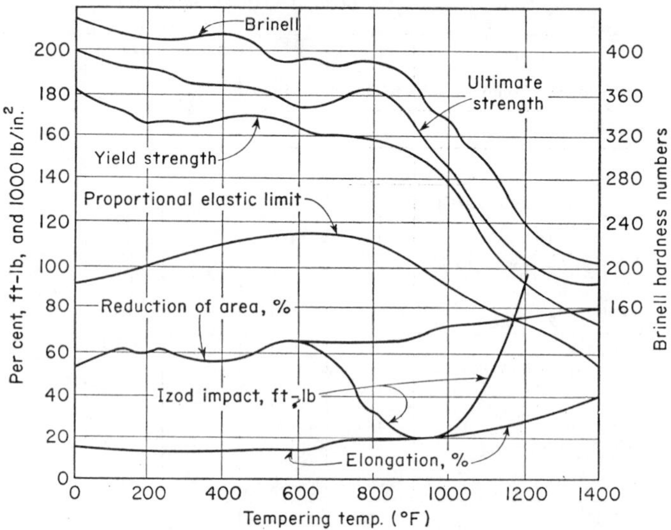

FIG. 14.10 Effect of tempering temperature on properties of 12.23% Cr, 0.09% C steel (Types 403 and 410), oil-quenched from 1750 F. (After Thum, *The Book of Stainless Steels*)

The corrosion- and scale-resistant steels are, in general, much more difficult to machine than the hardness would indicate when compared with the plain carbon steels. Machinability may be improved considerably by the addition of sulfur or selenium in the amounts given for Type 416. About 0.50 per cent molybdenum is also beneficial to machinability. Selenium in amounts smaller than about 0.05 per cent has less effect upon the corrosion resistance of these steels than does sulfur. The greater difficulty of machining may be accounted for on the basis of the higher frictional characteristics of these alloys. The free-machining grades, i.e., those containing sulfur, selenium, or other elements of this type, tend to reduce these frictional characteristics and, furthermore, decrease the tendency to seize and to gall when they are used in contact with each other as wearing surfaces.

14.10 Chromium Steel—High Carbon. In order to develop greater hardness in the high-chromium steels for use in cutlery and other applications involving wear, the carbon content must be higher than that of the previous class. One of the first cutlery steels contained between 11 and 14 per cent chromium and 0.30 and 0.40 per cent carbon. At present, a modified cutlery steel contains between 0.30 and 1.0 per cent carbon, and the chromium content is between 12 and 18 per cent as in Types 420, 440A, and 440B. With large percentages of chromium, the carbon content must be increased if hardness is required.

The cutting properties and ability to retain a cutting edge are reflected in the hardness values. The cutting qualities of the corrosion-resistant cutlery steels compare favorably with those of the plain carbon steels and are superior to the cheap grades of plain carbon cutlery steels. A comparison of the hardness of plain carbon and corrosion-resistant cutlery steels is given in Table 14-IV.

TABLE 14-IV. COMPARATIVE HARDNESS OF STAINLESS AND CARBON STEEL KNIVES.

(From Thum, *The Book of Stainless Steels*)

Type of Knife	Brinell Hardness Number
Good shear steel carver	505-525
Good shear steel table blade	510-530
Common steel carver	510-525
Common steel blade	510-525
Best cast steel pocket blade	560-580
Best cast steel razor	625-640
Stainless carvers	500-520
Stainless steel table blades	500-550
Stainless pocket blades	550-600

Although these steels will harden with air-cooling, better results are usually obtained with an oil quench. Maximum hardening can be obtained by quenching in water or brine, but there is greater liability of warping and breaking. The high-chromium steels must be held at the hardening temperature about twice as long as the plain carbon steels because of the low rate of diffusion. The effect of tempering Type 420 steel is shown in Fig. 14.11.

Steels containing between 1.0 and 1.2 per cent carbon and 16 to 20 per cent chromium have a very good combination of strength, hardness, toughness, and resistance to abrasion or corrosion. Therefore, they are used

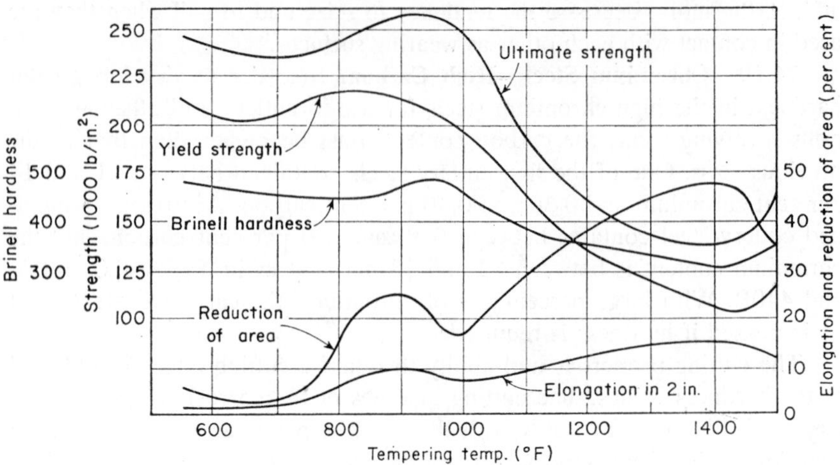

FIG. 14.11 Effect of tempering temperature on properties of 13.0% Cr, 0.30% C steel (Type 420), oil-quenched from 1825 F. (After Thum, *The Book of Stainless Steels*)

for ball bearings, chemical equipment, oil pumps, etc. Hardening is done by heating to a temperature between 1900 and 1925 F (1038 and 1052 C) for double the time required for plain carbon steels and quenching in oil. Fig. 14.12 shows the hardness and impact values for a 1 per cent carbon-17.5 per cent chromium steel. The best combination of hardness and toughness is obtained by tempering at about 800 F (427 C). The tempering range of 750 to 1050 F (399 to 566 C) is to be avoided. Higher tempering tem-

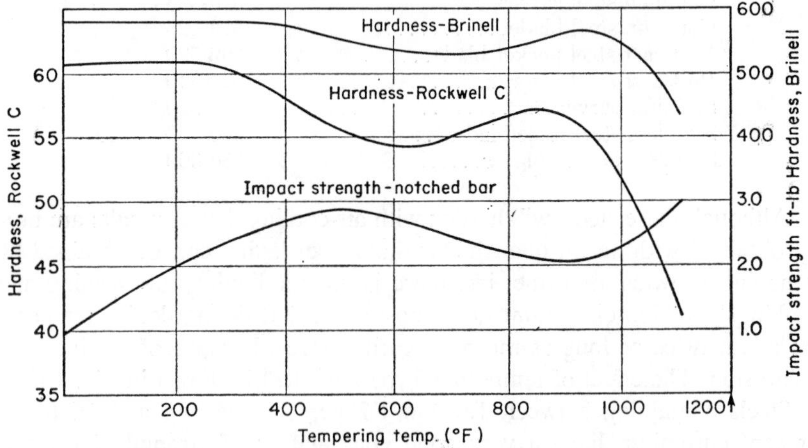

FIG. 14.12 Effect of tempering temperature on hardness and impact values of 17.5% Cr, 1% C steel, oil-quenched. (After Thum, *The Book of Stainless Steels*)

peratures make the steel machinable. The mechanical properties of the martensitic stainless steels are compared in Table 14-III.

14.11 Chromium Steel—Ferritic. The structure of those steels that contain in excess of about 14 per cent chromium with small amounts of carbon, such as Type 430, will consist of chrome ferrite with some chromium carbide if the carbon content is greater than about 0.10 per cent. With a ferritic structure, these steels are not susceptible to heat treatment or grain refinement. When the carbon content is on the low side of the range, the corrosion resistance is very good. With a greater percentage of carbon, the corrosion resistance may not be as good. The ferritic steels may be cold-formed with facility; therefore, they are used extensively for deep-drawn articles, such as vessels for the chemical and food industries and for architectural and automotive trim. Steels of this range of composition are resistant to oxidation at temperatures up to about 1550 F (845 C). The properties of this type of steel are shown in Table 14-V.

An alloy containing about 27 per cent chromium and not more than 0.35 per cent carbon, Type 446, is used where resistance to oxidation at

TABLE 14-V. PROPERTIES OF FERRITIC STAINLESS STEELS.
(Krivobok, *Metal Progress,* Vol. 34, 1938, pp. 47-52)

Composition	16-20% Cr (0.12% C max.)		23-30% Cr (0.30% C max.)	
Microstructure	Ferrite with a little pearlite		Ferrite	
Effect of heat treatment	Somewhat modified		Not modified	
Effect of cold work	Modified		Modified*	
Room temperature properties	Annealed	Cold-Worked	Annealed	Cold-Worked
Tensile strength lb/in.²	75,000-90,000	90,000-190,000	75,000-95,000	85,000-175,000
Yield point lb/in.²	40,000-55,000	65,000-130,000	45,000-60,000	55,000-135,000
Elongation in 2 in. (%)	30-20	20-2	30-20	25-2
Reduction of area (%)	55-40	40-20	60-50	50-20
Izod impact, (ft-lb)	75-5	30-2	Very low	Very low
Brinell hardness	140-180	175-275	140-180	150-250
Strength when hot				
1000 F	47,500-54,000		49,000-52,000	
1100 F	37,000-41,000		20,000-34,000	
1300 F	15,000-23,000		10,000-15,000	
1500 F	8,000-12,000		6,000- 7,500	
Creep strength (1% in 10,000 hr)				
1000 F	5,500- 8,500		6,000- 7,000	
1100 F	2,300- 5,000		2,700- 3,300	
1200 F	1,300- 2,100		1,600- 1,800	
1300 F	1,000- 1,400		500- 800	

* Age hardening and similar treatments are excluded.

temperatures between 1500 and 2100 F (815 and 1150 C) is required. As in other ferritic steels, these high-chromium alloys cannot be hardened by heat treatment, and grain refinement cannot be accomplished by heating; in fact, heating to about 2000 F (1095 C) generally produces coarsening. The presence of nitrogen may reduce the tendency toward grain coarsening. The 27 per cent chromium steels are not as easy to cold-form as the 17 per cent chromium steels.

When the ferritic chromium steels, containing in excess of about 20 per cent chromium, are exposed to a temperature of about 900 F (485 C) for a long period, a precipitation of a second phase from the ferrite is likely to occur, thus reducing their ductility at room temperature. This constituent is an iron-chromium compound forming at 50 atomic per cent chromium and is known as "sigma phase." The ductility can be restored by cooling rapidly from a temperature of between 1400 and 1600 F (760 and 870 C).

Welding of the 27 per cent chromium steels is difficult because of the marked grain coarsening that occurs at the welding temperature, thus reducing the ductility. The use of a chromium-nickel-iron welding rod tends to decrease this embrittlement.

14.12 Chromium Steel—Low Nickel. The addition of nickel to the chromium steels extends the austenite region and thus increases the air-hardening tendency of these steels. The addition of about 2 per cent nickel to the 16 to 18 per cent chromium, low-carbon alloys, renders them heat-treatable, as in the case of Type 431. The effect of carbon content on the properties of these steels cooled in air from 1740 F (950 C) and tempered at 1200 F (649 C) is given in Table 14-VI. The heat treatment of these steels is complicated by the possible presence of delta ferrite at the austenitizing temperature, thus requiring close control of composition and quenching temperature to assure effective hardening.

TABLE 14-VI. EFFECT OF CARBON CONTENT ON PROPERTIES OF CHROMIUM STEEL WITH LOW NICKEL; BARS, $1\frac{1}{8}$ IN. D, AIR-HARDENED FROM 1740 F, TEMPERED AT 1200 F.

(From Thum, *The Book of Stainless Steel*)

C %	Cr %	Ni %	Ult. Strength (lb/in.²)	Yield Point (lb/in.²)	Elongation (%)	Izod Impact (ft-lb)
0.11	18.0	2.18	101,500	80,600	27.5	87,83,88
0.16	17.2	2.30	109,800	86,900	24.0	68,76,68
0.21	17.6	2.46	118,500	88,300	22.0	50,60,56
0.27	17.9	2.20	133,300	101,900	20.0	46,42,47
0.35	17.4	2.01	150,800	127,700	17.0	20,20,18

These steels have been employed for various high-strength fittings on seaplanes because of their excellent combination of strength and corrosion resistance. They are also serviceable for valve parts and pumps handling hot or cold salt water or other brines.

14.13 18 Per Cent Chromium-8 Per Cent Nickel Steel. The structure of an 18 per cent chromium steel can be made austenitic at room temperature by the addition of 8 per cent nickel. In discussing the phase relations of the iron-chromium-nickel-carbon alloys, it was noted that, under conditions that were as near to equilibrium as possible, the structure would be composed of austenite, ferrite, and carbide. However, with most practical methods of cooling, these phases are not easily produced. To insure the retention of austenite at room temperature, rapid cooling is recommended. Since they are usually in the austenitic state, they are commonly referred to as the austenitic steels.

The corrosion resistance of the 18 per cent chromium-8 per cent nickel alloys (commonly referred to as 18-8 stainless steel) is superior to that of the ferritic or martensitic steels, since the presence of nickel extends the passive range and improves the stability of the film formed. The cold-forming and welding characteristics of these steels are very satisfactory. The austenitic structure provides high creep resistance, toughness, and strength, particularly above 1100 F (595 C). Because of the greater strength at elevated temperatures, corrosion resistance, and forming characteristics, they are well-adapted for use in engine manifolds, collector rings, food equipment, chemical equipment, still tubes, etc., in spite of the higher cost. It is desirable to maintain the carbon content as low as possible so that there will be the least possibility of the existence of carbides at the grain boundaries to reduce corrosion resistance. The carbon content of Types 301, 302, and 302B may vary between 0.08 and 0.2 per cent. The retention of carbide in solution with the austenite is insured by rapid cooling from a temperature of about 1900 F (1040 C). Types 304 and 308 are low-carbon grades containing a maximum of 0.08 per cent carbon, whereas Type 304L with the carbon controlled to a maximum of 0.03 per cent is practically immune to carbide precipitation.

If the carbon content of this steel is on the high side of the permissible range, and the alloy is heated in the range between about 800 and 1600 F (425 and 870 C) for any appreciable length of time, carbides will precipitate at the grain boundary, greatly reducing corrosion resistance at ordinary temperatures. This subject is discussed later under carbide precipitation.

The development of Types 201 and 202, low-nickel, high-manganese stainless steels, was stimulated to satisfy a need for conservation of nickel. These steels are comparable to Types 301 and 302 respectively with regard to rate of work hardening, mechanical properties and for most corrosion resisting applications.

14.14 High-Chromium, High-Nickel Steels. In order to obtain greater resistance to oxidation at elevated temperature, the chromium and nickel content may be about 25 per cent and 12 or 20 per cent, respectively. These steels are illustrated in Types 309, 309S, 310, 310S, and 314. Because of their greater strength and resistance to oxidation at elevated temperatures, these materials are used for furnace parts, such as conveyor chains, rails, shafts, rollers, etc.

Alloys containing about 29 per cent chromium and 9 per cent nickel have been used with success in resisting the action of hot sulfur dioxide, salt water, and other corrosive agents. The strength is somewhat higher and the ductility is a little lower than the 18-8 steel. These materials are also cast and may have the following properties:

Yield strength will vary between 55,000 and 60,000 lb/in.^{2}
Tensile strength may be between 96,000 and 105,000 lb/in.^{2}
Elongation will be about 28 per cent in 2 in.
Reduction of area will be about 30 per cent.

These alloys are usually austenitic, although with certain analyses the structure may be composed of chromium ferrite and austenite, as shown in the phase diagram.

The maximum resistance to high-temperature oxidation is obtained with about 30 per cent chromium. The addition of nickel greatly improves the creep characteristics of the chromium steels by increasing the high-temperature strength.

14.15 Mechanical Properties of 18-8 Stainless Steel. Since the 18-8 stainless steels are austenitic, they cannot be hardened by heat treatment. Increased strength can be attained, however, by cold working. In the soft condition, the tensile strength is about 100,000 lb/in.^{2}, and the elongation is about 65 per cent in 2 in. By cold working, the tensile strength may be raised very appreciably as shown in Fig. 14.13. With 60 per cent reduction by cold working, the strength of the 18-8 stainless steel may be increased to approximately 190,000 lb/in.^{2} These steels are quite susceptible to relatively small variations in composition, as is apparent if Figs. 14.13 and 14.14 are compared. An increase of carbon content from 0.05 to 0.14 per cent increases the strength and response to cold working.

Again referring to Figs. 14.13 and 14.14, it will be noted that the nickel content is also of considerable importance. The elongation of these alloys is favored by the higher nickel content. The carbon content is maintained as low as possible to reduce the tendency for carbide precipitation.

Although the tensile strength of the 18-8 stainless steel is quite high, the proportional limit is low, thus rendering this material unsuitable for applications in which permanent changes of dimension are not permitted with

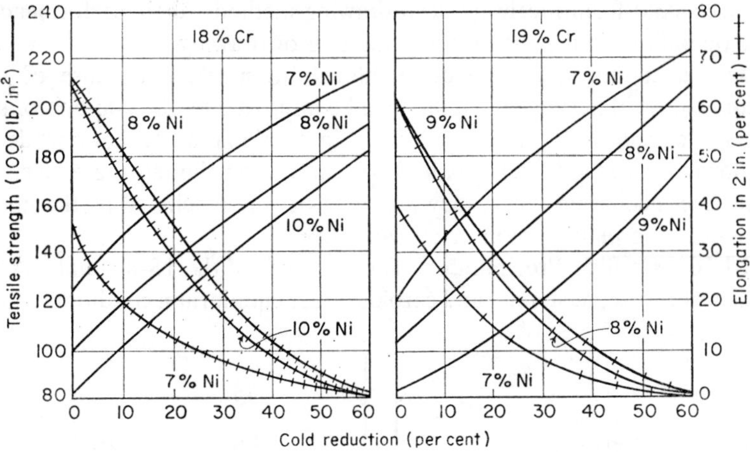

FIGS. 14.13 and 14.14 Properties of stainless steel after various amounts of cold work. *Left:* for 18% Cr, 0.05% C with 7, 8, and 10% Ni. *Right:* for 19% Cr, 0.05% C with 7, 8, and 9% Ni. (Krivobok and Lincoln, *Trans. ASM* **25,** 1937, p. 646)

high stresses. The yield strength of the 18-8 stainless steel is about the same as many other steels with similar tensile strength. Some specifications require a proof test in which the stress required to produce a given amount of permanent deformation is determined. Such tests require careful technique, accurate instruments, and considerable time. The data used in the determination of the proof stress of an 18-8 stainless steel are shown in Fig. 14.15. In this case, the proof stress of 101,000 lb/in.² is that stress corresponding to a permanent deformation of about 0.0001 in./in. Although the 18-8 stain-

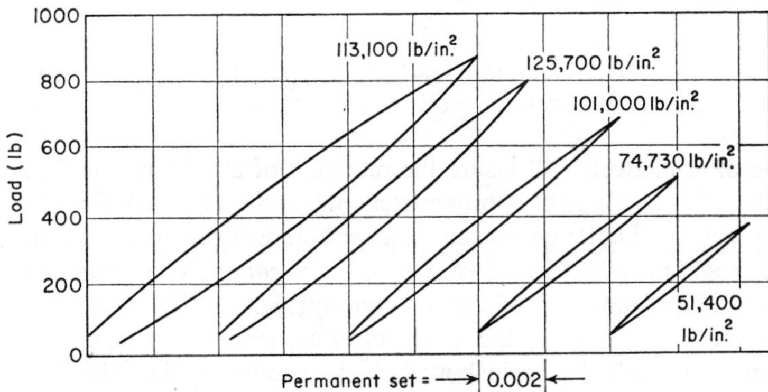

FIG. 14.15 Stress-strain curves, by optical methods, during the determination of proof stress for 18.27% Cr, 7.57% Ni, and 0.11% C alloy. (Krivobok, Lincoln, and Patterson, *Trans. ASM* **25,** 1937, p. 664)

less steels cold-form nicely by deep-drawing methods, their performance in these operations is dependent upon the rate of drawing.

The austenitic steels are quite difficult to machine because of their gummy nature. Their machinability can be improved by the addition of such elements as sulfur or selenium, usually in combination with molybdenum or phosphorus, respectively, as in Type 303. Even though these steels are called free-machining types, they are not as easily machined as the normal free-machining plain carbon or alloy steel.

14.16 Carbide Precipitation. The structure of the 18-8 stainless steels consists of ferrite, austenite, and carbide under equilibrium conditions. Rapid

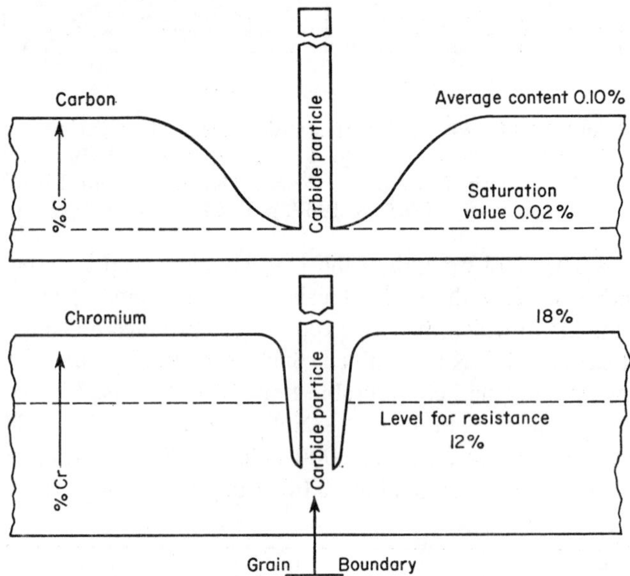

FIG. 14.16 Schematic diagram showing probable distribution of carbon and chromium in vicinity of carbide particle. (Thum, *The Book of Stainless Steels*)

cooling of such steels will insure the retention of austenite, but if they are heated within the range of approximately 800 and 1600 F (425 and 870 C) for any appreciable length of time, the carbide will precipitate in the grain boundaries. This action is referred to as *sensitization*. The effect of sensitization on the chromium and carbon concentration is shown in Fig. 14.16. The upper diagram shows the variation in carbon content in passing from one grain, through the grain boundary, to another grain. The chromium content varies in the manner shown in the lower portion of this figure. There is a narrow region at the grain boundary which contains less than 12 per cent chromium. This is below the 12 per cent chromium minimum required for good corrosion resistance. If there were no carbon present in the alloy, this condition would not be possible. The condition shown in Fig. 14.16 is a transient condition; therefore, exposing the steel to a temperature of about 1200 to 1400 F (650 or 760 C) for a long time will tend to reduce the chromium deficiency. This treatment, although it lowers the chromium content in the grains, will raise the chromium content at the edge of the grains to about 12 per cent, thus restoring corrosion resistance.

The sensitization of an austenitic steel permits corrosive attack to start at the grain boundary where there is a deficiency of chromium. Since the grains are more resistant than the boundaries, the corrosion follows the boundaries, which is typical of intergranular corrosion. The resistance to corrosion at elevated temperatures is reduced somewhat by cold working. This, however, has been attributed to the precipitation of the carbide on the many slip planes within the grain rather than at the grain boundaries. The general precipitation of carbide is not as detrimental as grain-boundary precipitation.

Once carbide precipitation has occurred at the grain boundary, resolution can be attained by heating the steel to between 1800 and 1900 F (980 and 1040 C) whereupon the carbide will be taken into solution with the austenite. Quenching the alloy from this temperature will produce homogeneous austenite.

Sensitization may occur in the welding of 18-8 stainless steel. The precipitation usually occurs not in the weld itself, but in that portion of the structure which is heated to within the sensitization range for a sufficient length of time. In spot welding, sensitization is not usually a problem because of the short time at temperature and the rapidity with which the material cools. After ordinary 18-8 stainless steels have been welded, they should be heated to 1800-1900 F (980-1040 C) and quenched to restore the corrosion-resistant qualities.

For many applications, the liability of sensitization is remote. When a part is to be used at temperatures within the sensitization range, as in exhaust manifolds or welded parts that cannot be heat-treated after welding, a modification of the analysis of the steel is recommended. These modified steels are called the stabilized 18-8 types.

14.17 Stabilization. Since intergranular corrosion is due principally to a deficiency of chromium at the grain boundaries, this deficiency having been brought about by the precipitation of the carbide, any means by which the carbon can be made to combine with some other element rather than chromium will tend to prevent sensitization. It is by this method that the 18-8 stainless steel can be stabilized or rendered insensitive to heating in the range of 800 to 1600 F (425 to 870 C). The elements employed are those that have stronger carbide-forming tendencies than chromium; these include columbium and titanium. If columbium is used, the amount present should be at least ten times the carbon content, as in Type 347; with titanium, there should be about five times the carbon content present, as in Type 321. Columbium is not only a strong carbide-former, but also tends to stabilize the ferrite. By maintaining the chromium content on the high side, the formation of ferrite may be prevented.

The action of these stabilizing elements is apparently to permit the retention of the chromium by the austenite. If carbon is precipitated from the grains, it will be in the form of titanium or columbium carbide. Apparently, the precipitation of a carbide does not reduce the corrosion resistance, provided no chromium is removed from the grains.

In welding of 18-8 stainless steel, it is preferable to use the columbium-stabilized type in preference to the titanium-stabilized type. Titanium is easily oxidized, and, although loss of this element may be only slight with a single bead, multiple bead welds and cross welds may decrease the titanium content sufficiently to reduce its stabilizing effect.

14.18 Applications of 18-8 Stainless Steel. The 18-8 stainless steels are used for a very wide variety of applications. The resistance of these steels to common corroding media is shown in Table 14-VII. They are very satisfactory at elevated temperatures and have been used with considerable success for steel tubes operating in the temperature range of 900 to 1000 F (480 to 540 C). If the 18-8 stainless steels are heated for a very long period at temperatures of the order of 1200 or 1300 F (650 or 705 C), they may become magnetic and less ductile at room temperature. When overheating occurs for any appreciable length of time, the material may be considerably embrittled, and failure may occur at elevated temperature by intercrystalline rupture. Overheating may result from the formation of a heavy deposit in the tube which prevents proper heat flow. This can be eliminated by proper cleaning at regular intervals.

The presence of ferrite tends to improve high temperature ductility of these steels and may be accomplished by adding titanium, molybdenum, or, to a lesser extent, vanadium, or by decreasing the nickel content. Finer-grained steels also tend to have improved ductility at high temperatures.

The 18-8 stainless steels are particularly susceptible to pit corrosion, which is accelerated by hydrochloric acid, chlorides, foreign particles, etc. It has been found that the addition of about 3 per cent molybdenum will prevent serious pitting even in a 10 per cent aqueous solution of ferric chloride. Types 316, 316L, and 317 are alloys of this kind.

As a group, the 18-8 austenitic stainless steels have a very low thermal conductivity with a coefficient of expansion about 50 per cent greater than carbon steel. This fact should be kept in mind where heat transfer is involved in design.

TABLE 14-VII. CORROSION OF 18-8 STAINLESS STEEL (TYPE 304) IN VARIOUS MEDIA.

(From Uhlig, *Corrosion Handbook*)

CORROSIVE MEDIUM	TEMPERATURE	DURATION OF TEST (HR)	WEIGHT LOSS (mdd*)	(ipy†)
20% Nitric acid	Room	1	Nil	—
20% Nitric acid	Boiling	18	Nil	—
3% Nitric acid	Boiling	18	Nil	—
1% Nitric acid	Boiling	18	Nil	—
Nitric acid fumes	230 F (110 C)	13	100	0.018
10% Hydrochloric acid	Room	1	360	0.065
10% Sulfuric acid	Room	1	432	0.079
1% H_2SO_4 + 2% HNO_3	Room	17	Nil	—
0.25% H_2SO_4 + 0.25% HNO_3	Room	17	Nil	—
10% Acetic acid, C.P.	Room	3	Nil	—
10% Acetic acid, C.P.	Boiling	12	Nil	—
Glacial acetic acid, U.S.P.	Room	276	0.1	0.000
Glacial acetic acid, U.S.P.	Boiling	167	130	0.024
Crude acetic acid	Boiling	166	375.5	0.068
10% Phosphoric acid, C.P.	Boiling	17	Nil	—
10% Carbolic acid, C.P.	Boiling	16	Nil	—
10% Chromic acid (tech.)	Boiling	41	204	0.037
Concentrated sulfurous acid	Room	22	Nil	—
0.5% Lactic acid	Boiling	16	4.1	0.001
1.0% Lactic acid	150 F (65 C)	16	Nil	—
2.0% Lactic acid	Boiling	16	5.1	0.001
50% Lactic acid	Boiling	16	12,240	2.23
85% Lactic acid	Boiling	16	1,560	0.284
10% Tartaric acid	Boiling	39	Nil	—
1% Oxalic acid	Boiling	39	177.6	0.032
10% Oxalic acid	Room	17	139.2	0.025
10% Formic acid	Boiling	1	3,240	0.590
10% Formic acid	Room	17	2.4	0.000
10% Malic acid	Room	17	Nil	—
10% Sodium sulfite	Boiling	16	Nil	—
10% Sodium bisulfite	Boiling	16	Nil	—
10% Ammonium sulfate	Boiling	16	Nil	—
10% Ammonium chloride	Boiling	16	Pitted	—
Lemon juice	Room	89	Nil	—
Orange juice	Room	91	Nil	—
Sweet cider	Room	23	Nil	—
Canned rhubarb	Boiling	16	Nil	—
Canned tomatoes	Boiling	16	Nil	—
10% Sodium hydroxide	Boiling	41	Nil	—

* mdd = milligrams per square decimeter per day.

† ipy = inch penetration per year.

Types 304L and 347 have been widely used to satisfy the special corrosion problems associated with liquid metal cooling systems in nuclear reactors. The mass transfer effect is a serious problem in handling molten sodium or lead. This is caused by the hot liquid metal dissolving the container alloy and the subsequent deposition at a critical point when it is cooled by the heat-exchanger. This results in a reduction of the wall thickness and plugging of the system. Intergranular corrosion also presents a problem in these applications. The grain boundaries are attacked by either the liquid metal or by impurities such as oxides that may be present.

14.19 Precipitation-Hardenable Stainless Steels. The requirements of the airframe and missile industries for suitable high-strength materials for skins, ribs, bulkheads, and other structural components has inspired the development of precipitation-hardenable stainless steels in competition with titanium alloys for applications at temperatures up to about 1000 F (540 C). Modification of the basic composition 17-7 austenitic stainless steel has yielded three new groups of alloys that are capable of being hardened in different ways. Hardening may be accomplished by (1) precipitation from a martensitic matrix, (2) precipitation from a martensitic matrix after the transformation of austenite to martensite, or (3) precipitation from an austenitic matrix. Typical compositions of alloys in each group are shown in Table 14-VIII.

Alloys of the first, or martensitic group, are normally solution-annealed

TABLE 14-VIII. COMPOSITION (%) OF PRECIPITATION-HARDENABLE STAINLESS STEELS.

Alloy	C	Cr	Ni	Mo	Al	Mn	Si	Other
				Martensitic				
17-4 PH	0.07 max	16.5	4.00	—	—	1.00 max	1.00 max	Cu 2.75
Stainless W	0.07	17.0	7.00	—	0.20	—	—	Ti 0.70
				Semi-Austenitic				
17-7PH	0.07	17.0	7.00	—	1.15	0.60	0.40	—
PH 15-7 Mo	0.09 max	15.0	7.00	2.50	1.00	1.00 max	1.00 max	—
AM-350	0.10	16.5	4.30	2.75	—	0.80	0.25	N 0.10
AM-355	0.13	15.5	4.30	2.75	—	0.95	0.25	N 0.10
				Austenitic				
17-10P	0.12	17.0	10.0	—	—	—	—	P 0.25
HNM	0.30	18.5	9.50	—	—	3.50	—	P 0.23

at a temperature of 1900 F (1040 C) and air-cooled with the resultant transformation of austenite to martensite. Subsequent aging in the temperature range of 850-1050 F (455-565 C) causes a precipitation effect. Typical mechanical properties after aging at a temperature of 900 F (480 C) are as follows: tensile strength 195,000 lb/in.^{2}, yield strength 180,000 lb/in.^{2}, elongation of 13 per cent in 2 in. These alloys have poor cold-forming and shearing characteristics in the annealed condition.

Alloys of the second, or semi-austenitic group, are of two types: the 17-7 PH and PH 15-7 Mo alloys which respond to precipitation hardening and the AM-350 and AM-355 alloys which are classified in this group because of the similarity of their heat treatments, even though they do not actually harden by the precipitation mechanism. All alloys in this group are generally furnished in the "mill-annealed" condition. This treatment consists of a solution-anneal at a temperature of 1950 F (1065 C) and provides maximum formability. A choice of two hardening cycles is possible, i.e., a subzero cooling treatment that results in higher strength or a double-aging treatment that produces improved ductility in the 17-7 PH and the PH 15-7 Mo alloys.

The subzero cooling treatment for the 17-7 PH and the PH 15-7 Mo alloys consists of an austenite conditioning treatment at a temperature of 1750 F (955 C), followed by cooling to a temperature of −100 F (−73 C) to produce the transformation to martensite and the subsequent aging at a temperature of 950 F (510 C) to produce Condition RH 950 material. Typical mechanical properties for the 17-7 PH alloy are as follows: tensile strength 235,000 lb/in.^{2}, yield strength 220,000 lb/in.^{2}, and elongation of 6 per cent in 2 in. The properties of the PH 15-7 Mo alloy are as follows: tensile strength 240,000 lb/in.^{2}, yield strength 215,000 lb/in.^{2}, and elongation of 6 per cent in 2 in.

The double-aging treatment for 17-7 PH and PH 15-7 Mo alloys consists of heating to a temperature of 1400 F (750 C) and cooling to below 60 F (15 C) to cause the austenite to transform to martensite, followed by aging at a temperature of 1050 F (565 C) to produce Condition TH 1050 material. This treatment produces a tensile strength of 200,000 lb/in.^{2}, a yield strength of 185,000 lb/in.^{2}, and an elongation of 9 per cent in 2 in. in the 17-7 PH alloy. This same treatment produces a tensile strength of 210,000 lb/in.^{2}, a yield strength of 200,000 lb/in.^{2}, and an elongation of 7 per cent in 2 in. in the PH 15-7 Mo alloy.

The treatments for the AM-350 and AM-355 alloys are similar to those given for the previous alloys with a slight modification of the aging temperature. Since these alloys are hardened by the austenite-martensite transformation, the aging treatment is in effect a tempering operation without secondary hardening. The subzero cooling treatment is capable of developing a tensile strength of 200,000-215,000 lb/in.^{2}, a yield strength of 175,000-195,000 lb/in.^{2}, and an elongation of 10-12 per cent in 2 in.

Alloys of the third, or austenitic group, are solution-annealed by rapid cooling from the temperature range of 2000-2250 F (1095-1230 C) to produce a supersaturated austenitic matrix with excellent ductility and forming characteristics. Aging in the temperature range of 1200-1400 F (650-760 C) will produce a tensile strength of 140,000-180,000 lb/in.^{2}, a yield strength of 100,000-140,000 lb/in.^{2}, and an elongation of 16-20 per cent in 2 in.

REFERENCES

Borgmann *et al.*, *Corrosion of Metals,* American Society for Metals, Metals Park, Ohio, 1946.

Metals Handbook, American Society for Metals, Metals Park, Ohio, 1961.

Samans, *Engineering Metals and Their Alloys,* Macmillan Co., New York, 1949.

Symposium on the Nature, Occurrence, and Effects of Sigma Phase, Technical Publication No. 110, American Society for Testing Materials, Philadelphia, Pa., 1950.

Thum, *The Book of Stainless Steels,* American Society for Metals, Metals Park, Ohio, 1935.

Uhlig, *Corrosion Handbook,* John Wiley & Sons, New York, 1948.

Woldman, *Engineering Alloys,* American Society for Metals, Metals Park, Ohio, 1954.

Zapffe, *Stainless Steels,* American Society for Metals, Metals Park, Ohio, 1949.

QUESTIONS

1. Why is chromium effective in increasing the corrosion resistance of steel?
2. What is the sigma phase, and what is its effect on properties of alloys?
3. How is the response to hardening of corrosion-resistant steels related to the constitution diagram of that particular system?
4. Draw a curve showing approximately the effect of different amounts of chromium upon the resistance to oxidation of steel at elevated temperature.
5. What is Invar, and what is its application?
6. What is Elinvar, and what is its application?
7. What is the approximate composition of corrosion- and heat-resistant steels that are called austenitic steels?
8. How may the corrosion- and scale-resistant steels be divided into three groups?
9. What types of steels are classified as martensitic stainless steels?
10. What types of steels are referred to as ferritic stainless steels?
11. What steels are included in the classification of austenitic stainless steels?
12. What steels are included in the classification of low-alloy, corrosion-resistant steels? What are their advantages?

13. What types of steels are used for cutlery? Indicate some typical compositions of steels used for this purpose.
14. What are some uses for the chrome steels of the ferritic type?
15. What advantages may be derived from the use of the austenitic steels in comparison with the ferritic and martensitic steels?
16. Indicate some specific applications of 18 per cent chromium, 8 per cent nickel steels.
17. What difficulties may be encountered with austenitic steels which contain too high a carbon content?
18. Can an 18 per cent chromium, 8 per cent nickel steel be hardened by heat treatment? Why?
19. What is the mechanism of sensitization of the 18-8 stainless steel?
20. If a piece of 18-8 stainless steel has been sensitized, how may it be returned to its normal condition?
21. How may 18-8 stainless steel be stabilized?
22. In what three ways may the precipitation-hardenable stainless steels be hardened?

XV

HIGH-STRENGTH, HEAT-RESISTANT ALLOYS

15.1 High-Temperature Characteristics of Metals. The growing demand for alloys to meet the requirements of high-temperature engineering applications presents a serious problem to the metallurgist. This demand resulted initially from the greater efficiency found possible in many process operations and heat engines, including cracking stills, steam turbines, and, more recently, in turbosuperchargers and gas turbines. Supersonic aircraft and missiles present new problems in oxidation, abrasion, and high-temperature strength because aerodynamic heating in high-speed flight causes skin temperatures to increase to high values. Aircraft power plants and rocket motors require improved materials to meet the needs of increased thrust.

A large number of the modern high-temperature alloys have been developed since 1940 as a result of extensive wartime research on "superalloys" sponsored by the Office of Scientific Research and Development and the National Aeronautics and Space Administration, formerly the National Advisory Committee for Aeronautics. It is important that the engineer recognize the inherent characteristics of metals at elevated temperatures and the limitations of the more important commercial materials available for such use.

The damage to metals and alloys by loss of strength at high temperature may be of three general types: (1) by oxidation or exposure to contaminating media with resultant loss of metal; (2) by incipient surface cracks brought about by cyclic thermal stresses; and (3) by changes in the properties of the metal with increasing temperature, with or without attendant phase changes.

Loss of strength by oxidation and scaling is an extreme case of corrosion and has already been discussed in a general way in Chapter XIV.

Thermal stresses produced in the surface of a metal by a sudden localized change in temperature will tend to produce incipient ruptures in the skin, which may develop into a well-defined, heat-check network with subsequent heating cycles. The thermal gradient is intensified in metals having an inherently low thermal conductivity and high thermal expansion at elevated temperatures. Heat-checking or thermal fatigue is aggravated by large volume changes in the transformation range in ferrous alloys, by the presence of corroding media which infiltrate the cracks, and by a relatively low strength at high temperatures. This type of failure is observed in ingot molds, large brake drums, locomotive driving wheels, glass molds, die-casting dies, etc.

The common structural and machinery steels exhibit a definite decrease of yield strength and ultimate strength with increasing temperature, as

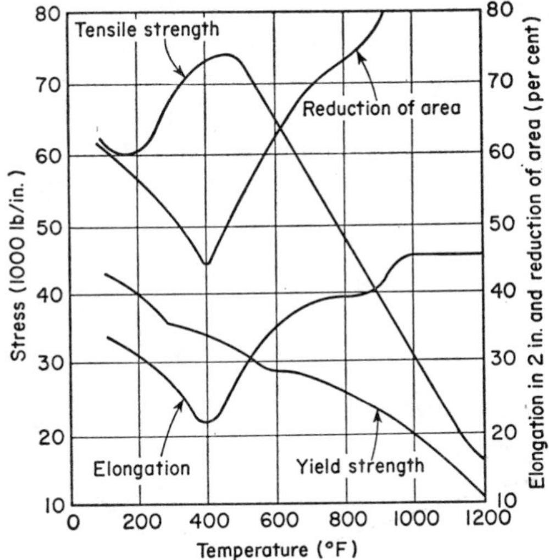

FIG. 15.1 Short-time tensile properties of forging steel. (After *Effect of Temperature on the Properties of Metals, Symposium ASME-ASTM,* 1931)

illustrated in Fig. 15.1. These curves result from short-time tensile tests obtained by heating standard tensile specimens in a furnace mounted on the testing machine and conducting the test as soon as the specimen reaches thermal equilibrium at a given temperature.

With increasing temperature, the stress-strain curve for a given metal changes in character and becomes flatter, indicating less resistance to slip. Above a certain temperature, hardening does not result from plastic deformation; hence, a metal will continually deform under a constant load. This characteristic is known as *creep*.

The type of actual fracture is influenced by the temperature and the rate of strain. At elevated temperatures, fractures are generally intercrystalline, compared to the transcrystalline failures at room temperature resulting from the slip mechanism in ductile metals. That temperature or range of temperature at which the grain boundary becomes weaker than the grain itself is referred to as the *equicohesive temperature* for a given metal or alloy.

15.2 Mechanism of Creep. In most cases, the problem of application of metals for elevated-temperature service is concerned with continuous flow at constant stress. At ordinary temperatures most metals and alloys exhibit hardening and strengthening as a result of plastic deformation. Such strain-hardenable metals will remain in a state of constant strain as long as the applied stress is constant. In the absence of strain hardening, a metal will continue to deform or flow while the applied stress is essentially constant. Furthermore, the higher the magnitude of the stress, the more rapid will be the rate of flow. Since little can be done to control this continuing deformation, a structure must be designed on the basis of a specified amount of deformation that can be tolerated in a given time and then must be removed from service when this time has elapsed.

The assembly of creep data on metals and alloys is of paramount importance for use in the design of structures to be used at elevated tempera-

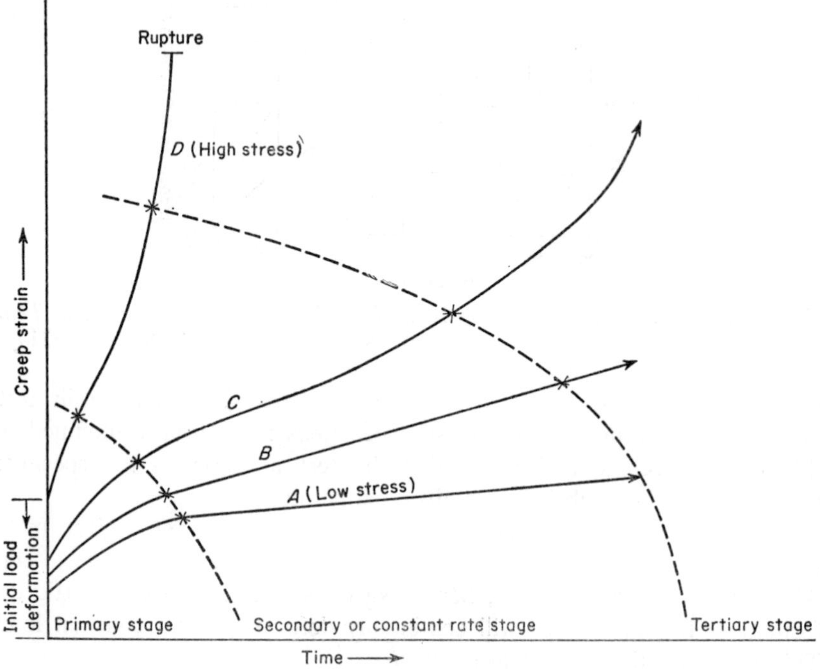

FIG. 15.2 Three stages of creep under different stresses at constant temperature.

tures. Creep data are usually secured by subjecting a specimen of the metal to a constant load at the temperature in question and by observing the change in length as a function of time. Creep tests are usually conducted at several temperature levels. The general character of the change of length with respect to time during creep is shown schematically in Fig. 15.2. Curve *A* represents a low constant load at constant temperature, whereas curves *B*, *C*, and *D* illustrate the trend of increasing constant load with different specimens.

Upon application of the load, some deformation takes place immediately as a result of elastic strain and some plastic strain. This is followed by a period in which the rate of creep gradually decreases; this period is known as the stage of *primary creep*. The creep rate then reaches a minimum, and deformation continues at a uniform rate for a period of time known as the stage of *secondary creep*. With relatively large amounts of deformation, the creep rate increases markedly, introducing the stage of *tertiary creep*, which leads to eventual rupture of the specimen if the test is carried to completion.

The stress and temperature at which the creep test is made will affect the character of the creep curve. High stresses and high temperatures will increase the minimum creep rate during secondary creep and will decrease the time required for rupture, as has been shown in Fig. 15.2. The creep rate of a material is also influenced by certain inherent characteristics of the metal and manufacturing variables including phase transformations, alloying elements, structure, grain size, precipitation effects, deoxidation practice, etc.

Creep tests are time-consuming and for economical reasons are seldom continued to final rupture of the specimen. Extrapolation of the rate of secondary creep has been employed to provide design data for engineering applications, providing the creep test has progressed for a sufficient period to define the slope of the creep curve in the constant rate or secondary stage, and assuming that the material will not pass into the tertiary stage during the useful life of the equipment. This usually requires a minimum of 1000 hr of testing time, depending on the material, stress, and temperature. The projected creep values are employed in several ways, a common usage being the stress that will produce 1 per cent elongation in 10,000 or 100,000 hr at a given temperature.

15.3 Stress-Rupture. It should be apparent from the foregoing discussion of creep that short-time tensile properties of metals and alloys at elevated temperatures are of very limited value to the engineer concerned with design for sustained high-temperature operation. Creep tests, on the other hand, require an excessive amount of time to provide even sufficient data to allow extrapolation for the life expectancy of a proposed design.

As an economical compromise for the preliminary evaluation of high-temperature materials, the stress-rupture or creep-rupture test is used for the screening of prospective materials for service in turbosuperchargers, gas turbines, and jet-engine parts. This test yields important information on the high-temperature characteristics of alloys where the intended operating life is limited or considerable deformation can be tolerated.

Stress-rupture specimens are tested at temperatures and under stress conditions that will assure early test failure, usually within 1000 hr. Creep characteristics of a metal under these conditions were illustrated by curve *D* in Fig. 15.2. A logarithmic plot of the stresses and rupture times for a series of tests at a given temperature usually results in a straight-line relationship, provided structural changes do not occur in the alloy to change the slope of the line. A family of stress-rupture curves at various temperatures for an 18 per cent chromium-8 per cent nickel, corrosion-resistant steel is shown in Fig. 15.3. Stress-rupture data are generally reported as the stress required to rupture a specimen in a period of 10, 100, or 1000 hr, often with an indication of the ductility expressed as the percentage elongation determined after rupture.

15.4 Classification of Heat-Resistant Materials. Materials for high-temperature service may be conveniently classified on the basis of the major element or component of which they are composed. In some of the more complex superalloys developed in recent years, several elements may exert an equal influence. The following groups provide a general classification for study purposes:

1. Iron-base alloys.
2. Nickel-base alloys.
3. Cobalt-base alloys.
4. Refractory metals and alloys.
5. Metal-ceramics.
6. Ceramics.

Certain alloys lack the ductility required in forming operations and, hence, are usually cast. This is particularly true of the cobalt-base alloys. Other alloys are melted with difficulty because of the ease with which they are oxidized, thus requiring special vacuum-melting techniques or preparation by powder-metallurgical methods. The refractory metal compounds and metal-ceramics are prepared by the latter method.

A comparison of some typical cast and wrought heat-resistant alloys is given in Tables 15-I and 15-II, respectively. These tabulations include some of the corrosion- and scale-resistant alloys for the purpose of comparison. High-temperature strength characteristics have been presented from available data for many of the alloys. With the exception of certain of the better-

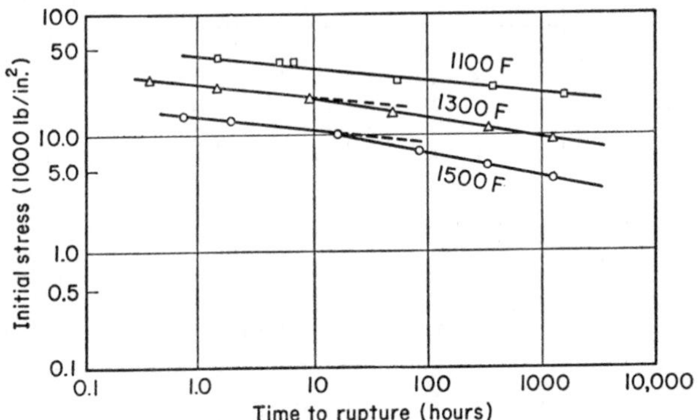

Fig. 15.3 Stress-rupture curves for 18-8 stainless steel. (By permission from *Properties of Metals at Elevated Temperatures*, by G. V. Smith, copyright 1950, McGraw-Hill Book Co.)

known alloys, data in the literature include incomplete information on the heat treatment and processing of test material which, in many cases, is based on single test results. Since manufacturing variables, heat treatment, impurities, test conditions, etc., have a strong influence on the high-temperature strength of metals, the engineer should employ a sizable factor of safety in the use of reported values in design calculations. The design problem in moving parts is further complicated by dynamic stresses acting at elevated temperatures. Very little reliable information is available on high-temperature fatigue and impact properties of alloys, and engineers are not in agreement on the actual value of such data. In the final analysis, the service testing of actual parts under operating conditions provides the best source of information on the selection of alloys and manufacturing methods.

15.5 Iron-Base Alloys. It has been shown in previous chapters that the carbon steels temper readily and have poor creep resistance above about 600 F (315 C). The low-alloy steels containing strong carbide-forming elements, such as molybdenum, vanadium, chromium, and tungsten, resist tempering and have satisfactory creep resistance up to about 800 F (425 C). The medium-alloy and hot work tool steels resist tempering up to a temperature of about 1000 F (540 C). The powerful influence of small quantities of molybdenum in improving the creep resistance of relatively pure iron is illustrated in Fig. 15.4.

High-alloy austenitic steels of the 18-8 type are required for satisfactory strength up to about 1300 F (705 C). Aside from strength requirements, all steels for high-temperature service contain varying amounts of chromium and, in some cases, silicon or aluminum to provide scale resistance. The high-chromium ferritic steels suffer a serious loss of ductility at room

Table 15-1. TYPICAL CAST HEAT-RESISTANT ALLOYS.

Alloy	Nominal Percentage of Principal Elements										Short-time T.S. (lb/in.²)	Stress (lb/in.²) for 1% Creep in 10,000 hr		Stress-Rupture in 1000 hr			
														1200 F (650 C)		1500 F (815 C)	
	C	Cr	Ni	Co	Mo	W	Ti	Al	Fe	Other	1500 F (815 C)	1200 F (650 C)	1500 F (815 C)	lb/in.²	% Elong.	lb/in.²	% Elong.
12 Cr	<0.15	12.0	1.0	—	—	—	—	—	Bal.	—	—	—	—	—	—	—	—
18 Cr	<0.30	20.0	2.0	—	—	—	—	—	Bal.	—	—	—	—	—	—	—	—
28 Cr	<0.50	28.0	4.0	—	—	—	—	—	Bal.	—	10,000	—	1,000	—	—	1,800	—
20-10 Cr-Ni	0.30	20.0	10.0	—	—	—	—	—	Bal.	—	25,000	8,500	4,500	—	—	5,900	—
29-10 Cr-Ni	0.35	28.0	9.5	—	—	—	—	—	Bal.	—	25,000	—	2,750	—	—	—	—
25-12 Cr-Ni	0.35	26.0	12.5	—	—	—	—	—	Bal.	—	25,000	9,000	4,500	15,000	—	6,000	—
28-15 Cr-Ni	0.35	28.0	16.0	—	—	—	—	—	Bal.	—	—	—	5,100	—	—	6,600	—
25-20 Cr-Ni	0.40	26.0	20.0	—	—	—	—	—	Bal.	—	—	—	5,500	—	—	7,000	—
35-15 Ni-Cr	0.55	15.0	35.0	—	—	—	—	—	Bal.	—	30,000	—	6,000	—	—	7,000	—
39-19 Ni-Cr	0.55	19.0	39.0	—	—	—	—	—	Bal.	—	—	—	6,700	—	—	—	—
N-155, Multimet	0.15	21.0	20.0	20.0	3.0	2.0	—	—	Bal.	1.0 Cb, 0.15 N	50,000	—	—	38,000	—	13,000	—
60-12 Ni-Cr	0.55	12.0	Bal.	—	—	—	—	—	25.0	—	25,000	—	4,000	—	—	6,000	—
67-17 Ni-Cr	0.55	17.0	Bal.	—	—	—	—	—	10.0	—	—	—	—	—	—	—	—
Hastelloy B	0.10	1.0	Bal.	—	28.0	—	—	—	5.0	—	58,500	—	—	40,500	—	12,500	—
Hastelloy C	0.10	16.0	Bal.	—	17.0	4.0	—	—	5.0	—	56,500	—	—	42,500	—	14,500	—
Hastelloy X	0.15	22.0	Bal.	—	9.0	0.6	—	—	18.5	—	52,000	—	—	—	—	10,000	—
Cosmoloy F	0.04	15.0	Bal.	—	4.0	2.0	3.5	4.5	—	0.08 B, 0.07 Zr	—	—	—	—	—	—	—
GMR-235	0.15	15.0	Bal.	—	5.0	—	2.0	3.0	10.0	0.06 B	85,000	—	—	—	—	25,000	8.0
Inconel 713C	0.12	13.0	Bal.	—	4.5	—	0.6	6.0	1.0	2.0 Cb	110,000	—	—	—	—	40,000	5.0
Nicrotung	0.10	12.0	Bal.	10.0	—	8.0	4.0	4.0	—	0.05 B, 0.05 Zr	115,000	—	—	—	—	—	4.0
HS-21 (Vitallium)	0.25	27.0	3.0	Bal.	5.0	—	—	—	1.0	—	59,000	24,000	9,000	44,000	—	14,000	5.0
HS-30 (422-19)	0.40	24.0	17.0	Bal.	6.0	—	—	—	1.0	—	—	—	—	—	—	21,700	—
HS-31 (X-40)	0.40	25.0	10.0	Bal.	—	8.0	—	—	1.0	—	59,500	—	13,500	46,000	—	23,000	7.0
HS-36 (L251)	0.40	19.0	10.0	Bal.	—	14.5	—	—	1.0	0.03 B	—	—	—	—	—	25,500	—
S-816	0.40	19.0	20.0	Bal.	4.0	4.0	—	—	3.0	4.0 Cb	70,000	—	—	46,000	—	22,000	1.0
60-25-15 Cr-Fe-Mo	0.03	Bal.	—	—	15.0	—	—	—	25.0	—	90,000	—	—	—	—	—	—
60-15-25 Cr-Fe-Mo	0.03	Bal.	—	—	25.0	—	—	—	15.0	—	—	—	—	—	—	32,000	9.0

Table 15-II. TYPICAL WROUGHT HEAT-RESISTANT ALLOYS.

Alloy	Nominal Percentage of Principal Elements										Short-time T.S. (lb/in.²)	Stress (lb/in.²) for 1% Creep in 10,000 hr		Stress-Rupture in 1000 hr			
														1200 F (650 C)		1500 F (815 C)	
	C	Cr	Ni	Co	Mo	W	Ti	Al	Fe	Other	1500 F (815 C)	1200 F (650 C)	1500 F (815 C)	lb/in.²	% Elong.	lb/in.²	% Elong.
5 Cr (501)	>0.10	5.0	—	—	—	—	—	—	Bal.	—	9,500	2,000	—	6,000	—	1,500	—
12 Cr (410)	<0.15	12.0	—	—	—	—	—	—	Bal.	—	8,000	2,000	500	4,500	—	—	—
17 Cr (430)	<0.12	17.0	—	—	—	—	—	—	Bal.	—	8,000	2,000	750	—	—	—	—
27 Cr (446)	<0.35	25.0	—	—	—	—	—	—	Bal.	—	7,000	1,500	250	4,000	—	1,000	—
18-8 (304)	<0.08	19.0	9.5	—	—	—	—	—	Bal.	—	28,000	7,000	750	15,000	—	3,500	—
18-8 Ti (321)	<0.08	18.0	9.5	—	—	—	—	—	Bal.	Ti 5 × % C	32,000	8,000	750	17,500	—	3,500	—
18-8 Cb (347)	<0.08	18.0	10.5	—	—	—	—	—	Bal.	Cb 10 × % C	32,000	9,500	1,000	17,000	—	4,500	—
18-8 Mo (316)	<0.10	17.0	12.0	—	2.5	—	—	—	Bal.	—	40,000	15,000	3,000	25,000	—	7,000	—
24-12 (309)	<0.20	23.0	13.5	—	—	—	—	—	Bal.	—	32,000	8,500	750	20,500	—	5,000	—
25-20 (310)	<0.25	25.0	20.5	—	—	—	—	—	Bal.	—	50,000	9,000	1,000	17,000	—	4,500	—
19-9 DL	0.30	19.0	9.0	—	1.25	1.2	0.30	—	Bal.	0.4 Cb	39,000	20,000	7,000	40,000	2.5	10,500	6.0
19-9 DX	0.30	19.0	9.0	—	1.50	1.2	0.55	—	Bal.	—	46,000	—	—	42,000	—	—	—
Timken 16-25-6	0.10	16.0	25.0	—	6.0	—	—	—	Bal.	0.15 N	47,000	20,000	5,500	34,000	4.5	9,000	20.0
A-286	0.08	15.0	25.0	—	1.25	—	2.10	0.35	Bal.	0.25 V	37,000	29,000	10,000	46,000	8.5	8,400	—
Discaloy 24	0.04	13.5	26.2	—	3.9	—	1.6	0.11	Bal.	—	36,000	35,000	—	41,000	—	—	—
Incoloy T	0.10	20.5	32.0	—	—	—	1.0	—	Bal.	—	—	—	—	26,100	—	7,000	—
N-155, Multimet	0.15	21.0	20.0	20.0	3.0	2.5	—	—	Bal.	1.0 Cb, 0.15 N	46,000	27,000	9,000	40,000	—	13,000	—
S-590	0.42	20.5	20.0	20.0	4.0	4.0	—	—	Bal.	4.0 Cb	52,000	28,000	9,500	38,000	37.0	16,000	10.0
Refractaloy 26	0.05	18.0	37.0	20.0	3.0	—	2.8	0.2	Bal.	—	71,000	45,000	17,000	63,000	—	18,000	16.0
Refractaloy 80	0.10	20.0	20.0	30.0	10.0	5.0	—	—	Bal.	—	69,000	—	—	—	—	—	—
Hastelloy X	0.15	22.0	Bal.	—	9.0	—	—	—	18.5	—	52,000	—	—	30,500	—	10,000	—
Inco 739	0.07	15.0	Bal.	—	—	—	1.7	2.7	1.0	—	—	—	—	—	—	20,000	—
Inconel	0.04	15.5	Bal.	—	—	—	—	—	1.0	—	—	—	—	14,500	—	3,700	—
Inconel W	0.04	15.0	Bal.	—	—	—	2.5	0.6	7.0	—	—	—	—	54,000	—	11,500	—
Inconel X	0.03	15.0	Bal.	—	—	—	2.3	0.9	6.5	0.6 Cb	64,000	—	18,000	62,000	1.5	18,500	2.5
Nimonic 75	0.12	20.0	Bal.	—	—	—	0.4	0.06	2.4	—	—	—	—	—	—	—	—
Nimonic 80A	0.05	20.0	Bal.	—	—	—	2.3	1.0	0.5	—	—	—	—	56,000	—	15,500	—
M-252	0.15	19.0	Bal.	10.0	10.0	—	2.5	0.75	5.0	—	100,000	—	—	70,000	—	18,000	—
René 41	0.10	19.0	Bal.	10.0	10.0	—	3.0	1.50	1.0	0.005 B	126,000	—	—	100,000	—	25,000	—
Udimet 500	0.15	17.5	Bal.	16.5	4.0	—	3.0	2.75	4.0	0.008 B	125,000	—	—	95,000	—	35,000	—
Udimet 700	0.15	15.0	Bal.	17.5	5.0	—	3.5	4.25	1.0	0.10 B	130,000	—	—	100,000	—	—	—
Waspaloy	0.05	19.5	Bal.	13.5	4.25	—	3.0	1.25	2.0	0.005 B, 0.06 Zr	100,000	—	—	85,000	—	20,000	—
Inco 700	0.10	15.0	Bal.	28.0	3.0	—	2.0	3.0	0.5	—	107,000	—	—	87,000	—	30,000	—
Nimonic 90	0.08	20.0	Bal.	16.0	—	—	2.3	1.4	0.5	—	—	—	—	63,000	—	17,900	—
K-42-B	0.05	18.0	Bal.	22.0	—	—	2.5	0.2	13.0	—	71,000	18,800	7,500	39,500	0.5	15,000	2.5
S-816	0.38	20.0	20.0	Bal.	4.0	4.0	—	—	4.0	4.0 Cb	70,000	42,000	11,500	50,000	4.0	22,000	10.0
V-36	0.31	25.0	20.0	Bal.	4.0	2.6	—	—	3.0	2.2 Cb	40,000	—	—	—	—	18,000	—
HS-25 (L605)	0.12	20.0	10.0	Bal.	—	15.0	—	—	1.0	—	47,000	—	14,000	58,000	—	18,000	—
Jetalloy 1570	0.22	20.0	28.0	Bal.	—	7.0	4.0	—	2.0	—	82,000	—	—	84,000	—	24,000	—
														(1800 F (980 C))		(2000 F (1095 C))	
Mo-Cb	0.03	—	—	—	Bal.	—	—	—	—	0.75 Cb	—	—	—	42,000	—	12,000	—
Mo-Zr	0.02	—	—	—	Bal.	—	—	—	—	0.09 Zr	—	—	—	45,000	—	—	—
Mo-Ti	0.02	—	—	—	Bal.	—	0.46	—	—	—	90,000	—	—	43,000	—	16,000	—

temperature, caused by the separation of the "sigma phase," when held in the range 900 to 1300 F (480 to 705 C) for an extended period of time. This phenomenon may also be noted at higher holding temperatures with the 18-8 austenitic steels unless the composition is carefully balanced and controlled to prevent separation of ferrite and embrittlement resulting from the formation of "sigma phase."

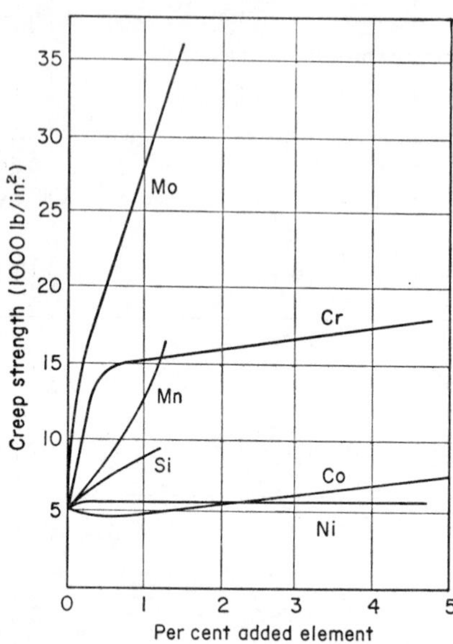

Fig. 15.4 Effect of certain elements on creep strength of iron at 800 F—0.0001% per hour. (Austin *et al.*, *Trans. AIME*, 1945)

Iron-base casting alloys are of two general types: (1) iron-chromium ferritic alloys and (2) chromium-nickel and nickel-chromium austenitic alloys. The 12, 18, and 28 per cent chromium alloys of the iron-chromium group are essentially corrosion- and scale-resistant irons having low strength and ductility at elevated temperatures. The 20-10, 29-10, 25-12, 28-15, and 25-20 per cent chromium-nickel alloys have greater strength and ductility at elevated temperatures and will also withstand greater fluctuations in temperature. These austenitic alloys are widely used for furnace parts, which must withstand creep under heavy loads and also resist attack from sulfur-containing gases.

Alloys in which the nickel exceeds the chromium content, such as in 35-15 and 39-19 per cent nickel-chromium alloys, are used where severe thermal shock is present, but they will not withstand atmospheres containing appreciable amounts of sulfur. They are employed for quenching fixtures for hardening steel and in furnace parts to resist the effects of carburizing and nitriding.

Wrought iron-base alloys for moderate and high-temperature service include: (1) low-alloy machinery steels such as the AISI 4XXX and 6XXX series, (2) medium-alloy special steels including AISI Type 501 and hot work tool steels and (3) high-alloy ferritic and austenitic steels of the AISI 400 and 300 series, respectively, including complex modifications of the 18-8 and 25-20 types that make up the iron-base superalloys for use at temperatures in excess of about 1000 F (540 C). Typical modifications include 19-9 DL, Timken 16-25-6, A-286, and Discaloy 24. Alloys 19-9 DL and Timken 16-25-6 alloys are strengthened by the alloying effect of carbide-forming elements and by "hot-cold-working" in the range of 1200-1400 F (650-760 C) rather than by heat treatment. Alloys A-286 and Discaloy 24 are hardened by solution treatment and aging. These alloys have been used for jet-engine turbine rotors at temperatures up to 1200 F (650 C). Superalloys N-155 (Multimet), S-590, Refractaloy 26, and Refractaloy 80 may be considered as borderline alloys under the iron-base classification, since they contain almost equivalent amounts of chromium, nickel, cobalt, and iron.

15.6 Nickel-Base Alloys. The nickel-base alloys may be classified as (1) nickel-chromium, (2) nickel-chromium-molybdenum, (3) nickel-chromium-cobalt, and (4) nickel-chromium-cobalt-molybdenum with other more complex variations of these basic compositions.

The nickel-chromium series of high-nickel alloys is highly resistant to corrosion at elevated temperatures, has excellent thermal shock resistance, and possesses electrical resistivity, which makes certain of the alloys particularly adaptable for electric heating elements. "Inconel" has been widely used for industrial applications that require corrosion- and oxidation-resistance at elevated temperatures, such as for retorts and fixtures for heat-treating furnaces. "Inconel X" is an age-hardening alloy possessing high strength and corrosion resistance up to temperatures of about 1500 F (815 C) and has been used for gas-turbine blades, combustion liners in jet engines, tail pipes, etc.

Nickel-chromium-molybdenum alloys are resistant to attack by a variety of chemical agents; they have good high-temperature strength and ductility. Alloys in this group include the "Hastelloys," Cosmoloy F, GMR-235, Inconel 713C, etc.

Nickel-chromium-cobalt alloys, such as Nimonic 90 and K-42-B, are age-hardenable turbine blade alloys.

Alloys of the nickel-chromium-cobalt-molybdenum series have been developed as substitutes for cobalt-base alloys, such as S-816, in order to conserve the available supply of cobalt. These alloys possess the high strength required in jet-engine turbines. Alloys of this type respond to precipitation hardening and include Inco 700, M-252, René 41, Udimet 500, Waspaloy, etc. Vacuum melting of these alloys produces high purity turbine blade material with a substantial increase in high-temperature strength and ductility.

15.7 Cobalt-Base Alloys. The cobalt-base alloys are noted for their excellent high-temperature strength characteristics and corrosion resistance. Red-hardness and wear-resistant qualities have caused certain of these alloys to be used as cutting-tool materials for a number of years prior to their use in engineering design for high-temperature service. The widespread use of these alloys is somewhat restricted by certain disadvantages, including the relatively high cost and limited availability of cobalt, combined with the very poor machinability and lack of ductility of the medium- and high-carbon varieties. The high-temperature strength increases with carbon content, but there is a corresponding decrease in ductility in most cases.

The original alloy of this group which continues in use is a combination of cobalt, chromium, nickel, and molybdenum known as Stellite or HS-21 ("Vitallium"). This has been employed extensively in the form of precision investment castings for gas-turbine blades in jet engines and aircraft superchargers. A variation of the cobalt-chromium-nickel-tungsten type, known as HS-31 (X-40), possesses even superior high-temperature characteristics at temperatures up to about 1700 F (925 C). Alloy HS-25 (L605), a low-carbon weldable wrought alloy, finds application in afterburners at temperatures of the order of 2000 F (1095 C). Alloy S-816 has been widely used for forged turbine blades in jet engines, but in view of its high strategic alloy content, it has been replaced by nickel-chromium-cobalt-molybdenum alloys which have equivalent properties at temperatures up to about 1600 F (870 C).

15.8 Refractory Metals and Alloys. The continuing demand for materials to operate at temperatures above 1600 F (870 C) has directed attention to the refractory metals and their alloys. The maximum temperature at which a metal possesses useful strength properties is directly related to its recrystallization temperature. Recrystallization temperatures for pure metals will approximate 40 per cent of the absolute melting temperatures and may be increased substantially by the addition of alloying elements. For example, cobalt with a melting point of 2723 F (1495 C) and a recrystallization temperature of approximately 810 F (435 C) can be alloyed to provide useful properties at temperatures up to about 1700 F (925 C). The recrystallization temperature and melting points of a number of pure metals are compared in Fig. 15.5. The strategic position of the refractory metals as a base for high-temperature alloys is illustrated in Fig. 15.5. From the standpoint of cost and availability, alloy research and development have been concentrated on the nonprecious refractory metals, chromium, molybdenum, and tungsten. These metals have melting points of 3430 F (1888 C), 4760 F (2727 C), and 6170 F (3410 C), respectively.

The difficulty of providing a crucible material to withstand these temperatures during any melting operation is apparent. The problem is further complicated by metal reactions with both the crucible and the atmosphere at the melting temperatures. Until recently, the preparation of these metals and their alloys has been limited to parts of relatively small size that could be compacted and sintered by powder metallurgical methods. The early development of tungsten filament wire from sintered compacts by hot swagging and drawing is an example of this method of preparation.

Extensive work on the vacuum melting of chromium-base alloys was conducted during World War II with concentrated study on the 60-25-15 and 60-15-25 chromium-iron-molybdenum alloys. Although these alloys possess good creep resistance at 1600 F (870 C), their application has been limited by brittleness at room temperature. Ductile chromium has been produced from high-purity electrolytic chromium, and continued research

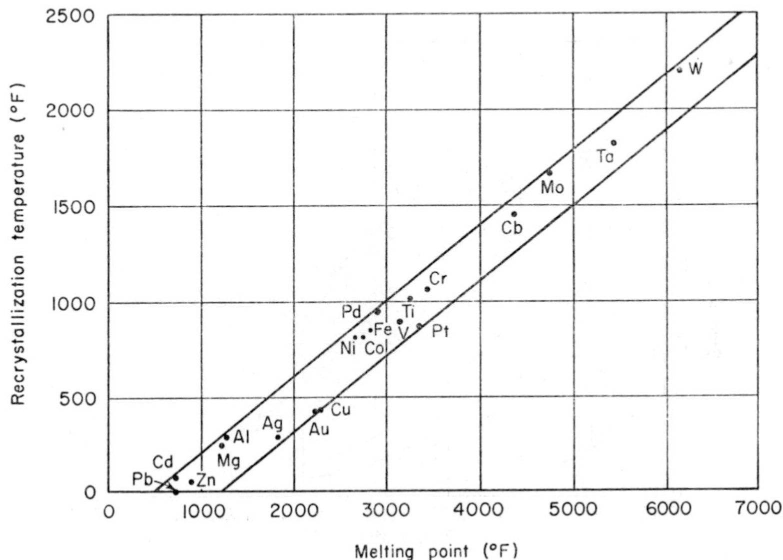

FIG. 15.5 Recrystallization temperature vs. melting point for cold-worked pure metals.

may develop useful alloys for applications at temperatures up to about 2000 F (1095 C).

Ductile molybdenum is a development made possible by the arc-cast process. This material shows considerable promise in the high-temperature field, particularly at temperatures above about 1800 F (980 C), as compared to the commercially available superalloys. Aside from a high melting point, this metal has the advantage of possessing a high thermal conductivity and a low coefficient of thermal expansion. A serious disadvantage in connection with its use is the ease of formation of an oxide film, which vaporizes readily above 1400 F (760 C). Several methods of protection have been proposed, including cladding with nickel-base alloys, vapor-phase deposition with silicon, and coating with ceramic compositions having similar thermal-expansion characteristics. Research on molybdenum alloys has been directed toward improving the oxidation resistance and increasing the hardness of the base metal. An alloy containing 0.5 per cent titanium shows considerable promise for the future after the oxidation problem has been solved.

In view of the fact that tungsten possesses the highest melting point of the metallic elements, it is only natural to expect attention to be directed toward the possibility of producing a ductile tungsten by the arc-cast

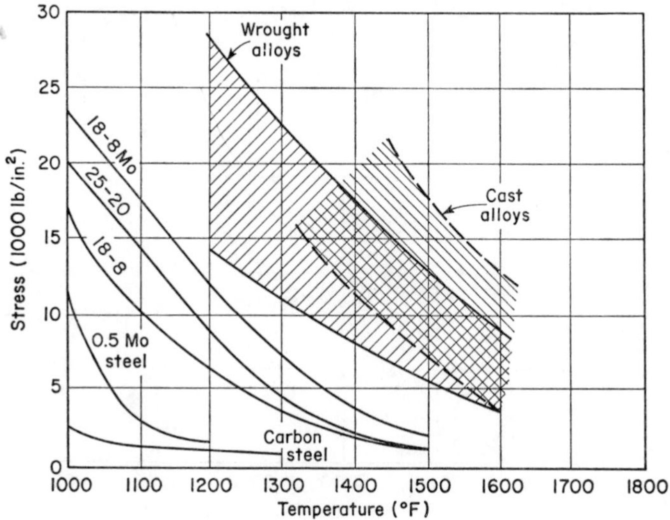

FIG. 15.6 Creep strength of superalloys vs. steel (creep rate of 0.0001% per hour). (By permission from *Properties of Metals at Elevated Temperatures,* by G. V. Smith, copyright 1950, McGraw-Hill Book Co.)

process. Current research indicates that forgeable tungsten ingots may be available for high-temperature application of the future.

15.9 Ceramics and Metal-Ceramics. The ceramics in general possess very low thermal conductivity with poor thermal and mechanical shock characteristics at room temperature. Studies made on relatively pure alumina, beryllia, and zirconia bodies, however, have indicated that these materials have better high-temperature strength characteristics than metals at temperatures above about 2000 F (1095 C).

To improve the thermal conductivity of the ceramics, a group of materials known as metal-ceramics, ceramals, or cermets has been developed by powder metallurgical methods or by metal infiltration of pressed ceramic shapes. These materials contain a nickel, cobalt, or refractory metal base combined with a carbide, boride, oxide, silicide, or nitride. Low impact strength and ductility restrict the application of the cermets as structural materials. Titanium carbide bonded with nickel, cobalt, or Inconel has been recognized as having the best high-temperature properties. Recent developments with titanium boride and chromium boride cermets indicate that these materials may possess even higher strength at 2000 F (1095 C).

15.10 Evaluation of High-Temperature Alloys. The relative position of the cast vs. wrought superalloys with respect to the creep-resistant steels is compared in Figs. 15.6 and 15.7 for creep and rupture strengths, re-

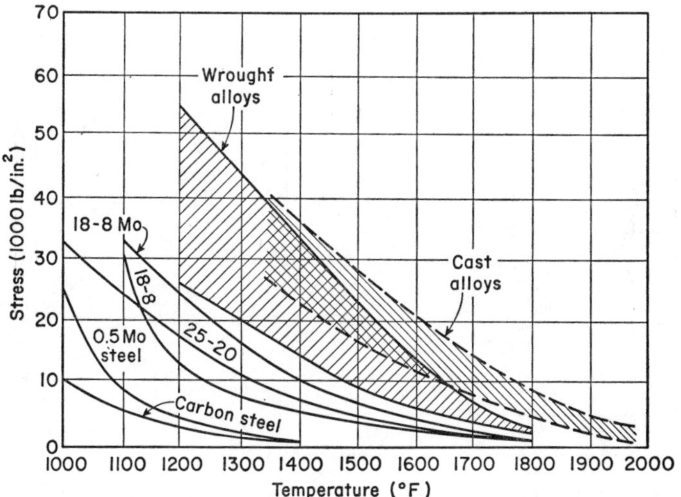

FIG. 15.7 Rupture strength of superalloys vs. steel (rupture in 1000 hr). (By permission from *Properties of Metals at Elevated Temperatures*, by G. V. Smith, copyright 1950, McGraw-Hill Book Co.)

spectively. A comprehensive comparison of the stress-rupture properties of various alloys is shown in Fig. 15.8.

In general, the wrought alloys appear to have better strength properties at 1200 F (650 C) than the cast alloys, whereas the cast alloys are superior at temperatures above about 1500 F (815 C). Since the high-temperature alloys must be cast in sand or ceramic molds, the subsequent slow rate of cooling produces a coarse and variable grain size. It has been indicated previously that high-temperature fractures above the equicohesive temperature range are intergranular. In a coarse-grain structure, the amount of grain boundary material to suffer from incipient softening would be less than in a fine-grained material. This may explain the higher strength properties normally experienced with cast materials at elevated temperatures.

It is recognized that high-temperature data on many of the alloys discussed in this chapter are meager and limited in usefulness by the absence of complete manufacturing information. The wrought alloys are generally sensitive to variations in finishing temperatures. Heat treatment also plays an important role in both the wrought and cast alloys. The austenitic struc-

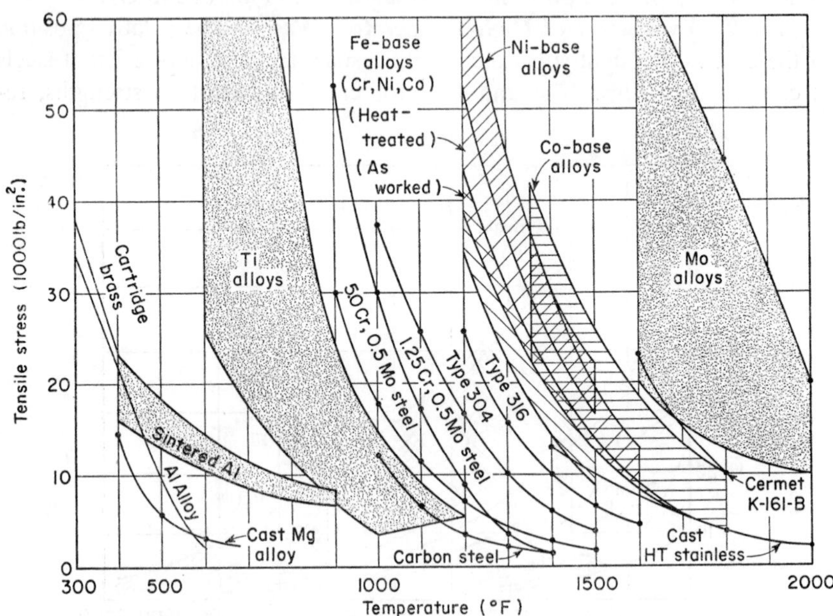

Fig. 15.8 Stress-temperature curves for various alloys (rupture in 1000 hr). (From *Metals Handbook, 1954 Supplement*)

ture is generally stronger at elevated temperatures when a second phase can be precipitated from solid solution. Most of the complex superalloys are susceptible to precipitation effects, either by actual heat treatment or as a result of high-temperature service. There is a limiting temperature and time at temperature for a given alloy beyond which loss of strength will occur, as a result of over-aging, as discussed in Chapter VI.

REFERENCES

Clark, *Metals at High Temperatures,* Reinhold Publishing Corp., New York, 1950.

Ductile Chromium, American Society for Metals, Metals Park, Ohio, 1957.

Goodwin and Greenidge, *Forgeable Arc-Melted Tungsten, Metal Progress* **59,** Metals Park, Ohio, 1951, pp. 812-814.

Grant, Frederickson, Taylor, *Heat Resistant Alloys from 1200° F to 1800° F, Iron Age,* March 18, April 8, April 15, 1948.

Haynes Alloys for High-Temperature Service, Haynes Stellite Co., Kokomo, Indiana, 1948.

Metals for Supersonic Aircraft and Missiles, American Society for Metals, Metals Park, Ohio, 1958.

Metals Handbook, American Society for Metals, Metals Park, Ohio, 1961.

Parker, *High Temperature Properties of the Refractory Metals, Trans. ASM* **42,** Metals Park, Ohio, 1950, pp. 399-404.

Parker *et al., High Temperature Properties of Metals,* American Society for Metals, Metals Park, Ohio, 1951.

Résumé of High Temperature Investigations, Timken Roller Bearing Co., Canton, Ohio, 1940-50.

Sheet Materials for High Temperature Service, American Society for Metals, Metals Park, Ohio, 1958.

Smith, *Properties of Metals at Elevated Temperatures,* McGraw-Hill Book Co., New York, 1950.

Symposium on Effect of Temperature on the Properties of Metals, American Society of Mechanical Engineers and American Society for Testing Materials, Philadelphia, Pa., 1931.

Symposium on Materials for Gas Turbines, American Society for Testing Materials, Philadelphia, Pa., 1946.

The Metal Molybdenum, American Society for Metals, Metals Park, Ohio, 1959.

Utilization of Heat Resistant Alloys, American Society for Metals, Metals Park, Ohio, 1954.

QUESTIONS

1. What three general types of damage may occur to metals and alloys resulting in loss of strength at high temperature?
2. What effect may sudden localized changes of temperature have upon the surface of metals?
3. What two characteristics in metals may intensify heat-checking?
4. What is the effect of temperature upon the yield strength and ultimate strength of steel?
5. What is creep in metals?
6. What is the equicohesive temperature? How is the equicohesive temperature related to the elevated temperature behavior of metals and alloys?
7. How does the stress-strain diagram of a metal at elevated temperature differ from that at ordinary atmospheric temperature?
8. Describe the three stages of creep with respect to the relation between deformation and time.
9. How are the creep characteristics of a material determined?
10. How are stress-rupture or creep-rupture tests made?
11. Into what groups may heat-resistant materials be classified? Give some examples of materials in each group.
12. What are the two general types of iron-base casting alloys used for heat resistance?

13. Give an example of the nickel-base, heat-resistant alloys.
14. Give a typical example of a cobalt-base, heat-resistant alloy.
15. What materials are included in the refractory metals and alloys?
16. What are some of the limitations of ceramics for use in temperature-resistant applications?

XVI

CAST IRONS

16.1 Classification. The general group of iron-carbon alloys referred to as cast iron includes a wide variety of alloys distinguished from one another by differences of composition and structure. The only characteristic common to all alloys of this group is that they are usually sand-cast to the desired shape. The general designation *cast iron* is meaningless, except in the sense of an iron-carbon alloy containing relatively large amounts of carbon either as graphite or as iron carbide. Therefore, it is necessary to make a more specific designation.

Alloys that come within this broad group include white cast iron, gray cast iron, malleable cast iron, and nodular cast iron. Each of these may be modified by the addition of alloy elements in order to obtain specific properties. Most cast iron requires the presence of elements other than iron and carbon to produce the type desired.

White cast iron is an iron-carbon alloy in which all of the carbon is combined in the form of iron carbide and in which the carbon content exceeds 1.7 per cent. An example of this structure is shown in Fig. 16.1.

Gray cast iron is an iron-carbon alloy consisting of a structure of flakes of graphite embedded in a matrix of steel. Any alloy of iron, carbon, and other elements that satisfies this structural requirement is included in this classification. In gray cast iron, graphite flakes nucleate from the melt and a portion of the compound, iron carbide, dissociates into ferrite and graphite. Fig. 16.2 shows a gray cast iron with flakes of graphite intermingled with pearlite. With a certain combination of composition and cooling rate, the structure of the cast iron may consist of a mixture of the structures found in gray and white cast iron. This is called *mottled cast iron*. The general range of composition of gray, mottled, and white cast irons is shown in Fig. 16.3.

Malleable cast iron (malleable iron) is defined as an iron-carbon alloy consisting of a structure of small particles (nodules) of graphite (temper carbon) formed by the dissociation of cementite in the solid state and embedded in a matrix of ferrite or pearlite. Such a structure is shown in Fig. 16.4.

Malleable and gray cast irons differ from each other in two respects. First, the iron carbide is either partially or completely dissociated in the malleable cast iron; second, the dissociation of iron carbide in the malleable cast iron occurs only while the alloy is solid. The dissociation in gray cast iron occurs during the early stages of solidification; hence, the difference in the character of the graphite in each. When the dissociation of iron carbide is incomplete, the structure may consist of temper carbon embedded in a matrix of steel. A material having this kind of structure is referred to as *pearlitic malleable cast iron.*

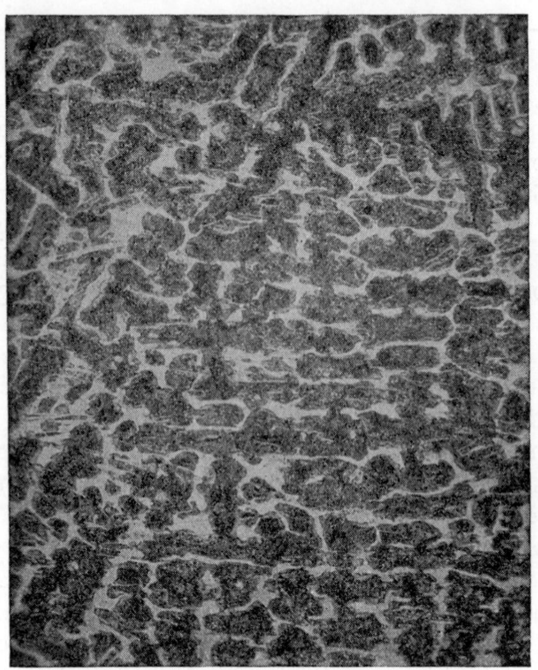

Fig. 16.1 White cast iron (200×).

Nodular cast iron is an iron-carbon alloy having a structure composed of nodules of graphite formed directly during the process of solidification and embedded in a matrix of steel. This structure differs from that of gray cast iron in that the graphite is in nodular form: furthermore, these nodules are produced upon solidification and subsequent cooling and not by a heat treatment as is required in the production of malleable cast iron. The structure of nodular cast iron is shown in Fig. 16.5.

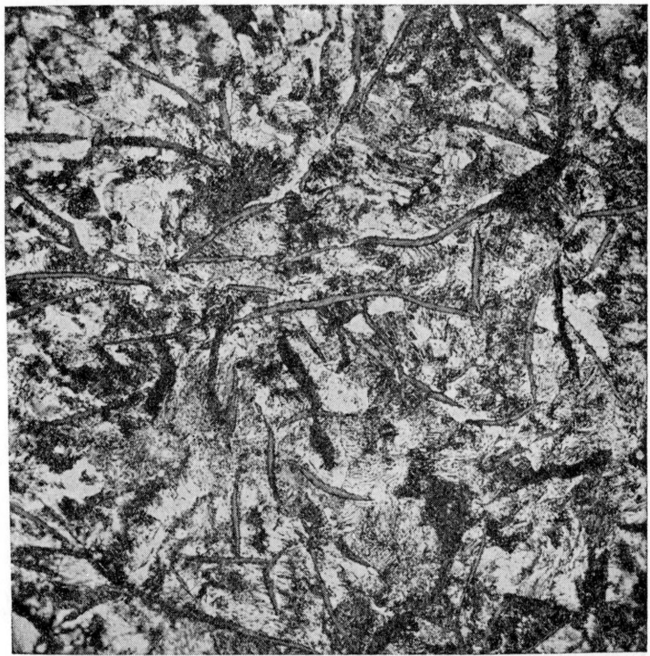

FIG. 16.2 Gray cast iron (250×).

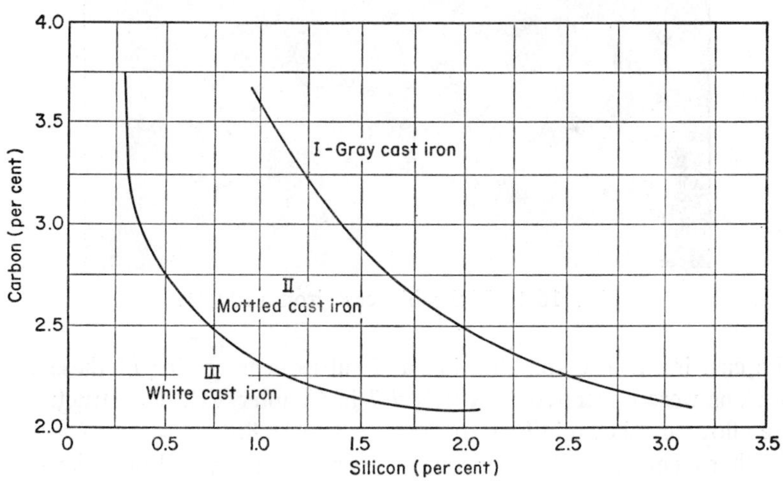

FIG. 16.3 Composition limits for gray, mottled, and white cast irons. (Tonimura, H., *Mem. Coll. Eng.*, Kyushu Imp. Univ., Vol. 6, No. 2, 1931, pp. 115-188)

16.2 White Cast Iron. The iron-carbon alloys containing more than 1.7 per cent carbon in the form of iron carbide (white cast iron) behave in accordance with the equilibrium diagram of the iron and iron carbide system. That portion of the diagram between 1.7 and 6.67 per cent carbon is of interest here. Those alloys containing between 1.7 and 4.2 per cent carbon are designated as hypoeutectic white cast irons; the alloy containing 4.2 per cent carbon is called a eutectic white cast iron; and those alloys containing from 4.2 to 6.67 per cent carbon are known as hypereutectic

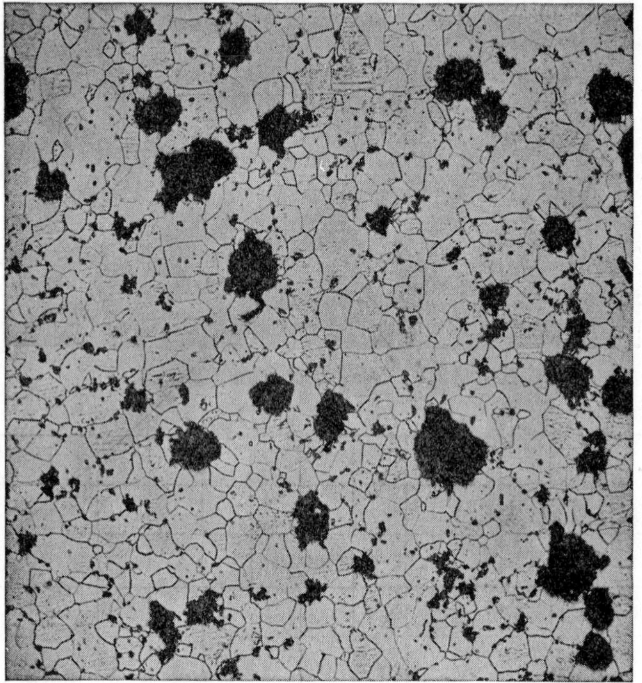

FIG. 16.4 Malleable cast iron (100×).

white cast irons. The solidification and subsequent cooling of these white cast irons were illustrated in Article 7.7 in Chapter VII. The structures of these alloys are shown in Figs. 16.6, 16.7, and 16.8.

Other elements such as silicon, manganese, sulfur, and phosphorus are present in white cast iron. While these elements alter the temperature of transformation and the composition of phases, they do not alter the general character of phase relations that exist in the simple iron-iron carbide diagram. If the alloy is cooled more rapidly, a higher percentage of silicon can be tolerated without the occurrence of graphitization.

White cast iron is not employed for structural parts because of its excessive brittleness; however, it is useful in resisting wear by abrasion. The dissociation of the iron carbide of a gray cast iron can be prevented by rapid cooling. Hence, the use of a chill in the mold will form white cast iron at the surface, thus producing a gray cast iron with a hard, wear-resistant surface. The properties of white cast iron vary widely. The hardness and brittleness increase as the carbon content increases. The tensile strength varies between 25,000 and 50,000 lb/in.^{2} and is usually about 35,000

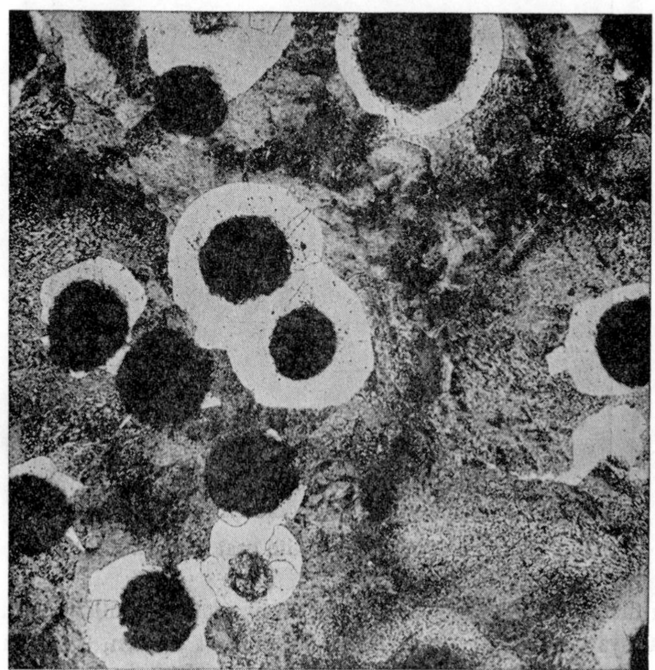

FIG. 16.5 Nodular cast iron as cast (250×).

lb/in.^{2} The hardness ranges between about 350 and 500 Brinell. The compressive strength varies between 200,000 and 250,000 lb/in.^{2}

In view of the hardness and wear resistance of white cast iron, it is used for wearing plates, pump liners, parts of abrasion machinery, parts of sand slingers, etc. The use of this alloy is restricted to parts that can be sand-cast and do not require very much machining. White cast iron can be machined with some tools, although with difficulty; hence, it is common practice to finish by grinding.

16.3 Dissociation or Graphitization of Cementite. The presence of carbon as graphite in cast iron has already been mentioned. Here again.

the versatility of the iron-carbon alloys is evident. Although the compound iron carbide is metastable, it is not difficult to retain it as such when in combination with pure iron. In alloys of iron, carbon, and silicon, cementite is unstable; hence, graphite may be present in the structure.

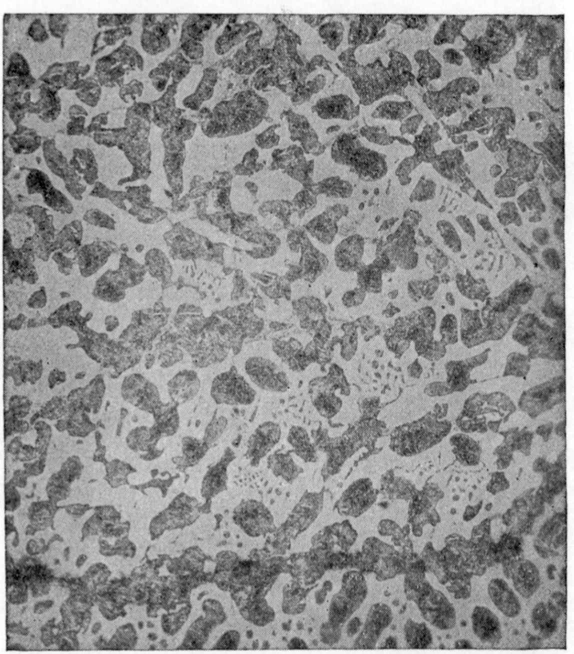

FIG. 16.6 Hypoeutectic white cast iron (200×).

With the addition of graphitizers, such as nickel or silicon, to the iron-carbon alloys containing more than 1.7 per cent carbon, the iron carbide dissociates in accordance with the following reaction:

$$Fe_3C = 3Fe + C$$

The extent of this dissociation is controlled by adjusting the amount of these elements in the alloy or by controlling the rate of solidification and subsequent cooling. A rapid rate of cooling will tend to retard the decomposition of the compound, whereas very slow cooling will give more opportunity for dissociation. High carbon also favors graphitization, as does inoculation of the melt with artificial nuclei. *Meehanite* is a grade of gray cast iron in which additions of calcium silicide are made to the melt to produce a fine distribution of graphite and resultant excellent mechanical properties.

Since graphite is relatively soft and without strength, graphitization of the cementite will tend to lower the hardness and improve machinability.

The several elements that are present in the iron-iron carbide alloys control the extent of dissociation or the stability of iron carbide. The elements that are normally present in these alloys, in addition to carbon, include silicon, manganese, sulfur, and phosphorus. When the cementite graphitizes at very high temperature it usually forms flakes which grow with continued graphitization. The structure resulting from this graphitization will consist of flakes or plates of graphite embedded in a matrix of iron and iron carbide. Several factors control the size and shape of the graphite flakes. The most

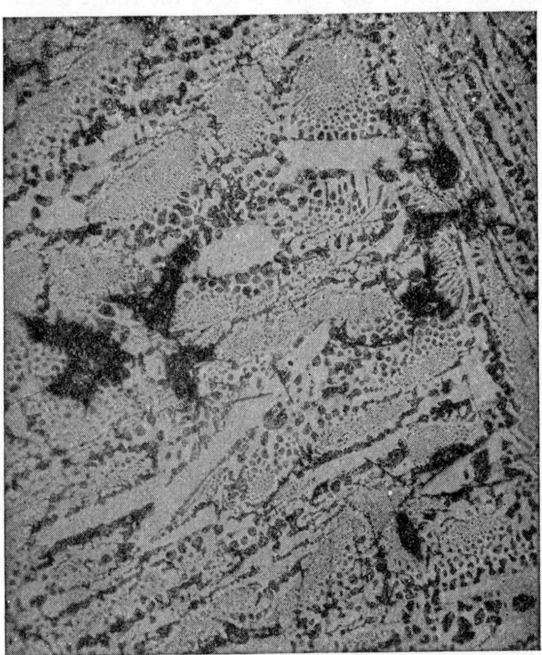

FIG. 16.7 Near-eutectic white cast iron (200×).

desirable structure consists of small flakes of graphite uniformly distributed with random orientation. If the iron carbide has been sufficiently graphitized such that the matrix is essentially a steel, the alloy is described as a gray cast iron.

By properly balancing the composition, no graphitization may occur on solidification and subsequent cooling, but by heating the alloy to a suitable temperature for a sufficient length of time, the iron carbide will dissociate. However, under this condition, the graphite will be in nodular form, producing the structure of malleable cast iron.

16.4 Effect of Silicon. Silicon is one of the most important elements present in gray cast iron. It is the element that exerts the greatest controlling effect on the relative proportion of combined and graphitic carbon. By increasing the silicon content in a given cast iron, a greater proportion of graphite will be formed in the structure at the expense of the combined carbon. This serves as a method by which the properties of the iron can be controlled.

The amount of silicon in cast iron varies between 1 and 3.5 per cent. It is customary to vary the characteristics of the iron by changing both the carbon and silicon content. For high-strength iron, it is usually desirable to keep the carbon content on the lower side of the range and the silicon con-

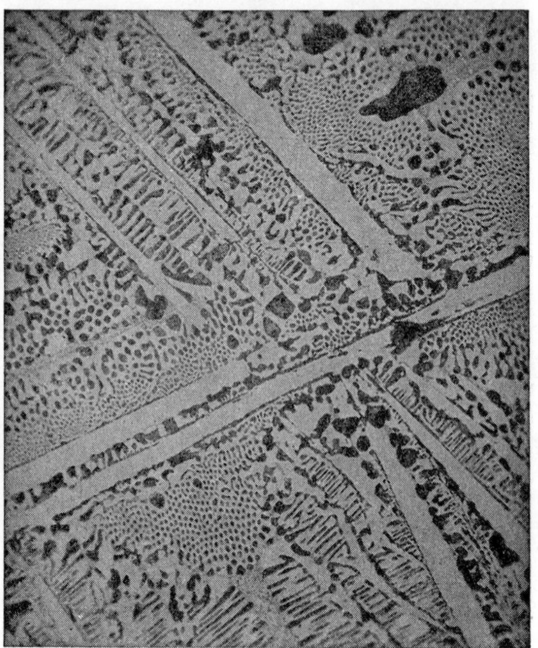

FIG. 16.8 Hypereutectic white cast iron (200×).

tent on the higher side. Care must be exercised, however, that the proper balance is maintained so that the cast iron will have the desired machinability.

The relation between the structure of thin sections of cast iron and the carbon and silicon content is shown in Fig. 16.9. In Region I, the iron carbide is stable; thus, the structure is that of white cast iron. In Region II, there is sufficient silicon to cause graphitization of all of the iron carbide except the eutectoid cementite. This leads to what is called a pearlitic gray cast iron, consisting of flakes of graphite embedded in a matrix of eutectoid steel. In Region III, the large amount of silicon promotes a more complete

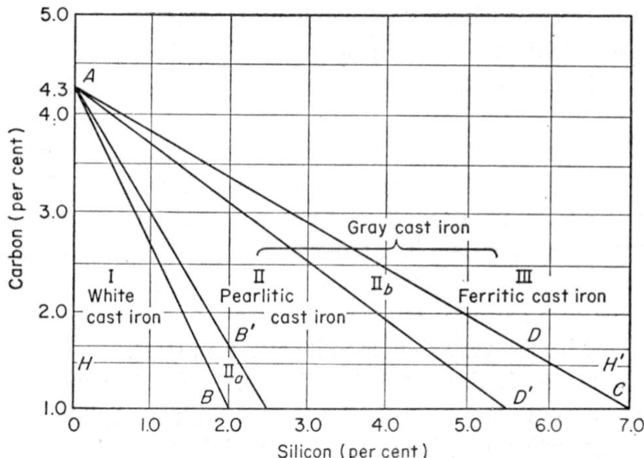

FIG. 16.9 Relation of structure to carbon and silicon content of cast iron. (Maurer)

dissociation of cementite to form graphite and ferrite. The dissociation is so complete that there is a negligible proportion of cementite remaining in the structure.

The relation between the tensile properties and the carbon and silicon content of several alloys is shown in Fig. 16.10. The greatest structural strength is derived with a carbon content of about 2.75 per cent and a silicon content of about 1.5 per cent. In comparing this region with the

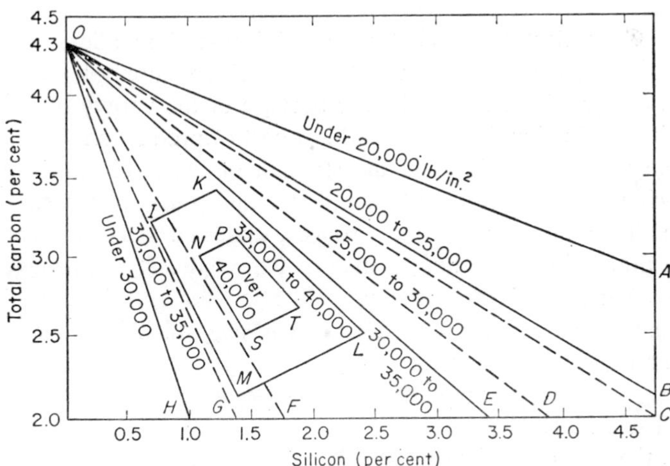

FIG. 16.10 Relation of tensile properties to carbon and silicon content of cast iron. (Coyle, *Trans. ASM* **12,** 1927, p. 446)

regions in Fig. 16.9, it will be observed that the structure should be pearlitic. Although these figures are based upon small sand-cast samples, they will serve as an indication of the most desirable analyses of cast iron for general work.

16.5 Effect of Manganese. The amount of manganese usually found in cast iron varies between 0.4 and 1.0 per cent. Less than 1.0 per cent has very little effect upon the properties of cast iron. With more than about 1.25 per cent manganese, strength, hardness, and resistance to wear are increased, whereas the machinability is decreased because of the greater stability of the iron carbide. The general effect of manganese on the matrix of the gray cast iron is very similar to its effect on steel.

16.6 Effect of Sulfur and Phosphorus. Sulfur is usually present in cast iron in amounts less than 0.12 per cent with negligible effect on the mechanical properties, provided there is a sufficient quantity of manganese. In the absence of the proper proportion of manganese, sulfur tends to stabilize the cementite. It is common practice to have a sufficient quantity of manganese present to combine with the sulfur to form manganese sulfide.

The most marked effect of phosphorus is to improve the fluidity of the cast iron. It is usually desirable to maintain the phosphorus content at less than about 0.30 per cent. However, some cast irons contain as much as 0.90 per cent phosphorus. For high-strength cast iron, it is desirable to maintain a low phosphorus content. Phosphorus in amounts less than 0.3 per cent is dissolved in the ferrite of gray cast iron, but, with more than this amount, a hard, brittle compound, iron phosphide (Fe_3P), is formed. With larger amounts of phosphorus, an iron-iron phosphide eutectic (steadite) is formed.

16.7 Gray Cast Iron. That group of iron-carbon alloys containing flakes of graphite embedded in a matrix of steel, commonly known as gray cast iron, enjoys wide commercial use because of the variety of properties that can be obtained and the ease and low cost of production. The strength of gray cast iron depends upon the strength of the steel matrix and the size and character of the graphite flakes making up the structure.

Since the composition of the strongest annealed steel corresponds to eutectoid proportion of carbon, it may be expected that the strongest gray iron will be produced when the graphitization is controlled so that the matrix contains eutectoid proportions of carbon, which is about 0.70 per cent in the iron-graphite system.

Some typical analyses of cast irons used for many different purposes are given in Table 16-I.

Since the rate of cooling of gray cast iron with a given silicon content controls to a large degree the extent of dissociation of iron carbide, it is of great importance to consider the effect of section thickness on the structural

TABLE 16-I. ANALYSES (%) OF GRAY CAST IRON FOR DIFFERENT APPLICATIONS.
(From *Metals Handbook*)

Application	Total C	Si	Mn	P	S	Ni	Cr	Mo
Automobile cylinder—plain	3.25	2.25	0.65	0.15	0.10	—	—	—
Automobile cylinder—heavy duty	3.25	1.90	0.65	0.15	0.10	1.75	0.45	—
Automobile pistons—plain	3.35	2.25	0.65	0.15	0.10	—	—	—
Automobile piston rings	3.50	2.90	0.65	0.50	0.06	—	—	—
Automobile brake drums	3.30	1.90	0.65	0.15	0.08	1.25	0.50	—
Machinery								
Light service or thin section	3.25	2.25	0.50	0.35	0.10	—	—	—
Medium service or heavy section	3.25	1.75	0.50	0.35	0.10	—	—	—
Heavy service with heavy section	3.25	1.25	0.50	0.35	0.10	—	—	—
High strength iron-plain	2.75	2.25	0.80	0.10	0.09	—	—	—
High strength iron-Ni	2.75	2.25	0.80	0.10	0.09	1.00	—	—
High strength iron-Mo	2.75	2.25	0.80	0.10	0.09	—	—	0.35
Car wheels	3.35	0.65	0.60	0.35	0.12	—	—	—
Light forming and stamping or forging dies	3.30	1.50	0.60	0.20	0.10	2.00	0.60	—

characteristics. The methods of controlling the dissociation of cementite were discussed in Article 16.3. Thick sections of castings will cool slower than thin sections; hence, whereas the thick sections may have a gray structure, the thin sections may be white, thus indicating a high "section sensitivity" which is undesirable. In this event, the thin sections are not readily machinable, and difficulties in production may be encountered. If the silicon content is adjusted so that there will be graphitization in the thin sections, the strength of the heavy sections may be greatly reduced because of the marked graphitization. Finer and more uniform graphitization can be obtained by the addition of nickel instead of by increasing the silicon content, thus reducing the "section sensitivity."

Fig. 16.11 shows the effect of the rate of cooling on the decomposition

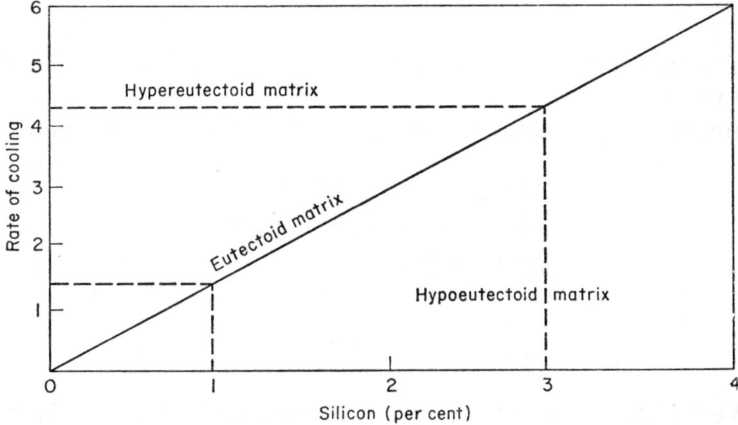

FIG. 16.11 Effect of rate of cooling and percentage silicon on decomposition of cementite.

of cementite for different amounts of silicon. With larger amounts of silicon, a faster rate of cooling must be employed in order to maintain a eutectoid matrix. As an example of the effect of chilling, consider a casting in which the thickness varies from $\frac{1}{4}$ to 1 in. With an ordinary cast iron containing approximately 2 per cent silicon, the heavy portions will be coarse-grained and soft compared to the thin sections. If another cast iron is used in which the silicon content is reduced to about 1.5 per cent, the heavy portion of the casting will possess a very good structure, but the thin section will be hard and unmachinable. If, in this case, two parts of nickel are added for each part of silicon that was removed, a dense structure will be obtained in the heavy section, and the light section will be machinable.

There is a wide variety of compositions used to obtain given properties of gray cast iron. These variables are so wide that there are very few specifications which call for specific analyses of cast iron. The American Society for Testing Materials, in cooperation with the American Foundrymen's Society, has established certain specifications for cast iron on the basis of strength and not of composition. This classification is as follows where the first three classes cover the ordinary grades and the remainder are considered high-strength cast irons:

Class	T.S. Min. (lb/in.^{2})
20	20,000
25	25,000
30	30,000
35	35,000
40	40,000
50	50,000
60	60,000

The relation between strength, hardness, structure, and section size is shown in Fig. 16.12.

Analyses of gray cast iron usually fall within the following limits:

Total carbon: 2.5-3.5%
Combined carbon: 0.6-0.9%
Silicon: 1.5-3%
Manganese: 0.5-0.8%
Sulfur: 0.1-0.2%
Phosphorus: 0.2%

Very high-strength gray cast iron is produced by very carefully controlled production. Castings with strength as high as 80,000 lb/in.^{2} have been produced. Since gray cast iron consists of a matrix of steel, it will be obvious that the properties of these alloys can be altered to a certain degree by heat treatment, which will be discussed in a subsequent article.

Gray cast iron finds extensive use in cylinders and pistons for internal combustion engines, locomotive cylinders, machine tool beds and frames, flywheels, engine frames, and other parts where the tensile strengths and ductility associated with steel are not necessary, and the high compressive strength and damping capacity of gray cast iron are desired.

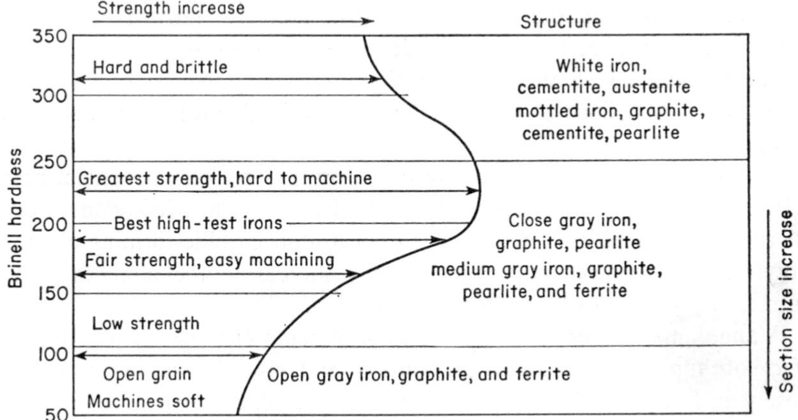

FIG. 16.12 Relation between strength, hardness, structure, and section size of cast iron. (*Cast Metals Handbook*)

16.8 Alloys in Gray Cast Iron. For most ordinary purposes, unalloyed gray cast iron is quite suitable. Where requirements, however, are more severe than can be met with the normal gray cast iron, modification of the properties may be desirable by the use of alloying elements. The alloying elements that are most commonly found in gray cast iron are nickel, chromium, molybdenum, vanadium, copper, and silicon.

It has already been shown that nickel can be used very effectively as a graphitizer to impart more uniform characteristics in variable sections of castings. It not only controls graphitization, but it also tends to reduce the size of the graphite flake. Furthermore, nickel does not tend to reduce the amount of carbon combined in the pearlite as markedly as silicon. As a graphitizer, it is considered half as powerful as silicon. By proper manipulation of the composition, the addition of nickel increases the strength of cast iron. Nickel is generally used in amounts up to 4 per cent in the usual grades of nickel cast iron. An alloy containing 4.5 per cent nickel, 1.5 per cent chromium, low silicon and high carbon contents is known as *Ni-Hard* and possesses outstanding wear-resistant properties. An alloy gray cast iron containing approximately 15 per cent nickel, 6 per cent copper, 2 per cent chromium, and normal amounts of the other common elements is known as *Ni-Resist.* It has an austenitic matrix with excellent growth-, corrosion-, and heat-resistant properties up to 1500 F.

Chromium acts in an opposite way to that of nickel and silicon in that it promotes the formation of combined carbon. This forms a complex iron-chromium carbide which is stable and imparts hardness and wear resistance to the cast iron. Chromium improves tensile strength when added in amounts up to about 0.5 per cent. It also reduces "growth" caused by graphitization in heat treatment or during service above about 700 F, as will be discussed in a subsequent article. The combination of chromium and nickel will produce an iron which is fine-grained and higher in strength, with graphite uniformly dispersed.

The addition of molybdenum to cast iron imparts very beneficial properties and is used in amounts up to about 1.2 per cent. Molybdenum does not promote graphitization, and it is only a mild carbide-forming element. Molybdenum dissolves in ferrite, thereby increasing the strength of the matrix, and promotes better distribution of graphite.

Vanadium is a very strong carbide-stabilizing element. It also has the effect of improving graphite distribution and size. Vanadium is usually present in amounts of less than about 0.25 per cent, and it acts somewhat in the manner of molybdenum as far as strengthening the cast iron.

Copper has been used to some extent in low-carbon cast iron for improvement of tensile strength. It is a slight graphitizer and is beneficial under certain circumstances. Copper is dissolved principally in the matrix and, therefore, acts as a strengthener.

High silicon contents increase the acid-corrosion resistance of cast iron, finding particular application in the chemical industries.

16.9 Properties of Gray Cast Iron. The properties by which gray cast iron is evaluated are tensile strength, hardness, and transverse strength as indicated by modulus of rupture. Other characteristics that distinguish gray cast iron are low cost, relatively low melting point, excellent fluidity, good machinability, high damping capacity, good compressive strength, abrasive hardness of as-cast surface, all combined with low impact resistance and lack of ductility.

Since the properties of gray cast iron are directly related to the rate of cooling and, therefore, to the section of a casting, it is important that test bars be of the proper size. Agreement between foundry and purchaser as to the controlling section of the casting determines the size of the transverse or arbitration test bar. These are shown in the following table:

Test Bar	Diameter (in.)	Length (in.)	Distance between Supports in Transverse Test (in.)	Controlling Section of Casting (in.)
A.	0.875	15	12	0.50 maximum
B.	1.20	21	18	0.50-1.00
C.	2.00	27	24	1.00-2.00

A comparison of the properties of different classes of gray cast iron is given in Table 16-II.

TABLE 16-II. PROPERTIES OF DIFFERENT CLASSES OF GRAY CAST IRON.

(Sisco, *Modern Metallurgy for Engineers*, Pitman Publishing Corp., New York, 1948)

Class No.	Minimum Tensile Strength (lb/in.²)	MINIMUM TRANSVERSE STRENGTH (LB)			Average Mod. of Rupture (lb/in.²)	Average Compressive Strength (lb/in.²)	Secant Modulus (lb/in.²) (10⁶)
		Bar A	Bar B	Bar C			
20	20,000	900	1,800	6,000	42,000	80,000	14.2
25	25,000	1,025	2,000	6,800	46,000	92,000	14.5
30	30,000	1,150	2,200	7,600	54,000	108,000	16.0
35	35,000	1,275	2,400	8,300	60,000	118,000	18.0
40	40,000	1,400	2,600	9,100	64,000	120,000	19.0
50	50,000	1,675	3,000	10,300	68,000	140,000	20.0
60	60,000	1,925	3,400	—	70,000	150,000	—

16.10 Growth in Gray Cast Iron. When gray cast iron parts are repeatedly heated in service to above about 700 F (370 C), a permanent expansion, known as *growth,* takes place. This phenomenon results in the loss of strength and increased brittleness, aside from the obvious dimensional changes. Several theories have been advanced to explain this phenomenon. Oxidation and growth apparently work together in lowering strength characteristics. Growth is probably caused by graphitization of proeutectoid and eutectoid cementite. The infiltration of corroding media, such as furnace gases, superheated steam, etc., into the metal along the path of graphite flakes may cause internal oxidation and consequent growth. Allotropic volume changes caused by heating through the critical range and the pressure of occluded gases may also contribute to the phenomenon. When gray cast iron is heated locally through many cycles to a temperature of the order of 1000 F (540 C), strong thermal gradients on the surface cause the formation of incipient cracks which eventually develop into well-defined *heat checks* or *fire cracks* with ultimate failure. Heat-checking occurs in a relatively few cycles where the thermal gradient is steep. Early formation of such checks is undoubtedly caused by surface stresses aggravated by the presence of weak graphite flakes and surface growth. The incipient check provides a path for penetration of oxidizing media. This condition is noted in large brake drums and ingot molds.

Growth tendencies can be reduced by dense, fine-grain structures as normally found in light sections. Growth in gray cast iron may be further reduced by carbide-stabilizing elements of which chromium is the most effective. The influence of chromium as a growth-reducing agent is shown in Fig. 16.13. Molybdenum or nickel is usually employed with chromium

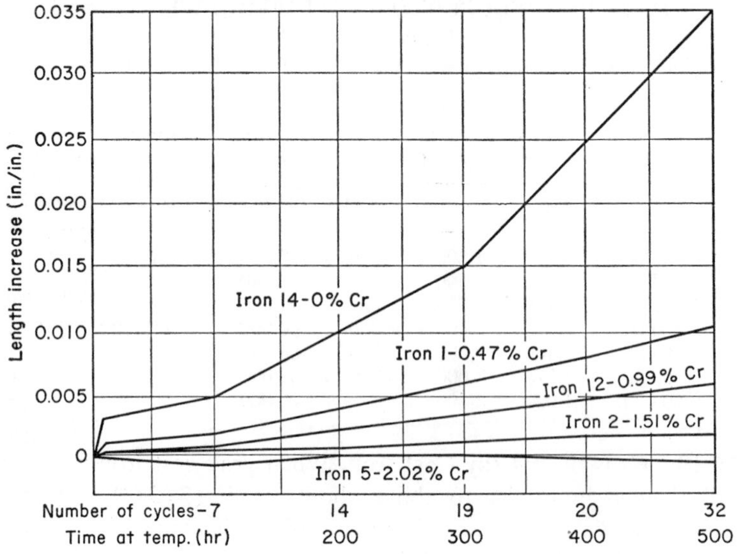

FIG. 16.13 Growth of gray cast iron after cyclic heating to 1472 F. (*Cast Metals Handbook*)

to improve elevated strength properties and resistance to rupture caused by growth. Reduction in volume changes in the thermal critical range is also possible by alloying. The complete elimination of the critical range by use of high-alloy austenitic cast iron (Ni-Resist) has been previously noted for high-temperature applications.

16.11 Malleable Cast Iron. Malleable cast iron derives its name from the fact that it is much more ductile than gray cast iron. Ordinary malleable cast iron consists of nodules of graphite embedded in the matrix of ferrite. A pearlitic malleable cast iron may be produced which consists of nodules of graphite embedded in a matrix of pearlite.

Malleable cast iron is usually composed of from 1.75 to 2.75 per cent carbon, 0.8 to 1.2 per cent silicon, less than 0.40 per cent manganese, less than 0.20 per cent phosphorus, and less than 0.12 per cent sulfur. An alloy, the composition of which falls within this range, when cast in sand molds will have the structure of white cast iron. By heating the casting to above the critical temperature for the proper length of time, the cementite can be dissociated. By very slowly cooling, the cementite in the pearlite can also be made to decompose to ferrite and graphite. Since no graphitization occurs while any of the metal is in the liquid form, the graphite will not be in the form of flakes as it is in gray cast iron, but in the form of nodules commonly referred to as *temper carbon*. When the amount of combined carbon is still quite high or graphitization has not been complete enough to produce a matrix of very nearly eutectoid composition, the iron is designated as *mottled cast iron* because of the appearance of the fracture.

One of the malleableizing or annealing processes that is used in this country consists of packing white iron castings of proper composition in a large metal box. Castings are packed in sand, cinders, or mill scale to support them during the annealing operation and also to exclude air, thus preventing scaling and decarburization. Usually about 45 hr are required to heat the furnace and its contents to about 1500 F (815 C). The charge is then held at a temperature between 1500 and 1650 F (815 and 900 C) for about 50 hr. Cooling from this temperature is very slow, requiring about 60 hr. It can be appreciated that this is a relatively slow process, requiring of the order of 150 hr for the complete cycle.

Through the development of controlled atmosphere furnaces, it has been possible to eliminate packing material, thereby decreasing the time required to heat the furnace to the desired temperature and decreasing the cost. There has been some tendency to decrease the annealing time by using somewhat higher annealing temperatures.

Malleable castings made by the short-cycle process have properties equal to those of castings made by the longer-cycle method. The short-cycle treatment requires something of the order of 48 hr.

Sometimes malleable iron produced by the method just described is referred to as *black-heart malleable cast iron*. In Europe, a considerable amount of malleable cast iron has been produced by packing the castings in an oxidizing material. In this treatment, a large proportion of the carbon is oxidized and, therefore, removed from the structure. Hence, the casting consists almost entirely of ferrite. Products of this treatment are referred to as *white-heart malleable cast iron*.

Pearlitic malleable iron may be produced with the same base alloy as would be used for the usual malleable iron, but with a variation of the treatment during or after annealing. It may be produced by including carbide stabilizing elements that will permit the development of the pearlitic structures by the standard anneal for regular malleable iron. The pearlitic irons include those with structures of coarsely spheroidized cementite, bainite, martensite, or a tempered matrix.

Malleable cast iron may be produced with a tensile strength varying between 50,000 and 60,000 lb/in.^{2} and a yield point between 30,000 and 40,000 lb/in.^{2} The elongation varies between 10 and 25 per cent in 2 in. Some of the properties of two grades of malleable cast iron are given in Table 16-III. As with gray cast iron, malleable cast iron is usually not purchased on the basis of composition, but in accordance with the mechanical properties that are required. This is because certain mechanical properties can be met with different compositions, foundry practice, and annealing practice. Malleable cast iron has been used extensively in agricultural implements and equipment, marine hardware, parts for locomotives and railroad cars, brackets and levers on automobiles and trucks, and various small machinery components. A more detailed list of uses of malleable cast iron is shown in Table 16-IV.

TABLE 16-III. MECHANICAL PROPERTIES OF TWO MALLEABLE CAST IRONS.
(After *Cast Metals Handbook*)

Property	ASTM Specification A47-52, Grade 35018	ASTM Specification A47-52, Grade 32510
Tensile strength (lb/in.^{2})	53,000-63,000	50,000-60,000
Yield point (lb/in.^{2})	35,000-42,000	32,500-41,000
Elongation in 2 in. (%)	18-25	10-20
Modulus of elasticity, tension (lb/in.^{2})	25×10^6	25×10^6
Poisson's ratio	0.17	0.17
Ultimate shearing strength (lb/in.^{2})	48,000	48,000
Yield point, shear (lb/in.^{2})	23,000	23,000
Modulus of elasticity, shear (lb/in.^{2})	12.5×10^6	12.5×10^6
Modulus of rupture, torsion (lb/in.^{2})	58,000	58,000
Hardness, Brinell	110-150	110-150
Charpy impact, V-notch, 0.197-in. depth (ft lb)	16.5	—
Izod impact, V-notch, 0.394-in. square bar, 0.079-in. depth (ft-lb)	16.5	—
Endurance limit (lb/in.^{2})	26,500-31,500	25,000-30,000

Some of the properties of pearlitic malleable cast iron are given in Table 16-V. It will be seen that very high strength can be attained by the partial-malleableizing treatment, at a sacrifice, of course, of ductility expressed by per cent elongation.

16.12 Nodular Cast Iron. A grade of cast iron known as *nodular* or *ductile* cast iron is produced by treating the molten alloy with magnesium,

TABLE 16-IV. TYPICAL USES OF MALLEABLE CAST IRON.
(After *Cast Metals Handbook*)

Agricultural Implements: Plows, tractors, harrows, reapers, mowers, binders, cultivators, rakes, spreaders, dairy and poultry equipment, presses, tools, pumps, wagon parts, fence parts.
Automobiles: Parts for frames, wheels, spring assembly, brakes, motor transmission, axles, steering gear assembly, body accessories.
Boilers, Tanks, and Engines: Boilers, tanks, fittings, engine parts, outboard motors, diesel engines.
Building Equipment: Hardware for windows, doors, garage equipment, awning hardware.
Conveyor and Elevator Equipment: Chains, buckets, pulleys, rollers, cranes, hoists, fittings.
Electrical and Industrial Power Equipment: Motor and generator parts, pumps, stokers, electrical locomotives and tractors, steam specialties, outlet and switch boxes.
Hardware and Small Tools: Pneumatic and portable tools, miscellaneous tools, saddlery, hardware, table and kitchen utensils.
Household Appliances: Stoves, sewing machines, refrigerators, washing and ironing machines, vacuum cleaners, dishwashing machines, oil burners, electric fans and toasters, radios.
Machine Tools: Lathes, planers, shapers, grinders, screw machines, gear cutters, drills.
Machinery for Special Uses: Textile, cement, rubber, shoe, mining and quarrying, grinding, forging, foundry, bakery, wood-working, bottling, ice, laundry equipment.
Marine Equipment: Anchors, chains, capstans, fastenings, towing bits, hardware.
Metal Furniture and Fixtures: Stoves, beds, desks, filing cabinets, shelving, hotel supply equipment.
Municipal, State, and Public Service: Manhole covers, street lamp standards, guard rail equipment, traffic markers, transmission line fittings, electric railway fittings.
Pipe Fittings and Plumbing Supplies: Elbows, unions, reducers, flanges, valves, bolts, nuts.
Railroads: Wide variety of parts for construction of locomotives, freight and passenger cars; guard rails, track accessories.
Road and Contractor's Machinery: Rollers, excavators, cranes, hoists, tractors, graders, scarifiers, mixers, pavers.
Toys and Specialties: Sleds, wagons, automobiles, bicycles, carriages, gun parts.

cerium, or a combination of the two elements. Other elements may have a similar effect. Magnesium is added in quantities of the order of 0.07 to 0.10 per cent, followed by an addition of ferro-silicon. When the alloy solidifies, graphitization occurs to form nodules of graphite. Elements such as titanium, tin, lead, and arsenic act as inhibitors. The addition of the cerium may overcome the effect of the inhibitors. The composition of nodular iron is usually in the following range:

Total carbon	3-4%
Si	1.80-2.80%
Mn	0.15-0.90%
P	0.10% max
S	0.03% max
Mg	0.01-0.10%

TABLE 16-V. PROPERTIES OF PEARLITIC MALLEABLE IRON ASTM TENTATIVE SPECIFICATION A220-55T.

Grade	Tensile Strength, Min lb/in.^{2}	Yield Point or Yield Strength Min lb/in.^{2}	Elongation in 2 in. Min Per Cent	Brinell
45010	65,000	45,000	10	163-207
45007	68,000	45,000	7	163-217
48004	70,000	48,000	4	163-228
50007	75,000	50,000	7	179-228
53004	80,000	53,000	4	197-241
60003	80,000	60,000	3	197-255
80002	100,000	80,000	2	241-269

Nodular iron castings may be used in the as-cast condition or heat-treated. The maximum toughness, impact value, and machinability is secured by a ferritizing anneal such as the following:

A. Primary carbide absent
   1. Heat to 1450 F (790 C).
   2. Hold for 1 hr/in. of section.
   3. Cool at rate of 20 F/hr to 1350 F (735 C).
   4. Cool at rate of 100 F/hr to 800 F (440 C).

B. Primary carbide present
   1. Heat to 1650 F (900 C).
   2. Hold for 2 hr/in. of section.
   3. Cool to 1450 F (790 C) in 1 hr.
   4. Cool at rate of 20 F/hr to 1350 F (735 C).
   5. Cool at rate of 100 F/hr to 800 F (440 C).

Other heat treatments to secure greater strength include air-cooling or oil-quenching followed by tempering. Some of the mechanical properties of nodular iron are compared with those of other cast ferrous materials in Table 16-VI. The structure of nodular cast iron will of course depend upon the treatment to which it has been subjected. The significant part of the structure that identifies the iron as the nodular type is the spheroid of graphite as shown in Fig. 16.5. The matrix may consist of a hypo- or hyper-

TABLE 16-VI. PROPERTIES OF NODULAR IRON AND OTHER FERROUS METALS.
(From Vennerholm *et al.*, *Trans. SAE* **4**, 1950, pp. 422-435)

PROPERTY	GRAY CAST IRON	MALLEABLE IRON	PEARLITIC MALLEABLE IRON	CAST STEEL 0.30 C	CAST STEEL 0.30 C	NODULAR IRON High Tensile	NODULAR IRON High Duct. As-cast	NODULAR IRON High Duct. Annealed
Tensile strength ($lb/in.^2$)	30,000	53,000	60,000	76,000	108,000	100,000	80,000	65,000
Yield point ($lb/in.^2$)	—	35,000	43,000	42,000	79,000	80,000	55,000	45,000
Elongation (%)	—	18	10	26	19	1.5	10	16
Reduction in area (%)	—	—	—	32	46	1.0	8	20
Brinell hardness	190	130	170	140	220	225	190	160
Modulus of elasticity ($lb/in.^2$)	19×10^6	25×10^6	25×10^6	30×10^6	30×10^6	24×10^6	24×10^6	24×10^6
Impact (ft-lb)	10	16	(12)	20	35	40	180	260
Heat treatment	none	anneal	anneal	anneal	quench temper	none	none	anneal

eutectoid structure consisting of ferrite and cementite or one of the nonequilibrium structures such as may be secured by heat treatment.

16.13 Heat Treatment of Cast Irons. The operations of heat treatment that can be employed to modify the properties or structures of cast irons are stress relieving, annealing, quenching, tempering, and surface hardening.

The relief of casting stresses is accomplished by heating within the range 800-1000 F (430-540 C), holding one hour per inch of section, and then following by furnace cooling. As a general rule, this treatment will not change the structure or promote graphitization. The temperature should, however, be reduced for gray cast irons containing appreciable amounts of graphitization agents and increased when strong carbide stabilizing elements are present. The operation promotes dimensional stability in castings prior to machining or service.

Annealing operations for cast iron may be classed as: (1) full annealing, (2) graphitizing, or (3) malleableizing, depending on the purpose. In the broad sense, the term *annealing* implies softening. Machinability of iron castings may be improved by annealing processes which cause graphitization of hard massive proeutectoid cementite or cause decomposition of the eutectoid cementite. The former is accomplished by *full annealing,* which consists of heating to above the upper critical temperature in the range 1500-1800 F (815-985 C). The exact time and temperature are governed by the presence of carbide stabilizing elements. After the proeutectoid cementite is broken up by graphitization, castings are furnace-cooled to allow formation of coarse pearlite. Decomposition of pearlite in gray cast iron by a process known as *graphitizing* consists of annealing below the lower critical temperature in the range 1200-1500 F (650-815 C) until the combined carbon is reduced by decomposition of the eutectoid cementite to form ferrite and graphite. The rate of graphitization will be great in high-silicon irons. *Malleableizing* is a specialized annealing operation in the manufacture of malleable cast iron, as described in a previous section.

Quenching and tempering increase the hardness and abrasion resistance of gray cast iron. However, other mechanical properties are not, in general, improved over the as-cast product. Tensile, transverse, and impact strength, together with deflection, are all lowered by quenching. Tempering permits recovery to values corresponding to the as-cast condition. Alloy cast iron with fine graphite flakes dispersed in a pearlitic matrix offers the best response to hardening, giving hardness values as high as 600 Brinell. Castings that are to be quenched should be designed with all precautions kept in mind as to drastic section changes. An austenitizing range of 1450-1600 F (790-870 C) is suitable for most gray cast iron grades, with the lower temperatures giving highest strength properties in a given composition. The usual quenching medium is oil, although many alloy grades will form a martensitic structure by air hardening. The effects of tempering at various temperatures are shown in Fig. 16.14.

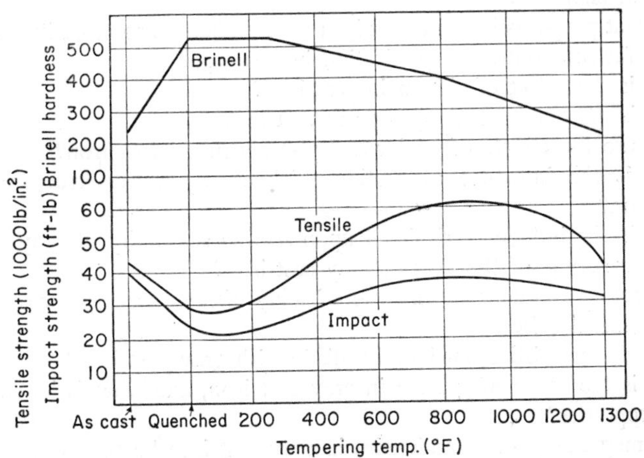

FIG. 16.14 Properties of quenched and tempered gray cast iron. (*Cast Metals Handbook*)

Surface hardening of regular grades of gray cast iron by means of flame or induction hardening is common for certain applications. Alloy irons containing aluminum as an essential alloying element may also be nitrided.

REFERENCES

Alloy Cast Iron Handbook, American Foundrymen's Society, Des Plaines, Ill., 1944.

Bolton, *Gray Cast Iron,* Penton Publishing Co., Cleveland, Ohio, 1937.

Boyles, *The Structure of Cast Iron,* American Society for Metals, Metals Park, Ohio, 1947.

Cast Metals Handbook, American Foundrymen's Society, Des Plaines, Ill., 1957.

Molybdenum Steels, Irons, Alloys, Climax Molybdenum Co., New York, 1948.

Sisco, *Alloys of Iron and Carbon,* Vol. II, McGraw-Hill Book Co., New York, 1937.

Symposium on Cast Iron, American Society for Testing Materials, Philadelphia, Pa., 1933.

Symposium on Malleable Iron Castings, American Society for Testing Materials, Philadelphia, Pa., 1931.

Vennerholm, Bogart, and Melmoth, "Nodular Cast Iron," *Trans. SAE* **4,** 1950, pp. 422-436.

QUESTIONS

1. Define each of the following: white cast iron, malleable cast iron, gray cast iron, mottled cast iron, and nodular cast iron.
2. Distinguish between the method of formation of graphite in gray cast iron and malleable cast iron.
3. For what purposes is white cast iron used? What are some of its limitations?
4. What factors control graphitization in cast iron?
5. What is the effect of silicon in cast iron?
6. What is the effect of manganese in cast iron?
7. What are the effects of sulfur and phosphorus in cast iron?
8. What is the influence of section thickness upon the structure of gray cast iron?
9. What is the influence of nickel upon section sensitivity of gray cast iron?
10. What are some of the uses of gray cast iron?
11. Give some examples of the use of alloy elements in gray cast iron and their general effect.
12. What characteristics particularly distinguish gray cast iron?
13. What is the cause of growth in gray cast iron, and what is its effect upon the properties?
14. How may the growth tendencies in gray cast iron be reduced?
15. Explain the process of producing malleable cast iron and indicate the structural changes that occur during the process.
16. Distinguish between black-heart malleable cast iron and white-heart malleable cast iron.
17. What are the particular characteristics of malleable cast iron that make it adaptable for certain engineering applications?
18. What is the structure of nodular cast iron, and how is nodular cast iron produced?
19. What heat treatments are commonly applied to cast iron?

XVII

COPPER- AND NICKEL-BASE ALLOYS

17.1 Copper. Copper is a very important engineering metal not only in the pure state, but also when combined with other elements to form alloys. In the pure state, its greatest application lies in the electrical industry because of its high electrical conductivity. In the annealed state, copper is very soft and ductile. Its tensile strength is approximately 35,000 lb/in.^{2} with an elongation of about 40 per cent in 2 in. The purest form of copper is commercially produced by electrolytic deposition, and the product is called *cathode copper*.

The properties of copper can be altered appreciably by cold working. The tensile strength may be raised to about 50,000 lb/in.^{2} with an elongation of about 4 per cent in 2 in. by rolling into sheet. Hard-drawn wire 0.102 in. in diameter may have a tensile strength as high as 65,000 lb/in.^{2} However, the electrical conductivity of copper is somewhat decreased by cold working. The presence of a relatively small proportion of other elements in the copper greatly decreases electrical conductivity if the second element dissolves in the copper. Nevertheless, in some applications where greater strength is required, alloying may be necessary, and a sacrifice of conductivity may be permitted.

Most copper for electrical purposes contains from 0.02 to 0.05 per cent oxygen which combines with copper to form cuprous oxide (Cu_2O). This is called electrolytic tough pitch copper and it has an electrical conductivity of approximately 101.6 per cent of that of pure copper. This type of copper should not be employed in applications where it may be subject to heating in a reducing atmosphere, because it is embrittled at a temperature as low as 750 F (400 C).

Cathode copper may be melted, cast, and solidified under a protective atmosphere to exclude oxygen to produce an improved quality of copper. The material is called oxygen-free, high-conductivity copper (OFHC[1]).

[1] Registered trade-mark, The American Metal Co., Ltd.

Copper sheet and wire may be purchased with different strengths, depending upon the amount of cold work. The designations for the different conditions and the minimum properties expected are shown in Table 17-I.

TABLE 17-Ia. CLASSIFICATION OF COPPERS.
(From ASTM Designation B224-58)

DESIGNATION	TYPE OF COPPER	FORMS AVAILABLE FROM FABRICATORS			
		Flat Products	Pipe and Tube	Rod and Wire	Shapes
	Tough Pitch Coppers				
ETP	Electrolytic tough pitch	X	X	X	X
FRHC	Fire-refined, high-conductivity tough pitch	X	X	X	X
FRTP	Fire-refined tough pitch	X	X	—	X
ATP	Arsenical, tough pitch	X	X	X	—
STP	Silver bearing tough pitch	X	X	X	X
SATP	Silver bearing arsenical, tough pitch	X	X	—	—
	Oxygen-Free Coppers				
OF	Oxygen-free without residual deoxidants	X	X	X	X
OFP	Oxygen-free, phosphorus bearing	X	X	X	—
OFTPE	Oxygen-free, phosphorus and tellurium bearing	—	—	X	X
OFS	Oxygen-free, silver bearing	X	—	X	X
OFTE	Oxygen-free, tellurium bearing	—	—	X	X
	Deoxidized Coppers				
DHP	Phosphorized, high residual phosphorus	X	X	X	X
DLP	Phosphorized, low residual phosphorus	X	X	—	X
DPS	Phosphorized, silver bearing	X	X	—	X
DPA	Phosphorized, arsenical	X	X	—	—
DPTE	Phosphorized, tellurium bearing	—	—	X	X

Copper is also the basis of a very important class of alloys. A large amount of copper is used for tubing in water lines, for fuel and oil lines in aircraft, and for conducting foodstuffs, beverages, pulp-mill pulp, chemicals, etc. In view of its excellent thermal conductivity, copper is used for kettles, etc. Copper is also very resistant to atmospheric corrosion and is, therefore, used for roofing, window screens, etc.

17.2 Alloys of Copper. Any elements that form a solid solution with copper have a marked effect in decreasing the electrical conductivity. The influence of several elements on the electrical conductivity of copper is shown in Fig. 17.1. Elements such as lead, tellurium, and selenium are rela-

TABLE 17-Ib. EXPECTED MINIMUM MECHANICAL PROPERTIES OF COPPER.

Form	Condition	Diameter Range (in.)	Tensile Strength Range (lb/in.²)		% Elong. in 10 in. min. Range	ASTM Spec.
Wire	Soft or annealed	0.46-0.0226	36,000-38,500 max.		35-25	B3-56
Wire	Med. hard-drawn	0.46-0.0403	49,000-60,000 max.		3.75-0.88*	B2-52
Wire	Hard-drawn	0.46-0.0403	49,000-67,000 min.		3.75-0.85*	B1-56
Sheet, strip, plate, and rolled bar			min.	max.		
	Light cold-rolled	—	32,000	40,000	—	B152-58
	Half hard	—	37,000	46,000	—	B152-58
	Hard	—	43,000	52,000	—	B152-58
	Spring	—	50,000	58,000	—	B152-58
	Extra spring	—	52,000	—	—	B152-58
	Hot-rolled	—	30,000	38,000	—	B152-58
	Hot-rolled and annealed	—	30,000	38,000	—	B152-58

* For smaller wire, gauge length 60 in.

tively insoluble in solid copper; therefore, they do not have any significant effect on conductivity, but they do improve machinability. Cadmium has a definite strengthening effect on copper, with a relatively small adverse effect on the electrical conductivity. The use of about 1 per cent cadmium

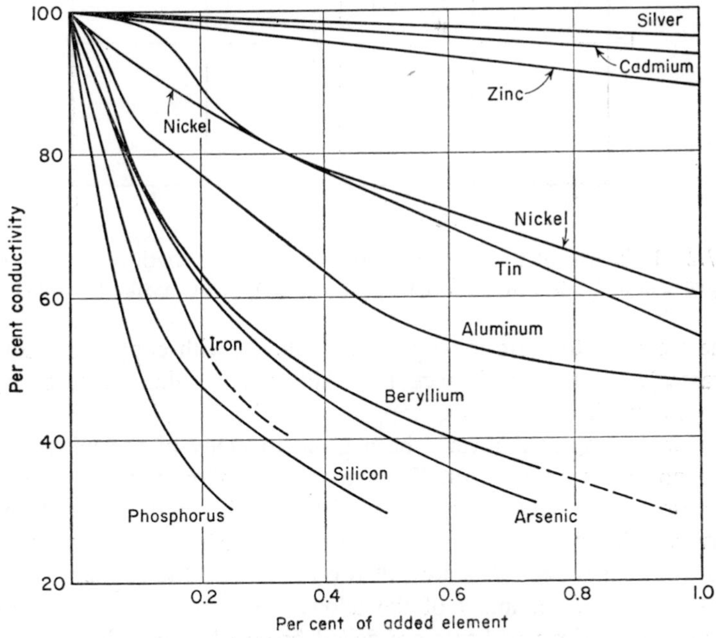

FIG. 17.1 Influence of some elements on the electrical conductivity of copper. (*Copper Data*. 1936 ed., British Copper Development Assn., London)

increases the Brinell hardness of copper from 100 to about 130 with a decrease of about 12 per cent in electrical conductivity. The influence of several elements on the strength of copper is shown in Fig. 17.2.

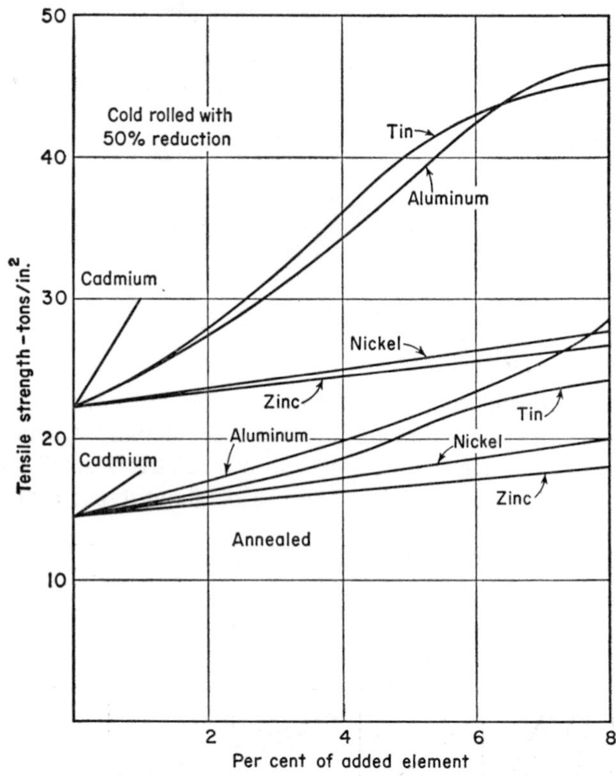

FIG. 17.2 Influence of some elements on the tensile strength of copper. (*Copper Data,* 1936 ed., British Copper Development Assn., London)

The elements that are most generally alloyed with copper are zinc (in the brasses), tin (in the bronzes), nickel, silicon, aluminum, cadmium, and beryllium. A study of the copper alloys is, indeed, rather confusing because of the names that have been given to many of the alloys, names that are not always consistent. For example, an alloy named *nickel silver* is composed principally of copper, nickel, and zinc and does not contain any silver. It has been given this name because of its silvery appearance. Most of the copper-tin alloys are called bronzes, and the copper-zinc alloys are called brasses, yet many of the alloys that probably conform to the common compositions of brass are called bronze because they look more like the copper-tin alloys.

The alloys of copper will be discussed from the standpoint of composition rather than name. The copper-zinc alloys will be considered as a group regardless of whether they are called brass or bronze. It is preferable that the engineer consider these alloys from the character of their composition and properties rather than from the name alone.

17.3 Copper-Zinc Alloys. The equilibrium diagram of the copper-zinc system is shown in Fig. 17.3. The region of the alpha solid solution

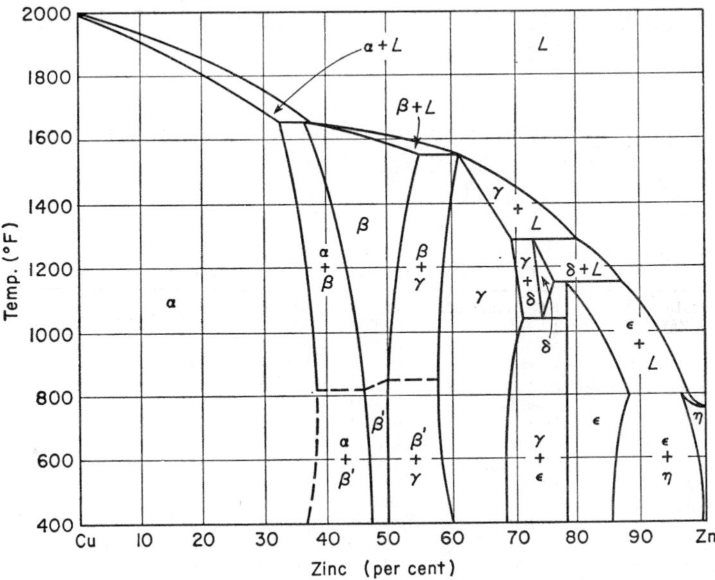

FIG. 17.3 Equilibrium diagram for copper-zinc system.

is quite wide, extending from zero to 38 per cent zinc. With a larger amount of zinc, a second solid solution, beta, is observed. Two forms of this solid solution are recognized, namely, beta existing above 815 F (435 C) and beta prime existing below this temperature. In alloys containing between about 38 and 46 per cent zinc, alpha and beta prime coexist at ordinary temperatures after slow cooling. With more than 50 per cent zinc, another solid solution called gamma is found.

The mechanical characteristics of alloys in this system are very closely related to the equilibrium diagram as in other alloys. The tensile strength and ductility of the copper-zinc alloys increase with increasing amount of zinc up to about 30 per cent. With the appearance of the beta solid solution, the strength continues to increase, but the ductility begins to decrease. The beta and gamma solid solutions are not as ductile as the alpha solid solution; therefore, very few alloys containing more than about 40 per cent zinc are of commercial importance unless they are for special purposes. If the structure of the alloy contains an appreciable amount of the gamma phase, the alloy will be quite brittle and of little engineering value.

Those alloys that contain only alpha solid solution can be easily hot- or cold-worked, whereas those alloys that contain both alpha and beta will not withstand an appreciable amount of cold working without rupture and, therefore, must be formed while hot. Alloys that contain mostly beta solid solution can be hot-rolled, forged, or extruded easily. When the gamma solid solution is present, both cold and hot working are difficult. Such alloys are restricted to use in the form of castings.

Table 17-II. COMPOSITION, PROPERTIES, AND USES OF SOME COPPER-ZINC ALLOYS.
(Compiled from *Metals Handbook*)

Common Name and Composition (%)	Condition	T.S. (lb/in.²)	Y.S. (lb/in.²)	Elong. in 2 in. (%)	Some Uses
Gilding metal (95 Cu, 5 Zn)	0.035 mm G.S.	35,000	11,000	45	Coins, fuse caps, emblems, jewelry.
	¼ hard	42,000	32,000	25	
	Hard	56,000	50,000	5	
	Spring	64,000	58,000	4	
Commercial bronze (90 Cu, 10 Zn)	0.035 mm G.S.	38,000	12,000	45	Grillwork, screen cloth marine hardware, primer caps, costume jewelry.
	¼ hard	45,000	35,000	25	
	Hard	61,000	54,000	5	
	Spring	72,000	62,000	3	
Red brass (85 Cu, 15 Zn)	0.035 mm G.S.	41,000	14,000	46	Architectural trim, electrical sockets, condenser tubes, costume jewelry.
	¼ hard	50,000	39,000	25	
	Hard	70,000	57,000	5	
	Spring	84,000	63,000	3	
Low brass (80 Cu, 20 Zn)	0.035 mm G.S.	46,000	15,000	48	Architectural ornaments, musical instruments.
	¼ hard	53,000	40,000	30	
	Hard	74,000	59,000	7	
	Spring	91,000	65,000	3	
Cartridge brass (70 Cu, 30 Zn)	0.035 mm G.S.	49,000	17,000	57	Radiator cores, lamp fixtures, springs.
	¼ hard	54,000	40,000	43	
	Hard	76,000	63,000	8	
	Extra spring	99,000	65,000	3	
Yellow brass (65 Cu, 35 Zn)	Annealed	49,000	17,000	57	Architectural grillwork, same as 70-30.
	¼ hard	54,000	40,000	43	
	Hard	74,000	60,000	8	
	Spring	91,000	62,000	3	
Muntz metal (60 Cu, 40 Zn)	Annealed	54,000	21,000	45	Architectural trim, large nuts & bolts, condenser tubes.
	½ hard	70,000	50,000	10	
Architectural bronze (57 Cu, 40 Zn, 3 Pb)	As extruded	60,000	20,000	30	Trim & hardware.
Admiralty metal (71 Cu, 28 Zn, 1 Sn)	Annealed	53,000	22,000	65	Condenser tubes.
Naval brass (60 Cu, 39¼ Zn, ¾ Sn)	Rod soft an.	57,000	25,000	47	Condenser plates, welding rod, marine hardware, propeller shafts, valve stems.
	Rod ¼ H (8%)	69,000	46,000	27	
	Rod ½ H (20%)	75,000	53,000	20	
Manganese bronze (58.5 Cu, 39 Zn, 1.4 Fe, 1 Sn, 0.1 Mn)	Rod soft an.	65,000	30,000	33	Pump rods, valve stems, welding rod.
	Rod ¼ H (10%)	77,000	45,000	23	
	Rod ½ H (20%)	84,000	60,000	19	
Aluminum brass (76 Cu, 22 Zn, 2 Al)	Tube 0.025 G.S.	60,000	27,000	55	Condenser tubes.

The useful engineering alloys are those containing less than about 40 per cent zinc. Some of the commercially important copper-zinc alloys contain small amounts of other elements. These are listed in Table 17-II.

The alloys containing approximately 90 per cent copper and 10 per cent zinc are used principally for screen wire, hardware, screws, rivets, and costume jewelry. These alloys are easily worked and sometimes, because of their characteristic color, are referred to as commercial bronze. The structure of this alloy is shown in Fig. 17.4 to consist of the solid solution

FIG. 17.4 Wrought commercial bronze, 90% Cu - 10% Zn (100×).

alpha. A large quantity of alloy containing about 30 per cent zinc (cartridge brass) is used in the form of sheets, wire, and tubes for all types of drawing and trimming operations. The structure of this alloy is also composed of the solid solution alpha as shown in Fig. 17.5. *High brass* is an alloy containing about 34 per cent zinc and is used extensively for drawing and spinning operations and for springs. A tensile strength of the order of 125,000 lb/in.^{2} may be attained in the spring temper.

The alloys containing between 30 and 40 per cent zinc are most susceptible to season cracking. (Season cracking is discussed in Article 17.5.) An alloy containing approximately 60 per cent copper and 40 per cent zinc is called *Muntz metal*. The structure of this alloy is shown in Fig. 17.6. This alloy contains a large proportion of alpha solid solution with some beta solid solution and is quite resistant to fresh water. For this rea-

FIG. 17.5 Cartridge brass, 70% Cu - 30% Zn cold worked and heated to 1200 F (100×).

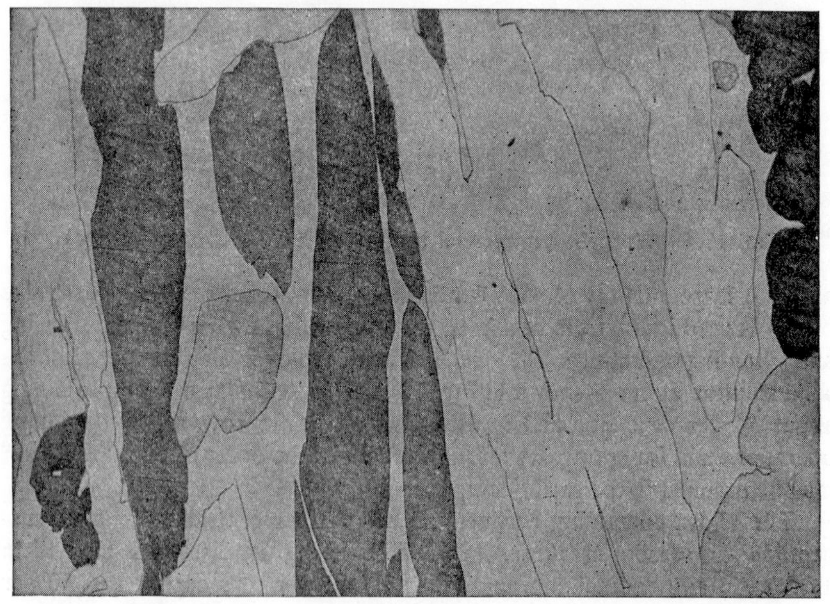

FIG. 17.6 Muntz metal, 60% Cu - 40% Zn hot worked (α-solid solution-white with some β-solid solution-dark). (100×)

son, it is used extensively for condenser tubes where corrosive conditions are not too severe. Its relatively low cost compared with other alloys makes its use economical.

The corrosion resistance of Muntz metal can be improved by the addition of about 1 per cent tin. In this form, the alloy is called *naval brass* or *tobin bronze* and is suitable for use in salt water. The structure of naval brass is shown in Fig. 17.7 to be composed primarily of solid solution alpha. Other elements, such as lead, nickel, manganese, and silicon, may be added to the copper-zinc alloys with beneficial results. These special alloys will be considered later.

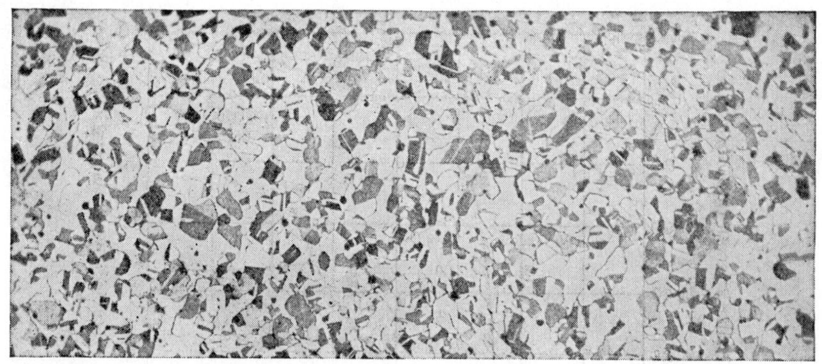

FIG. 17.7 Naval brass, 60%Cu - 39%Zn - 1%Sn half hard (100×).

17.4 Recrystallization of Brass. As in all alloys that do not possess an allotropic transformation, the physical properties of brass can be altered only by cold working and recrystallization. The dendritic structure of cast brass can be broken up by forging, rolling, cold working followed by recrystallization, or somewhat by diffusion heat treatment.

The inferior characteristics imparted to castings by the existence of dendritic structure have been discussed. Homogenization can be accomplished most thoroughly by hot working or by cold working followed by heating to above the recrystallization temperature. Neither of these procedures is possible with castings, because the working will alter the shape of the casting. A certain amount of homogenization can be accomplished by soaking the alloy at a sufficiently high temperature for a sufficient length of time to permit the elimination of segregation by diffusion of the constituents throughout the structure. The change in grain size and properties of a brass by cold working followed by heating to different temperatures is shown in Fig. 17.8. At the recrystallization temperature, the tensile strength begins to decrease markedly and, by heating above the recrystallization temperature, the grain size increases. Structural changes in brass by recrystallization were shown in the photomicrographs in Fig. 6.11.

17.5 Season Cracking. Brasses that contain more than 20 per cent zinc are susceptible to spontaneous cracking and to corrosion at grain boundaries in the cold-worked condition. After severe cold work, as in deep-drawing operations, high internal stresses are established in the material. Cracks may develop under this condition after a time; hence, the term *season cracking*. Season cracking can be prevented in these alloys by a low-temperature anneal which tends to decrease the residual or internal stresses in the material.

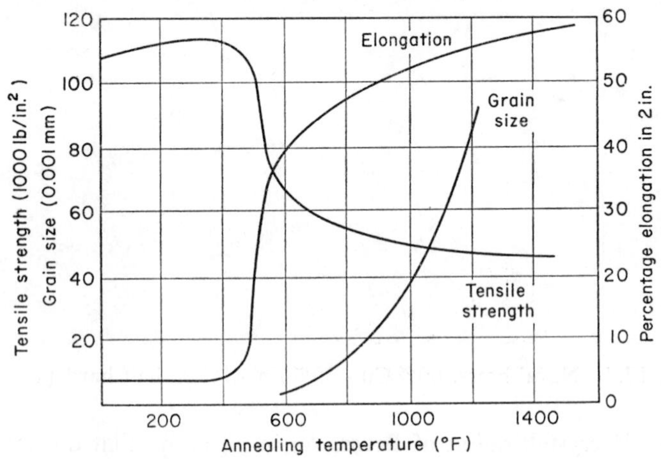

FIG. 17.8 Characteristics of yellow brass wire drawn 63% hard and heated for 1 hr at indicated temperatures. (*Metals Handbook*)

17.6 Leaded Brasses. The machinability of most of the nonferrous metals and alloys can be improved by the addition of lead. The addition of up to about 4 per cent lead improves the machinability of brass quite appreciably. Such an addition tends to reduce the ductility and, hence, the cold-forming characteristic of brass.

Lead is relatively insoluble in copper and, therefore, exists as particles distributed throughout the structure of the alloy. Under certain conditions, lead may effectively be forced from the alloy as a lead sweat at temperatures above 325 F (165 C).

17.7 Special Brasses. The addition of a small amount of tin to the brasses considerably improves their resistance to salt water corrosion. One of these alloys, as has already been mentioned, is naval brass or tobin bronze, composed of approximately 60 per cent copper, 0.5 to 1.5 per cent tin, with the balance zinc.

The addition of manganese to brass accomplishes considerable improvement in mechanical properties and corrosion resistance. In many instances where the manganese furnishes a cleansing or deoxidizing action, very little manganese, if any, may appear in the final analysis. These alloys, called manganese bronzes even though they are essentially brasses, are usually of the Muntz metal composition basically, i.e., 60 per cent copper, 40 per cent zinc, with the addition of from 0.5 to 1.5 per cent tin, 0.8 to 2.0 per cent iron, and a maximum of 0.50 per cent manganese. Manganese bronze is best worked in the hot state. It is not particularly well adapted to cold working, although sometimes it is cold-worked in a mild way. The strength of different forms of wrought manganese bronze is shown in Table 17-III.

TABLE 17-III. MECHANICAL PROPERTIES OF WROUGHT MANGANESE BRONZE.
(After *Metals Handbook*)

Rod Diameter	Temper	Tensile Strength (lb/in.^{2})	Yield Strength* (lb/in.^{2})	Elongation in 2 in. (%)	Hardness Rockwell B	Shear Strength (lb/in.^{2})
1 in.	Soft anneal	65,000	30,000	33	65	42,000
1 in.	$\frac{1}{4}$ hard (10%)	77,000	45,000	23	83	47,000
1 in.	$\frac{1}{2}$ hard (20%)	84,000	60,000	19	90	48,000
2 in.	$\frac{1}{4}$ hard (10%)	72,000	42,000	27	77	44,000

* 0.5% extension under load.

In casting applications, the composition of manganese bronze is somewhat variable. The composition and mechanical properties of three types are given in Tables 17-IVa and 17-IVb, respectively. Castings of this material are quite resistant to the action of sea water, and it is used extensively

TABLE 17-IVa. NOMINAL COMPOSITION (%) OF CAST MANGANESE BRONZE.

Commercial Designation	Cu	Zn	Sn	Pb	Mn	Fe	Al
Leaded manganese bronze	59	37.0	0.75	0.75	0.5	1.25	0.75
No. 1 manganese bronze	58	38.5	—	—	1.0	1.25	1.25
High-strength manganese bronze	60-68	Bal.	0.20 max.	0.20 max.	2.5-5.0	2.0-4.0	3.0-7.5

TABLE 17-IVb. TENSILE PROPERTIES OF CAST MANGANESE BRONZE.

Commercial Designation	Tensile Strength (1000 lb/in.^{2})	Yield Strength (1000 lb/in.^{2})	Elong. in 2 in. (%)	ASTM Spec.
Leaded manganese bronze	60-80	20-32	15	B132-52
No. 1 manganese bronze	65	25	20	B147-52
High-strength manganese bronze	90-110	45-60	18-12	B147-52

for marine-engine pumps, ship propellers, gears, and other purposes for which a corrosion-resistant, high-strength casting is required.

The addition of appreciable quantities of silicon to the basic yellow brasses improves strength, hardness, machinability, and salt water corrosion resistance. These alloys find use in the die-casting industry because their increased fluidity allows accurate reproduction of fine detail, and their lower casting temperatures permit longer die life. They may also be sand-cast, forged, rolled, or drawn, and vary in composition from about 70 to 80 per cent copper, 15 to 30 per cent zinc, 1 to 4 per cent silicon, plus other incidental elements. Their tensile strengths are in the range of 60,000 to 90,000 lb/in.^{2}; yield strengths 25,000 to 40,000 lb/in.^{2}; and

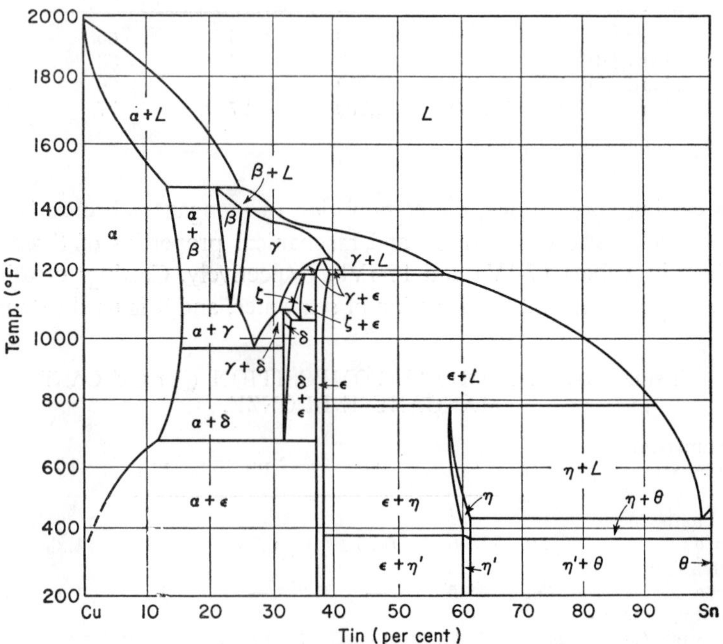

FIG. 17.9 Equilibrium diagram for copper-tin system.

elongation in 2 in. 5 to 20 per cent. Die castings with certain compositions may develop a tensile strength of as high as 115,000 lb/in.^{2}

17.8 Copper-Tin Alloys. Alloys containing principally copper and tin are considered as true bronzes. These materials possess desirable properties of strength, wear resistance, and salt water corrosion resistance. The equilibrium diagram for the copper-tin system is shown in Fig. 17.9. This system has a more limited region of homogeneous solid solution than any found in the copper-zinc system. The homogeneous alpha phase exists within the limits of between zero and about 16 per cent tin. However, at room temperature, the solubility of tin in copper is restricted. With larger proportions of tin, the hard compound Cu_3Sn, epsilon, may appear in the structure, decreasing ductility. Relatively few of the true bronzes are pure copper-tin alloys, since it is usually desirable to make certain additions that improve the characteristics of the plain two-component alloys. The useful engineering alloys in this system are those containing less than about 20 per cent tin.

For the sake of clarity, the general range of compositions of bronzes with respect to copper and tin content may be divided into four groups as follows:

1. Those alloys containing up to about 8 per cent tin which are used for sheets and wire as well as for coins. These alloys are readily cold-worked.
2. Those alloys containing between 8 and 12 per cent tin which are principally used for gears and other machine parts, for bearings that are heavily loaded, and for marine fittings to resist impingement attack by sea water.
3. Those alloys containing between 12 and 20 per cent tin which are used to considerable extent for bearings.
4. Those alloys containing between 20 and 25 per cent tin which are used principally for bells. Alloys of this group are very hard and relatively brittle and are used mostly in the cast condition.

17.9 Phosphor Bronze. *Phosphor bronze* is a term applied to the straight alpha tin bronzes containing between 1.25 and 10 per cent tin in which phosphorus is used as a deoxidizer, generally without the addition of other alloying elements. When phosphorus is added solely as a deoxidizer, no phosphorus is retained in the alloy; but because of the greater freedom from oxide, the strength and ductility are somewhat improved. If some of the phosphorus is retained, the hardness and strength are considerably increased. Phosphorus is a stronger hardener than tin due to the formation of the hard compound Cu_3P. Treatment with phosphorus tends to improve fluidity in casting and the general soundness. The various grades find applications including springs, wire brushes, electrical contacts, gears, shaftings, bushings, and a number of other uses. Table 17-V gives the composition and properties of some of the commercial phosphor bronze wrought alloys.

17.10 Bearing Bronzes. By the addition of zinc and lead to the basic phosphor-bronze alloys in amounts equal to or less than the tin content, considerable improvement in machinability, plasticity, and adaptability to bearing surfaces can be attained. One of the alloys that is used extensively is referred to as 88-10-2. The designation refers to the approximate percentage of copper, tin, and zinc, respectively. This alloy has also been referred to as government bronze, gun metal, or composition "G." It has excellent salt water corrosion resistance, particularly under impingement conditions, and is used for marine castings, such as pumps, valve bodies, bearings, bushings, etc.

The mechanical properties of alloys of the 88-10-2 or 88-8-4 group are given in Table 17-V. Castings of this analysis can be heat-treated by an-

TABLE 17-Va. COMPOSITION (%) OF PHOSPHOR BRONZE, BEARING BRONZE, LEADED TIN BRONZE, AND 85-5-5-5.
(After *Metals Handbook* and *Metals and Alloys Data Book*)

Common Name	Form	Cu	Sn	P	Pb	Zn	Trade Name
1 Phosphor bronze (5 Sn)	Wrought	Bal.	3.5-5.8	0.03-0.35	—	—	
2 Phosphor bronze (8 Sn)	Wrought	Bal.	7.0-9.0	0.03-0.25	—	—	
3 Phosphor bronze (10 Sn)	Wrought	Bal.	9.0-11.0	0.03-0.25	—	—	
4 Leaded tin bronze	Cast	86-90	5.5-6.5	—	1.0-2.0	3.0-5.0	Navy M
5 Leaded tin bearing bronze	Cast	85-89	7.5-9.0	—	1.0	3.0-5.0	Bearing bronze
6 Gun metal	Cast	87-89	7.5-10.5	—	0-2.0	1.5-4.5	
7 High-leaded tin bronze	Cast	81-85	6.25-7.5	—	6.0-8.0	2.0-4.0	Bearing bronze
8 High-leaded tin bronze	Cast	78-82	9.0-11.0	—	8.0-11.0	—	Bushing and bearing bronze
9 High-leaded tin bronze	Cast	75-79	6.25-7.5	—	13.0-16.0	—	Anti-acid bronze
10 High-leaded tin bronze	Cast	68.5-73.5	4.5-6.0	—	22.0-25.0	—	Semi-plastic bronze
11 85-5-5-5	Cast	84-86	4.0-6.0	—	4.0-6.0	4.0-6.0	Ounce metal

nealing at about 1400 F (760 C) for 1 hr per in. of thickness followed by cooling in air. This treatment may be used to increase ductility, and it is particularly useful in reducing leakage due to lack of structural soundness in pressure applications. These alloys may be modified by the addition of up to 2 per cent lead to improve machinability, in which case they are known as leaded tin bronzes. The alloy known as 80-10-10, or high-leaded tin bronze, containing approximately 80 per cent copper, 10 per cent tin, and 10 per cent lead is used quite extensively for high-speed, heavy-pressure bearings and bushings because of the improved frictional characteristics. The mechanical characteristics of alloys of this type are given in Table 17-V.

TABLE 17-Vb. PROPERTIES OF PHOSPHOR BRONZE, BEARING BRONZE, LEADED TIN BRONZE, AND 85-5-5-5.
(After *Metals Handbook* and *Metals and Alloys Data Book*)

Common Name	Form	Section (in.)	Tensile Strength (lb/in.^{2})	Yield Strength (lb/in.^{2})	Elong. (%)	Hardness	Comp. Str. (lb/in.^{2}) for 0.001 in. set
1 Phosphor bronze (5% Sn)	Flat annealed	0.040	47,000	19,000	64	RB 26	—
	½ hard flat	0.040	68,000	55,000	28	RB 78	—
	Flat hard	0.040	81,000	75,000	10	RB 87	—
	Spring flat	0.040	100,000	80,000	4	RB 95	—
	Flat extra spr.	0.040	107,000	80,000	3	RB 97	—
	Wire annealed	0.080	50,000	20,000	58	—	—
	Wire ¼ hard	0.080	68,000	60,000	24	—	—
	Wire ½ hard	0.080	85,000	80,000	8	—	—
	Wire hard	0.080	110,000	—	5	—	—
	Wire spring	0.080	140,000	—	2	—	—
	Rod ½ hard	0.500	75,000	65,000	25	RB 80	—
	Rod ½ hard	1.0	70,000	58,000	25	RB 78	—
2 Phosphor bronze (8 Sn)	Flat annealed	0.040	55,000	—	70	RF 75	—
	Flat ½ hard	0.040	76,000	55,000	32	RB 84	—
	Flat hard	0.040	93,000	72,000	10	RB 93	—
	Flat spring	0.040	112,000	—	3	RB 98	—
	Flat extra spr.	0.040	120,000	—	2	RB 100	—
	Wire annealed	0.080	60,000	24,000	65	—	—
	Wire ¼ hard	0.080	81,000	—	—	—	—
	Wire ½ hard	0.080	105,000	—	—	—	—
	Wire hard	0.080	130,000	—	—	—	—
	Wire spring	0.080	140,000	—	—	—	—
	Rod ½ hard	0.50	80,000	65,000	33	RB 85	—
3 Phosphor bronze (10 Sn)	Flat annealed	0.040	66,000	—	68	RB 55	—
	Flat ½ hard	0.040	83,000	—	32	RB 92	—
	Flat hard	0.040	100,000	—	13	RB 97	—
	Flat spring	0.040	122,000	—	4	RB 101	—
	Flat extra spr.	0.040	128,000	—	3	RB 103	—
	Wire annealed	0.080	66,000	—	70	—	—
	Wire ¼ hard	0.080	93,000	—	—	—	—
	Wire ½ hard	0.080	118,000	—	—	—	—
	Wire hard	0.080	147,000	—	—	—	—
4 Leaded tin bronze (Navy M)	Sand cast	0.505	38,000	16,000*	35	66 Brin.**	13,000
5 Leaded tin bronze (bearing bronze)	Sand cast	0.505	36,000	18,000*	30	68 Brin.**	13,000
6 Gun metal	Sand cast	—	40,000	20,000*	30	75 Brin.**	15,000
7 High-leaded tin bronze (bearing bronze)	Sand cast	0.505	34,000	17,000*	20	60 Brin.**	—
8 High-leaded tin bronze (bushing & bearing bronze)	Sand cast	—	32,000	17,000	12	65 Brin.**	14,500
9 High-leaded tin bronze (anti-acid bronze)	Sand cast	—	30,000	16,000	15	55 Brin.**	15,000
10 High-leaded tin bronze (semi-plastic bronze)	Sand cast	—	21,000	—	10	48 Brin.**	13,000
11 85-5-5-5 (ounce metal)	Sand cast	—	34,000	17,000	25	60 Brin.**	11,000

* 0.5% elongation under load.
** 500-kg load.

Another alloy which is used very widely for bearings, low-pressure valves, pipe fittings, small gears, ornamental fixtures, as well as many other purposes is called the 85-5-5-5 alloy or Ounce Metal. This alloy contains approximately 85 per cent copper, 5 per cent tin, 5 per cent zinc, and 5 per cent lead. It is readily machinable, casts quite well, and has reasonably good mechanical properties at low cost. The properties of this alloy are given in Table 17-V.

17.11 Brazing Alloys. The alloys used for brazing are composed principally of copper and zinc and sometimes contain in addition tin or nickel. In some instances, a copper-zinc-phosphorus alloy is employed. The compositions and characteristics of several of the brazing alloys are considered in detail in Chapter XXI.

17.12 Copper-Nickel Alloys. The equilibrium diagram of the copper-nickel system, Fig. 17.10, shows that copper and nickel are completely soluble in both the liquid and solid states. The corrosion resistance of these alloys increases with the nickel content. The relation between nickel content and some of the mechanical properties of the alloys is shown in Fig. 17.11. The principal high-copper-nickel alloys are listed in the following table:

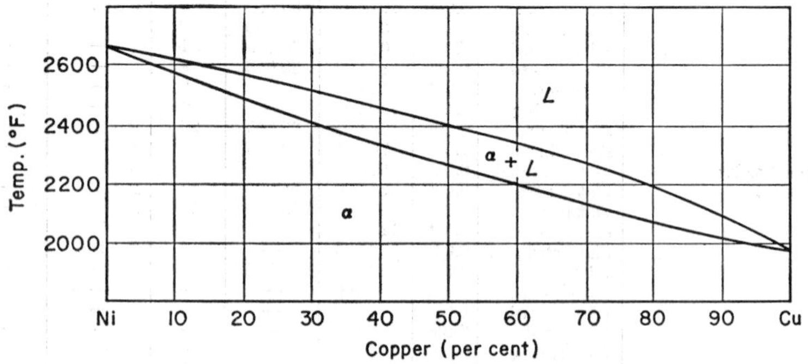

FIG. 17.10 Equilibrium diagram for copper-nickel system.

Cupronickel	2 to 30 per cent nickel; remainder, copper.
Nickel silver	10 to 30 per cent nickel; 5 per cent zinc; remainder, copper.
U. S. five-cent coin	25 per cent nickel; remainder, copper.
Constantan	45 per cent nickel; remainder, copper.

The addition of nickel to copper has a marked effect upon the color of the alloy, the alloy becoming progressively whiter in appearance until, with

20 per cent nickel, it is practically white. The profound effect of nickel upon the electrical characteristics of copper is shown in Fig. 17.12. An alloy containing about 45 per cent nickel with the remainder copper has a very high resistivity and an extremely low temperature coefficient of resistivity which make it particularly useful for certain types of resistors. Used extensively in thermocouples, this alloy is commonly called "constantan."

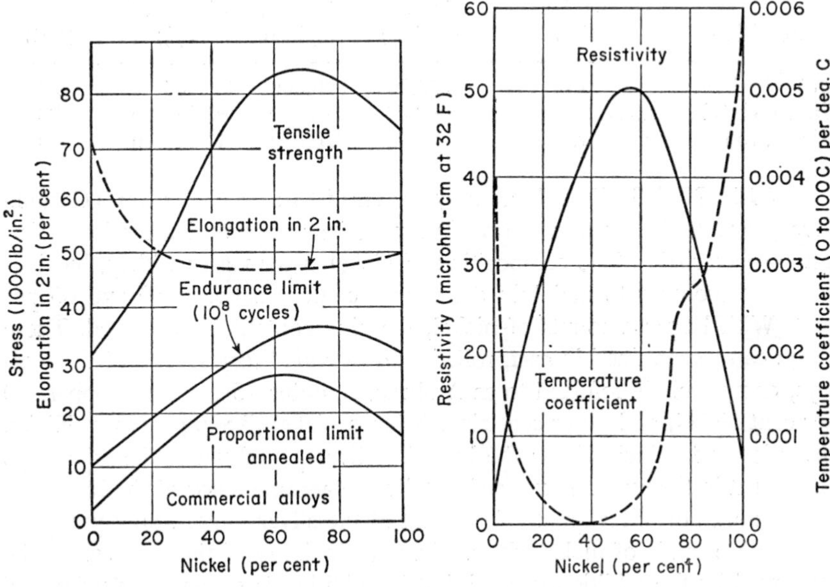

FIG. 17.11 Effect of nickel on the mechanical properties of copper-nickel alloys. (Sisco, *Modern Metallurgy for Engineers,* Pitman Pub. Corp., New York, 1948)

FIG. 17.12 Effect of nickel on the electrical properties of copper-nickel alloys. (Sisco, *Modern Metallurgy for Engineers,* Pitman Pub. Corp., New York, 1948)

17.13 Cupronickel. Cupronickel is a term that is not necessarily applied to any specific alloy, although it has been used principally in connection with the alloys containing 15, 20, and 30 per cent nickel with the remainder copper. Others, however, have been employed containing $2\frac{1}{2}$, 5, 10, and 25 per cent nickel. The first group of three alloys is most commonly employed and is useful particularly because of its strength in combination with resistance to corrosion. The 30 per cent cupronickel alloy has been used with considerable success for condenser tubes and tubing in conducting salt water and other corroding substances. The alloys containing a lower nickel content, such as the 20 per cent cupronickel alloy, are used for turbine blades and for parts requiring resistance to corrosion and erosion.

These alloys can be formed by hot-forging or cold-working operations. Some of the characteristics of 30 per cent cupronickel are given in Table 17-VI.

TABLE 17-VI. MECHANICAL PROPERTIES OF 30% CUPRONICKEL.
(From *Metals Handbook*)

Condition	Section (in.)	Tensile Strength (lb/in.^{2})	Yield Strength (lb/in.^{2})	Elongation (%)	Hardness (RB)
Hot-rolled, flat	1.0	55,000	20,000	45	35
Annealed, tube	1.0 O.D. × 0.065	60,000	25,000	45 in 2 in.	36
$\frac{1}{2}$ hard, rod	1.0	75,000	70,000	15 in 2 in.	80

17.14 Nickel Silver. The alloys referred to as nickel silver (German silver) owe their importance principally to their color and corrosion-resistant characteristics. The composition of nickel silver varies widely, but it is usually customary to maintain the copper content between 60 and 65 per cent. With larger amounts of nickel, the zinc content is decreased. These alloys can be obtained in cast, rolled, and extruded forms. One of these alloys, commercially referred to as Ambrac, contains 75 per cent copper, 20 per cent nickel, and 5 per cent zinc. It has a nickel-like appearance. Nickel silver is used principally as the base for silver-plated ware, the advantage being that in the event of wear of the silver plate, the color of the base substance does not differ greatly from that of the silver plate. Many plumbing fixtures are made of nickel silver. The machinability of these alloys can be improved by the addition of lead, but its addition is somewhat detrimental to deep-drawing characteristics and, furthermore, may be responsible for cracking when the material is heated to elevated temperatures, as in hot-working or annealing processes. The properties of characteristic nickel silvers in different conditions are shown in Table 17-VII.

17.15 Copper-Nickel-Tin Alloys. The addition of nickel to the copper-tin alloys produces some improvement in tensile properties and may permit precipitation hardening as well as somewhat improved corrosion resistance. The properties of a heat-treated cast alloy containing 5 per cent nickel, 5 per cent tin, 2 per cent zinc, known as high-strength bronze, may have a yield strength of about 60,000 lb/in.^{2}, a tensile strength of 80,000 lb/in.^{2}, with an elongation of from 5 to 25 per cent in 2 in.

An alloy containing about 70 per cent copper, 29 per cent nickel, and 1 per cent tin is used quite extensively for condenser tubes, plumbing fittings, tableware for restaurants, laundry machinery, etc. This alloy can be hot- and cold-rolled quite readily. In the hot-rolled state, such an alloy may have a yield strength of about 25,000 lb/in.^{2}, a tensile strength of 54,000

Table 17-VII. COMPOSITION (%) AND PROPERTIES OF NICKEL SILVER.

(Compiled from *Metals Handbook* and Hoyt, *Metals and Alloys Data Book*)

Form	Section	Condition	Cu	Ni	Zn	Other	Tensile Strength ($lb/in.^2$)	Yield Strength ($lb/in.^2$)	Elong. in 2 in. (%)	Hardness
Sheet	—	Soft	75	20	5		55,000	20,000	40	71 Brin.*
Sheet	—	Hard	75	20	5		130,000	95,000	2	160 Brin.*
Flat	0.04 in.	Annealed	63 -66.5	17 -19.5	Bal.		58,000	25,000	40	RB 40
		¼ hard					65,000	50,000	20	RB 73
		½ hard					74,000	62,000	8	RB 83
		Hard					85,000	74,000	3	RB 87
Rod	0.5 in.	Annealed					56,000	25,000	42	—
		Hard					70,000	60,000	20	RB 78
Wire	0.08 in.	Annealed	63 -66.5	16.5-19.5	Bal.		58,000	25,000	45	—
		¼ hard					73,000	65,000	16	—
		½ hard					86,000	80,000	7	—
		Hard					103,000	90,000	3	—
Sheet	—	Soft	65	10	25		50,000	15,000	55	RB 32
		Hard					90,000	65,000	3	RB 90
Flat	0.04 in.	Annealed	53.5-56.5	16.5-19.5	Bal.		60,000	27,000	40	RB 55
		Hard					100,000	85,000	3	RB 91
		Spring					115,000	93,000	2.5	RB 99
Wire	0.08 in.	Annealed					60,000	—	40	—
		Spring					145,000	—	2	—
Cast	—	Sand cast	65	25	Bal.	5 Sn, 1.5 Pb	50,000	24,000	15	130 Brin.*
Cast	—	Sand cast	64	20	Bal.	4 Sn, 4 Pb	40,000	25,000	15	105 Brin.*
Cast	—	Sand cast	57	12	Bal.	2 Sn, 9 Pb	34,000	15,000	20	60 Brin.*
Cast	—	Sand cast	60	16	Bal.	3 Sn, 5 Pb	38,000	17,000	25	75 Brin.*

* 500-kg load

lb/in.², an elongation of 46 per cent in 2 in., and a reduction of area of 72 per cent. By cold drawing into wire, a tensile strength of the order of 130,000 lb/in.² may be attained. Some tests conducted at temperatures up to about 800 F (430 C) have shown that this alloy has characteristics commensurate with plain carbon steel used at a temperature of 1000 F (540 C).

17.16 Copper-Aluminum Alloys. The equilibrium diagram of the copper end of the copper-aluminum system is shown in Fig. 17.13. Alloys

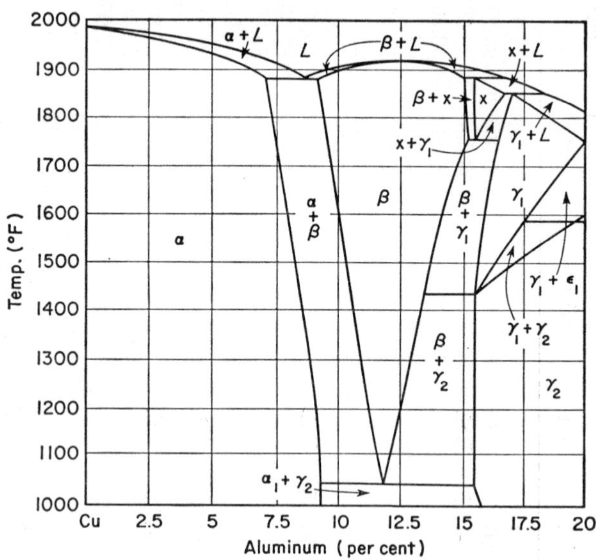

FIG. 17.13 Copper-rich end of the copper-aluminum equilibrium diagram.

in this system usually contain up to about 10 or 11 per cent aluminum and up to about 4 per cent iron with small amounts of tin. The alpha-phase alloys may be hot- or cold-worked but are difficult to machine. Typical compositions of commercial copper-aluminum alloys, sometimes referred to as aluminum bronze, are given in Table 17-VIII. These alloys are quite hard, have a high tensile strength, and resist wear, corrosion, impact, and fatigue. They are used particularly for gears in heavy machinery, feed nuts, bearings, pump parts, valve guides, nonsparking hand tools, cold-working dies, electrical contacts, etc. The 89-10-1 alloy is heat-treated by quenching in water from a temperature of about 1500 or 1600 F (820 or 870 C) and tempering between 700 and 1100 F (370 and 600 C). Typical properties of these alloys are given in Table 17-VIII.

Aluminum bronze is not an easy alloy to cast in view of the marked shrinkage during solidification and a tendency to form shrinkage cavities.

TABLE 17-VIII. COMPOSITION (%) AND PROPERTIES OF ALUMINUM BRONZE.

(Compiled from *Metals Handbook*)

Form	Cu	Al	Fe	Ni	Zn	Tensile Strength (lb/in.^{2})	Yield Strength (lb/in.^{2})	Elong. in 2 in. (%)	Hardness
Sheet, 0.041 in. annealed	94.88	5.02	0.04	—	0.06	60,300	25,600	65.8	RB 48.5
0.041 in. C.R. 44%						100,000	63,900	8.0	RB 93.5
Rod, <1 in., annealed	95.19	4.66	—	—	—	55,000	20,000	60.0	RB 45
<1 in., drawn 50%						110,000	65,000	15.0	RB 92
Sheet, 0.02 in. annealed	91.74	8.10	0.04	0.02	0.10	78,200	42,200	41.8	—
0.02 in. 37% red.						91,000	65,700	12.8	—
Rod, <1 in., annealed	91.73	8.01	Traces of Pb and Fe			60,000	15,000	70	RB 33
<1 in., 50% red.						140,000	65,000	10	RB 89
Rod, annealed	89.25	9.25	0.6	0.5	Sn 0.4	80,000	40,000	22	—
Com. temper						95,000	55,000	16	—
Hot-rolled	87.45	5.62	—	6.93	—	109,900	103,300	20.0	241 Brin.*
C.R. an. 572 F						117,600	116,800	12.0	—
Heat-treated	89.0	10-11	1	—	—	90,000-100,000	70,000-80,000	2-6	RC 20-25
Sand cast	86.9 min	9.5	0.75-1.50	—	—	62,000	26,000	25	122 Brin.**
Sand cast, ht.tr. A[1]						80,000	32,000	25	145 Brin.**
Sand cast, ht.tr. B[2]						82,000	40,000	22	160 Brin.**
Sand cast	83.0 min	10	3-5	2.5 max	—	75,000	35,000	18	155 Brin.**
Sand cast, ht.tr. C[3]						100,000	40,000	15	190 Brin.**
Sand cast, ht.tr. D[4]						105,000	52,000	10	230 Brin.**

* 500-kg load.

** 3000-kg load.

[1] Heat treatment A: Solution treatment at 1600-1650 F, quenched in water (rapid agitation), aged at 1100-1150 F for 1 hr, quenched in water.

[2] Heat treatment B: Same as A, aged at 1000 F for 1 hr, quenched in water.

[3] Heat treatment C: Solution treatment at 1600-1650 F, quenched in oil or water.

[4] Heat treatment D: Same as C, aged at 1000 F for 1 hr, quenched in water.

The design of parts in which this alloy is to be used must be simple. Dry-sand molds and very careful foundry practice must be employed.

17.17 Copper-Cadmium Alloys. Cadmium is sometimes alloyed with copper in order to produce increased strength without great decrease in electrical conductivity. The addition of most elements to copper markedly decreases the electrical conductivity. The tensile strength of copper containing about 1 per cent cadmium is approximately 83,000 lb/in.^{2} in the hard-drawn form as compared with about 64,000 lb/in.^{2} for hard-drawn pure copper. With this amount of cadmium, the electrical conductivity is about 80 per cent that of pure soft copper.

17.18 Copper-Silicon Alloys. A portion of the copper-silicon equilibrium diagram is shown in Fig. 17.14. Silicon is soluble in copper up to

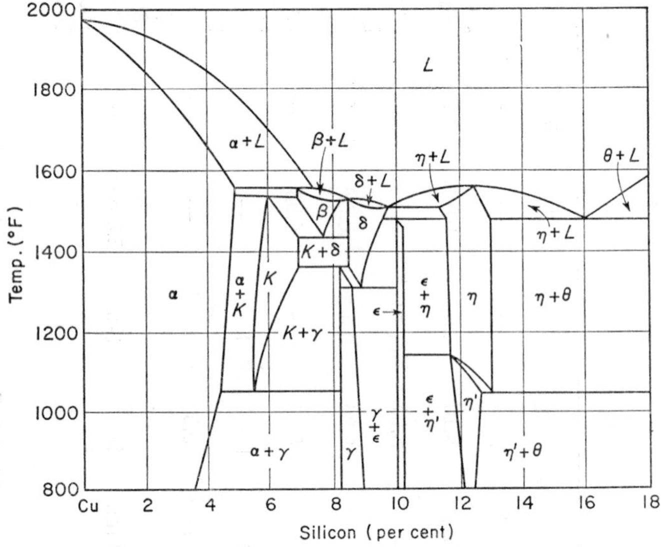

FIG. 17.14 Copper-rich end of the copper-silicon equilibrium diagram.

about 5.3 per cent at a temperature of about 1550 F (845 C). This solubility decreases so that, at room temperature, it is somewhat less than 4 per cent silicon. There is a wide variety of copper-silicon alloys that contain other elements in small proportions. The majority of these are usually referred to by the common term, silicon bronzes, or by trade names, such as Everdur, Herculoy, etc. The silicon content usually varies between about 1.0 and 5.5 per cent silicon.

The value of the silicon bronzes lies primarily in their resistance to corrosion, combined with the high strength and toughness comparable with that of the low-carbon steels. In many applications, they are suitable low-cost substitutes for the tin bronzes since they have good resistance to salt water corrosion, except under impingement attack. They likewise possess strategic importance in periods of tin shortage. These materials are cast and worked hot or cold. Some reluctance to their use has been caused by the need for sound melting practice. They are obtainable in the form of strip, plate, wire, rod, tube, pipe, casting ingots, etc. The other elements that are commonly found in these alloys are iron, tin, and lead for improved machinability. In some applications, manganese may be employed up to about 1.5 per cent. A few typical analyses of these alloys are given in Table 17-IX.

TABLE 17-IX. TYPICAL ANALYSES (%) OF SOME SILICON BRONZES.
(After *Metals and Alloys Data Book*)

Trade Name	Cu	Si	Sn	Other
Cusiloy	95	3	1	Fe 1
Duronze I	97	1	2	—
Duronze III	91	2	—	Al 7
Everdur 1000	95	4	—	Mn 1
Everdur 1015	98.25	1.5	—	Mn 0.25
Herculoy 418	96.25	3.25	0.5	—
Olympic Br. A	96	3	—	Zn 1
PMG 94	93	2.5	—	Zn 4, Fe 0.5

The tensile strength of these alloys varies in the annealed condition between 50,000 and 70,000 lb/in.^{2} and in the hard condition between 70,000 and 110,000 lb/in.^{2}, with a yield strength of 20,000 and 100,000 lb/in.^{2}, respectively. The elongation is about 50 and 5 per cent in 2 in., respectively. By more severe cold working, as may be attained in producing spring temper, the tensile strength may reach a value of the order of 145,000 lb/in.^{2} These alloys are not hardenable by precipitation methods. The tensile strength of castings will vary between about 40,000 and 70,000 lb/in.^{2}, and the elongation will vary between 20 and 35 per cent in 2 in. The silicon bronzes find application in high-strength bolts, bells, rivets, springs, propeller shafts, etc.

17.19 Copper-Beryllium Alloys. A portion of the copper-beryllium equilibrium diagram is shown in Fig. 17.15. Beryllium is soluble in copper up to about 2.1 per cent beryllium at a temperature of approximately 1600 F (870 C). Below this temperature, the solubility decreases; as a result, alloys in this system containing up to about 3 per cent beryllium exhibit precipitation hardening characteristics. It is customary to keep the beryllium content below 3 per cent because of alloy cost and difficulty in cold working. The alloys most commonly used contain between 1.0 and 2.25 per cent beryllium and are available in cast or wrought form. Some of the characteristic properties of these alloys are given in Table 17-X.

Heat treatment of these alloys is accomplished by quenching from a solution temperature of about 1450 F (790 C) into water and reheating to a precipitation temperature between 480 and 620 F (250 and 330 C) for 2 or 3 hr, the exact time and temperature depending on the properties desired. Cold rolling or cold drawing greatly increases the strength of these alloys, as can be seen from the table. By a combination of cold work and precipitation, it is possible to obtain a tensile strength of 200,000 lb/in.^{2}, an elongation of 2 per cent, and a Brinell hardness of 400.

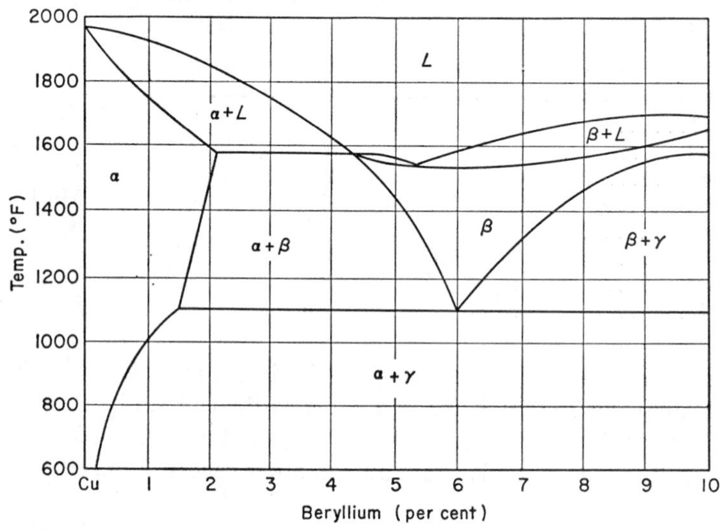

FIG. 17.15 Copper-rich end of the copper-beryllium equilibrium diagram.

Beryllium-copper alloys have been used quite successfully for certain tools which require hardness and nonsparking characteristics. Their corrosion and high fatigue resistance have made them very useful for springs, gears, diaphragms, bearings, and other purposes in which corrosion resistance is required in addition to high strength. They are also used for electrical contacts and molds for forming plastics. Even though these alloys contain a relatively small proportion of beryllium, their cost is rather high; therefore, their use is justified only when other alloys of lower cost will not fulfill the requirements of the particular application.

17.20 Nickel. Nickel is one of the most important metals used in engineering. Pure nickel finds considerable application where resistance to corrosion is required under certain conditions. It finds extensive use in the

TABLE 17-X. SOME PROPERTIES OF BERYLLIUM COPPER.
(From *Metals Handbook*)

Form	Condition	Section (in.)	Tensile Strength (lb/in.^{2})	Yield Strength 0.01% offset (lb/in.^{2})	Elongation (%)	Hardness
Strip	A	0.032	72,000	25,000	50	RB 50
Strip	½ H	0.032	95,000	62,000	12	RB 93
Strip	H	0.032	107,000	70,000	6	RB 97.5
Strip	AT	0.032	175,000	80,000	5	RC 41
Strip	½ HT	0.032	188,000	105,000	4	RC 42
Strip	HT	0.032	195,000	110,000	3	RC 42.5
Rod	AT	0.75	190,000	140,000*	3.7	380 Brin.
Rod	HT	0.75	200,000	171,000*	2.8	400 Brin.

A—Solution treated and quenched.
AT—Solution treated, quenched, and precipitation-hardened.
½ H—Solution treated, quenched, cold-rolled 20.7% (if strip) or cold-drawn 37.1% (if wire).
½ HT—Same as ½ H and precipitation-hardened.
H—Solution treated, quenched, cold-rolled 37.1% (if strip) or cold-drawn 60.5% (if wire).
HT—Same as H and precipitation-hardened.

* True stress for 0.1% offset.

chemical industry and in the production of caustic soda. It serves as an excellent base coating for the electroplating of chromium. Nickel-clad steel is used for the construction of heavy tanks and kettles. Probably the greatest quantity of nickel, however, is used in alloys. Reference has already been made to the stainless steels and the heat-resistant steels and other alloys that contain considerable nickel. The nominal composition of the different types of nickel is given in Table 17-XI.

"A" nickel is a commercially pure nickel which is obtainable in wrought form and contains 99.4 per cent nickel with the remainder principally cobalt. An alloy of nickel containing about 4.5 per cent manganese is used where improved resistance to attack by sulfur compounds at temperatures below about 1000 F (540 C) is desired. This alloy is called "D" nickel. An alloy called "E" nickel, which is very similar to "D" nickel and which contains about 2 per cent manganese, is used for similar purposes. The "L" nickel with a low carbon content is used for applications in which large plastic deformation is involved in forming. This material does not work-harden as much as the other materials.

Another alloy of nickel contains about 4.5 per cent aluminum and is called "Z" nickel. This material is subject to precipitation hardening and, therefore, provides a good combination of high strength and resistance to

TABLE 17-XI. NOMINAL COMPOSITION OF NICKEL AND NICKEL-BASE ALLOYS.

	Ni	Cu	Fe	Mn	Si	C	S	Other Elements
Nickel Grade A	99.4	0.1	0.15	0.2	0.05	0.1	0.005	
"D" Nickel	95.2	0.05	0.15	4.5	0.05	0.1	0.005	
"E" Nickel	97.7	0.05	0.10	2.0	0.05	0.1	—	
"L" Nickel	99.4	0.1	0.15	0.2	0.05	0.02	0.005	
"Z" Nickel (Duranickel)	94.0	0.05	0.25	0.25	0.40	0.16	0.005	0.33 Ti
"Z" Nickel Type B (Permanickel)	98.75	0.03	0.10	0.17	0.13	0.26	0.005	0.30 Mg
Electronic Grade "A"	99.5	0.05	0.10	0.15	0.10	0.1	0.005	
599 Alloy (Electronic)	99.5	0.02	0.07	0.05	0.20	0.05	0.005	
Monel	67.0	30.0	1.4	1.0	0.1	0.15	0.01	
R-Monel	67.0	30.0	1.7	1.1	0.05	0.10	0.035	
K-Monel	66.0	29.0	0.9	0.75	0.5	0.15	0.005	2.75 Al
H-Monel	65.0	29.5	1.5	0.9	3.0	0.1	0.015	
S-Monel	63.0	30.0	2.0	0.9	4.0	0.1	0.015	
Inconel (HR)*	77.0	0.2	7.0	0.25	0.25	0.06	0.007	15.0 Cr
Inconel W (HR)	75.0	—	7.0	0.6	0.25	0.04	—	15 Cr, 2.5 Ti, 0.6 Al
Inconel "X" (HR)	73.0	0.05	7.0	0.5	0.4	0.04	0.007	0.9 Al, 15 Cr, 1.0 Cb, 2.5 Ti
Inconel X, 550 (HR)	73	—	6.5	0.5	0.3	0.03	—	15 Cr, 0.6 Cb, 2.5 Ti, 1.1 Al
Nichrome (HE)	80	—	0.5	0.1	0.2	0.05	—	20 Cr
80-20 Ni Cr (HE) (Nichrome)	80	—	—	—	—	—	—	20 Cr
Chromel (HE)	90.0	—	—	—	—	—	—	10.0 Cr
Hy Mu "80"	79	—	17	—	—	—	—	4 Mo
Mumetal	77	5	17	—	—	—	—	1.5 Cr
78.5 Permalloy	78.5	—	21.5	—	—	—	—	
4-79 Permalloy	79	—	17	—	—	—	—	4 Mo
Supermalloy	79	—	16	—	—	—	—	5 Mo
60-12 Ni-Cr (HE)	Bal.	—	25.0	—	—	0.55	—	12 Cr
67-17 Ni Cr (HR)	Bal.	—	10.0	—	—	0.55	—	17 Cr
Hastelloy A (CR)	57	—	20	—	—	—	—	20 Mo
Hastelloy B (HR + CR)	62	—	5	—	—	—	—	30 Mo
Hastelloy C (HR + CR)	58	—	5	—	—	—	—	17 Mo, 15 Cr, 5 W
Hastelloy D (CR)	85	3	—	—	10	—	—	
Hastelloy F (CR)	47	—	17	—	—	0.05	—	2.5 Co, 22 Cr, 6.5 Mo, 0.6 W
Hastelloy X (HR)	47	—	18	—	—	0.10	—	1.5 Co, 22 Cr, 9 Mo, 0.6 W
Illium G (CR)	58	6	6	—	—	—	—	22 Cr, 6 Mo
K-42-B (HR)	42	—	13	—	—	—	—	22 Co, 18 Cr, 2.5 Ti
Inco 700 (HR)	49	—	0.5	0.05	0.2	0.1	—	28 Co, 15 Cr, 3 Mo, 2 Ti, 3 Al
Inco 739 (HR)	77	—	1.0	0.05	0.2	0.07	—	15 Cr, 1.7 Ti, 2.7 Al
M-252 (HR)	Bal.	—	5.0 max	1.0	0.65	0.15	—	19 Cr, 10 Co, 10 Mo, 2.5 Ti, 0.87 Al
Nimonic 80A (HR)	76	—	0.5	0.7	0.5	0.05	—	20 Cr, 2.3 Ti, 1 Al
Nimonic (HR)	76	—	2.4	0.4	0.6	0.12	—	20 Cr, 0.4 Ti, 0.06 Al
Nimonic 90 (HR)	58	—	0.5	0.5	0.4	0.08	—	20 Cr, 16 Co, 2.3 Ti, 1.4 Al
Waspaloy (HR)	Bal.	—	2.0	1.0 max	0.75 max	0.1 max	—	19.5 Cr, 13.5 Co, 4.25 Mo, 2.5 Ti, 1.25 Al

* Type of application:
(HR) = Heat-resisting alloy
(CR) = Corrosion-resisting alloy
(HE) = Heating element alloy

corrosion. This alloy is satisfactory for application in sulfidizing atmospheres at temperatures below 600 F (320 C) and is utilized for springs, pump rods, shafts, etc. The heat treatment consists of heating to a solution temperature of 1600 F (870 C) for $\frac{1}{2}$ hr, followed by a water quench and then heating to a precipitation temperature of 1100 F (595 C) for a period of from 8 to 16 hr, followed by furnace cooling. Some of the properties of nickel and the low alloys of nickel are given in Table 17-XII.

TABLE 17-XII. SOME PROPERTIES OF NICKEL.
(Compiled from *Metals Handbook*)

Material	Condition	Tensile Strength (lb/in.^{2})	Yield Strength (lb/in.^{2})	Elong. in 2 in. (%)	Reduction of Area (%)	Hardness
Nickel Electrolytic	Annealed	46,000	8,500	30	—	—
"A" Nickel	Annealed	50,000- 80,000	10,000- 30,000	50-35	75-60	RB 40-65
	Cold-drawn rod	65,000-115,000	40,000- 90,000	35-15	70-50	RB 70-100
	Hot-rolled rod	55,000- 80,000	15,000- 45,000	65-30	75-60	RB 40-65
	Cast	53,000	25,000	25	—	—
"D" Nickel	Hot-rolled	86,000	34,000	40	60	RB 70
"Z" Nickel	Annealed rod	90,000-120,000	30,000- 60,000	50-25	65-45	RB 75-90
	Drawn rod	100,000-150,000	50,000-130,000	35-15	60-40	RB 8-31
	Drawn and age-hardened rod	60,000-190,000	120,000-150,000	20- 7	30-15	RC 31-40

There are many special (proprietary) nickel-rich alloys that are employed in electron tube parts. An example of these alloys is Electronic Grade "A" Nickel and 599 Alloy, indicated in Table 17-XI. Control of the composition of these materials is important to the proper functioning of tube cathodes. An improper amount of certain elements may cause the cathode to be slow in attaining a high emission level.

17.21 Nickel-Base Alloys. The alloys that contain a substantial proportion of nickel are of great commercial importance. Many of the corrosion-resistant steels containing nickel have been discussed in Chapter XIV. Aside from the so-called stainless steels that contain nickel, there are the low-expansion alloys containing about 36 per cent nickel, the glass-sealing alloys with from 30 to 50 per cent nickel, the alloys with low-temperature coefficient of modulus of elasticity, and the magnetic alloys containing up to about 65 per cent nickel.

Those alloys that contain more than 50 per cent nickel are classed as the nickel-base alloys. The nominal composition of this group of alloys is given in Table 17-XI.

17.22 Monel. Monel nominally contains 67 per cent nickel, 30 per cent copper, and small amounts of iron and manganese. The Monels are particularly useful in applications that require resistance to acids, alkalies, brines, water, and food products. The Monel family of alloys consists of five compositions: Monel, R-Monel, K-Monel, H-Monel, and S-Monel. Typical analyses of these alloys are given in Table 17-XI.

The alloy known as R-Monel possesses the same general characteristics as Monel, but it is a free-machining alloy intended for processing in automatic screw machines. The improved machining qualities are derived from the addition of from 0.025 to 0.060 per cent sulfur.

K-Monel is a precipitation-hardenable Monel which contains about 3 per cent aluminum, and its strength approaches that of heat-treated alloy steels. The corrosion resistance is similar to that of Monel. The heat treatment consists of heating to a solution temperature of 1600 F (870 C) for ½ hr, followed by quenching in water. This is followed by heating to a precipitation temperature of 1100 F (595 C) for 8 to 16 hr, and then furnace cooling.

S-Monel is used primarily in castings and contains about 4 per cent silicon. This alloy is responsive to precipitation hardening. A hardness of about 350 Brinell makes it suitable for use when resistance to galling and erosion is important, as in valve seats, and where sliding contact is involved under corrosive conditions. This alloy is softened by heating to a temperature of 1600 F (870 C) for 1 hr, followed by cooling in air to 1200 F (650 C) and then quenching in oil or water. Hardening is accomplished after the softening treatment by heating to 1100 F (595 C) for 4 to 6 hr, followed by furnace cooling.

H-Monel is similar to S-Monel, but contains only 2.75 to 3.25 per cent silicon. This alloy cannot be treated to as high a strength and hardness as the S-Monel, but for many applications, it is adequate.

Typical properties of these alloys are given in Table 17-XIII.

TABLE 17-XIII. SOME PROPERTIES OF NICKEL-COPPER ALLOYS.
(Compiled from *Metals Handbook* and *Nickel and Nickel Alloys*, International Nickel Co.)

Material	Condition	Tensile Strength (1000 lb/in.^{2})	Yield Strength (1000 lb/in.^{2})	Elong. in 2 in. (%)	Reduction of Area (%)	Hardness
Monel......	Annealed rod	70- 85	25- 40	50-35	75-60	RB 60-75
	Hot-rolled	80- 95	40- 65	45-30	75-60	RB 75-90
	Cold-drawn	85-125	55-120	35-10	70-50	RB 85-RC 23
R-Monel....	Annealed rod	70- 85	25- 40	50-35	70-55	RB 60-75
	Hot-rolled	75- 90	35- 60	45-25	70-50	RB 70-85
	Cold-drawn	80-115	50-100	35-15	65-50	RB 80-100
K-Monel....	Annealed rod	90-110	40- 60	45-35	70-50	RB 75-90
	Precipitation-hardened	130-150	90-110	30-20	50-30	RC 21-28
	Cold-drawn and precipitation-hardened	140-170	100-130	30-15	50-25	RC 27-33
H-Monel....	As-cast	90-115	45- 75	20-10	—	175-250 Brin.
S-Monel....	As-cast or annealed and					275-350 Brin.
	precipitation-hardened	110-145	80-115	4- 1	—	325-375 Brin.

17.23 Other Nickel-Base Alloys. There is a large group of essentially nickel-base alloys that are particularly serviceable for resistance to elevated temperatures or to corrosion. Many of these alloys are utilized for both applications. These alloys are indicated in Table 17-XI. A few of these alloys are particularly well adapted for use as electrical heating elements, as indicated in the table.

Some of the high-nickel alloys are especially useful as magnetic materials. The initial permeability (permeability at low-magnetizing force) increases with increasing amounts of nickel added to iron, attaining a high value with about 80 per cent nickel. The alloy with about 80 per cent nickel also has very high maximum permeability and is called "Permalloy." The composition of some of the alloys is given in Table 17-XI. The magnetic characteristics of 78.5 Permalloy are sensitive to the cooling rate to which it is subjected in heat treatment. Additions of chromium and molybdenum eliminate this sensitivity. The 4-74 Permalloy is one of these modified alloys.

REFERENCES

Book of Standards, Part II, Non-Ferrous Metals, American Society for Testing Materials, Philadelphia, Pa.

Cast Metals Handbook, American Foundrymen's Society, Chicago, Ill., 1957.

Ellis, *Copper and Copper Alloys,* American Society for Metals, Metals Park, Ohio, 1948.

Hoyt, *Metal Data,* Reinhold Publishing Corp., New York, 1952.

Hull, *Casting of Brass and Bronze,* American Society for Metals, Metals Park, Ohio, 1950.

Metals Handbook, American Society for Metals, Metals Park, Ohio, 1961.

Wilkins and Bunn, *Copper and Copper-Base Alloys,* McGraw-Hill Book Co., New York, 1943.

QUESTIONS

1. What is the greatest use of pure copper?
2. Indicate the general relationship between properties and the phase diagram of the copper-zinc system.
3. What are the particular characteristics of alloys of the copper-zinc system which contain only alpha solid solutions?
4. What is commercial bronze, and what are some of its applications?
5. What is cartridge brass, and for what type of applications is it best suited?
6. What is Muntz metal? For what is it used?
7. How does naval brass differ from Muntz metal?
8. By what process may the cast structure of a brass be broken up to produce a more homogeneous structure?
9. What is season cracking? Indicate its cause and remedy.
10. What is the purpose of adding lead to brass?
11. What advantages may be secured by the addition of such elements as manganese and silicon to brass?
12. What types of alloys are included under the classification of bronzes?
13. Into what four groups may bronzes be divided? Indicate the general application of each group.
14. What benefits may be secured by the inclusion of phosphorus in bronzes?

15. What benefits may be derived by the addition of zinc and lead to bronzes?
16. Give some examples of bronzes that are frequently used for bearings.
17. Give some examples of alloys that are used for brazing.
18. What alloy is commonly used because of its high resistivity and low-temperature coefficient of resistivity?
19. What is the range of composition of cupronickel? What are some of its applications?
20. Indicate the approximate composition and characteristics of nickel silver.
21. What characteristics and uses are possessed by the copper-nickel-tin alloys?
22. What is the general composition of aluminum bronze? Indicate some of its characteristics and uses.
23. What is the effect upon the properties when cadmium is added in copper?
24. Indicate the approximate composition and some of the applications of the copper-silicon alloys.
25. Indicate the approximate compositions, properties, and uses of the copper-beryllium alloys.
26. Distinguish between "A" nickel and "Z" nickel, as to composition, properties, and uses.
27. What is included in the Monel family of alloys?
28. Indicate some of the differences in the composition, properties, and use of the different Monel metals.

XVIII

LIGHT METALS AND THEIR ALLOYS

18.1 Scope. Light alloys have become of great importance in engineering for the construction of transportation equipment. Many of these lightweight alloys have sufficiently high strength to warrant their use for structural purposes, and, as a result of their use, the total weight of transportation equipment has been considerably decreased. Probably the greatest application of the light alloys is in the construction of aircraft. The extensive use of light alloys began with the discovery of precipitation hardening in these alloys.

At present, the metals that serve as the base of the principal light alloys are aluminum and magnesium. Titanium and its alloys are included in this group since they have a density much lower than that of steel. This chapter will be concerned with a discussion of the principal alloys of these elements—their properties and uses.

18.2 Aluminum. Commercially pure aluminum with a specific gravity of 2.71 is 99 per cent pure, the remainder consisting principally of iron and silicon. In the annealed state, it has a tensile strength of 13,000 lb/in.^{2} with an elongation of about 40 per cent in 2 in. It can be cold-worked with ease and may be work-hardened to a strength of about 24,000 lb/in.^{2}. Its resistance to the action of the atmosphere and to several chemicals gives it further advantages, although commercially pure aluminum is not usually resistant to strong alkalies or to some weak alkaline solutions. Its corrosion resistance under oxidizing conditions depends upon the natural development of an aluminum oxide surface. In commercially pure form, aluminum is used principally for cooking utensils and chemical equipment. It is obtainable in the form of sheet, plate, tubing, wire, rivets, rod, bar, etc. Aluminum is also used extensively in paint as a paint pigment. It is also used for the deoxidation of steel. On a weight basis, the electrical conductivity of aluminum is about 200 per cent that of copper; on a volume basis, it is about 61 per cent that of copper. The aluminum alloys are used most extensively for structural purposes.

18.3 Alloying Elements in Aluminum. The alloying elements commonly used in commercial aluminum alloys include copper, silicon, magnesium, manganese, and occasionally zinc, nickel, and chromium. The overall effect of alloy additions is to raise the tensile strength, yield strength, and hardness with corresponding reduction of percentage elongation. Alloying elements are added extensively to aluminum castings to improve casting qualities as well as mechanical properties. In general, the aluminum alloys may secure their higher mechanical properties by:

(1) Solid solution strengthening, or
(2) Responding to precipitation hardening, or
(3) Strain hardening by cold work.

Copper has been the principal alloying element in aluminum for many years. It is employed in amounts up to about 4 per cent in wrought alloys and up to about 8 per cent in castings. Its effect is to decrease shrinkage and hot shortness and to provide a basis for age hardening in many aluminum alloys.

Silicon is probably second to copper in its importance as an alloying element, principally in casting alloys. It is used in amounts ranging from about 1 to 14 per cent as a primary or secondary alloying element. Silicon improves casting qualities, such as fluidity and freedom from hot shortness, in addition to providing corrosion resistance, low thermal expansion, and high thermal conductivity. Casting alloys containing silicon are noted for good impact toughness and pressure tightness. Silicon usually appears as an impurity in all commercial aluminum alloys. It is employed as a secondary alloying element in aluminum-magnesium alloys which depend on the formation of magnesium silicide as a hardening agent.

Magnesium is alloyed with aluminum in amounts usually ranging from about 1 to 10 per cent. Such alloys are lighter than aluminum, possess good mechanical properties, are easily machined, and probably possess, with increasing magnesium content, the best salt water and alkaline corrosion resistance of the aluminum-base alloys.

Zinc in amounts up to about 10 per cent, usually associated with other elements, may be added to improve mechanical properties through the formation of hard intermediate phases, such as Mg_2Zn.

Manganese and chromium are added in small amounts to increase both the strength and the corrosion resistance of aluminum alloys. Nickel improves the strength at elevated temperatures with some decrease in corrosion resistance. In the absence of silicon, iron decreases the hardening capacity of the aluminum-copper alloys by removing copper from the solid solution. Lead and bismuth are sometimes added to produce free-machining grades of aluminum alloys. Small amounts of titanium or columbium have been used as grain-refining agents in certain alloys.

18.4 Aluminum-Copper Alloys. The equilibrium diagram of the aluminum-copper system is shown in Fig. 18.1. A detailed diagram for the

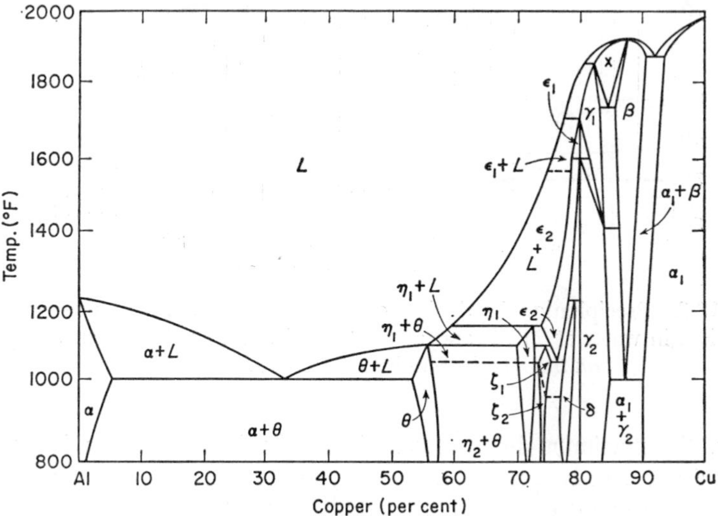

FIG. 18.1 Equilibrium diagram for aluminum-copper system.

aluminum-rich portion was shown in Fig. 6.1 of Chapter VI under the discussion of precipitation hardening. The solubility of copper in aluminum decreases from 5.65 per cent at 1018 F (548 C) to less than 0.25 per cent at room temperature. A eutectic is formed at 33 per cent copper. The useful aluminum-rich copper alloys are those containing less than about 10 per cent copper. The mechanical properties of both the wrought 4 per cent copper and the cast 8 per cent copper alloys are improved by precipitation of the theta phase ($CuAl_2$) from the solid solution. The mechanism of precipitation hardening in such alloys was discussed in detail in Chapter VI. The influence of the copper content and the time of aging or over-aging on the hardness and corresponding strength properties is shown in Fig. 18.2. The structure of a commercial alloy containing 4 per cent copper is shown in Fig. 18.3. The particles that appear are not precipitate, but impurities. This alloy has been solution-treated and aged.

The true binary alloys of this system have limited commercial usefulness and are usually found combined with small percentages of other elements, such as silicon, manganese, magnesium, zinc, etc. In the heat-treated con-

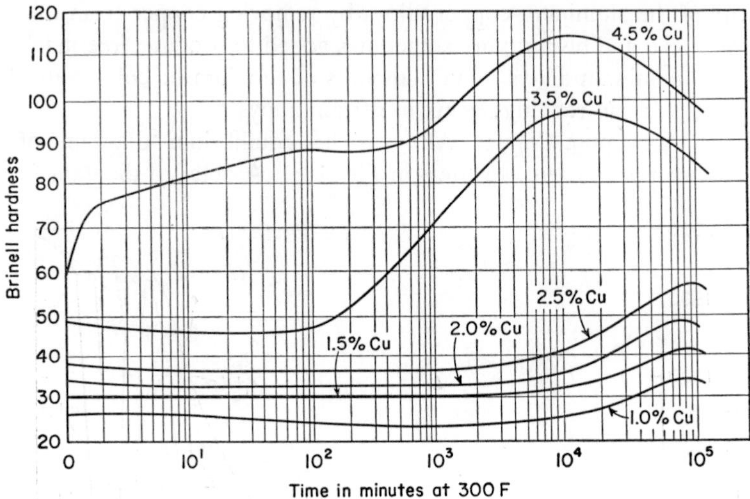

FIG. 18.2 Precipitation hardening curves for some aluminum-copper alloys quenched in water at 212 F and aged at 300 F. (After Hunsicker, *Symposium on the Age Hardening of Metals, ASM,* 1939)

dition, the maximum improvement in mechanical properties is noted at about 6 per cent copper in binary alloys. The corrosion resistance of aluminum is greatly decreased by the addition of copper and may, in general, be said to be the poorest of the conventional aluminum-base alloys.

18.5 Aluminum-Silicon Alloys. The aluminum-silicon equilibrium

FIG. 18.3 Aluminum alloy 2017; 95% Al, 4% Cu, 0.5% Mn; solution-treated and aged at room temperature (100×).

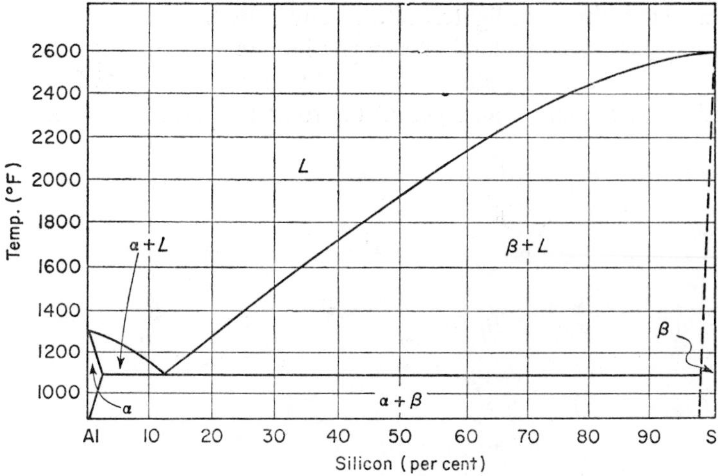

FIG. 18.4 Equilibrium diagram for aluminum-silicon system.

diagram is shown in Fig. 18.4. The solubility of silicon in aluminum decreases from 1.65 per cent at the eutectic temperature of 1070 F (577 C) to a negligible amount at room temperature. A eutectic forming at 11.6 per cent silicon is largely responsible for the hardness of the cast aluminum-silicon alloys. The silicon content of the binary alloy seldom exceeds 14 per cent. The addition of other elements, such as iron, zinc, copper, etc., may be made to form special constituents in the ternary system with age-hardening effects. Silicon is being substituted for a large portion of the copper in the high-copper casting alloys. The effect of silicon on the mechanical properties of the cast binary alloy is shown in Fig. 18.5. The structure of a commercial alloy containing 7 per cent silicon is shown in Fig. 18.6. The aluminum-silicon eutectic is clearly shown in the photomicrograph.

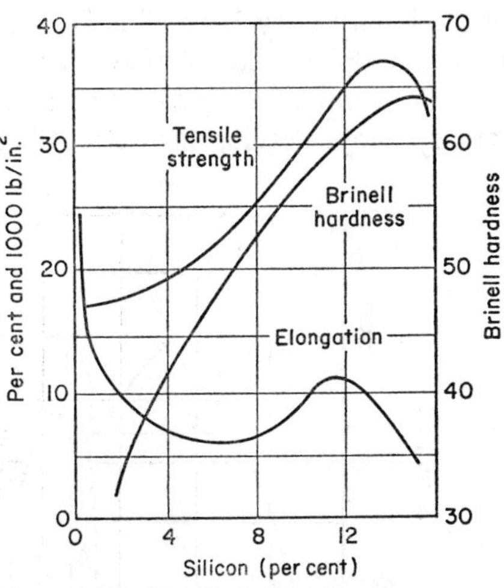

FIG. 18.5 Mechanical properties of aluminum-silicon alloys, chill-cast test bars. (Kempf, *Metals Handbook*)

18.6 Aluminum-Magnesium Alloys. Alloys of the aluminum-magnesium series constitute the lightest of the conventional aluminum-base alloys. The equilibrium system, aluminum-magnesium, is shown in Fig. 18.7. Magnesium has a maximum solubility of 14.9 per cent at 845 F (452 C), decreasing to less than about 2 per cent at room temperature. Age hardening

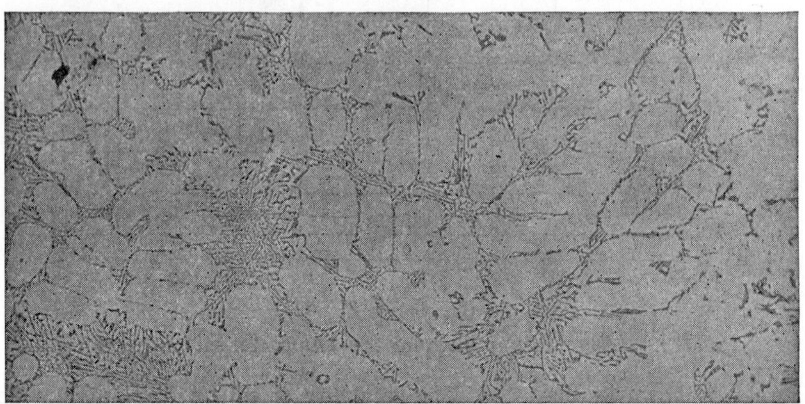

FIG. 18.6 Aluminum alloy 356—as cast; 92% Al, 7% Si, 0.3% Mg (100×).

of the binary alloy is possible by precipitation of the beta phase (AlMg) from the alpha solid solution. In the presence of silicon, the hard, intermetallic compound Mg_2Si forms as an important hardening agent. These ternary alloys develop high strength through precipitation hardening.

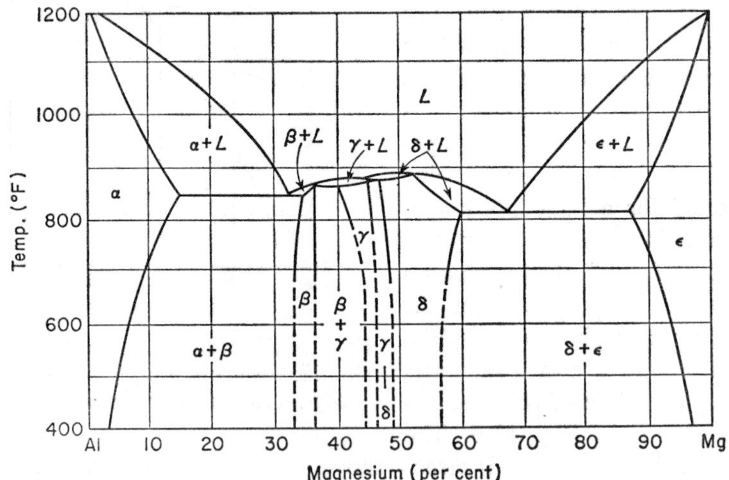

FIG. 18.7 Equilibrium diagram for aluminum-magnesium system.

Magnesium contents up to about 10 per cent are commonly found in the binary alloys. With a high proportion of magnesium, casting becomes difficult in that excessive drossing or surface oxidation of the bath and inability to produce pressure-tight castings are encountered. With proper foundry technique, however, satisfactory commercial castings are possible.

In general, aluminum-magnesium alloys offer maximum resistance to atmospheric, salt water, and alkaline corrosion of any of the aluminum-base alloys. This resistance naturally decreases with the presence of impurities.

A comparison of the mechanical properties of the sand-cast and aged binary alloys is shown in Figs. 18.8 and 18.9, respectively. The marked improvement produced by aging in cast alloys containing more than 6 per cent magnesium is quite clear.

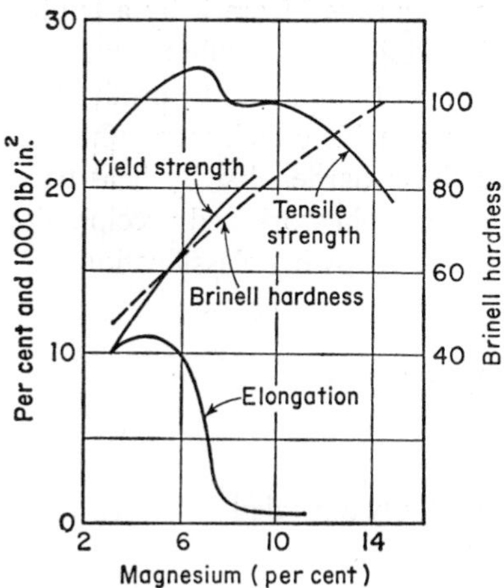

FIG. 18.8 Mechanical properties of aluminum-magnesium alloys, sand castings. (Sicha and Keller, *Metals Handbook, ASM,* 1948)

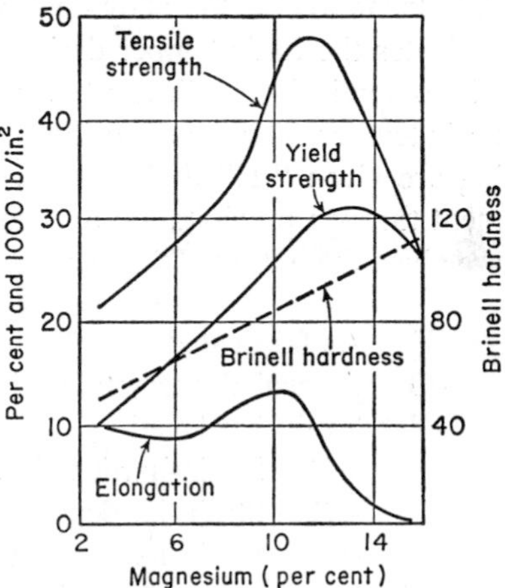

FIG. 18.9 Mechanical properties of aluminum-magnesium alloys, heat-treated sand castings. (Sicha and Keller, *Metals Handbook, ASM,* 1948)

18.7 Classification of Aluminum Alloys. The aluminum alloys are broadly classed as (1) casting or (2) wrought alloys. The casting alloys are usually designated by a number assigned to each composition such as 43, 195, etc. The wrought alloys are designated by a four-digit designation system established by the Aluminum Association. The first digit identifies the alloy type. The second digit indicates the specific alloy modification. The last two digits identify the specific aluminum alloy or indicate the aluminum purity. The meaning of the first digit is as follows:

Designation	*Type of Alloy*
1xxx	Aluminum 99.00% min and greater
2xxx	Copper
3xxx	Manganese
4xxx	Silicon
5xxx	Magnesium
6xxx	Magnesium and silicon
7xxx	Zinc
8xxx	Other element
9xxx	Unused series

A letter following the alloy designation and separated from it by a hyphen indicates the basic-temper designation. The addition of a subsequent digit, where applicable, indicates the specific treatment employed to produce the basic temper. Those compositions that are hardenable only by strain hardening are given "-H" designations; whereas those hardenable by heat treatment through precipitation or a combination of cold work and precipitation are given the letter "-T" in accordance with the following classification:

- -F As fabricated
- -O Annealed, recrystallized (wrought only)
- -H Strain-hardened
  - -H1 Strain-hardened only
  - -H2 Strain-hardened and then partially annealed
  - -H3 Strain-hardened and then stabilized
- -W Solution heat-treated—unstable temper
- -T Heat-treated to stable tempers
  - -T2 Annealed (cast only)
  - -T3 Solution treated and cold-worked
  - -T4 Solution treated followed by natural aging at room temperature
  - -T5 Artificially aged only after an elevated-temperature, rapid-cool fabrication process such as casting or extrusion
  - -T6 Solution treated and artificially aged
  - -T7 Solution treated and stabilized to control growth and distortion
  - -T8 Solution treated, cold-worked, and artificially aged
  - -T9 Solution treated, artificially aged, and cold-worked

A second numeral following the basic temper letter indicates the degree of hardness produced by the specific processing operation. The numeral "8" designates the full hard commercial temper, whereas the numeral "4" indicates a hardness midway between the fully annealed (O) and 8 temper. The extra hard temper is designated by the numeral "9." As an illustration, the designation 1100-H14 refers to commercially pure wrought aluminum, indicated by 1100, which has been strain-hardened, indicated by -H1, to a tensile strength midway between the hardest and softest commercial tempers, indicated by 4. Recommended treatments to obtain specific tempers are given in subsequent sections.

Prior to the present system of temper designations, it was the practice to indicate the various tempers as $\frac{1}{4}$H, $\frac{1}{2}$H, $\frac{3}{4}$H, and H. Such terminology may still be found in the literature.

18.8 Aluminum Casting Alloys. The nominal compositions and typical mechanical properties of common aluminum sand-casting alloys are given in Tables 18-I and 18-II, respectively. Strength and hardness values are representative of standard half-inch diameter, tensile-test specimens, as cast in green-sand molds without subsequent machining.

Aluminum alloys are not only cast in sand molds, but certain compositions are ideally suited to the permanent-mold and die-casting processes for

TABLE 18-I. NOMINAL COMPOSITION OF ALUMINUM SAND-CASTING ALLOYS.
(* Alloys also used for permanent-mold castings)

	% ALLOYING ELEMENTS					
ALLOY	Cu	Fe	Si	Mg	Zn	Ni
* 43	0.1	0.8	5.0	0.05	0.2	—
108	4.0	1.0	3.0	0.03	0.2	—
112	7.0	1.5	1.0	0.07	2.2	0.3
* 113	7.0	1.4	2.0	0.07	2.2	0.3
* 122	10.0	1.5	1.0	0.2	0.5	0.3
* 142	4.0	0.8	0.6	1.5	0.1	2.0
195	4.5	1.0	1.2	0.03	0.3	—
212	8.0	1.4	1.2	0.05	0.2	—
214	0.1	0.4	0.3	4.0	0.1	—
B214	0.1	0.4	1.8	4.0	0.1	—
F214	0.1	0.4	0.5	4.0	0.1	—
220	0.2	0.3	0.2	10.0	0.1	—
319	3.5	1.2	6.3	0.5	1.0	0.5
* 355	1.3	0.6	5.0	0.5	0.2	—
A355	1.5	0.6	5.0	0.5	0.1	0.8
* 356	0.2	0.5	7.0	0.3	0.2	—
A612	0.5	0.5	0.15	0.7	6.5	—
* 750	1.0	0.7	0.7	—	—	1.0
*A750	1.0	0.7	2.5	—	—	0.5
*B750	2.0	0.7	0.4	0.75	—	1.2

TABLE 18-II. TYPICAL MECHANICAL PROPERTIES OF ALUMINUM SAND-CASTING ALLOYS.*

Alloy	Tensile Strength (lb/in.²)	Yield Strength (Offset = 0.2%) (lb/in.²)	Elongation in 2 in. (%)	Brinell Hardness (500-kg Load 10-mm Ball)	Shear Strength (lb/in.²)	Endurance Limit† (lb/in.²)
43-F	19,000	8,000	8.0	40	14,000	8,000
108-F	21,000	14,000	2.5	55	17,000	11,000
112-F	24,000	15,000	1.5	70	20,000	9,000
113-F	24,000	15,000	1.5	70	20,000	9,000
122-T61	41,000	40,000	<0.5	115	32,000	8,500
142-T21	27,000	18,000	1.0	70	21,000	8,000
142-T571	32,000	30,000	0.5	85	26,000	11,000
142-T77	30,000	23,000	2.0	75	24,000	10,500
195-T4	32,000	16,000	8.5	60	26,000	7,000
195-T6	36,000	24,000	5.0	75	30,000	7,500
195-T62	41,000	32,000	2.0	90	33,000	8,000
212-F	23,000	14,000	2.0	65	20,000	9,000
214-F	25,000	12,000	9.0	50	20,000	7,000
B214-F	20,000	13,000	2.0	50	17,000	8,500
F214-F	21,000	12,000	3.0	50	17,000	8,000
220-T4	48,000	26,000	16.0	75	34,000	8,000
319-F	27,000	18,000	2.0	70	22,000	10,000
319-T5	30,000	26,000	1.5	80	24,000	11,000
319-T6	36,000	24,000	2.0	80	29,000	11,000
355-T51	28,000	23,000	1.5	65	22,000	8,000
355-T6	35,000	25,000	3.0	80	28,000	9,000
355-T7	38,000	36,000	0.5	85	28,000	10,000
355-T71	35,000	29,000	1.5	75	26,000	10,000
A355-T51	28,000	24,000	1.5	70	22,000	8,000
356-T51	25,000	20,000	2.0	60	20,000	8,000
356-T6	33,000	24,000	3.5	70	26,000	8,500
356-T7	34,000	30,000	2.0	75	24,000	9,000
356-T71	28,000	21,000	3.5	60	20,000	8,500
A612-F	35,000	25,000	5.0	75	26,000	8,000
750-T5	20,000	11,000	8.0	45	14,000	—
A750-T5	20,000	11,000	5.0	45	14,000	—
B750-T5	27,000	22,000	2.0	65	18,000	10,000

* Tensile properties and hardness based on unmachined cast test specimens, 0.5 in. diameter. Modulus of elasticity, average 10.3×10^6 lb/in.²

† 500×10^6 cycles with R. R. Moore-type specimen.

the quantity production of close tolerance parts. The composition of some alloys for permanent-mold casting is the same as that of some of the sand-casting alloys. Additional alloys for permanent-mold castings are given in Table 18-III. The typical properties of permanent-mold casting alloys are indicated in Table 18-IV. The composition of aluminum alloys for die-

TABLE 18-III. NOMINAL COMPOSITION OF ALUMINUM PERMANENT-MOLD CASTING ALLOYS.
(Other alloys indicated by * in Table I)

Alloy	Cu	Fe	Si	Mg	Zn	Ni
A108	4.5	1.0	5.5	0.1	0.5	—
C113	7.0	1.5	3.5	0.3	2.5	0.5
A132	1.0	1.3	12.0	1.0	0.1	2.5
F132	3.0	1.2	9.5	1.0	1.0	0.5
138	10.0	1.5	4.0	0.25	0.5	0.5
152	7.0	1.5	5.5	0.3	1.0	0.5
B195	4.5	1.0	2.5	0.03	0.3	—
A214	0.1	0.4	0.3	4.0	1.8	—
333	3.75	1.2	9.0	0.6	1.0	0.5
C355	1.25	0.2	5.0	0.5	0.1	—
A356	0.2	0.2	7.0	0.3	0.1	—
C612	0.5	1.4	0.3	0.35	6.5	—

casting is given in Table 18-V. The typical properties of the aluminum die-casting alloys are indicated in Table 18-VI.

The original binary aluminum-copper, sand-casting alloy containing 8 per cent copper has been largely replaced by alloys containing other elements to improve casting and machining properties, such as 112, 113, and 212. The mechanical properties of these general-purpose alloys are practically identical, with 112 having somewhat better machinability.

Aluminum-silicon casting alloys, such as 43 and 356, are noted for excellent casting characteristics, being employed for pressure-tight hydraulic castings and thin-wall designs where fluidity in pouring is desired to reproduce fine detail accurately. This property, together with their atmospheric corrosion resistance, makes them particularly desirable for ornamental work. Such alloys are more ductile and shock-resistant than the aluminum-copper alloys. They find use in aircraft and marine fittings, cooking utensils, and miscellaneous thin-sectioned, intricate castings.

Aluminum-magnesium alloys, such as 214 and 220, are noted for their corrosion resistance even in marine atmospheres. Additions of silicon are sometimes included to improve casting and mechanical properties. Such alloys are used for food-handling equipment, chemical and sewage fittings, aircraft-structural fittings, and miscellaneous castings requiring strengths with resistance to corrosion and shock. These alloys are unsuitable for pressure-tight castings or elevated-temperature applications.

Aluminum-copper-silicon alloys, such as 108 and 319, combine certain properties of the aluminum-copper and aluminum-silicon alloys to produce

TABLE 18-IV. TYPICAL MECHANICAL PROPERTIES OF ALUMINUM PERMANENT-MOLD CASTING ALLOYS.*

Alloy	Tensile Strength (lb/in.^{2})	Yield Strength (Offset = 0.2%) (lb/in.^{2})	Elongation in 2 in. (%)	Brinell Hardness (500-kg Load 10-mm Ball)	Shear Strength (lb/in.^{2})	Endurance Limit† (lb/in.^{2})
43-F	23,000	9,000	10.0	45	16,000	8,000
A108-F	28,000	16,000	2.0	70	22,000	13,000
113-F	28,000	19,000	2.0	70	22,000	9,000
C113-F	30,000	24,000	1.5	85	24,000	9,500
122-T551	37,000	35,000	0.5	115	30,000	8,500
122-T65	48,000	36,000	0.5	140	36,000	9,000
A132-T551	36,000	28,000	0.5	105	28,000	13,500
A132-T65	47,000	43,000	0.5	125	36,000	—
F132-T5	36,000	28,000	1.0	105	—	—
138-F	30,000	24,000	1.5	100	24,000	—
142-T571	40,000	34,000	1.0	105	30,000	10,500
142-T61	47,000	42,000	0.5	110	35,000	9,500
152-T524	29,000	16,000	1.0	95	—	—
B195-T4	37,000	19,000	9.0	75	30,000	9,500
B195-T6	40,000	26,000	5.0	90	32,000	10,000
B195-T7	39,000	20,000	4.5	80	30,000	9,000
A214-F	27,000	16,000	7.0	60	22,000	—
333-F	34,000	19,000	2.0	90	27,000	14,500
333-T5	34,000	25,000	1.0	100	27,000	12,000
333-T6	42,000	30,000	1.5	105	33,000	15,000
333-T7	37,000	28,000	2.0	90	28,000	12,000
355-T51	30,000	24,000	2.0	75	24,000	—
355-T6	42,000	27,000	4.0	90	34,000	10,000
355-T62	45,000	40,000	1.5	105	36,000	10,000
355-T71	36,000	31,000	3.0	85	27,000	10,000
C355-T61	46,000	34,000	6.0	100	32,000	14,000
356-T6	38,000	27,000	5.0	80	30,000	13,000
356-T7	32,000	24,000	6.0	70	25,000	11,000
A356-T61	41,000	30,000	10.0	90	28,000	13,000
C612-F	35,000	18,000	8.0	70	—	11,000
750-T5	23,000	11,000	12.0	45	15,000	9,000
A750-T5	20,000	11,000	5.0	45	14,000	9,000
B750-T5	32,000	23,000	5.0	70	21,000	11,000

* Tensile properties and hardness based on unmachined specimens 0.5 in. diameter individually permanent mold cast. Modulus of elasticity, average 10.3×10^6 lb/in.^{2}

† 500×10^6 cycles with R. R. Moore-type specimen.

good casting characteristics, weldability, moderate strength, and pressure tightness. They are used for manifolds, valve bodies, automotive cylinder heads, etc.

Certain compositions for castings have been developed for improved mechanical properties through heat treatment. Recommended treatments for the precipitation hardening of several such alloys are given in Table

TABLE 18-V. NOMINAL COMPOSITION OF ALUMINUM DIE-CASTING ALLOYS.

Alloy	Cu	Si	Mg	Zn	Ni	Sn		Mn
13	—	12.0	—	—	—	—	—	
43	—	5.0	—	—	—	—	—	
A214	—	—	3.8	1.8	—	—	—	
218	—	—	8.0	—	—	—	—	
360	—	9.5	0.5	—	—	—	—	
364	—	8.5	0.3	—	—	—	—	0.35% Cr 0.03% Be
380	3.5	9.0	—	—	—	—	—	
384	3.8	12.0	—	—	—	—	—	

18-VII for various temper designations. Directly following the solution heat treatment, castings should be quenched in water at 150-212 F (66-100 C) to minimize distortion and quenching stresses. The heating times will vary with furnace conditions.

The most popular heat-treatable casting alloy in general use is probably 195, basically an aluminum-copper alloy with the addition of a small quantity of silicon. It is used where a combination of high tensile properties and machinability is required. Where pressure tightness, resistance to corrosion, fluidity, and weldability are needed, an aluminum-silicon alloy con-

TABLE 18-VI. TYPICAL MECHANICAL PROPERTIES OF ALUMINUM DIE-CASTING ALLOYS.*

Alloy	Tensile Strength (lb/in.^{2})	Yield Strength (Offset = 0.2%) (lb/in.^{2})	Elongation in 2 in. (%)	Shear Strength (lb/in.^{2})	Endurance† Limit (lb/in.^{2})
13	42,000	21,000	2.5	28,000	19,000
43	33,000	16,000	9.0	21,000	17,000
A214	40,000	22,000	10.0	26,000	18,000
218	45,000	27,000	8.0	29,000	20,000
360	47,000	25,000	3.0	30,000	19,000
A360**	46,000	24,000	5.0	29,000	18,000
364	43,000	23,000	7.5	26,000	18,000
380	48,000	24,000	3.0	31,000	21,000
A380**	47,000	23,000	4.0	20,000	30,000
384	47,000	25,000	1.0	21,000	30,000

* Finished properties and hardness based on unmachined ASTM standard round specimens, 0.25-in. diameter, die-cast by cold-chamber process. Modulus of elasticity, average 10.3×10^6 lb/in.^{2}

† 500×10^6 cycles with R. R. Moore-type specimen.

** Components, primarily iron, controlled to lower limits.

TABLE 18-VII. RECOMMENDED HEAT TREATMENT OF ALUMINUM ALLOY CASTINGS.

	Sand Casting						Permanent-Mold Casting					
	Solution Heat Treatment			Precipitation Heat Treatment			Solution Heat Treatment			Precipitation Heat Treatment		
Alloy and Final Temper	(°F)	(°C)	(hr)	(°F)	(°C)	(hr)	(°F)	(°C)	(hr)	(°F)	(°C)	(hr)
122-T551	—	—	—	—	—	—	—	—	—	340	171	18-22
122-T61	950	510	12	310	154	10-14	—	—	—	—	—	—
122-T65	—	—	—	—	—	—	950	510	8	340	171	7-9
A132-T551	—	—	—	—	—	—	—	—	—	340	171	14-18
A132-T65	—	—	—	—	—	—	960	516	8	340	171	14-18
F132-T5	—	—	—	—	—	—	—	—	—	400	204	7-9
142-T21	—	—	—	650	343	2-4	—	—	—	—	—	—
142-T571	—	—	—	340	171	22-26	—	—	—	340	171	22-26
142-T61	960	516	6	450	232	1-3	960	516	6	400	204	3-5
142-T77	960	516	6	650	343	1-3	—	—	—	—	—	—
152-T524	—	—	—	—	—	—	—	—	—	525	274	5-7
195-T4	960	516	12	—	—	—	—	—	—	—	—	—
195-T6	960	516	12	310	154	3-6	—	—	—	—	—	—
195-T62	960	516	12	310	154	12-20	—	—	—	—	—	—
B195-T4	—	—	—	—	—	—	950	510	8	—	—	—
B195-T6	—	—	—	—	—	—	950	510	8	310	154	5-7
B195-T7	—	—	—	—	—	—	950	510	8	500	260	4-6
319-T5	—	—	—	400	204	7-9	—	—	—	—	—	—
319-T6	940	504	12	310	154	2-5	—	—	—	—	—	—
333-T5	—	—	—	—	—	—	—	—	—	400	204	7-9
333-T6	—	—	—	—	—	—	940	504	8	310	154	2-5
333-T7	—	—	—	—	—	—	940	504	8	500	260	4-6
355-T51	—	—	—	440	227	7-9	—	—	—	440	227	7-9
355-T6	980	527	12	310	154	3-5	980	527	8	310	154	3-5
355-T62	—	—	—	—	—	—	980	527	8	340	171	14-18
355-T7	980	527	12	440	227	3-5	980	527	12	440	227	7-9
355-T71	980	527	12	475	246	4-6	980	527	8	475	246	4-6
A355-T51	—	—	—	440	227	7-9	—	—	—	—	—	—
C355-T61	—	—	—	—	—	—	980	527	12	350	177	10-12
356-T51	—	—	—	440	227	7-9	—	—	—	—	—	—
356-T6	1000	538	12	310	154	3-5	1000	538	8	310	154	3-5
356-T7	1000	538	12	400	204	3-5	1000	538	8	440	227	7-9
356-T71	1000	538	12	475	246	2-4	—	—	—	—	—	—
A356-T61	—	—	—	—	—	—	1000	538	12	310	154	6-10
750-T5	—	—	—	430	221	7-9	—	—	—	430	221	7-9
A750-T5	—	—	—	430	221	7-9	—	—	—	430	221	7-9
B750-T5	—	—	—	430	221	7-9	—	—	—	430	221	7-9

taining some magnesium, such as 356, is employed. This alloy finds use in transmission cases, aircraft fittings and pump parts, water-cooled cylinder blocks, etc. For the highest combination of tensile properties, elongation, and impact resistance, the aluminum-magnesium alloy 220-T4 is used. The statement has been made previously that this composition is unsuitable for pressure-tight castings and generally requires somewhat specialized foundry technique. Certain compositions have been developed to retain strength and hardness properties at somewhat elevated temperatures of the order of 500-600 F (260-315 C). Alloys, such as 142 and 355, find use in air-cooled, internal-combustion engines and for Diesel pistons and cylinder heads.

18.9 Aluminum Wrought Alloys. The nominal composition and typical mechanical properties of standard wrought aluminum alloys are given in Tables 18-VIII and 18-IX, respectively. The modulus of elasticity of both

Table 18-VIII. Nominal Composition of Wrought Aluminum Alloys.

Alloy	Cu	Si	Fe	Mn	Mg	Zn	Cr	Ni	Other
EC	Maximum impurities 0.40%								
2EC	0.05	0.4	0.3	0.01	0.6	0.05	0.01	—	
1050	0.05	0.25	0.4	0.05	0.05	0.05	—	—	Ti 0.03
1060	0.05	0.25	0.35	0.03	0.03	0.05	—	—	Ti 0.03
1100	0.2	Si + Fe =	1.0	0.05	—	0.10	—	—	
1130	0.2	Si + Fe =	0.7	—	—	—	—	—	
1175	0.1	Si + Fe =	0.15	—	—	—	—	—	
1260	0.04	Si + Fe =	0.40	—	0.01	—	—	—	
2011	5.5	0.4	0.7	—	—	0.3	—	—	Pb 0.4 Bi 0.4
2014	4.4	0.9	1.0	0.8	0.5	0.25	0.10	—	Ti 0.15
2017	4.0	0.8	1.0	0.7	0.5	0.25	0.10	—	
2018	4.0	0.9	1.0	0.2	0.7	0.25	0.10	2.0	
2024	4.5	0.5	0.5	0.5	1.5	0.25	0.10	—	
2025	4.5	0.8	1.0	0.8	0.05	0.25	0.10	—	Ti 0.15
2117	2.6	0.8	1.0	0.2	0.3	0.25	0.10	—	
2218	4.0	0.9	1.0	0.2	1.5	0.25	0.10	2.0	—
2618	1.3	0.25	1.1	—	1.5	—	—	1.0	Ti 0.07
3003	0.2	0.6	0.7	1.2	—	0.10	—	—	
3004	0.25	0.3	0.7	1.2	1.0	0.25	—	—	
4032	0.9	12.2	1.0	—	1.0	0.25	0.10	0.9	
4043	0.3	5.0	0.8	0.05	0.05	0.10	—	—	Ti 0.20
4343	0.25	7.5	0.8	0.10	—	0.20	—	—	
5005	0.2	0.4	0.7	0.2	0.8	0.25	0.10	—	
5050	0.2	0.4	0.7	0.1	1.4	0.25	0.10	—	
5052	0.1	Si + Fe =	0.45	0.1	2.5	0.10	0.25	—	
5056	0.1	0.3	0.4	0.12	5.0	0.10	0.12	—	
5083	0.1	0.4	0.4	0.6	4.5	0.25	0.15	—	Ti 0.15
5086	0.1	0.4	0.5	0.5	4.0	0.25	0.15	—	Ti 0.15
5154	0.1	Si + Fe =	0.45	0.1	3.5	0.20	0.25	—	Ti 0.20
5254	0.05	Si + Fe =	0.45	0.01	3.5	0.20	0.25	—	Ti 0.05
5356	0.1	Si + Fe =	0.50	0.12	5.0	0.10	0.12	—	Ti 0.15
5357	0.07	0.12	0.17	0.32	1.0	—	—	—	
5454	0.1	Si + Fe =	0.40	0.75	2.7	0.25	0.12	—	Ti 0.20
5456	0.2	Si + Fe =	0.40	0.75	5.0	0.25	0.12	—	Ti 0.20
5554	0.1	Si + Fe =	0.40	0.75	2.7	0.25	0.12	—	Ti 0.12
5556	0.1	Si + Fe =	0.40	0.75	5.0	0.25	0.12	—	Ti 0.12
5652	0.04	Si + Fe =	0.40	0.01	2.5	0.10	0.25	—	
6053	0.1	0.6	0.35	—	1.2	0.10	0.25	—	
6061	0.27	0.6	0.7	0.15	1.0	0.25	0.25	—	Ti 0.15
6062	0.27	0.6	0.7	0.15	1.0	0.25	0.09	—	Ti 0.15
6063	0.1	0.4	0.35	0.10	0.67	0.10	0.10	—	Ti 0.10
6151	0.35	0.9	1.0	0.20	0.62	0.25	0.25	—	Ti 0.15
6253	0.1	0.6	0.5	—	0.25	2.0	0.25	—	
6463	0.2	0.4	0.15	0.05	0.67	—	—	—	
6951	0.27	0.3	0.8	0.10	0.6	0.20	—	—	
7072	0.1	Si + Fe =	0.7	0.10	0.10	1.0	—	—	
7075	1.6	0.5	0.7	0.30	2.5	5.6	0.3	—	Ti 0.20
7076	0.65	0.4	0.6	0.5	1.6	7.5	—	—	Ti 0.20
7079	0.6	0.3	0.4	0.2	3.3	4.3	0.17	—	Ti 0.10
7178	2.0	0.5	0.7	0.3	2.7	6.8	0.3	—	Ti 0.20
7277	1.2	0.5	0.7	—	2.0	4.0	0.25	—	Ti 0.10

Table 18-IX. TYPICAL MECHANICAL PROPERTIES OF WROUGHT ALUMINUM ALLOYS.

Alloy	Tensile Strength (lb/in.^{2})	Yield Strength (offset = 0.2%) (lb/in.^{2})	Elongation % in 2 in.		Brinell Hardness (500-kg. Load 10-mm Ball)	Shear Strength (lb/in.^{2})	Endurance Limit* (lb/in.^{2})
			Sheet Spec. $\frac{1}{16}$ in. Thick	Round Spec. $\frac{1}{2}$ in. Dia.			
EC-O	12,000	4,000	Wire approx. 23% in 10 in.		—	8,000	—
EC-H19	27,000	24,000	Wire approx. 1-$\frac{1}{2}$% in 10 in.		—	15,000	7,000
2EC-T6	32,000	29,000	—	19	—	—	—
2EC-T64	17,000	9,000	—	24	—	—	—
1060-O	10,000	4,000	43	—	19	7,000	3,000
1060-H18	19,000	18,000	6	—	35	11,000	6,500
1100-O	13,000	5,000	35	45	23	9,000	5,000
1100-H18	24,000	22,000	5	15	44	13,000	9,000
2011-T3	55,000	43,000	—	15	95	32,000	18,000
2011-T6	57,000	39,000	—	17	97	34,000	18,000
2011-T8	59,000	45,000	—	12	100	35,000	18,000
2014-O	27,000	14,000	—	18	45	18,000	13,000
2014-T4	62,000	42,000	—	20	105	38,000	20,000
2014-T6	70,000	60,000	—	13	135	42,000	18,000
2017-O	26,000	10,000	—	22	45	18,000	13,000
2017-T4	62,000	40,000	—	22	105	38,000	18,000
2018-T61	61,000	46,000	—	12	120	39,000	17,000
2024-O	27,000	11,000	20	22	47	18,000	13,000
2024-T3	70,000	50,000	18	—	120	41,000	20,000
2024-T36	72,000	57,000	13	—	130	42,000	18,000
2024-T4	68,000	47,000	20	19	120	41,000	20,000
2024-T81	70,000	65,000	6	—	128	43,000	18,000
2024-T86	75,000	71,000	6	—	135	45,000	18,000
2025-T6	58,000	37,000	—	19	110	35,000	18,000
2117-T4	43,000	24,000	—	27	70	28,000	14,000
2218-T72	48,000	37,000	—	11	95	30,000	—
3003-O	16,000	6,000	30	40	28	11,000	7,000
3003-H18	29,000	27,000	4	10	55	16,000	10,000
3004-O	26,000	10,000	20	25	45	16,000	14,000
3004-H38	41,000	36,000	5	6	77	21,000	16,000
4032-T6	55,000	46,000	—	9	120	38,000	16,000
5005-O	18,000	6,000	30	—	28	11,000	—
5005-H18	29,000	28,000	4	—	—	16,000	—
5005-H32	20,000	17,000	11	—	36	14,000	—
5005-H38	29,000	27,000	5	—	51	16,000	—
5050-O	21,000	8,000	24	—	36	15,000	12,000
5050-H38	32,000	29,000	6	—	63	20,000	14,000
5052-O	28,000	13,000	25	30	47	18,000	16,000
5052-H38	42,000	37,000	7	8	77	24,000	20,000
5056-O	42,000	22,000	—	35	65	26,000	20,000
5056-H38	60,000	50,000	—	15	100	32,000	22,000
5083-O	42,000	21,000	22	—	—	25,000	—
5083-H113	46,000	33,000	16	—	—	—	23,000
5086-O	38,000	17,000	22	—	—	23,000	—
5086-H34	47,000	37,000	10	—	—	27,000	—
5086-H112	39,000	19,000	14	—	—	—	—
5154-O	35,000	17,000	27	—	58	22,000	17,000
5154-H38	48,000	39,000	10	—	87	28,000	21,000
5154-H112	35,000	17,000	25	—	63	—	17,000
5254-O	35,000	17,000	27	—	58	22,000	17,000
5254-H38	48,000	39,000	10	—	80	28,000	21,000
5254-H112	35,000	17,000	25	—	63	—	17,000
5357-O	19,000	7,000	25	—	32	12,000	—
5357-H38	32,000	30,000	6	—	55	18,000	—
5454-O	36,000	17,000	22	—	62	23,000	—
5454-H34	44,000	35,000	10	—	81	26,000	—
5454-H112	36,000	18,000	18	—	62	23,000	—
5454-H311	38,000	26,000	14	—	70	23,000	—
5456-O	45,000	23,000	24	—	—	—	—
5456-H321	51,000	37,000	16	—	90	30,000	—
5456-H112	45,000	24,000	22	—	—	—	—
5456-H311	47,000	33,000	18	—	—	—	—
5652-O	28,000	13,000	25	30	47	18,000	16,000

TABLE 18-IX. (*Continued*)

Alloy	Tensile Strength (lb/in.^{2})	Yield Strength (offset = 0.2%) (lb/in.^{2})	ELONGATION % IN 2 IN.		Brinell Hardness (500-kg. Load 10-mm Ball)	Shear Strength (lb/in.^{2})	Endurance Limit* (lb/in.^{2})
			Sheet Spec. $\frac{1}{16}$ in. Thick	Round Spec. $\frac{1}{2}$ in. Dia.			
5652-H38	42,000	37,000	7	8	77	24,000	20,000
6053-O	16,000	8,000	—	35	26	11,000	8,000
6053-T6	37,000	32,000	—	13	80	23,000	13,000
6061-O	18,000	8,000	25	30	30	12,000	9,000
6061-T4	35,000	21,000	22	25	65	24,000	14,000
6061-T6	45,000	40,000	12	17	95	30,000	14,000
6062-O	18,000	8,000	—	30	30	12,000	9,000
6062-T4	35,000	21,000	—	25	65	24,000	14,000
6062-T6	45,000	40,000	—	17	95	30,000	14,000
6063-O	13,000	7,000	—	—	25	10,000	8,000
6063-T4	25,000	13,000	22	—	—	16,000	—
6063-T5	27,000	21,000	12	—	60	17,000	10,000
6063-T6	35,000	31,000	12	—	75	22,000	10,000
6063-T835	48,000	43,000	8	—	105	30,000	—
6151-T6	48,000	43,000	—	17	100	32,000	11,000
6463-O	13,000	7,000	—	—	25	10,000	8,000
6463-T4	25,000	13,000	22	—	—	16,000	—
6463-T5	27,000	21,000	12	—	60	17,000	10,000
6463-T6	35,000	31,000	12	—	73	22,000	10,000
7075-O	33,000	15,000	17	16	60	22,000	—
7075-T6	83,000	73,000	11	11	150	48,000	22,000
7079-T6	78,000	68,000	—	14	145	45,000	22,000
7178-O	33,000	15,000	15	16	60	22,000	—
7178-T6	88,000	78,000	10	11	160	52,000	22,000

* 500 × 10^6 cycles with R. R. Moore-type specimen.

cast and wrought aluminum alloys varies from 10×10^6 to 11.4×10^6 lb/in.^{2}, a value of 10.3×10^6 lb/in.^{2} usually being taken for most calculations.

The wrought alloys of aluminum may be classified as (1) those strain-hardenable by cold work and (2) those hardenable by means of precipitation hardening with or without cold work following the solution treatment.

In the first group are found such grades as EC, 1100, 3003, 3004, 4043, 5052, 5056, and 7072. These alloys are relatively low in cost, are formed and welded easily, and, because of their low alloy content, possess high corrosion resistance. The forms in which these alloys are available are indicated in Table 18-X. EC is a high-purity aluminum known as electrical-conductor grade. The strain-hardenable group of alloys is employed for formed parts in which the stresses are of low order, such as for railway and bus bodies, airplanes, shipping containers, cooking utensils, architectural designs, and miscellaneous forgings where stress conditions are not severe.

The second, or precipitation-hardenable, group is known as the *duralumin* type or simply *dural* when 4.0-5.5 per cent copper plays the dominant part in the precipitation-hardening process.

Alloys 2014, 2018, 2025, 4032, and 6151 are forging alloys, the one

Table 18-X. FORM OF PRODUCTS FOR ALUMINUM ALLOYS.

Alloy	Sheet	Plate	Wire	Rod	Bar	Rolled Shapes	Extruded Shapes, Tube and Pipe	Drawn Tube and Pipe	Rivets	Forgings and Forging Stock
EC			x	x	x	x				
2EC							x			
1050	Cladding on 2024									
1060	x	x								
1100	x	x	x	x	x				x	x
1130	Nos. 31 and 41 reflector sheet									
1175	Cladding on Nos. 12, 22, 13 & 23 reflector sheet									
1260										
2011			x	x	x					
2014	A	A		x	x	x	x			x
2017			x	x	x				x	
2018										x
2024	xA	xA	x	x	x		x	x	x	
2025										x
2117			Rivet	Rivet and redraw						
2218										x
2618										
3003	xA	xA	x	x	x		x	xA*		x
3004	xA	xA								
4032										x
4043			x	Redraw						
4343	Cladding on brazing sheet and No. 713 brazing sheet									
5005	x	x								
5050	x	x						Tube		
5052	x	x	x	x	x			Tube		
5056			xA	Rivet and redraw A					x	
5083										
5086	x	x								
5154	x	x	Welding				x			
5254										
5356			Welding							
5357	x	x								
5454										
5456		x								
5554	Welding electrode									
5556	Welding electrode		x							
5652										
6053			Rivet	Rivet					x	
6061	xA	xA	x	x	x	x	x	x	x	x
6062							Shapes and tube	Tube		
6063							x	x		
6151										x
6253	Cladding on 5056									
6463							x			
6951										
7072	Cladding on 3003, 3004, 5050, 6061, 7075									
7075	xA	xA	x	x	x		x			x
7076										
7079										x
7178	xA	xA					x			
7277									x	

A = Alclad
* coating on inside only

most commonly used being 2014. A modification of this composition in the form of 2025 is employed in large tonnages for aircraft-propeller forgings. Alloys 2018 and 4032 are used for forged pistons in internal combustion engines because of superior elevated temperature properties derived from nickel. Alloy 6151 has been used principally for aircraft radial-engine crankcases.

Alloy 2017 is the original composition of dural as employed in early dirigible construction. It has been largely superseded by 2024 and 7075 and newer alloys for aircraft construction. Alloy 2017 is still available in rod, wire, and bar form, finding use in rivets and for screw-machine stock when heat-treated to 2017-T4 for maximum machinability in the latter case.

Alloy 2024 is a variation of the basic 2017 composition and develops higher mechanical properties. It has become very popular and is used extensively in the form of sheet, plate, bar, rod, wire, extrusions, and rivets. Typical applications include aircraft structural members, hardware, etc.

Alloy 2011 is a free-machining grade, employing additions of lead and bismuth. It is available as wire, rod, and hexagonal bar stock. Maximum machinability is obtained when heat-treated to 2011-T3.

Alloys 6053 and 6063 are primarily extrusion alloys, although the former has been largely superseded by 6063, except for rivets. The ability of 6063 to be extruded into complicated shapes makes it particularly desired for architectural work. It is also used for portable irrigation tubing in 6063-T6.

Alloy 6061 combines the corrosion-resistant properties and ease of formability of 1100 with the precipitation-hardening characteristics of the dural types. This alloy is fast becoming popular as a general-purpose alloy for structural and architectural purposes, other than aircraft, where strength properties somewhat greater than 1100 or 3003 are required. It is available in the usual wrought forms exclusive of forgings. It is particularly well adapted for corrosion-resistant roofing nails for aluminum or asbestos shingles, canoes, furniture, transportation equipment, vacuum-cleaner tubing, etc.

Alloy 7178 provides the highest strength in this group. Because this alloy ages naturally at room temperature, it must be refrigerated, unless forming is to follow the solution treatment directly. It is available as sheet, plate, and extrusions. Present use is for high-strength aircraft structural parts.

Recommended heat treatments for the several alloys already discussed are shown in Table 18-XI. It should be realized that heating-time recommendations serve only as a guide in that heating time will vary with the furnace equipment and the load. Quenching from the solution temperature should be as rapid as possible to permit subsequent hardening through most

Table 18-XI. RECOMMENDED HEAT TREATMENT OF SOME WROUGHT ALUMINUM ALLOYS.

	Annealing Treatment			Solution Heat Treatment		Precipitation Heat Treatment		
Alloy	(°F)	Hours Approx.	Temper Design.	(°F)	Temper Design.	(°F)	Hours Approx.	Temper Design.
EC	650	*	—	—	—	—	—	—
1100	650	*	-0	—	—	—	—	—
2011	775	2-3	-0	950	-T4	320	12-16	-T6
2014	775	2-3	-0	940	-T4	340	8-12	-T6
2017	775	2-3	-0	940	-T4	—	—	—
2018	775	2-3	-0	950	-T4	340	8-12	-T61
2024	775	2-3	-0	920	-T4	375 375 375	11-13 7-9 11-13	-T81 -T86 -T6
2025	775	2-3	-0	960	-T4	340	8-12	-T6
2117	775	2-3	-0	940	-T4	—	—	—
2218	775	2-3	-0	950	-T4	460	5-8	-T72
3003	775	*	-0	—	—	—	—	—
3004	650	*	-0	—	—	—	—	—
4032	775	2-3	-0	950	-T4	340	8-12	-T6
4043	650	*	-0	—	—	—	—	—
5005	650	*	-0	—	—	—	—	—
5050	650	*	-0	—	—	—	—	—
5052	650	*	-0	—	—	—	—	—
5056	650	*	-0	—	—	—	—	—
5083	650	*	-0	—	—	—	—	—
5086	650	*	-0	—	—	—	—	—
5154	650	*	-0	—	—	—	—	—
5357	650	*	-0	—	—	—	—	—
5454	650	*	-0	—	—	—	—	—
5456	650	*	-0	—	—	—	—	—
6053	775	2-3	-0	970 —	-T4	350 450	6-8 1-2	-T6 -T5
6061	775	2-3	-0	970	-T4	320 350	16-20 6-10	-T6
6062	775	2-3	-0	970	-T4	320 350	16-20 6-10	-T6
6063	775	2-3	-0	—	—	450 365 350	1-2 4-6 6-8	-T5 -T6
6151	775	2-3	-0	960	-T4	340	8-12	-T6
6463	775	2-3	-0	—	—	450 365 350	1-2 4-6 6-8	-T5 -T6
7072	650	*	-0	—	—	—	—	—
7075	775	2-3	-0	880	-W	250	24-28	-T6
7079	—	—	—	830	-W	Room temp. 230-250	5 days 48	-T6
7178	775	2-3	-0	870	-W	250	24-28	-T6
7277	—	—	—	890	-W	210 315	4 8	-T6

* Time only sufficient to bring material to annealing temperature—cooling rate unimportant.

effective precipitation, and also to prevent grain-boundary separation of the insoluble phase with attendant decrease in corrosion resistance. Common practice is to quench in cold water; however, it is desirable to quench large sections in water at 140 to 212 F (60 to 100 C) to avoid excessive quenching strains. The cooling rate after precipitation is not important, provided it is not delayed to the point where over-aging occurs.

18.10 Alclad. The addition of appreciable amounts of copper to aluminum to form precipitation-hardening alloys materially reduces the corrosion resistance of the aluminum. This is particularly true for alloys 2014, 2024, 7075, and modifications of these alloys. If a surface layer of pure aluminum or copper-free aluminum alloy is applied to these alloys by a process known as cladding, a composite product is formed which provides the high strength of the core alloy protected by a corrosion-resistant skin. This layer constitutes approximately 5 per cent of the total thickness, and, since it is anodic to the core alloy, it also provides protection on sheared edges. Clad products are produced by Alcoa under the general name *Alclad*. Alclads 2014, 2024, and 7075 are available as sheet and plate. Alclad 2024 is clad with pure aluminum, whereas Alclads 2014 and 7075 have 6053 and 7072 coatings, respectively. The latter coatings are required to provide an electrode potential higher than the core. Alclad 3003 sheet is produced with a coating of 7072 for use in tank manufacture to prevent localized pit corrosion. Alclad 3003 is also produced as tubing with an inside layer of 7072 for use in heat-exchangers. Alclad 5056 clad with 6053 is available as rod and wire only and has found use in wire screen cloth for protection against insects. The alloys and cladding used for brazing sheet and reflector sheet are indicated in Table 18-XII.

TABLE 18-XII. ALLOYS AND CLADDING FOR BRAZING SHEET AND REFLECTOR SHEET.

Name	Alloy	Cladding
No. 11 Brazing sheet	3003	4343
No. 12 Brazing sheet	3003	4343
No. 21 Brazing sheet	6951	4343
No. 22 Brazing sheet	6951	4343
Nos. 12 and 22 reflector sheet	1100	1175
Nos. 13 and 23 reflector sheet	3003	1175

Clad products are heat-treated to develop mechanical properties in the core alloy; hence, the tensile and yield strengths usually are 1000-10,000 lb/in.^{2} below those for the unclad alloy.

18.11 Magnesium. Commercially pure magnesium has a specific gravity of 1.74 and is 99.8 per cent pure. In the annealed condition, the wrought metal has a tensile strength of about 27,000 lb/in.^{2} and an elongation of 15 per cent in 2 in. It can be cold-rolled to a strength of about 37,000 lb/in.^{2}, but is cold-formed with difficulty. The pure metal is used largely in magnesium-base alloys, as a deoxidizer and alloying agent in nonferrous metals, in vacuum-tube and dry-rectifier manufacture, and in pyrotechnics. In wartime, the largest tonnage goes into the manufacture of flares and incendiary bombs. Alloys of magnesium, rather than the pure metal, are employed for structural purposes. The desirable low specific gravity of magnesium is offset by two disadvantages: namely, lack of stiffness and ease of oxidation. The modulus of elasticity of magnesium is only 6.5×10^6 lb/in.^{2} compared with 10.3×10^6 and 29.5×10^6 lb/in.^{2} for aluminum and steel, respectively.

Magnesium and its alloys have poor resistance to corrosion, particularly in salt water and salt atmospheres, thus requiring a protective surface coating. A protective coating may be applied to magnesium and its alloys by dipping the material into a dichromate bath or by electrolytic anodizing. The electrolytic anodizing treatment has the advantage of abrasion resistance and improved resistance to salt spray, and provides an excellent base for paint. The coating is applied in an aqueous acidic electrolyte containing phosphate, fluoride, and chromate ions.

The ease of oxidation and ignition of magnesium when in a finely divided form has deterred the industrial acceptance of the metal and its alloys. Good housekeeping, reasonable precautions, and suitable fire protection, however, have reduced fire hazards to the point where melting and machining of magnesium alloys are commonplace. Magnesium powder and fine chips should be kept dry and not allowed to accumulate openly in production operations.

18.12 Alloying Elements in Magnesium. The alloying elements commonly added to magnesium are aluminum, zinc, manganese, and, for special purposes, tin, zirconium, cerium, thorium, and beryllium. Copper, iron, and nickel are considered impurities and must be kept at a minimum to provide the best corrosion resistance in the alloy.

Aluminum in amounts ranging from 3 to 10 per cent is the principal alloying element in most magnesium alloys. It increases the strength, hardness, and castability of magnesium. In excess of about 10 per cent, the alloy becomes brittle.

Zinc is used together with aluminum in magnesium-base alloys in amounts up to about 3 per cent to increase salt water corrosion resistance and to offset the harmful effects of iron and copper impurities. It also improves casting properties. An excessive amount of zinc produces porosity and brittleness because of the formation of the compound $MgZn_2$.

Manganese has very limited solubility in magnesium, particularly in the presence of aluminum. It is used in magnesium-aluminum and magnesium-aluminum-zinc alloys in quantities less than about 0.50 per cent to improve corrosion resistance and weldability without affecting strength properties. A binary alloy of magnesium with 1.20 per cent manganese is used for best weldability and hot-forming characteristics with a sacrifice of strength.

Silicon is not soluble in magnesium, but it forms the compound Mg_2Si

which increases the hardness of the alloy. It is usually held below 0.30 per cent to avoid extreme brittleness.

Tin is soluble in magnesium up to about 15 per cent at 1200 F (649 C), decreasing rapidly in solubility to room temperature with the precipitation of the beta phase (Mg_2Sn). A magnesium-aluminum-manganese alloy containing 5 per cent tin has good hammer-forging properties.

18.13 Magnesium-Aluminum Alloys. The magnesium-rich end of the equilibrium diagram of the magnesium-aluminum system is shown in Fig. 18.10. The solubility of aluminum in magnesium decreases from about 12.6

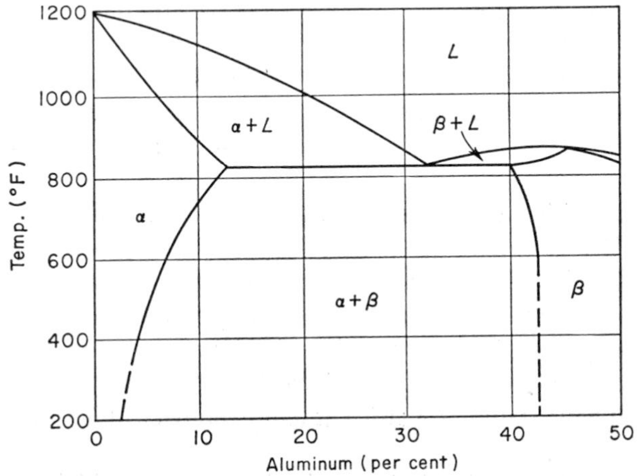

FIG. 18.10 Magnesium-rich end of magnesium-aluminum equilibrium diagram.

per cent at 819 F (437 C) to 2.3 per cent at room temperature. A eutectic forms at 32.2 per cent aluminum. With more than about 45 per cent aluminum, the beta solid solution is formed. The physical properties of the magnesium-aluminum casting alloys are improved by precipitation of the beta phase from the supersaturated alpha solid solution. Although the useful alloys of the system are those containing less than 10 per cent aluminum, usually with small amounts of zinc and manganese, the binary diagram is useful as a basis for studying such alloys.

18.14 Classification of Magnesium Alloys. Magnesium alloys may be broadly classified as (1) casting or (2) wrought alloys. The designations of the magnesium base alloys follow the recommendations of the American Society for Testing Materials.[1] These consist of not more than two letters representing the alloying elements, specified in the greatest amount and arranged in order of decreasing percentages or in alphabetical order if they

[1] ASTM Designation B275-59.

are of equal percentage. These letters are followed by the respective percentages rounded off to whole numbers and a serial letter to differentiate otherwise identical designations. The letters representing the alloying elements are as follows:

A—Aluminum	L—Lithium
B—Bismuth	M—Manganese
C—Copper	N—Nickel
D—Cadmium	P—Lead
E—Rare Earth	Q—Silver
F—Iron	R—Chromium
G—Magnesium	S—Silicon
H—Thorium	T—Tin
K—Zirconium	Y—Antimony
	Z—Zinc

The composition of the common magnesium-base alloys are given in Table 18-XIII.

18.15 Magnesium Casting Alloys. Representative mechanical properties of common magnesium casting alloys are given in Table 18-XIV.

The first group of alloys contains aluminum and zinc as the alloying elements. These alloys possess good casting characteristics, develop high strength, and have stable properties at temperatures as high as about 200 F (95 C).

Alloy AZ63A is desirable in applications requiring high yield strength with maximum toughness.

Alloy AZ81A is applicable for conditions requiring high ductility with a yield strength similar to that obtained with Alloy AZ63A-T4.

Alloy AZ91C is particularly useful where pressure tightness and good weldability of the casting are important. The mechanical properties of this alloy are intermediate to those of Alloys AZ92A and AZ63A.

The magnesium alloys containing rare earths and zirconium, EK30A, EK41A, and EZ33A, have relatively little difference in mechanical properties. All three of the alloys produce excellent pressure-tight castings and possess good creep strength at temperatures up to 500 F (260 C).

The two alloys, HK31A and HZ32A, containing thorium, are particularly useful at temperatures as high as 650 or 700 F (345 or 370 C). Alloy HK31A is best for application under short-time elevated temperature conditions, while Alloy HZ32A is applicable under long-time lower stress conditions.

Alloys ZH62A and ZK51A are primarily magnesium, zinc, zirconium alloys with the first alloy containing thorium. Alloy ZH62A possesses about

Table 18-XIII. COMPOSITION OF MAGNESIUM ALLOYS.

Alloy	Form	Aluminum	Mn min	Zinc	Zirconium	Rare Earth	Thorium	Ca max	Si max	Cu max	Ni max	Fe max	Other imp. max
AM 100A	PMC	4.3-10.7	0.10	0.30 max	—	—	—	—	0.30	0.10	0.01	—	0.30
AZ 31B	W	2.5-3.5	0.20	0.6-1.4	—	—	—	0.04	0.10	0.05	0.005	0.005	0.30
AZ 31C	W	2.4-3.6	0.15	0.5-1.5	—	—	—	—	0.10	0.10	0.03	—	0.30
AZ 61A	W	5.8-7.2	0.15	0.4-1.5	—	—	—	—	0.10	0.05	0.005	0.005	0.30
AZ 63A	PMSC	5.3-6.7	0.15	2.5-3.5	—	—	—	—	0.30	0.25	0.01	—	0.30
AZ 80A	W	7.8-9.2	0.12	0.2-0.8	—	—	—	—	0.10	0.05	0.005	0.005	0.30
AZ 81A	PMSC	7.0-8.1	0.13	0.4-1.0	—	—	—	—	0.30	0.10	0.01	—	0.30
AZ 91A	DC	8.3-9.7	0.13	0.35-1.0	—	—	—	—	0.50	0.10	0.03	—	0.30
AZ 91B	DC	8.3-9.7	0.13	0.35-1.0	—	—	—	—	0.50	0.35	0.03	—	0.30
AZ 91C	PMSC	8.1-9.3	0.13	0.4-1.0	—	—	—	—	0.30	0.10	0.01	—	0.30
AZ 92A	PMSC	8.3-9.7	0.10	1.6-2.4	—	—	—	—	0.30	0.25	0.01	—	0.30
EK 30A	PMSC	—	—	0.3 max	0.20 min	2.5-4.0	—	—	—	0.10	0.01	—	0.30
EK 41A	PMSC	—	—	0.3 max	0.40-1.0	3.0-5.0	—	—	—	0.10	0.01	—	0.30
EZ 33A	PMSC	—	—	2.0-3.1	0.50-1.0	2.5-4.0	—	—	—	0.10	0.01	—	0.30
HK 31A	WPMSC	—	—	0.3 max	0.40-1.0	—	2.5-4.0	—	—	0.10	0.01	—	0.30
HM 21A	W	—	0.45-1.1	—	—	—	1.5-2.5	—	—	—	—	—	0.30
HM 31XA	W	—	0.45-1.1	—	—	—	2.5-4.0	—	—	—	—	—	0.30
HZ 32A	PMSC	—	—	1.7-2.5	0.50-1.0	0.10 max	2.5-4.0	—	—	0.10	0.01	—	0.30
K 1A	SC	—	—	—	0.40-1.0	—	—	—	—	—	—	—	0.30
M 1A	W	—	1.20	—	—	—	—	0.30	0.10	0.05	0.01	—	0.30
TA 54A	W	3.0-4.0	0.20	0.3 max	—	—	—	—	0.30	0.05	0.01	—	0.30*
ZE 10A	W	—	—	1.0-1.5	—	0.12-0.22	—	—	—	—	—	—	0.30
ZE 41A	PMSC	—	0.15 max	3.5-5.0	0.40-1.0	0.75-1.75	—	—	—	0.10	0.01	—	0.30
ZH 62A	PMSC	—	—	5.2-6.2	0.50-1.0	—	1.4-2.2	—	—	0.10	0.01	—	0.30
ZK 20A	W	—	—	2.0 approx	0.45 approx	—	—	—	—	—	—	—	0.30
ZK 51A	PMSC	—	—	3.6-5.5	0.50-1.0	—	—	—	—	0.10	0.01	—	0.30
ZK 60A	W	—	—	4.8-6.2	0.45 min	—	—	—	—	—	—	—	0.30
ZK 60B	W	—	—	4.8-6.2	0.45 min	—	—	—	—	—	—	—	—
ZK 60XB	W	—	—	4.8-6.5	0.45 min	—	—	—	—	—	—	—	0.30
ZK 61A	PMSC	—	—	5.5-6.5	0.6-1.0	—	—	—	—	0.10	0.01	—	0.30

* Sn = 4.0-6.0

PMC Permanent-mold casting.
PMSC Permanent-mold and sand casting.
W Wrought.
DC Direct cast.

TABLE 18-XIV. TYPICAL MECHANICAL PROPERTIES OF CAST MAGNESIUM ALLOYS.

Sand and Permanent-Mold Castings

Alloy	Temper	Tensile Strength (lb/in.²)	Yield Strength (0.2% Offset) (lb/in.²)	Elongation in 2 in. (%)	Shear Strength (lb/in.²)	Hardness Rockwell E
AZ63A	-F	29,000	14,000	6	18,000	59
	-T4	40,000	14,000	12	18,000	66
	-T5	29,000	14,000	5	19,000	—
	-T6	40,000	19,000	5	21,000	83
AZ81A	-T4	40,000	14,000	12	18,000	66
AZ91C	-F	24,000	14,000	2	18,000	62
	-T4	40,000	14,000	11	18,000	66
	-T6	40,000	19,000	4	21,000	83
AZ92A	-F	24,000	14,000	2	19,000	77
	-T4	40,000	14,000	10	19,000	75
	-T5	24,000	14,000	2	20,000	—
	-T6	40,000	21,000	2	22,000	90
EK30A	-T6	23,000	16,000	3	—	49
EK41A	-T5	23,000	16,000	1	—	49
	-T6	25,000	18,000	3	—	59
EZ33A	-T5	23,000	15,000	3	—	59
HK31A	-T6	30,000	15,000	8	—	66
HZ32A	-T5	29,000	14,000	7	—	68
ZE41A	-T5	30,000 min	20,000	3.5	—	—
ZH62A	-T5	40,000	26,000	8	—	—
ZK51A	-T5	40,000	26,000	8	22,000	77
ZK61A	-T6	40,000 min	25,000 min	5 min	—	—
			Permanent-Mold Castings			
AM100A	-F	22,000	12,000	2	17,000	65
	-T4	40,000	13,000	10	19,000	62
	-T6	40,000	16,000	4	21,000	80
	-T61	40,000	22,000	1	—	—
			Die Castings			
AZ91A	-F	33,000	22,000	3	20,000	72
AZ91B	-F	33,000	22,000	3	20,000	72

Temper designations:
- -F = As fabricated (cast).
- -T4 = Solution treated.
- -T5 = Artificially aged.
- -T6 = Solution treated and artificially aged.
- -T61 = Solution treated and artificially aged for longer period.

the same mechanical properties as Alloy ZK51A, including creep and fatigue characteristics; it also has excellent weldability and the castings have good pressure tightness. Alloy ZK51A has limited weldability.

Alloy AM100A, containing aluminum and manganese as the alloying elements, is used only for permanent-mold castings. This alloy has very good mechanical properties, and, by aging artificially for a longer period (-T61), a higher yield strength is developed than with the -T6 treatment. However, this treatment further reduces the elongation.

The two alloys used for die-casting are AZ91A and AZ91B; they differ only slightly in the amount of copper.

The fatigue behavior of the cast magnesium alloys is similar to the aluminum alloys. The endurance strength for Alloys AZ63A, AZ91C, and AZ92A is about 22,000 lb/in.^{2} on the basis of the minimum stress being equal to one-quarter of the maximum stress under axial loading conditions and for 500×10^6 cycles.

18.16 Magnesium Wrought Alloys. Selected magnesium alloys may be rolled as sheet and plate, extruded as bars, shapes, or tubing, and forged by press or hammer forging, depending on the specific alloy in each case. Such alloys are readily hot-worked and formed at temperatures of 400-700 F (204-371C), usually by the use of heated rolls, dies, and punches. Cold forming is restricted to large radius bends.

Representative mechanical properties of common magnesium wrought alloys are shown in Table 18-XV.

Alloy AZ31B is probably the most widely used for extrusions. It possesses good forming characteristics and has moderately good strength and good ductility.

Alloy AZ31C is quite similar to Alloy AZ31B but has higher limits of impurities.

Higher strength can be attained with Alloy AZ61A with a ductility about the same as the AZ31 alloys. The highest strength is secured with Alloy AZ80A in the artificially aged (-T5) condition. However, better elongation with almost the same strength can be obtained with Alloy ZK60A in the artificially aged (-T5) condition.

Alloy ZK60XB is a very slight modification of ZK60 which is extruded from pellets of the alloy rather than from an extrusion ingot. This alloy is capable of producing maximum properties in large extruded shapes.

Alloy M1A is employed in cases wherein the strength requirements are not severe and where the somewhat lower ductility can be tolerated.

The form of the endurance curves of the magnesium alloys is similar to that of the aluminum alloys. The endurance strength of Alloys AZ80A and ZK60A in the forged or extruded condition is about 27,000 lb/in.^{2} on the basis of 500×10^6 cycles with the minimum axial stress being one-quarter the maximum axial stress. The endurance strength of Alloys AZ31B and AZ61A extruded and the Alloy AZ31B sheet is about 18,000 lb/in.^{2}

18.17 Titanium. Titanium has a relatively high melting point of 3140 F (1727 C), a low specific gravity of 4.5, and excellent resistance to

TABLE 18-XV. TYPICAL MECHANICAL PROPERTIES OF WROUGHT MAGNESIUM ALLOYS.

Alloy	Temper	Form of Product	Tensile Strength lb/in.²	Yield Strength (0.2% Offset) lb/in.²	Elongation in 2 in. %	Shear Strength lb/in.²	Hardness Rockwell E
AZ31B	-F	Tooling plate	35,000	19,000	12	—	—
	-F	Extruded hollow shapes and tube	36,000	24,000	16	—	51
	-F	Extruded bars, rods, solid shapes	38,000	29,000	15	19,000	57
	-O	Sheet and plate	37,000	22,000	17-23	—	—
	-H10	Sheet and plate	35,000	20,000	17	—	—
	-H11	Sheet and plate	37 000	20,000	20	—	—
	-H23	Sheet and plate	42,000	28,000	15	—	—
	-H24	Sheet and plate	38-42,000	24-32,000	15-19	—	—
	-H26	Sheet and plate	38-40,000	25-28,000	10	—	—
AZ31C	-F	Tread plate	35,000	19,000	14	—	63
	-F	Extruded bars, rods, solid shapes	38,000	29,000	15	19,000	57
	-F	Extruded hollow shapes and tube	36,000	24,000	16	—	51
AZ61A	-F	Extruded bars, rods, solid shapes	45,000	33,000	16	20,000	72
	-F	Extruded hollow shapes and tube	41,000	24,000	14	—	60
AZ80A	-F	Extruded bars, rods, shapes	49,000	36,000	11	22,000	77
AZ80A	-T5	Extruded bars, rods, and shapes	55,000	40,000	7	24,000	88
HK31A	-O	Sheet and plate	33,000	21,000	23	—	—
	-F	Extrusion	30,000	19,000	35	—	—
	-H24	Sheet and plate	31-37,000	23-29,000	8-15	20-21,000	68
HM21A	-T8	Sheet and plate	34,000	21,000	10	—	—
ZE10A	-O	Sheet and plate	31-33,000	16-23,000	18	—	—
	-H24	Sheet and plate	34-38,000	19-28,000	8-12	—	—
ZK20A	-F	Extrusion	40,000	31,000	7	—	—
ZK60A	-F	Extruded bars, rods, shapes (solid)	49,000	37,000	14	27,000	84
	-F	Extruded hollow shapes and tube	46,000	34,000	12	—	84
	-T5	Extruded bars, rods, solid shapes	52,000	43,000	11	26,000	88
ZK60B	-B	Extrusion	49,000	38,000	17	—	—
ZK60XB	-B	Extruded bars, rods, solid shapes	49,000	38,000	17	—	—

Tempers:
-F = As fabricated.
-O = Annealed.
-H10 = Slightly strain hardened.
-H11 = Slightly strain hardened.
-H23 = Strain hardened and partially annealed.
-H24 = Strain hardened and partially annealed.
-H26 = Strain hardened and partially annealed.
-T5 = Artificially aged only.
-T8 = Solution heat-treated, cold worked, and then artificially aged.
-TB = Pellet extrusions.

corrosion at temperatures below approximately 800 or 1000 F (425 or 540 C). The strength of alloys of titanium is from two to three times that of the aluminum alloys and equal to some of the alloy steels. The modulus of elasticity of titanium is 16×10^6 lb/in.^{2}, which means that it possesses greater stiffness than the aluminum alloys. These properties make the use of titanium alloys very attractive. However, in spite of the abundance (fourth most abundant in the earth's crust) of titanium ores, the cost is very high. In marine atmospheres and sea water, titanium is superior to the austenitic stainless steels and monel metal.

The metal and its alloys have the disadvantage of reacting with other elements at temperatures above about 800 F (425 C), which restricts its use at elevated temperatures. The reaction characteristics of titanium at elevated temperatures cause considerable difficulties in forming operations as well as in its initial production.

Titanium exists in two allotropic forms, alpha at temperatures up to 1625 F (885 C) and beta above this temperature. Alpha titanium is of the hexagonal structure while beta is body-centered cubic. Most alloying elements decrease the alpha to beta transformation temperature. Oxygen, nitrogen, and aluminum raise the transformation temperature. However, oxygen and nitrogen increase hardness and strength with a decrease in ductility and, hence, formability. The stabilizing effect of aluminum on the alpha phase promotes stability at higher temperatures, which makes aluminum an important element in many of the titanium alloys.

The alloying elements, iron, manganese, chromium, molybdenum, vanadium, columbium, and tantalum, stabilize the beta phase, thus decreasing the alpha-beta transformation temperature. Additions of columbium and tantalum produce improved strength and help in preventing the embrittlement produced by the presence of compounds of titanium and aluminum. The elements nickel, copper, and silicon are active eutectoid-formers, while manganese, chromium, and iron are sluggish in the formation of a eutectoid. The elements tin and zirconium are soluble in both the alpha and beta structures. There are three general types of titanium alloys depending upon the structures; namely, all alpha, alpha and beta, and all beta. The all-alpha alloys are not responsive to heat treatment and hence do not develop the strength possible in other alloys. The alpha-beta alloys are heat-treatable and possess good ductility. The beta alloys have relatively low ductility for strengths comparable to the other alloys. The alpha-beta alloys are age-hardenable. The composition and mechanical properties of the titanium alloys are presented in Table 18-XVI.

The alpha alloys in general have the highest strength at temperatures in the range 600-1100 F (315-595 C) and have the best resistance to oxidation in this temperature range. Their room temperature strength is not as good as the alpha-beta alloys. The strongest alloys are the alpha-beta type which respond to heat treatment. These alloys are more formable than the alpha alloys, but they are not weldable. When they are heated above the alpha-beta transformation temperature, they lose ductility as a result of grain coarsening.

TABLE 18-XVI. NOMINAL COMPOSITION AND PROPERTIES OF TITANIUM ALLOYS.

Unalloyed Titanium	Forms Available* Condition**	Test Form	Tensile Strength lb/in.²	Yield Strength lb/in.²	Elong. %
High Purity (99.9%)	Annealed	—	34,000	20,000	54
ASTM Grade 2 (99.2%)	Annealed BSTW	—	59,000	40,000	28
ASTM Grade 3 (99.0%)	Annealed BSTW	—	79,000	63,000	27
ASTM Grade 4 (99.0%)	Annealed BSTW	—	95,000	80,000	25
Alpha Titanium Alloys					
6% Al, 4% Zr, 1% V	Annealed S	Sheet	143,000	138,000	17
5% Al, 2.5% Sn	Annealed BSTW	—	125,000	120,000	18
8% Al, 1% Mo, 1% V	1800 F 5 min AC B, S 1100 F 8 hr AC	Sheet	147,000	135,000	16
8% Al, 2% Cb, 1% Ta	Annealed 1650 F, 1 hr AC BSW	Bar	126,000	120,000	17
8% Al, 8% Zr, 1% (Cb + Ta)	Vac. Ann. 1600 F, 8 hr FC B to 600 F	Bar	135,000	125,000	16
Alpha-Beta Titanium Alloys					
3% Al, 2.5% V	Annealed S, T	Strip	100,000	85,000	15
5% Cr, 3% Al	Annealed B, S	Bar	155,000	145,000	14
8% Mn	Annealed S	Sheet	138,000	125,000	15
2% Fe, 2% Cr, 2% Mo	Annealed B, S, T	Bar	137,000	125,000	18
	1480 F 1 hr WQ 900 F 24 hr AC	Bar	179,000	171,000	13
4% Al, 4% Mn	Annealed B, S, W	Bar	148,000	133,000	16
	1450 F 2 hr WQ 900 F 24 hr AC	Bar	162,000	140,000	9
4% Al, 3% Mo, 1% V	1625 F 2½ min WQ S	Sheet	140,000	95,000	15
	1625 F 2½ min WQ 925 F 12 hr AC	Sheet	195,000	167,000	6
4% Al, 4% Mo, 4% V	1550 F 1 hr WQ B 1000 F 6 hr AC	Bar	170,000	150,000	6
5% Al, 2.75% Cr, 1.25% Fe	Annealed B, S	Sheet	155,000	135,000	14
	1475 F 6 min WQ 900 F 5 hr AC	Sheet	195,000	165,000	6
5% Al, 1.5% Fe, 1.4% Cr, 1.2% Mo	Annealed BS	Bar	154,000	145,000	16
	1650 F 1 hr WQ 1000 F 24 hr AC	Bar	195,000	184,000	9
6% Al, 4% V	Annealed B, S, T, W	Sheet	135,000	120,000	11
	1700 F 20 min WQ 975 F 8 hr AC	Sheet	170,000	150,000	7
7% Al, 4% Mo	Annealed BT	Bar	160,000	150,000	15
	1650 F 20 min WQ 900 F 16 hr AC	Bar	190,000	175,000	12
2.5% Al, 16% V	1380 F 20 min WQ BSW	Sheet	105,000	45,000	16
	1380 F 20 min WQ 960 F 4 hr AC	Sheet	180,000	165,000	6
Beta Titanium Alloys					
13% V, 11% Cr, 3% Al	1400 F 30 min AC BSW	Sheet	135,000	130,000	16
	1400 F 30 min AC 900 F 16 hr AC	Sheet	180,000	170,000	6

* Forms available B = bars and billets; S = rolled flat products (sheet, strip, plate); W = wire; T = tubes and extrusions.
** Heat-treated AC = air-cooled; WQ = water-quenched; FC = furnace-cooled.
Frost, "Titanium Alloys Today," *Metal Progress*, March 1959, p. 95.

18.18 Beryllium. The specific gravity of beryllium is 1.85, which is almost as low as that of magnesium. The melting point is 2345 F (1285 C). Beryllium has a hexagonal close-packed structure. An outstanding property is the modulus of elasticity which is 40×10^6 lb/in.^{2} The strength of beryllium is not unusually high, between 36,000 and 78,000 lb/in.^{2} with an elongation of between 1.5 and 7 per cent. There is some evidence that an ultimate strength of 120,000 lb/in.^{2} may be secured with an elongation as high as 18 per cent. However, beryllium exhibits marked differences in properties in the longitudinal and transverse directions. The very low ductility of this metal has been a serious problem in its utilization. Beryllium fumes or dust particles are toxic when inhaled and therefore require care in handling.

The very low neutron absorption cross section of beryllium has stimulated a search for an improvement of its mechanical properties. This metal is an excellent moderator of neutrons and is useful as a neutron reflector. However, its brittleness limits its use in connection with reactor fuels.

Beryllium is used extensively as an alloying element in copper, nickel, and magnesium. These alloys are subject to precipitation hardening. The copper-beryllium alloys have excellent applications where high strength and corrosion resistance are required. The characteristics of these alloys are discussed in Chapter XVII.

18.19 Zirconium. Zirconium and its alloys have many characteristics in common with titanium, although its density (6.4 sp gr) is closer to that of steel. However, the alloys of zirconium that have been developed so far are too brittle for most engineering applications. The tensile strength of high purity zirconium is about 36,000 lb/in.^{2}, the yield strength is about 16,000 lb/in.^{2}, and the elongation is about 36 per cent in 2 in. Like titanium, zirconium exists in two allotropic forms, hexagonal close-packed structure (alpha) at temperatures up to 1590 F (865 C) and body-centered cubic (beta) up to the melting point of 3380 F (1860 C). An alloy under the name of Zircoloy 2 contains 1.5% tin, 0.12% iron, 0.10% chromium, 0.05% nickel, and zirconium as the balance. This alloy is employed as a cladding for fuel elements of reactors. It has considerably greater strength than zirconium and only slightly greater thermal-neutron cross section.

The corrosion resistance of zirconium is one of the outstanding characteristics that makes this metal highly attractive. Its nuclear properties have made it particularly useful in the field of atomic energy. While beryllium and magnesium have a lower cross section for neutron-capture than zirconium, these metals are reactive and lose strength rather seriously with increasing temperature. Zirconium has a thermal-neutron cross section of 0.18 barns. The nearest structural material to this is aluminum with a cross

Table 18-XVII. TENSILE PROPERTIES OF LIGHTWEIGHT METALS AND ALLOYS COMPARED WITH STEEL. (Strength/Sp Gr Basis)

Material	Specific Gravity	Mod. of Elast. 10^6/lb/in.^{2}	E/sp gr × 10^6	Tensile Strength lb/in.^{2}	T.S./sp gr	Yield Strength (0.2%) lb/in.^{2}	Y.S./sp gr	End. Lim. R. R. Moore Spec. lb/in.^{2}	E.L./sp gr	Elong. % in 2 in.	Brinell Hard-ness
Steel-SAE 1035	7.8	30	3.8	88,000	11,300	55,000^{PL}	7,000	41,000^{L}	5,300	31	169[1]
-SAE 4130	7.8	30	3.8	199,000	25,500	168,000	21,500	114,000^{L}	14,600	10	432[1]
Steel-stainless 17-7PH	7.9	28	3.5	230,000	29,100	217,000	27,500	—	—	7	46 R_C
Steel-stainless-302-30% cold worked	7.9	28	3.5	175,000	22,200	140,000	17,700	80,000	10,100	15	315
Zirconium (0.02% O_2)	6.4	12	1.9	30,000	4,700	12,000	1,900	—	—	30	26 R_A
Zircaloy 2	6.5	—	—	69,000	10,600	45,000	6,900	—	—	22	—
Titanium (99.9%)	4.5	16	3.6	34,000	7,600	20,000	4,500	—	—	54	—
Ti-7% Al, 4% Mo	—	—	—	190,000	—	175,000	—	—	—	12	—
Aluminum 1060-O	2.7	10.4	3.9	10,000	3,700	4,000	1,500	3,000^{S}	1,100	43	19[2]
Aluminum 7178-T6	2.8	10.4	3.7	88,000	31,200	78,000	27,800	22,000^{S}	7,900	11	160[2]
Beryllium—extruded flake	1.9	40	21.1	59-78,000	31,000-40,000	20-30,000^{PL}	10,500-15,800	—	—	4-7	—
Magnesium	1.7	6.5	38.2	37,000	21,600	27,000	15,800	—	—	9	50
Magnesium AZ80A-T5	1.8	6.5	36.1	55,000	30,500	40,000	22,200	24,000	13,300	7	88 R_E

L = Endurance limit
PL = Proportional limit
S = Endurance strength
[1] 3000 kg load
[2] 500 kg load

section of 0.22, but here again aluminum is not suitable at elevated temperatures (500-800 F) (260-425 C). The next material with good structural qualities is iron with a cross section of 2.4 barns. Hence the great interest in zirconium.

18.20 Evaluation of Light Alloys as Structural Materials. A comparison of the specific gravities of the light metals with respect to steel is indicated in Table 18-XVII. In considering materials for structural purposes, stiffness and strength-weight ratio may be of as much or more importance than the density. The stiffness is of course indicated by the modulus of elasticity. In general, the lightweight materials, with the exception of beryllium, have moduli lower than that of steel. In order to offset the greater deflection associated with the use of these materials, structural members such as beams made of these materials must be designed deeper or reinforced by ribs for stiffening purposes. The moduli of several materials are compared in Table 18-XVII.

The various properties may be compared by dividing the value of the property by the specific gravity. Such a comparison is included in Table 18-XVII. Factors other than strength and weight, such as corrosion resistance, formability, availability, cost, etc., may control the selection of a given material for a specific application. The characteristics of each of the light metals and alloys have been discussed in the preceding articles.

REFERENCES

Alcoa Aluminum Handbook, Aluminum Company of America, Pittsburgh, Pa., 1959.

Magnesium, American Society for Metals, Metals Park, Ohio, 1946.

Metals Handbook, American Society for Metals, Metals Park, Ohio, 1961.

Physical Metallurgy of Aluminum Alloys, American Society for Metals, Metals Park, Ohio, 1949.

Raynor, *The Physical Metallurgy of Magnesium and Its Alloys,* Pergamon Press, London, 1959.

Symposium on Titanium, American Society for Testing Materials, Philadelphia, Pa., 1957.

Technical Information on Titanium Metal, Remington Arms Co., Inc., Bridgeport, Conn., 1949.

von Zeerleder, *Technology of Light Metals,* Elsevier Publishing Co., Inc., New York, 1949.

White and Burke, *The Metal Beryllium,* American Society for Metals, Metals Park, Ohio, 1955.

Zirconium and Zirconium Alloys, American Society for Metals, Metals Park, Ohio, 1953.

QUESTIONS

1. What types of alloys are included under the general heading, light alloys?
2. What are some of the uses of commercially pure aluminum?
3. What alloying elements are commonly used in the commercial aluminum alloys?
4. In what ways may the mechanical properties of aluminum and its alloys be increased?
5. What is the effect of adding copper to aluminum?
6. What is the effect of the addition of silicon to aluminum?
7. What characteristics does magnesium confer upon aluminum?
8. What is the effect of the addition of zinc to aluminum?
9. What is the purpose of adding manganese and chromium to aluminum?
10. What is the influence of nickel on the elevated temperature strength and corrosion resistance of aluminum?
11. What elements may be added to aluminum to improve the machining qualities?
12. What elements are employed to act as grain-refining agents in aluminum alloys?
13. What are some of the characteristics of aluminum-copper alloys?
14. What is the effect of the addition of copper to the corrosion resistance of aluminum?
15. How may the greater hardness of aluminum-silicon alloys be explained?
16. What are some of the characteristics of aluminum-magnesium alloys?
17. Explain the method of designating wrought aluminum alloys as established by the Aluminum Association.
18. Indicate some of the common aluminum alloys that are used in castings.
19. Into what two classes may wrought aluminum alloys be placed?
20. Which of the two classes of wrought alloys have the better corrosion resistance, and why?
21. Give some examples of the composition and uses of some of the wrought aluminum alloys.
22. What is Alclad? What are its advantages?
23. What are some of the advantages and disadvantages of the use of magnesium and its alloys for engineering purposes?
24. How may magnesium alloys be given a surface protection?
25. What alloying elements are commonly used in magnesium alloys?
26. What element improves the salt water corrosion resistance of magnesium alloys?
27. By what method are the magnesium alloys classified?
28. What are the characteristics of titanium that make it attractive for certain engineering applications?
29. What are the principal elements that are utilized in alloying with titanium?
30. Why are some of the titanium alloys heat-treatable?
31. What particular property makes beryllium a particularly interesting material in engineering?

32. What are the advantages and uses of zirconium alloys?
33. Why are zirconium and its alloys of particular interest in the field of atomic energy?
34. Compare the light alloys from the general standpoint of application in engineering.

XIX

METALLURGY OF CASTING

19.1 Inherent Characteristics of Castings. The casting process has certain inherent advantages and disadvantages as compared with other methods of manufacture. It is possible to cast large or intricate shapes that could not be economically formed in one piece by any other process. Small quantities of production or experimental parts can be produced in sand molds at a fraction of the cost of preparing expensive dies for forging. On the other hand, inherent weaknesses in castings result from certain characteristics associated with the solidification process. These include (1) shrinkage; (2) segregation; (3) gas porosity; and (4) low *hot* strength.

Shrinkage

In molten metal, shrinkage occurs in three stages: (1) liquid shrinkage; (2) solidification shrinkage; and (3) solid-state contraction.

Molten metal is usually cast with a certain degree of superheat above the actual melting point. For steel, this value amounts to approximately 100 F (40 C) above the liquidus. If a mold were suddenly filled and cooling progressed without the further addition of molten metal, it would be observed that the volume of molten metal would progressively shrink as the temperature decreased. For steel, this *liquid shrinkage* amounts to about 0.9 per cent per 100 F (1.6 per cent per 100 C) of the volume at room temperature for a 0.35 per cent carbon steel.

As the metal solidifies, a further reduction in volume occurs. For steel, the value of *solidification shrinkage* is of the order of 3 to 4 per cent. It should be recognized that, for steel, this value is about double that caused by superheat and is largely the cause of shrinkage cavities found in castings. It can be corrected by providing proper feeding of molten metal to heavy sections. Certain alloys suffer from a condition known as *shrinkage porosity, microporosity, interdendritic shrinkage,* etc. This is particularly noted in the copper-tin alloys and certain aluminum-copper compositions which have a wide liquidus-solidus range. The cause of this type of microporosity is believed to be due to a dendritic growth tendency which seals off and prevents feeding between the dendrite branches or cells. Rapid cooling and the application of pressure, as in die casting, are means of reducing microporosity.

With further cooling below the solidification temperature, a *solid-state* *contraction* occurs. This contraction is compensated for by the patternmaker's shrinkage allowance in order that the final casting will have the required dimensions. For steel, this contraction is of the order of 6.9 to 7.4 per cent by volume at room temperature. The volume changes recorded for the cooling of a 0.35 per cent carbon steel are shown in Fig. 19.1.

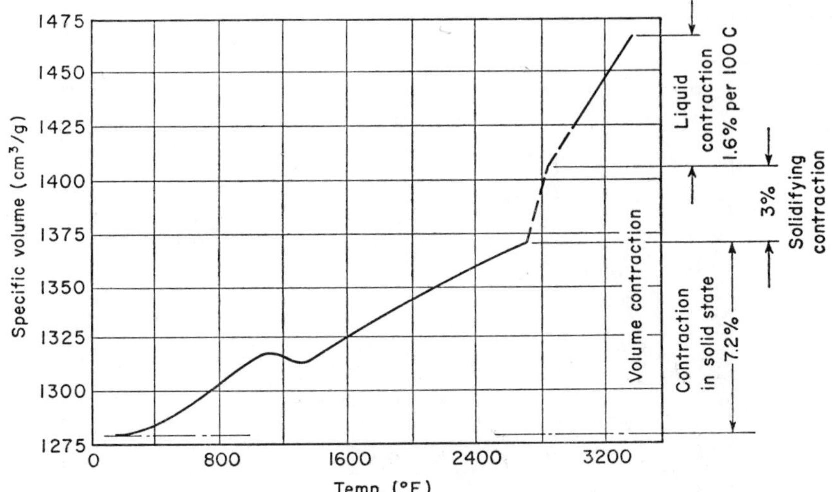

FIG. 19.1 Volume change on cooling a 0.35% carbon cast steel. (Briggs and Gezelius, *Trans. Am. Foundrymen's Society* **42,** 1950, p. 449)

Segregation

This weakness in castings may be divided into (1) gross, or chemical, segregation and (2) micro-, or dendritic, segregation. When molten metal solidifies in a mold, the initial crystals that freeze adjacent to the walls will be the lowest in alloy content, as indicated in Chapter III. Subsequent crystals will be progressively richer in alloy content and impurities until the metal that finally freezes at the center of the section has a composition somewhat different from the skin. This phenomenon is known as *chemical segregation* and may be somewhat altered by the rather slow process of diffusion in subsequent heat treatment.

Columnar crystals nucleate and grow from the mold wall into the melt in the form of dendrites, as discussed in Chapter II and illustrated in Fig. 2.9. A variation in composition occurs between the dendrite proper and the matrix that forms the boundary of the crystalline grain. This mechanism is known as *dendritic segregation*. It may be corrected somewhat in castings by a heat treatment designed to promote diffusion. It is completely broken up by hot working as found in wrought products. The dendritic structure in a steel casting is shown in Fig. 19.2.

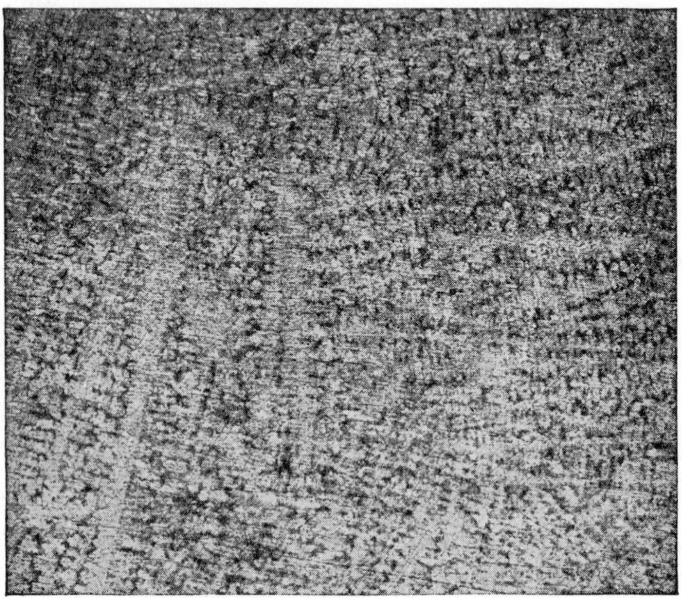

FIG. 19.2 Dendritic structure in steel casting.

The degree of segregation may be controlled by the rate of cooling from the melt. With slow cooling, as usually found in sand casting or in large sections, a coarse, weak dendritic structure tends to form. In die casting, where the operation is rapid, or in thin sections, where cooling is accelerated, the grain size is exceedingly fine, and the dendritic pattern is microscopic in nature.

Gas porosity

Metals possess greater affinity to dissolve gases in the molten state than in the solid state. Gases may be dissolved from products of combustion, from the atmosphere, or from reactions within the melt. With decreased solubility at lower temperatures, these gases come out of solution and are entrapped in the solidified metal as voids known as *gas porosity* or *pinholes*.

Several of the nonferrous alloys are particularly noted for this characteristic. By proper degasification technique, prior to pouring, this condition may be minimized or eliminated. Large voids, known as *blowholes,* may also result from the presence of moisture, volatile matter, or poor venting in sand molds.

Low hot strength

The strength of metals at temperatures near the freezing point is very low. Stresses imposed at these temperatures may lead to incipient flaws which later develop into well-defined cracks or hot tears upon cooling. Precautions may be taken to avoid stress concentrations through proper design in order to minimize shrinkage stresses at weak points. This will be discussed in the next article.

Other defects

Casting defects that are more directly the result of poor molding or foundry practice, rather than inherent in the solidification process, include cold shuts, dirt spots, fins, misruns, penetration, scabs, swells, washes, etc.

Cold shuts appear as irregular laps on the surface of a casting, resulting from low casting temperature or an oxide film on previously solidified metal. *Dirt spots* are caused by particles of molding sand, refractories, or slag being entrapped by the molten metal. *Fins* result from metal entering the spaces between loosely fitting components of a mold. *Misruns* result from poor metal fluidity and may be corrected by higher casting temperatures. *Penetration* is a term describing a rough casting surface where the metal has penetrated between the coarse grains of a sand mold. *Scabs* are projections on the surface of a casting caused by the cutting action of hot metal on the sand mold, whereas *swells* are bulges on the casting resulting from insufficient ramming of the sand. *Washes* result from erosion or the cutting action of hot metal on a mold cavity and are noted particularly in die-casting dies operating with high melting-point alloys.

19.2 Casting Design. Proper casting design should minimize the inherent weaknesses of the casting process. Close cooperation between designer, patternmaker, and foundryman in the early stages of planning is valuable in creating a design that can be economically molded to produce a satisfactory cast part. In addition to meeting certain foundry requirements, the part must also satisfy service needs as governed by the mechanical properties of the metal.

To compensate for solid-state shrinkage, a correction must be made in the pattern or mold design. *Shrinkage allowances* for common metals are approximately as follows:

Aluminum alloys	5/32 in. per ft
Brass	3/16 in. per ft
Bronze	1/4 in. per ft
Gray cast iron	1/8 in. per ft
Magnesium alloys	5/32 in. per ft
Malleable cast iron	3/32 in. per ft
Steel	1/4 in. per ft

These values should be used only as a guide since shrinkage varies with many factors, such as design complexity, relative volume of cores and metal, metal thickness and mass, pouring temperatures, mold material, etc. With experience gained through production, it may be found necessary to vary shrinkage allowances on different dimensions to maintain close control.

To permit removal of the pattern from a sand mold, or the ejection of the part from a permanent mold, requires a compensation known as *draft*. The amount of draft required is seldom less than one degree of taper and will vary with the molding process.

Dimensions that are to be finished by machining operations will require an additional allowance known as *machine-finish allowance* or *finish*. For reasons of economy, the amount of metal to be removed by machining should be held to the minimum required to clean up the surface. A more important reason for minimum removal is the fact that the soundest metal is that which forms the skin of the casting. This is usually overlooked in casting design with the result that the best metal is removed in the form of chips. A dense skin is particularly important for pressure tightness in hydraulic castings.

Where distortion is a problem with a given part, it may be necessary to distort intentionally or "fake" the pattern to provide a *distortion allowance*.

The effect of mass, or section size, has a strong influence on the mechanical properties of a cast structure. Briggs and Gezelius have noted a loss in strength and ductility at the center of steel castings with increase in section size. This has been attributed to microstructure, segregation, and decreased density as the mass increases. Since heavy sections cool slowly, more time is available to allow grain growth, segregation, and shrinkage cavities. To avoid mass effects, castings should be designed with uniform section thickness and a minimum of adjoining sections to cause *hot spots* or points of slow cooling and resultant low strength. The five basic methods by which sections may be joined are designated L, T, V, X, and Y, as illustrated in Fig. 19.3. Inscribed circles indicate the comparative mass between section and junction, with the mass varying as the square of the radii of the inscribed circles. The location of hot spots and resultant shrinkage cavities are shown by the shaded circles. Recommended designs are given to minimize the mass effect at junction points.

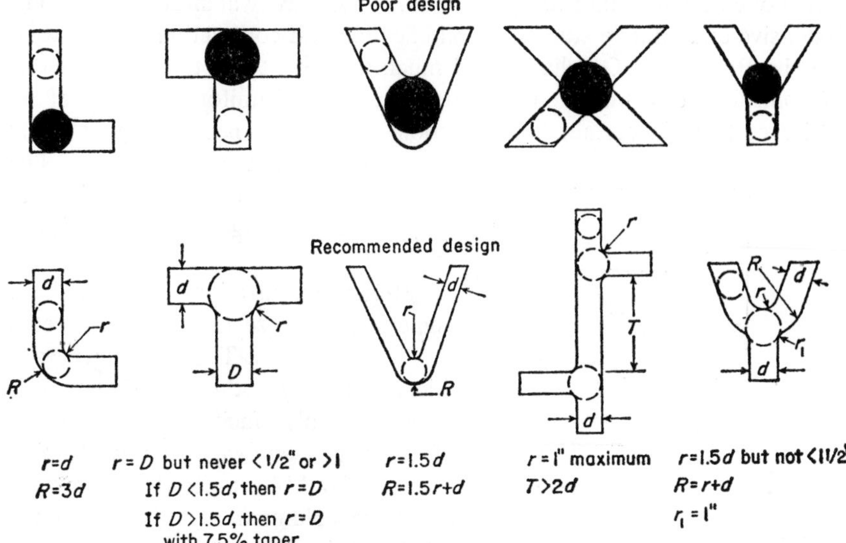

FIG. 19.3 Methods of joining cast sections to reduce mass effects.

Where heavy sections are required, provision must be made to feed hot metal to such points by means of appropriate risers; otherwise, a shrinkage cavity will result. Proper design should incorporate the principle of directional solidification where those portions of the mold farthest removed from the source of hot metal are caused to solidify first and to freeze progressively toward the source of liquid metal. This action may be obtained by gradually increasing the section size toward one or more locations where risers may be attached, or by causing a chilling action to be initiated by the use of in-

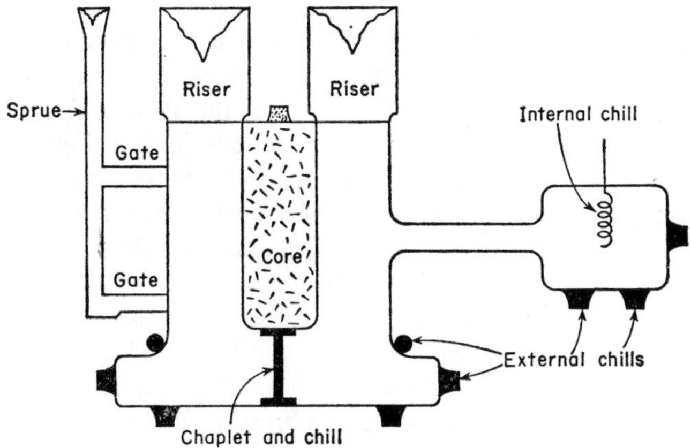

FIG. 19.4 Application of risers, chills, and cores to reduce shrinkage cavities.

ternal or external metal *chills* placed in the mold cavity at strategic locations. Extensive quantitative studies of solidification and contraction in steel castings by Briggs and Gezelius added much to present-day thinking in terms of controlled directional solidification in place of uniform cooling of an entire casting. In certain heavy sections, it may be necessary to employ cores to reduce the mass effect. The application of risers, chills, and cores to prevent shrinkage cavities is shown in Fig. 19.4.

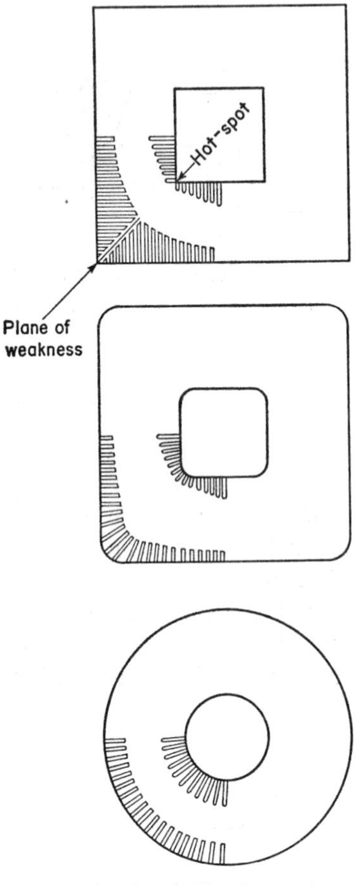

Stress concentrations, such as sharp angles, drastic section changes, etc., should be avoided. The use of generous fillets not only reduces serious stress raisers, but also facilitates the removal of patterns from sand molds. Section changes should be tapered. The growth of coarse columnar dendrites from the mold wall into the molten metal causes planes of weakness at all external angles. Hot spots formed at re-entrant angles also cause structural weakness. The columnar crystal pattern, shown in Fig. 19.5, represents points of constant temperature within molds of different shapes after a given time has elapsed subsequent to pouring. It should be noted that external corners cooled from two sides are quickly chilled compared to the hot spots formed at re-entrant angles. The use of generous radii and fillets improves the dendritic pattern and, hence, the structural design.

Projections, such as lugs and bosses, that seriously restrict natural contraction of the metal in the mold should be avoided where possible; otherwise, *hot tears* may occur. Ribs should be added where necessary to reinforce thin sections and prevent undue warpage. Opposing ribs should be staggered to avoid undesirable hot spots. The tendency toward hot tears can be reduced in sand castings by providing easily collapsible cores and properly rammed sand that will allow some compression under the stress of the contracting metal.

FIG. 19.5 Crystallization pattern governed by mold shape.

Large flat surfaces should be avoided in the interest of good appearance. Curved surfaces are more pleasing to the eye and tend to hide distortion and surface imperfections.

Proper design should also take into account the possibility of heat treatment to bring out the best mechanical properties in the casting. High stress concentrations should be avoided in order to prevent distortion and possible cracking caused by quenching.

19.3 Steel Castings. Steel castings are classified as carbon or alloy grades. Carbon steel castings are classed as low-carbon when the carbon content is less than 0.20 per cent, medium-carbon when between 0.20 and 0.50 per cent, and high-carbon when in excess of 0.50 per cent carbon.

Low-carbon steel castings have limited use in low-strength carburized parts, weldments, furnace accessories, and for electrical applications where a low magnetic hysteresis loss is desired. They are seldom heat-treated except for stress relief.

Medium-carbon steel castings are the backbone of the steel foundry industry, accounting for the bulk of the commercial output. This group offers competition to forging and, in general, possesses a lower ductility for the same carbon content with absence of directional properties. Castings are usually heat-treated to increase strength, ductility, impact, and endurance values.

High-carbon steel castings find use where high hardness and wear resistance are required, as in rolls, wheels, dies, etc. Such steels have poor weldability and are sensitive to thermal shock in heat treatment. Care must be exercised in design and heat treatment to avoid brittle failure.

Specifications for carbon steel castings based on composition or minimum strength properties have been prepared by several agencies and technical societies representing large tonnage consumers of steel castings, such as the American Society for Testing Materials, the American Railway Association, the Federal government, etc.

Alloy steel castings are classed as low- or high-alloy, with the division point arbitrarily taken at 8 per cent total alloy content. The low-alloy cast steels are pearlitic by nature and show improved strength, ductility, impact and endurance properties. The influence of the common alloying elements was discussed in previous chapters. Many compositions of low-alloy cast steel are produced by various foundries throughout the country, exceeding by far the number of standard AISI-SAE grades previously discussed. The low-alloy steels must be heat-treated to make effective use of the alloys present. For a comprehensive survey of the many complex low-alloy grades, the student is referred to the *Cast Metals Handbook* of the American Foundrymen's Society or the *Steel Castings Handbook* of the Steel Founders' Society. A few typical carbon and low-alloy cast steels are shown in Table

19-I, together with their mechanical properties in various conditions of heat treatment.

TABLE 19-I. TYPICAL COMPOSITIONS (%) AND MECHANICAL PROPERTIES OF CAST STEELS.
(After *Metals Handbook*)

Type of Steel	HEAT TREATMENT (°F)				Tensile Strength (lb/in.²)	Yield Strength (lb/in.²)	Elong. in 2 in. (%)	Red. of Area (%)
	Anneal	Norm.	Quench	Temper				
Low C								
0.17 C, 0.74 Mn	1650	—	—	—	65,900	35,000	34	61
0.40 Si	—	1650	1550	1250	67,000	38,000	34	61
Med. C								
0.30 C, 0.60 Mn	1650	—	—	—	74,200	37,600	29	45
0.36 Si	—	1650	1550	1100	89,400	62,500	26	51
High C								
0.42 C, 0.72 Mn	1650	—	—	—	84,250	41,500	26	40
0.41 Si	—	1650	—	1250	83,900	48,700	27	42
	—	1650	1550	1100	105,500	74,200	21	41
1330								
0.30 C, 1.50 Mn	—	1650	—	1200	95,000	55,200	28	54
0.35 Si	—	1650	1575	1000	126,200	102,000	18	41
8030								
0.30 C, 1.25 Mn	—	1650	—	1200	97,000	68,000	24	48
0.35 Si, 0.15 Mo	—	1650	1575	1000	160,000	145,000	12	22
8430								
0.30 C, 1.50 Mn	—	1650	—	1200	105,000	75,000	24	47
0.35 Si, 0.35 Mo	—	1650	1575	1000	165,000	148,000	14	35
8630								
0.30 C, 0.80 Mn, 0.35 Si	—	1650	—	1200	97,000	68,000	24	48
0.50 Ni, 0.50 Cr, 0.15 Mo	—	1650	1575	1000	160,000	145,000	12	22
2% Ni								
0.19 C, 0.75 Mn	—	1650	—	1200	82,000	55,000	30	55
0.35 Si, 2.0 Ni	—	1650	1650	1000	110,000	87,000	21	55
Ni-Mo								
0.20 C, 0.75 Mn, 0.35 Si	1750	—	—	—	78,000	46,000	26	50
1.9 Ni, 0.35 Mo	—	1750	—	1200	84,000	60,000	27	52
	—	1750	1550	1200	90,000	70,000	26	53
0.30 C, 0.30 Mo	—	1650	—	1200	90,000	60,000	22	45
0.40 C, 0.28 Mo	—	1650	—	1200	110,000	82,000	18	40
Ni-V								
0.30 C, 0.75 Mn, 0.35 Si	1650	—	—	—	97,500	62,250	24	47
1.50 Ni, 0.15 V	1650	1525	—	—	97,000	68,000	27	55
Cr-Mo								
0.35 C, 0.75 Mn, 0.35 Si	—	1700	—	1200	96,000	63,500	22	42
0.75 Cr, 0.35-0.40 Mo	—	1650	1600	1250	111,500	85,800	17	36

High-alloy cast steels are specialized compositions possessing distinctive corrosion-, heat-, or wear-resistant properties. Their structure may be ferritic, austenitic, or martensitic, depending on the specific composition and treatment. The iron-chromium, iron-chromium-nickel, and iron-nickel alloys making up the bulk of this group have been previously discussed in Chapter XIV.

An important high-alloy steel, known as high manganese, or Hadfield's austenitic manganese steel, accounts for a large tonnage of steel in the high-alloy class. Although available in some wrought forms, this steel is used largely in the form of castings with composition limits of 1.00 to 1.40 per cent carbon, 10.0 to 14.0 per cent manganese, 0.30 to 1.00 per cent silicon, 0.06 per cent maximum sulfur, and 0.10 per cent maximum phosphorus. As cast, it is very brittle because of the separation of free cementite during slow cooling in the mold, together with the presence of some martensite and transformation products of austenite. To render castings useful as engineering products, they are given a special heat treatment consisting of heating to 1900 F (1040 C) to dissolve the free cementite, followed by a water-quench to preserve a tough austenitic structure at room temperature. The mechanical properties for typical heat-treated castings are as follows:

Tensile strength.	118,000 lb/in.^{2}
Proportional limit.	42,900 lb/in.^{2}
Elongation in 2 in.	44%
Reduction of area.	39%
Endurance limit.	39,000 lb/in.^{2}
Brinell hardness.	190

In service, austenitic manganese steel develops a hard surface if subjected to conditions of impact wear, and the surface hardness may attain a value of 500 Brinell. It is used extensively for railroad switches and crossings, crusher parts, dredge buckets, dipper teeth, etc. It is not suitable in applications where pure abrasive wear is acting alone, such as in sand-blast nozzles. A hardened surface can be developed by the foundry on castings of this alloy by prepeening, but this layer is soon worn away in service unless conditions of impact are continuously present. The material is considered commercially unmachinable and is finished by grinding. It is nonmagnetic as quenched and, therefore, finds use as wear plates in large electromagnets for handling ferrous scrap.

If a quenched, austenitic manganese steel is reheated to about 700 F (370 C), it becomes embrittled. This phenomenon restricts its use to service temperatures below about 650 F (345 C) and complicates the possibility of fusion welding. At somewhat higher temperatures, the structure becomes magnetic, is very brittle, and increases in hardness to 500 Brinell because of the partial transformation of austenite.

A similar alloy, but with a modified composition, containing 0.60 to 0.80 per cent carbon, 10.0 to 15.0 per cent manganese, and 3.0 to 5.0 per cent nickel, exhibits similar toughness and wear-resistant properties as a result of a normalizing treatment and withstands somewhat higher service temperatures than alloys of the original analysis. An alloy of similar composition is available in rod form for satisfactory welding.

19.4 Heat Treatment of Steel Castings. The heat-treating operations generally employed with steel castings are full annealing, normalizing, stress relieving, quenching, and tempering. Other processes occasionally employed in specialized instances include spheroidizing, carburizing, nitriding, martempering, and precipitation hardening of copper cast steels.

Full annealing of castings consists of slow heating at a rate not exceeding 200 F/hr (95 C/hr) to well above the transformation range (Ac_3 plus 100-200 F (40-95 C)) and holding 1 hr/in. of heaviest section, followed by furnace cooling down to at least 1000 F (540 C). Certain alloy steels require a longer time at temperature to effect solution and diffusion of alloying elements. The purpose of the treatment is to relieve casting stresses and promote homogenization by diffusion of microsegregated regions. The coarse grain size resulting from this treatment improves machinability. If possible, castings should be removed from their molds and charged directly into the annealing furnace before their temperature falls below about 500 F (260 C). This practice is particularly desirable with high-carbon and alloy steel castings having air-hardening characteristics because of high hardenability and should also be followed in other processes requiring heating of these alloys to above the critical range.

Normalizing as applied to castings involves heating to above the Ac_3 or A_{cm} temperature and holding for a period of time to allow diffusion, followed by air cooling. The strength of castings which are normalized is higher than that of castings which are annealed because of a finer microstructure. If the normalizing temperature and time have been excessive, the resulting austenitic grain size from such a single heating will be coarse. A modification, known as *double normalizing,* employs a second heating just to the upper critical temperature for the purpose of causing grain refinement through allotropic recrystallization. When the double normalizing treatment is planned, the initial heating is intentionally at a high temperature. Since air cooling may impose serious stresses in complicated cast structures, the optimum mechanical properties may be brought out by a stress relief between 1000 and 1200 F (540 and 650 C), followed by furnace cooling down to at least 800 F (425 C). Since the latter operation is below the transformation range, no change in microstructure will occur. Normalizing is often employed for the purpose of producing a uniformly fine structure prior to subsequent quenching and tempering operations.

Quenching and tempering of castings develop mechanical properties not obtained by annealing or normalizing, provided the composition of the steel and the casting design will permit the more drastic rate of cooling without cracking. The details of these operations have been covered in Chapter VIII. When applied to castings, the general tendency is to employ somewhat higher tempering temperatures for the most effective stress relief and maximum ductility, followed by furnace cooling down to about 800 F (425 C).

The preceding treatments improve the mechanical properties of cast steel in increasing order as discussed. In general, the yield strength, reduction of area, and elongation are increased. This is particularly noticeable in the medium-carbon, high-carbon, and low-alloy grades. The increase is shown graphically for the medium-carbon steels in Figs. 19.6, 19.7, 19.8,

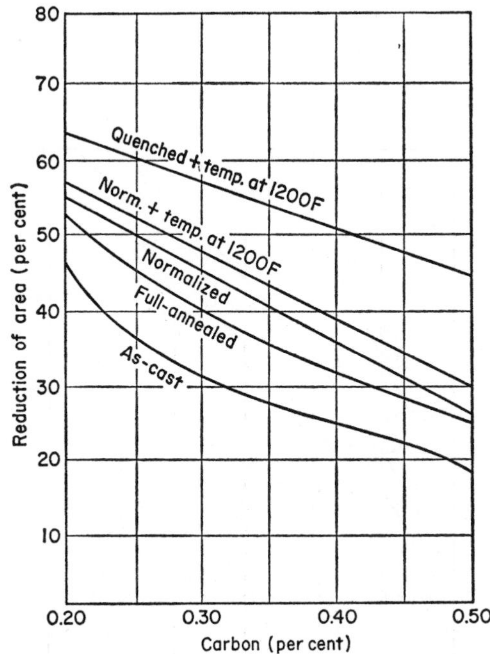

FIG. 19.6 Relation of reduction of area to heat treatment for steel castings of different carbon content. (After *Steel Castings Handbook*)

and 19.9. The mechanical properties of representative carbon and low-alloy cast steels were previously shown in Table 19-I. The ratio of endurance limit to tensile strength for a representative medium-carbon steel also increases from 0.40 as cast, 0.44 as annealed to 0.50 as quenched and tempered. The latter value is comparable to the ratio for a wrought product of similar composition.

The greatest improvement resulting from heat treatment is noted in the increase in toughness. The coarse, dendritic, as-cast structure would be expected to fail easily by brittle fracture and to have low impact values. The increase in notch impact properties for medium-carbon cast steels when subjected to the above treatments is shown in Fig. 19.10.

19.5 Gray Cast Iron. The metallurgy, heat treatment, and properties of the various grades of cast iron were discussed in Chapter XVI. Certain features of interest to the engineer from the standpoint of design and use of gray cast iron were noted, including fluidity, mass effect, growth, wear resistance, machinability, etc.

The proper design of gray cast iron parts follows the basic rules given in Article 19.2 with particular emphasis on uniform sections, since the mass effect is further complicated by the volume changes occurring during graphitization. Large sections cooling at a slower rate than thin adjoining sections will permit the separation of more graphite with a consequent marked expansion and a tendency toward hot tearing. The use of nickel to control

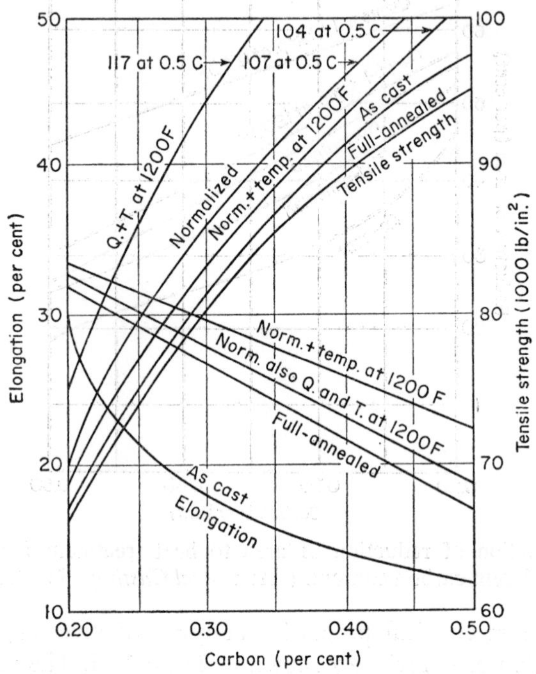

Fig. 19.7 Relation of strength and elongation to heat treatment for steel castings of different carbon content. (After *Steel Castings Handbook*)

section sensitivity and the use of chromium to reduce growth have been noted in Chapter XVI.

Gray cast iron possesses properties of wear resistance which are often incorporated into the design of a part. The lubricating value of the graphite flakes makes the material an ideal choice for applications in which metal-to-metal surfaces exposed to sliding friction might gall. Selected surfaces may be hardened by the formation of white or chilled cast iron to a predetermined depth by adjustment of the composition and control of the cooling rate. Metal chills incorporated in the mold to cause rapid cooling in certain areas should not be confused with the previous mention of chills to control directional solidification. Chilled cast iron provides excellent resistance to abrasive wear in applications where high unit pressures would normally deform gray cast iron, such as in grinding balls, sand-blast nozzles, steel-mill rolls, wire-drawing dies, car wheels, etc.

The ability to finish gray cast iron by machining varies inversely with the strength of the cast material. Graphite particles aid in lubrication and serve as effective chip breakers, making it possible to cut the material with

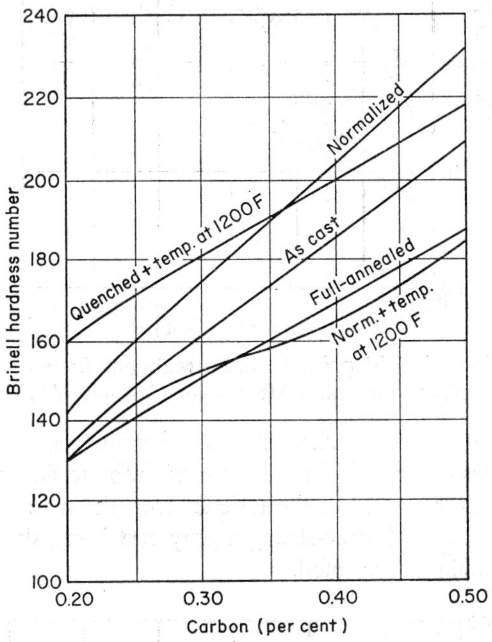

FIG. 19.8 Relation of hardness to heat treatment for steel castings of different carbon content. (After *Steel Castings Handbook*)

ease but with a rough finish. Grinding or polishing may be necessary where a high degree of finish is required. The presence of burnt-in sand on the surface of sand castings is particularly abrasive to cutting tools. Green sand molds also tend to produce a thin surface layer of chilled cast iron, and heavy roughing cuts may have to be made in order to penetrate below this layer. For this reason, it is desirable to consult with both the foundry and the machine shop before a design is completed. White cast iron is not considered to be commercially machinable and must be finished to size by grinding.

The endurance limit of gray cast iron is not as sensitive to the effect of

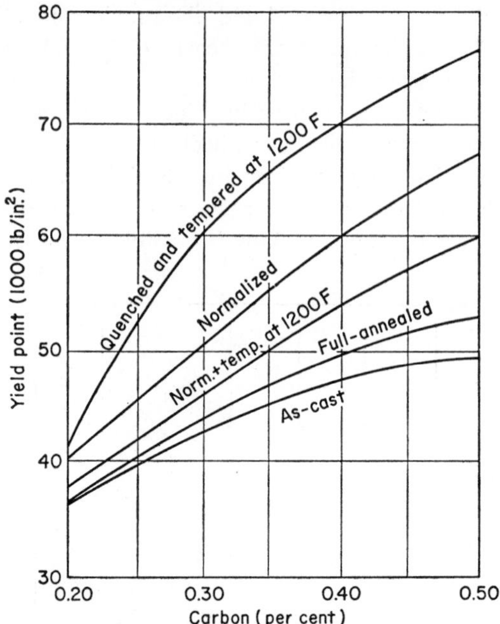

FIG. 19.9 Relation of yield point to heat treatment for steel castings of different carbon content. (After *Steel Castings Handbook*)

stress raisers, such as tool marks, as most other metals. Since the material consists of many stress raisers in the form of graphite flakes, an additional notch does not materially lower the fatigue strength. With the fine graphite dispersions found in the higher-strength gray cast irons, the notch sensitivity increases as might be expected.

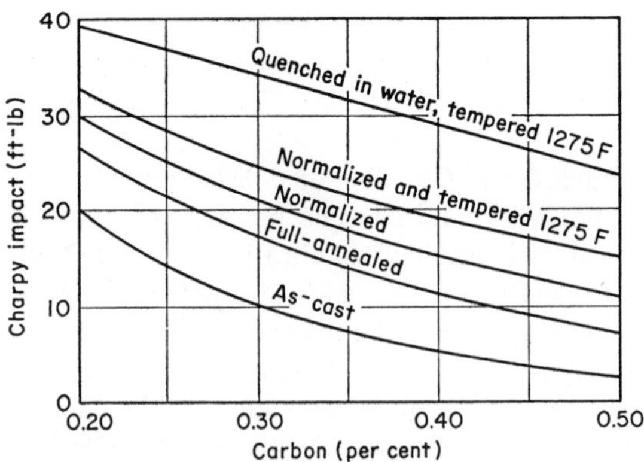

FIG. 19.10 Charpy keyhole notch impact properties of medium-carbon cast steels. (From *Steel Castings Handbook*)

19.6 Nonferrous Castings. The design of nonferrous castings follows the basic recommendations for casting design given in Article 19.2. Many nonferrous alloys have a strong tendency to oxidize and to dissolve gases from products of combustion or from the atmosphere. The problem of protecting the melt from undue drossing and gas absorption is one for the foundryman to control by use of protective slag on the bath and proper deoxidizing or degasifying techniques. The engineer is concerned, however, with the prospect of oxide inclusions and gas porosity in the final product. Early liaison between the engineer and foundryman is again emphasized as a wise precaution for successful part design.

The nonferrous casting alloys are classified according to their major elements, as copper-, aluminum-, magnesium-, zinc-, lead-, tin-, nickel-, and cobalt-base alloys. The compositions, properties, and applications of typical copper-, aluminum-, and magnesium-base casting alloys have been treated in previous chapters. Zinc-base alloys are essentially die-casting alloys and will be discussed under that section. The other groups represent a small portion of the production of nonferrous castings and are employed in specialized applications. Lead- and tin-base casting alloys are used largely for bearing babbitts and solders. The tin-base babbitts, although more costly than the lead-base babbitts, are used where higher bearing pressures, shock, and increased operating temperatures are encountered. The important nickel-base casting alloys include the various grades of Monel, Inconel, Hastelloy, and nickel-chromium and find use as corrosion- or heat-resistant alloys. The cobalt-base casting alloys are of the stellite type and are ideally suited to applications requiring hardness and strength at elevated temperatures.

Heat treatment of nonferrous castings is generally limited to the precipitation hardening of those alloys responding to such treatment and the aging of castings to promote dimensional stability.

19.7 Permanent-Mold Castings. The permanent-mold casting process is strongly competitive to sand casting for small parts when the quantity exceeds about 500 pieces. Dimensional tolerances of the order of ± 0.020 in./in. with smooth finish are possible, compared to about $\pm 1/32$ in./in. for the rough sand casting of small parts. Mold life varies from about 15,000 castings of cast iron to about 100,000 castings of aluminum. These figures are relative and approximate, since the actual life of a mold is usually determined by customer acceptance of variations and irregularities in a given part arising from heat checking or washing away of the mold cavity by hot metal.

The process offers a rapid rate of production, high yield, freedom from dirt, porosity, and shrinkage, together with the higher mechanical properties resulting from the fine grain structure produced. Aluminum alloy castings too large for die casting are commonly manufactured by this process.

19.8 Shell-Mold Castings. The shell-mold casting process is a development that is particularly adapted to the high production of castings with surface finish and dimensional tolerance superior to those produced by sand casting. The process of *shell-mold* casting consists of forming a thin shell of fine silica sand mixed with a thermosetting resin, such as phenol or urea formaldehyde, in contact with a preheated metal pattern. The sand-resin mixture is dumped upon the pattern maintained at a temperature in the range of 375 to 600 F (190 to 315 C) and held until polymerization of the resin has produced a shell of the desired thickness in contact with the heated pattern. The excess molding material is removed by inverting the pattern, after which the shell and pattern are placed in a baking oven to cure the shell. When cured, the shell is ejected from the pattern, and the molding process is repeated. The finished mold cavity is prepared by joining two half shells by clamping or by the use of adhesives. Generally, the assembled mold is backed up by sand or metal shot. Aside from a relatively smooth surface finish, the process is capable of maintaining tolerances of 0.002 to 0.006 in./in. Thin sections may be cast without chilling or misruns because of the lower cooling rate present in shell-molds. Molding costs are lowered by mechanization, decreased skilled-labor requirements, and the smaller quantity of molding sand involved compared to sand molding.

19.9 Plaster-Mold Castings. The plaster-mold casting process, as in sand casting, is adapted to any quantity of castings. Surface finish and tolerances are equivalent to that produced by permanent-mold casting and, in some instances, approach that of die-casting. Since the cooling rate is decreased by the refractory mold, the grain size is coarser than in processes using metal molds. Advantages claimed for the process include less porosity because of absence of moisture, slower pouring rate to allow escape of entrapped gases, less superheat with reduced oxidation of molten metal, and less casting stress and tendency toward hot tearing made possible by the ease of yielding of the plaster mold. The process is best adapted to the molding of bronze castings, although this does not restrict its use with the other nonferrous metals.

19.10 Die Castings. The die-casting process is capable of producing castings to dimensional tolerances of the order of ±0.001 to 0.003 in./in., depending on the casting alloy and shape of the part. The high cost of complicated metal dies requires long runs of several thousand pieces to make the process economical. Although 10,000 pieces is often considered the minimum commercial quantity for the process, this figure is subject to variation for a particular part. Die life varies from up to a million parts with zinc-base alloys, to about 100,000 parts with aluminum and magnesium alloys, and from 10,000 to 50,000 parts with the low-melting brasses. As stated before, the limits vary with the customer's acceptance of rough surfaces and the economics of salvage on parts produced in badly heat-checked dies.

Zinc alloys make up the largest quantity of die-cast production parts, with aluminum becoming extensively used for aircraft and related work. Magnesium and brass die-casting alloys occupy a small but important place in the die-casting industry. The compositions and mechanical properties of typical die-casting alloys are shown in Table 19-II. The process is, at present, limited to the casting of the yellow brasses and special silicon brasses which melt at temperatures below about 1600 F (870 C). Higher casting temperatures cause rapid heat checking and washing in metallic dies with die life reduced to below 1000 parts. Lower-melting alloys of lead or tin are occasionally die-cast for special purposes.

It should be noted that the zinc-base alloys do not have recognized yield strengths or elastic moduli. Creep occurs under constant loading at room temperature and increases with temperature. Such alloys should not be loaded continuously above about 200 F (95 C) in applications where dimensional change caused by creep is important.

Zamak 3 and Zamak 5 show a maximum shrinkage of 0.0007 and 0.0009 in./in., respectively, within about a month after casting. A stabilizing heat treatment for close tolerance work to accelerate these changes consists of heating for 3 to 6 hr at 212 F (100 C) or for longer periods at lower temperatures.

The zinc-base alloys also suffer further dimensional change through aging. After the above contraction is completed, an expansion occurs of the order of 0.0005 in./in. after ten years of aging at room temperature with no loss of impact strength.

In designing parts for die-casting, sections must be kept thin to permit solidification within the cycle of operation. Ribs may be added for necessary strengthening. Undercuts requiring slides or loose pieces in the die are costly. The part must also be adapted to the removal of necessary metal cores prior to the ejection of the piece from the die. Sharp external corners should be avoided and generous fillets should be provided, together with sufficient draft to permit ease of ejection from the die cavity. Design of the part must be closely coordinated with design of the die for the best commercial operation. Details of die design, including parting line location, arrangement of parts in multiple cavity molds, gating, vents, wells, cores, etc., should be worked out with a competent die-caster. Close liaison between part design and die design is more important in die casting than in any of the other casting processes. It is poor economy to attempt to save money on die cost. The die-caster should be allowed a free hand in designing a die for the most economical production of the part in his plant even though some modification of the original part design may be required.

Table 19-II. COMPOSITION (%) AND PROPERTIES OF TYPICAL DIE-CASTING ALLOYS.

ASTM No.	Alloy	Nominal Composition: Zn	Al	Cu	Mg	Si	Other	Tensile Strength (lb/in.^{2})	Typical Mechanical Properties of $\frac{1}{4}''$ Die-Cast Specimen: Yield Strength (lb/in.^{2})	Elongation in 2 in. (%)	Endurance (lb/in.^{2})	Impact (ft-lb)
Zinc Base												
AG40A[1]	Zamak 3	Bal.	4.0	—	0.04	—		41,000	—	10.0	6,900	43
AG41A[1]	Zamak 5	Bal.	4.0	1.0	0.04	—		47,500	—	7.0	8,200	48
Aluminum Base												
S12A[2]	Alcoa 13	—	Bal.	—	—	12.0		37,000	18,000	2.0	15,000	2
S5C[2]	Alcoa 43	—	Bal.	—	—	5.0		30,000	14,000	7.0	17,000	4
—	Alcoa 85	—	Bal.	4.0	—	5.0		40,000	22,000	3.5	17,000	2
G8A[2]	Alcoa 218	—	Bal.	—	8.0	—		42,000	23,000	7.0	18,000	
SG100A[2]	Alcoa 360	—	Bal.	—	0.5	9.5		42,000	23,000	2.0	19,000	3
SC84A[2]	Alcoa 380	—	Bal.	3.5	—	8.5		45,000	25,000	2.0	20,000	3
Copper Base												
ZS331A[3]	Yellow brass	32.0	—	Bal.	—	1.0	—	60,000	30,000	25.0	—	36
—	Doler No. 4	15.0	—	Bal.	—	3.5	—	85,000	50,000	8.0	—	36
—	Doler No. 5	10.0	1.0	Bal.	—	5.0	1.0 Mn	105,000	60,000	5.0	—	30
Magnesium Base												
AZ91[4]	Dowmetal R	1.0	9.0	—	Bal.	—	0.2 Mn	33,000	22,000	3.0	14,000	2
—	Eclipsaloy 130	—	1.3	—	Bal.	—	1.0 Mn	32,000	21,000	9.0	—	10

[1] ASTM Designation: B86-57T
[2] ASTM Designation: B85-58T
[3] ASTM Designation: B176-57
[4] ASTM Designation: B94-58

The die-casting process is noted for high production, close dimensional tolerances, high degree of surface finish, accuracy in reproducing fine detail, and higher strength properties made possible by the fine-grain structure produced. Casting size is limited to the capacity of available machines and over-all dimensions of die blocks. Sections as thin as 0.015 in. may be produced in small castings, and cores as small as $\frac{1}{32}$ in. in diameter can be used. Metal studs or inserts may be cast as an integral part of some castings. Threading costs may be reduced by casting accurate threads in the part.

19.11 Precision Investment Castings. The precision investment casting process is particularly adapted to the production of small, intricate, and close-tolerance parts of nonmachinable and nonforgeable alloys, such as stellite turbine blades and supercharger buckets. There is, however, no limitation on the metal used, provided it can be melted. Depending on the shape, section thickness, and alloy used, dimensional tolerances of about ± 0.003 in./in. can usually be maintained. Although any number of parts can be produced by the process, it has been found that between 500 to 2000 pieces constitute the economical minimum production run. Obviously, however, there is no quantity consideration involving parts which can be produced only by this method. At present, castings manufactured commercially by the process range in weight from less than 1 oz to about 5 lb, although castings up to 30 lb have been produced.

The process is particularly noted for accurate reproduction of fine detail, being limited in this respect by the fluidity of the metal being cast. Knife edges have been cast as thin as 0.0015 in. The thinnest preferred section is 0.020 to 0.040 in., whereas the thickest preferred section is 0.250 to 0.375 in. Heavier sections may be poured with attendant problems of expansion and contraction to increase the cost for acceptable parts. Long dimensions also present problems of holding close tolerances. The longest preferred dimension is about 2 in., although it is practical to cast up to about 7 in., as limited by the size of the investment flask. The practical minimum diameter of cored holes is 0.040 in., although holes 0.010 in. have been cast with the more fluid alloys. Internal or external threads can be accurately cast as small as 0.25 in. in diameter with a maximum of about 12 threads per in.

19.12 Centrifugal Castings. The centrifugal casting process is a somewhat specialized process having the advantage of consistently producing dense castings, since the metal freezes under pressure, accompanied by directional solidification from the mold wall. Centrifugal force causes the denser metal to be thrown to the periphery of the mold while the lighter nonmetallic particles and gases migrate toward the axis of rotation. With relatively thick tubular sections, solidification may also progress from the inner face of the casting, tending to produce center-line shrinkage.

Dimensional tolerances vary with the shape and size of the part and the choice of sand molds or permanent molds. The process is adapted to ferrous and most nonferrous alloys. Aluminum and magnesium alloys are not centrifugally cast to advantage because of their low density. Reasonably long production runs are usually required to justify set-up costs, and the process is further limited by the relatively few foundries having the available equipment and experience.

True centrifugal casting is restricted to parts that are cylindrical or symmetrical around an open center and capable of being revolved about a central axis. Machining of inner surfaces is generally required because of the tendency of inclusions to segregate in the bore. Typical parts adapted to the process include cylinder sleeves, gun barrels, hollow propeller shafts, pipe, and tubing. Tubing up to 50 in. in diameter and 16 ft in length has been produced with a wall 0.25 in. in thickness.

Semicentrifugal casting, employing a core to form the contour of the bore, is a variation of centrifugal casting and is useful in the production of parts, such as brakedrums, flywheels, pulleys, gears, etc.

Centrifuged or pressure casting is used for unsymmetrical parts which may be gated to a central sprue located at the axis of rotation. This method is often employed in connection with the precision investment casting process as covered in the previous article.

REFERENCES

Alloy Cast Iron Handbook, American Foundrymen's Society, Chicago, Ill., 1944.

American Malleable Iron Handbook, Malleable Founders' Society, Cleveland, Ohio, 1944.

Boyles, *The Structure of Cast Iron,* American Society for Metals, Metals Park, Ohio, 1947.

Briggs, *The Metallurgy of Steel Castings,* McGraw-Hill Book Co., New York, 1946.

Briggs and Gezelius, *Studies on Solidification and Contraction in Steel Castings, Trans. American Foundrymen's Society* **41,** 1933, p. 385; **42,** 1934, p. 449; **43,** 1935, p. 274; **44,** 1936, p. 1; and **45,** 1937, p. 61; also *The Effect of Mass Upon the Mechanical Properties of Cast Steel, ibid.* **26,** 1938, pp. 367-386.

Cady, *Precision Investment Castings,* Reinhold Publishing Corp., New York, 1948.

Campbell, *Casting and Forming Processes in Manufacturing,* McGraw-Hill Book Co., New York, 1950.

Cast Metals Handbook, American Foundrymen's Society, Chicago, Ill., 1944.

Clark, *Engineering Materials and Processes,* International Textbook Co., Scranton, Pa., 1959.

Die Casting for Engineers, New Jersey Zinc Co., New York, 1942.
Doehler, *Die Casting,* McGraw-Hill Book Co., New York, 1951.
Hull, *Casting of Brass and Bronze,* American Society for Metals, Metals Park, Ohio, 1950.
Marek, *Fundamentals in the Production and Design of Castings,* John Wiley and Sons, New York, 1950.
Metals Handbook, American Society for Metals, Metals Park, Ohio, 1961.
Practical Considerations in Die Casting Design, New Jersey Zinc Co., New York, 1948.
Steel Castings Handbook, Steel Founders Society of America, Cleveland, Ohio, 1960.

QUESTIONS

1. What are some of the inherent weaknesses in castings?
2. What is the significance of shrinkage in the production of castings?
3. What are the three stages of shrinkage?
4. How can the harmful effects of solidification shrinkage be reduced in castings?
5. What types of segregation are encountered in castings?
6. Discuss each type of segregation as to cause and its effect upon properties.
7. What defects may be found in castings which are not necessarily inherent in the process of solidification?
8. What factors must be considered in the design of a casting? Discuss each.
9. Why is it important to avoid marked changes in section in a casting?
10. What methods may be used to avoid localized heavy sections in a casting?
11. What is a hot tear in a casting? How may it be prevented?
12. What is the purpose of annealing and normalizing a steel casting?
13. What are some of the advantages of the production of castings by the permanent-mold process?
14. What are the advantages of the shell-mold casting process?
15. What advantages may be derived from the use of the plaster-mold method of casting?
16. Discuss the advantages and disadvantages of die casting.
17. What alloys are commonly die-cast?
18. Discuss the advantages and disadvantages of the production of castings by the precision investment process.
19. Discuss the classifications of the centrifugal-casting process.

XX

METALLURGY OF MECHANICAL WORKING

20.1 Plasticity. The forming of many parts for engineering structures is possible because the material employed is capable of considerable permanent deformation in the solid state. The curve of engineering stress vs. engineering strain for a metal is shown in Fig. 20.1. This type of stress-strain curve is characteristic of such materials as copper or brass. The property of a material by virtue of which it may be permanently deformed is called *plasticity*. The extent to which a material can be plastically deformed is indicated by the amount of permanent strain it can sustain before fracture occurs. A material for which the stress-strain diagram is of the type shown in Fig. 20.2 possesses less plasticity than the material represented by the stress-strain diagram of Fig. 20.1.

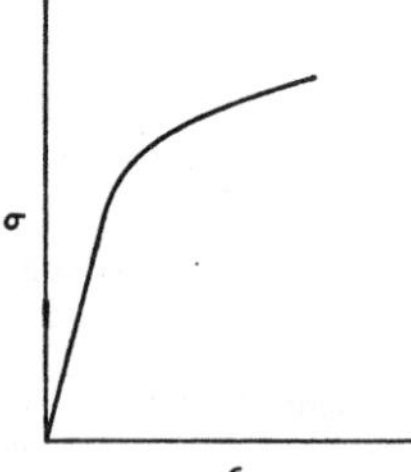

FIG. 20.1 Engineering stress vs. engineering strain for a ductile metal.

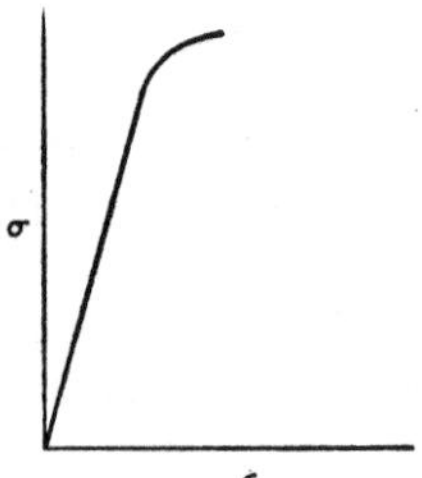

FIG. 20.2 Engineering stress vs. engineering strain for a metal with less plasticity than shown in Fig. 20.1.

Ductility and malleability are descriptive terms related to the ability of the material to be plastically deformed without fracturing in tension or compression, respectively. The orders of decreasing ductility and malleability in

TABLE 20-I. RELATIVE DUCTILITY AND MALLEABILITY OF COMMON METALS.

ORDER OF DECREASING DUCTILITY		ORDER OF DECREASING MALLEABILITY	
Metal	Crystal Lattice	Metal	Crystal Lattice
Gold	FCC	Gold	FCC
Silver	FCC	Silver	FCC
Platinum	FCC	Copper	FCC
Iron	BCC	Aluminum	FCC
Nickel	FCC	Tin	BCT
Copper	FCC	Platinum	FCC
Aluminum	FCC	Lead	FCC
Zinc	HCP	Zinc	HCP
Tin	BCT	Iron	BCC
Lead	FCC	Nickel	FCC
Magnesium	HCP	Magnesium	HCP

several metals are given in Table 20-I. The orders of decreasing ductility and decreasing malleability are not the same. The more ductile metals have crystal lattices of the face-centered cubic type, whereas the less ductile metals have the hexagonal close-packed or body-centered tetragonal type of lattice.

The mechanism of plastic deformation was discussed in Chapter V. While the deformation occurs by the operation of dislocations, the result is essentially the same as slip by block movement on certain crystallographic planes called *slip planes* and in a certain direction known as a *slip direction.* The combination of a slip plane and a slip direction forms a *slip system.* The availability of many potential slip systems in the cubic lattice may account for the relatively high ductility of metals that crystallize in this system.

The slip in grains is complicated by the large number of grains of differing orientation. These differently oriented grains cause slip interference, thus making plastic flow more difficult. The slip phenomenon in a polycrystalline metal can be revealed with the aid of a microscope. A polished piece of the metal is plastically deformed. When the surface is observed through the microscope, the grains are observed to contain parallel lines called *slip lines,* which are the intersections of slip planes with the surface of the specimen. Slip lines produced on an electrolytically polished surface of brass are shown in Fig. 20.3.

Those metals, such as low-carbon steel, that exhibit a distinct yield point in the stress-strain diagram are likely to experience surface irregularities during cold-forming operations. This surface irregularity, called *orange peel* or *stretcher strains,* is caused by the discontinuous yielding of the ma-

FIG. 20.3 Slip lines on polished surface of brass (500×).

terial in the form of Luder's bands. This effect is eliminated by subjecting the material to a small amount of plastic deformation by light rolling before using it in the cold-forming operation.

20.2 Strain Hardening. A metal that possesses a stress-strain diagram of the type shown in Fig. 20.4 can be continuously plastically deformed at a constant stress. The resistance to continued deformation does not change with the amount of plastic deformation. On the other hand, a metal for

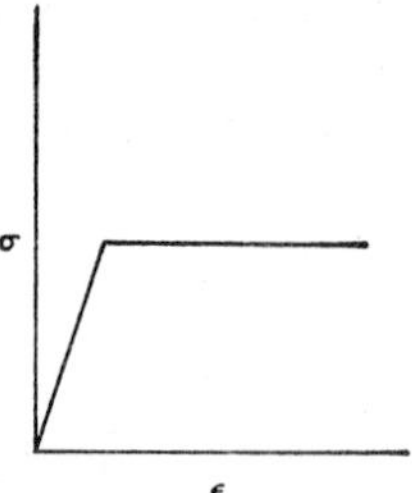

FIG. 20.4 Engineering stress vs. engineering strain for a metal with plastic strain at constant stress.

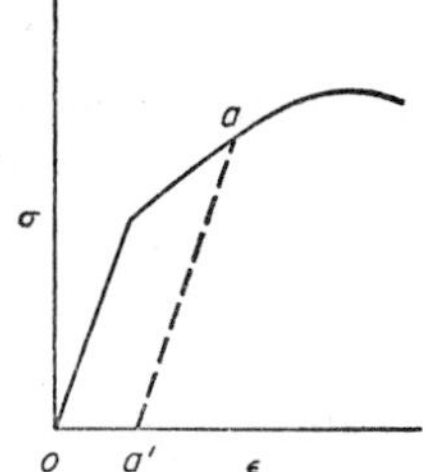

FIG. 20.5 Engineering stress vs. engineering strain illustrating strain hardening.

which the stress-strain diagram is of the type shown in Fig. 20.1 exhibits increasing resistance to plastic deformation with continued plastic deformation. If such a metal is plastically deformed to an amount corresponding to point *a* in the diagram of Fig. 20.5 and then the deforming force is released, the net amount of plastic deformation that has occurred is indicated by the strain *oa'*. If the load is again applied, elastic strain will occur up to a stress corresponding approximately to the point *a*, and then plastic deformation takes place. By this plastic deformation, the so-called proportional limit, or the elastic limit of the metal, has been increased, and the resistance to plastic deformation has been increased. Increase of resistance to plastic deformation is called *strain hardening*. Any material for which the stress-strain diagram exhibits increasing resistance to plastic strain can be strain-hardened.

In certain forming operations, strain hardening is a hindrance, since it makes plastic deformation of the material more difficult and may make necessary unusually high forces in order to secure the desired deformation. Although a metal that has a stress-strain diagram of the type shown in Fig. 20.4 would be easy to form, very few metals or alloys of practical importance have this type of diagram. Strain hardening, however, may be advantageous because of its effect of increasing the elastic limit of the material. The elastic limit of an alloy in which a phase change does not occur can be raised only by strain hardening.

It is of most interest in this discussion for the engineer to recognize that strain hardening produces an increase of elastic limit and a decrease of ductility. The extent of strain hardening in different metals varies considerably. An indication of the extent of strain hardening can be secured by comparing the stress-strain diagrams of the different materials.

20.3 Distinction between Cold Working and Hot Working. The relief of internal stress by recrystallization was discussed in Chapter V. If a metal or alloy is plastically deformed at a temperature somewhat above its recrystallization temperature, little if any strain hardening can occur. In this case, recrystallization takes place during plastic deformation, thus eliminating any strain-hardening effect. Under this condition, plastic deformation is relatively easy to accomplish. In view of the difference in behavior of materials above and below the recrystallization temperature, a means is provided by which cold working and hot working can be distinguished. *Cold working* is defined as plastic deformation at temperatures below the recrystallization temperature. *Hot working* is defined as plastic deformation at temperatures above the recrystallization temperature. In accordance with this definition, the plastic deformation of lead at room temperature is a hot-working process.

20.4 Effects of Cold Work. The advantages of cold working a metal over hot working include:

1. Smoother finish.
2. Closer dimensional tolerances.
3. Higher strength properties.
4. More uniform structure.

A smoother finish is possible at the lower temperatures employed for cold working since surface oxidation, or scaling, of the work is absent. All traces of scale caused by previous hot working or interoperational annealing must be removed from the surface prior to cold working to assure the best finish. Cold working also permits the maintenance of a better finish on rolls, dies, mandrels, and other cold-working tools. Closer dimensional tolerances are possible in the absence of thermal expansion and contraction, both in the work and in the equipment used in cold-working operations.

Cold working is generally followed by some form of annealing. Internal stresses are relieved without loss of strength or hardness by heating the metal to a temperature below the recrystallization temperature. This process is known as *stress-relief annealing.* Plasticity may be completely restored after cold working by heating to a temperature above the recrystallization temperature for a sufficient time to allow complete recrystallization. This is known as *process annealing.* After plasticity has been restored, the sequence of strain hardening and recrystallization, often referred to as the *plastic cycle,* may be continued almost indefinitely until the metal is formed into its final shape. The intermediate process annealing serves to produce a more uniform structure than is possible by hot working alone.

20.5 Combination of Cold Working and Precipitation Hardening. A proper combination of cold work and precipitation hardening can result in strength and hardness unattainable by either process alone. This is shown for copper-beryllium sheet in Table 20-II.

TABLE 20-II. TYPICAL EFFECTS OF COLD-WORKING COPPER-BERYLLIUM SHEET.

Condition	Tensile Strength (lb/in.^{2})	Elongation in 2 in. (%)	Brinell Hardness (3000 kg)
As annealed	70,000	45	110
Annealed and cold-worked	120,000	4	220
Annealed and aged	170,000	6	340
Annealed, cold-worked and aged	200,000	2	365

Severe cold working increases the number of available slip planes upon which potential nucleation and precipitation of a second phase may take place. The precipitating phase may also serve to inhibit recrystallization and growth, tending to raise the recrystallization temperature. When over-aging occurs, the inhibiting phase becomes ineffective, and recrystallization proceeds in the usual manner but at a higher temperature. These effects are illustrated in Fig. 20.6 with various combinations of severe cold work and aging. Several alloys are treated in this way to secure the benefits of both cold work and precipitation hardening.

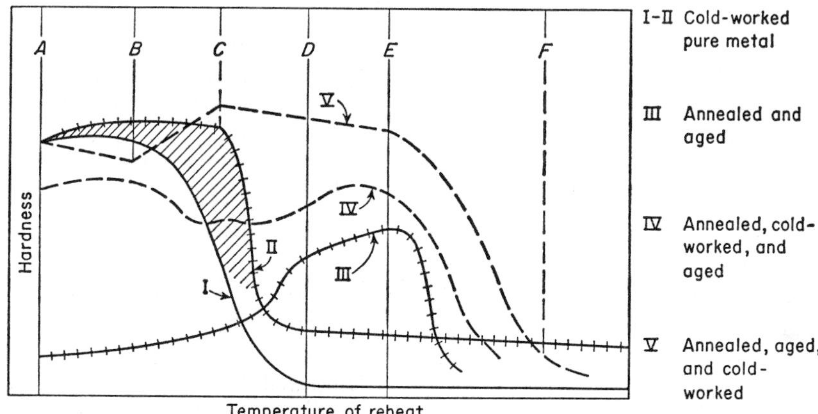

FIG. 20.6 Hardness vs. temperature curves for the recrystallization of alloys in different states. (After Harrington, *Symposium on the Age Hardening of Metals*)

20.6 Characteristics of Cold-Rolled Sheet. In view of the extensive use of cold-rolled sheet and strip for forming and deep-drawing operations, a knowledge of the characteristics that govern formability is valuable to the engineer.

Cold-rolled sheet and strip are manufactured with a specified stiffness known as *temper*. Since the degree of strain hardening varies widely from metal to metal, there is no direct correlation between the temper designations and the degree of deformation for different metals. The approximate mechanical properties for steel, brass, and aluminum sheet when cold-rolled to the standard temper designations employed by each industry are shown in Tables 20-III, 20-IV, and 20-V, respectively.

The formability or deep-drawing quality of sheet is governed by a combination of yield point or yield strength, hardness, grain size, elongation, reduction of area, etc. The hard tempers are useful as spring stock or where blanking operations are employed. The softer grades have better deep-drawing qualities, but with a tendency toward surface roughness or *orange peel* when composed of coarse grains. Deep-drawing qualities are affected by the directional properties characteristic of rolled metal. The difference

TABLE 20-III. APPROXIMATE MECHANICAL PROPERTIES FOR COLD-ROLLED STEEL SHEET.

Steel Temper Desig.	Rockwell Hardness	Depth of Cup for 0.050 in. Thick Strip in mm	Tensile Strength (lb/in.^{2})	Elongation for 0.050 in. Thick Strip (% in 2 in.)	Use
No. 5 dead soft	B 45	10-11	45,000	40	Deep drawing permitting slight stretcher strains. Bends 180° both ways.
No. 4 soft	B 60	9-10	50,000	30	Fairly deep drawing without stretcher strains. Bends 180° both ways.
No. 3 ¼ hard	B 70	8-9	55,000	20	Shallow drawing with very smooth finish. Bending 180° across the grain. Bending 90° along the grain.
No. 2 ½ hard	B 80	7-8	65,000	10	Bending up to 90° across the grain only.
No. 1 hard	B 90	6-7	80,000	3	Flat blanking only.

TABLE 20-IV. APPROXIMATE MECHANICAL PROPERTIES FOR COLD-ROLLED HIGH BRASS SHEET.

Brass Temper Designation	Reduction by Cold Work: B & S No.	Reduction by Cold Work: Per Cent	Rockwell Hardness	Tensile Strength (lb/in.^{2})	Elongation (% in 2 in.)
Soft	Annealed	0	F 60	48,000	67
¼ hard	1	10	B 50	52,000	46
½ hard	2	20	B 69	58,000	28
¾ hard	3	30	B 74	67,000	17
Hard	4	40	B 82	73,000	11
Extra hard	6	50	B 87	85,000	10
Spring	8	60	B 90	96,000	9
Extra spring	10	70	B 93	104,000	8

TABLE 20-V. APPROXIMATE MECHANICAL PROPERTIES FOR COLD-ROLLED ALUMINUM SHEET.

Aluminum Temper Designation		Reduction by Cold Work		Brinell Hardness (500 kg Load)		Tensile Strength (lb/in.^{2})		Elongation (% in 2 in.)	
Old	New	B & S No.	Per Cent	EC	3003	EC	3003	EC	3003
Soft	−O	Annealed	0	23	28	14,000	16,000	40	25
¼ Hard	−H12	2	20	—	—	16,000	18,000	8	7
½ Hard	−H14	4	40	32	40	18,000	21,000	5	4
¾ Hard	−H16	8	60	—	—	21,000	26,000	3	3
Full Hard	−H18	14	80	44	55	24,000	29,000	3	3

of elongation in different directions may cause the formation of *ears* in drawn products.

The character of the stress-strain diagram of an annealed low-carbon steel will lead to difficulties when this material is used in deep-drawing operations. A typical stress-strain diagram for an annealed low-carbon steel is shown in Fig. 20.7. The elongation associated with yielding (*a* to *b* in Fig. 20.7) leads to localized stretching of the metal surface which appears as surface blemishes known as *stretcher strains*.

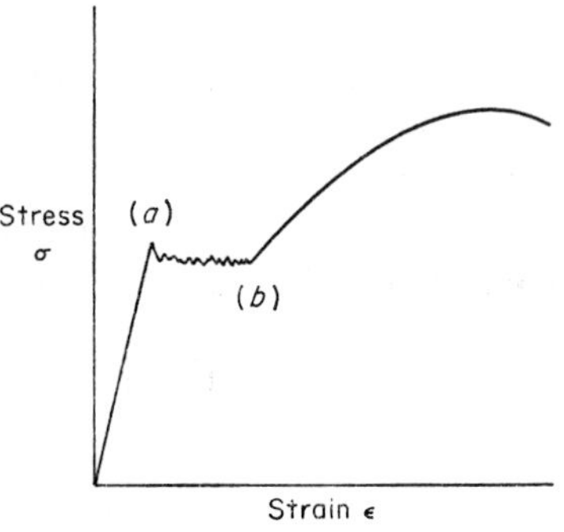

FIG. 20.7 Engineering stress vs. engineering strain for low-carbon steel illustrating yielding.

The yield point of a low-carbon steel can be eliminated by a certain amount of plastic deformation. The effect of cold work in the elimination of the yield point in a low-carbon steel is illustrated in Fig. 20.8. To improve the deep-drawing quality of a low-carbon steel, the sheet is given a reduction in thickness of about 1 per cent by *temper-rolling* or *stretcher leveling*. This treatment does not reduce the ductility of the annealed sheet. In some cases, particularly in a rimmed steel, the yield point may return during storage after this treatment. The return of the yield point occurs by a phenomenon known as *strain aging*. This is illustrated in Fig. 20.9. The effect of strain aging can be removed

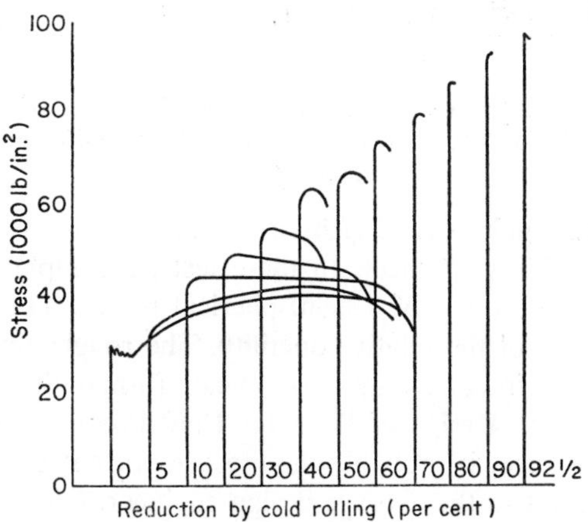

FIG. 20.8 Engineering stress vs. engineering strain for low-carbon steel plastically deformed different amounts. (Kenyon and Burns, *Trans, ASM* **21,** 1933, p. 577)

by an additional 1 per cent cold reduction prior to deep-drawing operations. Aluminum- or titanium-stabilized killed steels which do not show strain-aging characteristics are available. The mechanism of the yield point and strain aging are discussed in Chapter V.

The alpha brasses are particularly sensitive to a condition known as *season cracking* which results from high internal stress. Season cracking may occur spontaneously while the metal is in storage or in service under atmospheric conditions. Cracks develop spontaneously, rendering the metal useless. Season cracking develops more rapidly in the presence of corroding media. Deep-drawn brass parts should be stress-relieved immediately after cold working in order to eliminate the possibilities of season cracking.

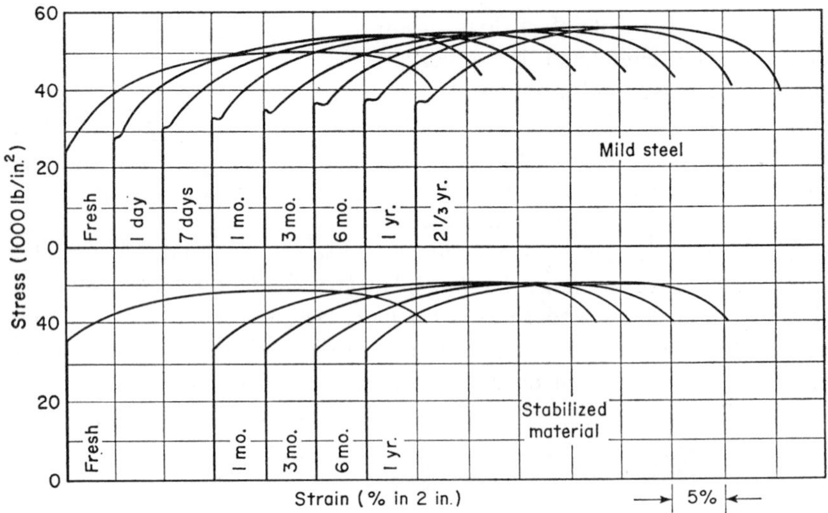

FIG. 20.9 Engineering stress vs. engineering strain, effect of aging after straining by rolling 1% for ordinary mild steel and non-aging mild steel. (*Metals Handbook*)

Cupping tests, such as the Erichsen or Olsen tests, are used to measure the relative formability of sheet. In these tests, a sample sheet is pressed into a cup form by the use of a hemispherical punch. The depth of cup is taken as a measure of the relative ductility. The roughening of the surface or character of any fracture may also indicate formability.

20.7 Factors in Design of Parts for Cold Press Forming. The press-working of metal sheet is probably the most important of the cold-working processes of interest to the average design engineer. Only the basic recommendations for the design of parts to be made by standard punching, forming, and drawing operations will be covered in this article.

Punching operations include blanking and perforating without a reduction in the thickness of the sheet. A blank should be designed to take into account the most economical use of standard size sheets, even though some modification of the original design is necessary to reduce waste. The diameter of perforated holes should not be less than the thickness of the sheet. To avoid tearing, holes should be separated from each other and from the edge of the sheet by a distance greater than the sheet thickness. Perforated slots and tabs should be as short as possible, no narrower than $1\frac{1}{2}$ times the sheet thickness for gauges heavier than $\frac{1}{16}$ in., and no narrower than $\frac{3}{32}$ in. for lighter sheet. The engineer should always consider the possible use of available standard stock-punching dies to reduce the die cost in punching operations.

In bending operations, a part is formed to shape without an intentional reduction in sheet thickness. It is preferable to make all bends at right angles to the direction of rolling in the sheet or to cross the flow lines at no less than a 45-degree angle for maximum strength. The radius of a bend should be no less than the sheet thickness, nor should any bend be located near a previously punched hole. Close angular tolerances are not economical because of *spring-back* resulting from elastic recovery when the applied bending force is removed. The amount of spring-back will vary with the temper of the sheet.

Drawing includes cupping and deep-drawing operations which are used to form cylindrical or irregular shells with or without reduction in sheet thickness. Cylindrical cups which may be formed in a single operation are the least expensive to produce. Flanged, irregular, or deep-drawn shapes will increase the cost of manufacture either by expensive die design, intermediate annealing requirements, or slow press operation.

The theoretical maximum reduction in the initial cupping operation is a cup that has a diameter 50 per cent of the diameter of the blank, with a height 75 per cent of the cup diameter. All radii in cupped and drawn parts should be at least 4 to 6 times the thickness of the sheet. Compressive stresses may be set up in the circumference of thin blanks during cupping operations, causing *wrinkles* which cannot be removed by subsequent drawing. Wrinkling can be prevented by the use of heavier sheet or by hold-down pressure on the edges of the blank during cupping.

Further reduction in cup diameter and increase in height is possible by deep drawing. Steel and brass require intermediate process annealing to restore plasticity between each redrawing operation. The reduction which may be permitted in drawing operations is less than that employed on the original cup and usually decreases with successive draws. In general practice, reductions decrease in the order of 30, 25, 16, 13, and 10 per cent of the preceding shell diameter for sheet having an initial thickness greater than 0.2 per cent of the blank diameter. The first redraw for lighter sheet should not exceed 20 per cent of the cup diameter. These general recommendations may vary slightly with local operating conditions.

Dimensional tolerances after the first redraw vary with the sheet thickness and shell diameter. A tolerance of ±0.005 in. can generally be held for $\frac{1}{32}$ in. sheet in shells up to 2 in. in diameter, increasing to about ±0.020 in. for $\frac{1}{4}$-in. sheet in shells up to 6 in. in diameter.

If the punch and die clearance is less than the thickness of the sheet, an intentional reduction of up to 50 per cent of the wall thickness is possible. This action is known as *ironing* and produces an elongated shell having a side wall thinner than the bottom.

20.8 Effects of Hot Work. The resistance of most metals to plastic deformation in the hot-working range varies directly with the rate of deformation and inversely with the temperature. The effect of temperature on the stress required to permit plastic flow in carbon steels for two different rates of compression is shown in Fig. 20.10. The breaks in these curves indicate the phase transformation, ferrite to austenite.

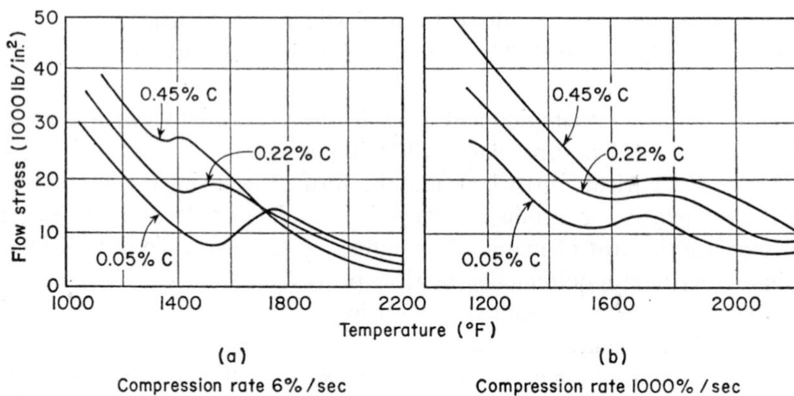

FIG. 20.10 Resistance of carbon steels to compression at different temperatures and different compression speeds. (Sachs and Van Horn, *Practical Metallurgy, ASM*)

The most economical method of converting a cast ingot to a wrought product in which maximum reduction of section is required is by hot working. Hot working may also provide the only means of forming certain shapes with some metals that have limited plasticity at room temperature.

The ease of plastic flow at temperatures above the recrystallization temperature is associated with the large number of available slip systems, the decrease of interatomic strength, and the continuous self-annealing and recrystallization that occurs. With increased rates of deformation, time may not permit the complete recrystallization of the strained lattice, and the process may resemble cold working in this respect. If hot working is done at temperatures close to the melting point, a coarse-grain structure may be produced, and there may be a strong tendency toward surface oxidation. A fine grain structure will be produced in a metal by continuing the hot-working process while the metal cools to a temperature near the recrystallization temperature. However, as the temperature approaches the recrystallization temperature, resistance to plastic flow increases, and greater power is required for the process. The temperatures for hot working are usually selected on a basis of compromising between the ease of working at high temperature and the better structure obtained at a lower temperature; this is a matter of economics. The temperature ranges generally used for forging various metals are given in Table 20-VI.

TABLE 20-VI. TEMPERATURE RANGES FOR FORGING DIFFERENT METALS.
(After Sachs and Van Horn, *Practical Metallurgy*)

Zinc alloys	550-425 F (290-220 C)
Magnesium alloys	750-400 F (400-205 C)
Aluminum alloys	900-750 F (480-400 C)
Aluminum	900-650 F (480-345 C)
Brasses	1475-1200 F (800-650 C)
Copper	1650-1200 F (900-650 C)
Monel	2100-1850 F (1150-1010 C)
Inconel	2300-1850 F (1260-1010 C)
Nickel	2300-1600 F (1260-870 C)
High-speed steels	2200-1900 F (1205-1040 C)
Stainless steels	2200-1900 F (1205-1040 C)
Carbon steels	2400 F (1315 C) for 0.10% C to 1900 F (1040 C) for 1.5% C.
Low-alloy steels	50 to 100 F (10 to 40 C) lower than carbon steels for the same carbon content.

Hot working of metals causes the segregated regions and impurities to be elongated in the direction of plastic flow, a condition which is revealed in the structure as *flow lines, fiber,* or *banded structure.* Fibering caused by the elongation of slag particles or nonmetallic inclusions, sometimes known as mechanical fibering, is not changed by heat treatment. The elongation of nonmetallic impurities is, in a large degree, responsible for the permanent directional properties found in wrought products. The ductility, impact, and fatigue properties are generally greater in the longitudinal direction, i.e., in the direction of plastic flow, than in the transverse direction. The yield strength and tensile strength differ only slightly in these two directions. The variation of directional properties increases with the magnitude of deforma-

TABLE 20-VII. DIRECTIONAL MECHANICAL PROPERTIES IN LOW-ALLOY STEEL FORGINGS.
(After Woldman, *Metal Process Engineering*, Reinhold Publishing Corp., New York, 1948)

Reduction Ratio Ingot Cross Section to Forged Cross Section	Direction Tested	Tensile Strength (lb/in.^{2})	Elong. (%)	Reduction of Area (%)	Impact (ft-lb)
1.7	Longitudinal	116,000	20.0	52.0	49
	transverse	116,000	18.0	64.0	40
3.2	Longitudinal	116,000	20.0	58.5	59
	transverse	115,000	16.0	61.0	29
6.1	Longitudinal	115,000	22.0	63.0	72
	transverse	115,000	12.0	55.0	25

tion, as shown for low-alloy steel forgings in Table 20-VII. Hot working also tends to close voids, blowholes, minor cracks, etc., causing increased density and generally improved uniformity of the structure.

20.9 Factors in Design of Forged Parts. Forgings are generally

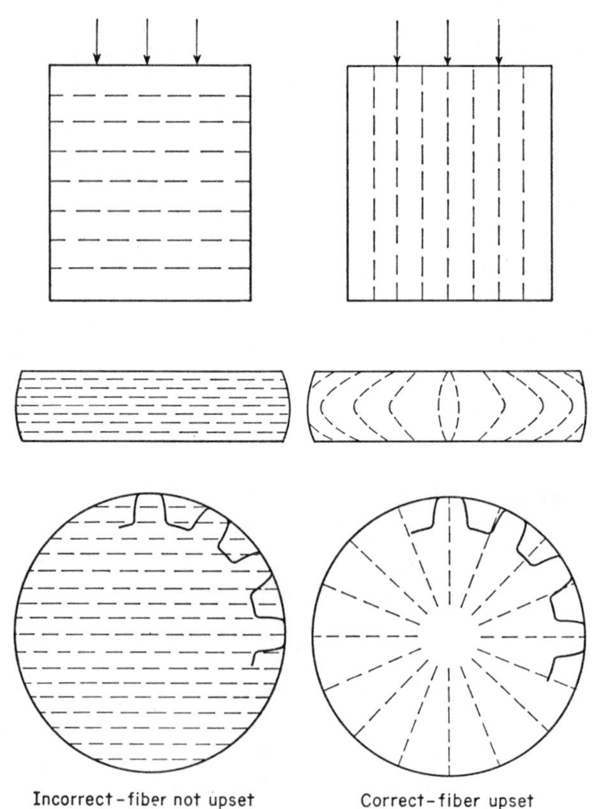

FIG. 20.11 Fiber flow patterns in forged gear blanks.

stronger in all directions than castings because of greater density and uniformity of structure. In a forged part, the tensile strength does not vary greatly with the direction of working. The impact and fatigue resistance are considerably higher in a longitudinal section than in a transverse section. Wherever these properties are important, the design engineer should work closely with the forging expert to make the most effective use of fiber flow in the design of a given part. The alteration of fiber pattern produced by upsetting a gear blank to secure better directional properties is shown in Fig. 20.11. The actual flow structure in a section of an engine crankshaft is illustrated in Fig. 20.12.

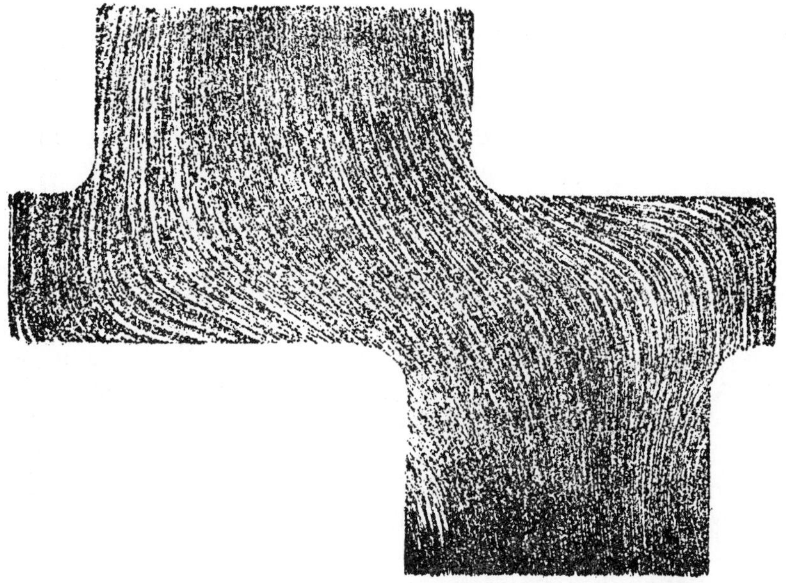

FIG. 20.12 Flow structure in a forged engine crankshaft.

Forgings are limited in size or weight and shape compared with castings. The economical limits on weight are about 800 lb for die forgings and about 100 tons for press and hammer forgings. Very simple shapes are often produced in any quantity by hammer forging between flat dies. The usual forged part requires the production of a large number of parts to justify the die-sinking and set-up costs. Forged parts should be designed for economical die preparation and maintenance. Undercuts, cored holes, and deep pockets should be avoided. Thin sections and sharp corners are difficult to produce by plastic flow in a die. Generous fillets and large radii are recommended not only for improved strength in the part but also for longer die life. The parting line should be on a plane centrally located with respect to the symmetry of the forging to assure ease of withdrawal of the metal from the die and subsequent removal of the flash. A draft allowance of about 7 degrees is required to permit withdrawal of drop forgings, with only 1 or 2 degrees for press forgings.

The tolerances in the production of drop forgings vary with size, shape, and allowable die wear from between about ±0.010 in. to ±0.030 in. Tolerance may be specified on width, thickness, draft, mismatch, fillets, and radii, but in no event should they be specified closer than absolutely necessary in the final part since frequent die rework or cold-sizing operations may be required. It is often more economical to add machining allowances of $\frac{1}{32}$ in. on small work and up to $\frac{1}{2}$ in. on large forgings where close tolerances appear necessary.

The volume of metal to be removed to clean up the surface of a forging is generally less than that required for a casting. The uniform structure permits good machinability with freedom from the abrasive action of burned-in sand found on sand castings. Freedom from blowholes results in lower scrap loss during machining operations.

20.10 Defects in Wrought Products. Most of the defects in wrought products originate in the ingot stage of production. These include pipe, segregation, blowholes, inclusions, and corner cracks. The *pipe* or major shrinkage cavity that forms in the top of the ingot accompanied by serious top *segregation* can be eliminated by shearing off that portion of the first semifinished rolled product, known as a *bloom,* which may contain the pipe. Piping is frequently reduced by the utilization of a hot top, which is a refractory ring placed upon the top of the ingot mold. Secondary piping is sometimes encountered in instances where solidification is completed at the top of the ingot before the lower portion has solidified, wherein the encased liquid metal shrinks. This may be prevented by the use of a big-end-up ingot mold. Secondary pipe cannot be readily removed and will be revealed as center porosity in bar products or as a lamination in plate or sheet.

Blowholes are voids caused by dissolved or occluded gases formed during the solidification of the ingot. In some instances, these blowholes will be welded shut and, thus, eliminated during the rolling operations. Sometimes, however, the blowholes may not be eliminated and will constitute a defect. Nonmetallic *inclusions* cannot be removed and may be elongated in the direction of working to form stringers. Inclusions are usually found in varying amounts in almost all steel. If they are uniformly distributed and in relatively small amounts, they will not cause any difficulty; but if they are coalesced in large stringers, they may cause serious trouble, such as pronounced directional characteristics, adversely affecting the strength. *Corner cracks* which are caused by planes of weakness in the ingot structure may open up upon rolling and cannot generally be eliminated except by chipping.

Internal cracks, known as *flakes, bursts,* or *shatter cracks,* are prevalent in large sections of high-carbon or high-alloy steels. They are believed to be caused by a combination of internal thermal stress and the evolution of hydrogen in the steel. These steels are particularly susceptible to this defect when charged into heating furnaces while cold or wet. The condition may be improved by very slow cooling immediately after the rolling of blooms or other large sections.

Overheating prior to hot working or heat treating causes excessive grain growth, oxidation of the grain boundaries, and serious surface oxidation or scaling. When a metal is badly overheated or burned, the oxidation at the grain boundaries renders the metal useless, and any part so treated must be scrapped.

Whenever steel is heated under oxidizing conditions, carbon may be removed from the surface layer, ultimately resulting in a soft, weak skin. This is called *decarburization* and can be prevented by the use of controlled atmosphere furnaces or molten salt baths.

Surface defects developed in hot-working operations include scabs, tears, slivers, seams, laps, blisters, rolled-in scale, scores, etc. *Scabs* are usually caused by elongation of metal splashes that have chilled and adhered to the wall of the ingot mold. *Tears* are transverse surface cracks caused by rupture of the skin of improperly heated metal which may develop into *slivers* with subsequent rolling. *Seams* may be caused by subsurface blowholes or longitudinal surface cracks that close but do not weld in the hot-working operation. *Laps* are seams caused by fins or corners that fold over without welding. *Blisters* result from dissolved gases that expand to raise the surface during heating or working. *Rolled-in scale* results from improper descaling of ingots or semifinished products allowing the scale particles to be embedded in the surface. *Scores* or scratches on the surface of wrought products may be traced to imperfections in rolls, guides, dies, etc.

The foregoing surface defects may be physically removed by scarfing, chipping, or grinding. Surface conditioning of ingots, blooms, slabs, billets, or other semifinished wrought products is expensive and justified only by the surface quality desired in the final product. The actual machining of the entire surface may be economical to assure complete elimination of surface and subsurface defects as in the case of some high-quality tool steel and piercing billets.

Other surface defects in wrought products may be related more directly to specific finishing operations. For example, drop forgings may contain sections where two surfaces fold together to cause *cold shuts;* relative shifting of dies will cause *mismatch;* insufficient metal to fill the die impression will cause *underfill.*

20.11 Wrought Products vs. Castings. The wrought products have certain inherent advantages over castings. In Chapter XIX, the process of the solidification of molten metal and the properties of castings were discussed. Cast materials which do not exhibit phase changes in the solid state do not lend themselves to any process by which grain refinement can be accomplished, assuming that the as-cast shape is to be maintained. The typical coarse-grained, dendritic, somewhat porous structure found in a cast ingot may be improved by mechanical working. The wrought products generally possess the following advantages over castings:

1. More uniform structure.
2. Greater density.
3. Higher strength properties.
4. Directional characteristics.
5. Closer dimensions permitting less machining.
6. Greater economy in mass production.

There are certain limitations that restrict the manufacture of products by mechanical working. These include the following:

1. Size is limited by the capacity of available equipment.
2. Shape must be relatively simple.
3. Quantity must justify set-up cost.
4. Material must be workable.

REFERENCES

Barrett, *Structure of Metals,* McGraw-Hill Book Co., New York, 1943.

Burke *et al., Grain Control in Industrial Metallurgy,* American Society for Metals, Metals Park, Ohio, 1949.

Campbell, *Casting and Forming Processes in Manufacturing,* McGraw-Hill Book Co., New York, 1950.

Chase, *Handbook on Designing for Quantity Production,* McGraw-Hill Book Co., New York, 1944.

Crane, *Plastic Working in Presses,* John Wiley and Sons, New York, 1944.

Hinman, *Pressworking of Metals,* McGraw-Hill Book Co., New York, 1941.

Jeffries and Archer, *The Science of Metals,* McGraw-Hill Book Co., New York, 1924.

Jevens, *Metallurgy of Deep Drawing and Pressing,* John Wiley and Sons, New York, 1942.

Naujoks and Fabel, *Forging Handbook,* American Society for Metals, Metals Park, Ohio, 1939.

Sachs and Van Horn, *Practical Metallurgy,* American Society for Metals, Metals Park, Ohio, 1940.

Smith, *Principles of Physical Metallurgy,* Harper & Brothers, New York, 1956.

Symposium on the Age Hardening of Metals, American Society for Metals, Metals Park, Ohio, 1939.

Symposium on Cold Working of Metals, American Society for Metals, Metals Park, Ohio, 1949.

QUESTIONS

1. What is plasticity?
2. How may the degree of plasticity be expressed?
3. In what type of crystal structure do the more ductile metals exist?
4. What is a slip plane?
5. What is a slip system?
6. Why is the slip mechanism more complicated in polycrystalline metals than in a single crystal of a metal?
7. What are stretcher strains?
8. How can the effect of orange peel be eliminated?
9. Show how the extent of strain hardening can be evaluated by means of stress-strain diagrams.
10. What type of stress-strain diagram would indicate the best forming characteristics of a metal?
11. Distinguish between cold working and hot working.
12. What are the advantages of cold working over hot working?
13. How may the effects of cold working be eliminated?
14. What is process annealing?
15. What are the advantages of combining cold working and precipitation hardening?
16. To what does the term temper refer in cold-rolled products?
17. What difficulties may be encountered in cold forming of a metal that possesses a distinct yield point?
18. How may the effects of a yield point be eliminated?
19. What is strain aging? How may it be eliminated?
20. What tests are commonly used to indicate the relative formability of materials?
21. What is the effect of spring-back in cold-forming operations?
22. What is the cause of wrinkles in cold-forming operations? How may they be prevented?
23. What are some of the advantages of hot-forming operations as compared with cold-forming operations? What are some of the disadvantages?
24. What is the character of flow lines that appear in hot-worked metals? How is this related to directional properties?
25. What precautions should be taken in the design of products to be produced by the forging process?
26. What defects may be found in wrought products?
27. What advantages do wrought products have over castings?
28. What are some of the limitations that restrict the manufacture of products by mechanical working?

XXI

METALLURGY OF SOLDERING, BRAZING, AND WELDING

21.1 Metallurgical Joining Methods. It is often more economical to produce metal parts and structures by joining simple components into a composite product rather than to employ a casting or wrought product. Metals may be permanently joined by soldering, brazing, or welding. These operations incorporate one or more of the principles of physical metallurgy discussed earlier in this text, including melting, deoxidation, degasification, casting, hot or cold working, heat treating, diffusion, etc.

Soldering may be defined as the joining of metals by means of another metal that has a melting point or range below about 1000 F (540 C) and lower than that of the metal to be joined. *Brazing* differs from soldering in that the added metal generally has a melting range above 1000 F (540 C) but below that of the base metal. *Welding* involves the joining of metals by heating to a suitable temperature above the recrystallization range or above the melting range with or without the application of pressure or added metal.

21.2 Soldering. Soldering alloys are basically composed of lead and tin and are known as "soft solders" in contrast to the brazing alloys or "hard solders" discussed in the next article. The high-tin solders tend to be hard and brittle compared to the high-lead solders which possess greater ductility but lower strength. The compositions and typical applications for the soft solders are given in Table 21-I.

Although, in general, the tin-lead solders are most satisfactory, in times of tin shortage it may be necessary to reduce or eliminate the amount of tin in the solder. A substitute has been developed[1] containing 87.5 per cent lead, 0.5 per cent arsenic, and 12 per cent antimony, which is satisfactory for soldering on steel and tin plate. However, this alloy is not satisfactory for soldering copper, brass, or galvanized iron because of lack of adherence

[1] Dowdell, Fine, Elliot, and Mattson, *A New Soft Solder: Lead-Arsenic-Antimony Alloy, Metal Progress* **43,** January 1943, p. 56.

Table 21-I. TYPICAL SOFT SOLDERING ALLOYS.

(After *Metals Handbook*)†

Nominal Composition (%)						Melting Range (°F)		Uses and Comments
Sn	Pb	Sb	Bi	Ag	Other	Solidus	Liquidus	
5-20	95-80	—	—	—	—	570-361	595-525	For coating and joining metals. For high temperature and differential soldering.
45	Bal.	—	—	—	—	361	439	For automobile radiator cores and hand-soldered electrical connections. Also a sweating solder.
50	Bal.	—	—	—	—	361	419	General-purpose solder.
38	Bal.	—	—	—	0.1 As (max)	361	458	Wiping solder for joining lead pipes and cable sheath.
35	Bal.	0.5	—	—	—	361	473	General-purpose solder. Also a sweating solder.
60	40	—	—	—	—	361	368	"Fine solder," used where temperature requirements are critical.
38-40	Bal.	—	—	—	—	361	464-458	Wiping solder. Also for automobile radiator cores and soldered electrical connections.
23	Bal.	—	—	—	9 Cd	294	455	Wiping solder.
15-35	Bal.	—	—	—	—	439-361	547-473	Low-tin solders. Automobile body solders.
32	Bal.	2	—	—	0.1 As (max)	367	466	Emergency wiping solder for joining lead pipes and cable sheath.
13	Bal.	0.5	23	—	0.1 As (max)	207	415	Wiping solder for joining lead pipes and cable sheath.
50	32	—	—	—	18 Cd	294	294	For fusibility and safety devices.
16	32	—	52	—	—	203	203	For fusibility, safety and fire prevention devices.
13	27	—	50	—	10 Cd	158	158	For fusibility, safety and fire prevention devices.
None*	Bal.	—	—	2.50	—	579	579	Emergency solder. For use where a high melting point is desirable. SAE E07.
2-5	Bal.	—	—	—	—	573-518	607-594	Dipping or coating metal.
4	Bal.	3.5	—	—	—	467	554	Automobile body solder.
1	Bal.	—	—	1.5	—	589	589	For sealing tin cans and for "high"-temperature use.
95	—	5	—	—	—	450	464	Sweating copper tubing joints.
—	Bal.	12	—	—	0.5 As	476	478	For joining ferrous parts. For filling joints in automobile bodies.
—	—	—	—	5	95 Cd	639	734	For "high" temperatures.
—	—	—	—	—	50 Cd, 50 Zn	508	619	For "high" temperatures.
—	—	—	—	—	82.5 Cd, 17.5 Zn	508	508	For "high" temperatures.

† Specification of soft solders are contained in ASTM Designation B32-58T.
* This solder is known to be sensitive to corrosion in outdoor environments.

and good flow characteristics. The strength of lap joints made with this alloy is comparable with that of those made with the usual 50 per cent tin, 50 per cent lead solder. The melting range is somewhat higher than for the 50-50 solder, but the range is only about 2 F. A few other solders have been developed as substitutes to reduce the use of tin. These are mostly alloys of lead and silver which have a higher melting range and require special techniques.

A concept of the strength of lead-tin soldered joints may be secured from results presented by Mample[2] in Table 21-II. For very short durations, loads much higher than those indicated in Table 21-II can be applied without failure.

TABLE 21-II. CONTINUOUS MAXIMUM SAFE LOAD ON SINGLE LAP JOINT FOR TWO SOLDERS AT 70 F.

Material	Condition	MAXIMUM SAFE LOAD (LB/IN.²) 62 Sn 38 Pb	20 Sn 80 Pb
Copper	Thin sheet, bend under load, shear angle 173°	570	275
Copper	Heavy members, no bend under load, shear angle 180°	785	375
Brass	Thin sheet, bend under load, shear angle 173°	630	300
Brass	Heavy members, no bend under load, shear angle 180°	730	350
Iron-black	Thin sheet, bend under load, shear angle 173°	520	250
Iron-black	Heavy members, no bend under load, shear angle 180°	680	325
Iron-tinned	Thin sheet, bend under load, shear angle 173°	—	210

The ultimate success of any soldering operation is governed primarily by surface cleanliness and freedom from oil, grease, or oxide films which prevent satisfactory adhesion of the solder. Fluxes are usually employed for this purpose. The function of a flux is fivefold:

1. To clean the surfaces.
2. To prevent oxidation of the surfaces when hot.
3. To lower surface tension or to allow normal action of metal surface tension.
4. To promote alloying of metal surface with the solder.
5. To promote wetting of the surface by the molten solder.

[2] A. Z. Mample, *An Engineering Approach to Soldering with Tin-Lead Alloys*, *Metals and Alloys* **21,** March 1945, pp. 702-707.

A flux must be of such a character that the residue is noncorrosive, nonhygroscopic, and nonconducting for electrical work. Furthermore, the flux should not readily volatilize, decompose, or carbonize during the soldering operation.

A variety of fluxes is available. Rosin is particularly effective for soldering electrical connections if it is properly used. When rosin is applied to the work, it must not be subjected to an excessive temperature or its fluxing qualities will be reduced and the residue will be carbonized. The particular advantage of rosin lies in its noncorrosive character which is of most importance for use in electrical circuits. Rosin is not so effective, however, as a flux for soldering sheet steel as some other materials.

The salt- or "acid"-type fluxes, composed of chlorides of ammonia, zinc, aluminum, etc., are excellent for many soldering operations, because they are very active and do not carbonize. However, these fluxes and their residues are highly corrosive, electrically conducting, and hygroscopic. These materials can be obtained in liquid or paste form and are particularly used in sheet-metal work. To prevent subsequent corrosion, the soldered joint should be thoroughly cleansed of the flux residue.

Some fluxes are used that contain organic acids, such as stearic, oleic, tartaric, benzoic, etc. Although these fluxes are corrosive, they are less corrosive than the inorganic-salt type. These fluxes are used on special occasion, as in soldering lead.

The application of heat to the surfaces to be joined may be by soldering iron, torch, molten bath, induction or resistance methods, or by wiping. The metal surfaces must be heated to the melting temperature of the solder and held at that temperature only long enough to allow uniform "wetting" or "tinning" of the previously prepared surface. An excessive soldering temperature or time at temperature promotes the formation of intermetallic compounds in many alloy systems which tend to weaken the resulting joint. Since little or no alloying takes place at the relatively low temperatures involved in soldering operations, the bond is only as strong as the soldering alloy. Thin solder films of the order of 0.003 to 0.005 in. thick are, therefore, preferred for the best joint strength. Where higher strength is required, it is often desirable to incorporate mechanical aids, such as crimping, riveting, seaming, etc.

Copper, iron, lead, nickel, tin, zinc, and many of their alloys may be soldered. Aluminum and the stainless steels, however, require special techniques in view of the stable and tenacious oxide films that are formed.

21.3 Brazing. Brazing alloys are of three general types: (1) copper and copper-base alloys; (2) silver brazing alloys, known as "silver solders"; and (3) aluminum brazing alloys.

Commercially pure copper is extensively used for production brazing of ferrous parts at a temperature of about 2150 F (1175 C). Copper-zinc alloys constitute the largest portion of the copper-base brazing alloys with melting ranges of from 1400 to 1700 F (760 to 925 C). The silver solders have been developed from binary copper-zinc alloys to which silver and other elements have been added in varying amounts to lower the melting range to within the limits of from 1100 to 1600 F (595 to 870 C). These alloys permit brazing at lower temperatures, which are necessary if damage to the base metal, such as structural change, annealing, or grain growth in cold-worked products and precipitation-hardenable alloys, is to be prevented. The compositions of common copper-base and silver brazing alloys are given in Tables 21-III and 21-IV respectively. Aluminum brazing alloys have been developed which consist principally of aluminum to which other elements have been added to decrease the melting range substantially below the melting point of aluminum.

TABLE 21-III. COPPER-BASE BRAZING ALLOYS.
(After *Metals Handbook*)

Brazing Alloy	Melting Range (°F)	Composition (%) Cu	Zn	Others	Remarks
ASTM-RBCaZn-A*...	1630-1650	57-61	Bal.	0.25-1.00 Sn	Common spelter solders, brass yellow color, in granulated and lump form for general brazing.
ASTM-RBCuZn-D*...	1600-1715	46-50	Bal.	9.0-11.0 Ni	
Spelter bronze........	1575	Bal.	45	3-5 Sn	Pale yellow in granulated or lump form for special work. Harder than ASTM grades.
Black button.........	1385-1440	Bal.	57-65	5-9 Sn	Gray to black, granulated, moderate strength, contains 1% Fe.
White spelter solder...	1600	Bal.	55-59	7-9 Ni	White, granulated form, match for nickel silver, cupronickel and the like.
Phos-copper..........	1304-1526	93	Bal.	Phosphorus	Rod or strip for self-fluxing brazes on copper, gray color.
White brazing rod....	1700	47	42	11 Ni	High-strength brazing rod.
Yellow bronze........	1595-1625	Bal.	42	0.5 Sn	General-purpose brazing or bronze-welding rod.
Copper..............	1981-2050	99.9	—	—	Tough-pitch or deoxidized copper in wire, granulated, or electrodeposited on steel for brazing in a hydrogen atmosphere.

* ASTM Designation: B260-56T.

Procedures have been developed for brazing 1100, 3003, 3004, 5052, 6053, and 6061 aluminum alloys. Although brazing wire is available, brazing sheet is frequently more economical. The sheet consists of 3003 alloy or some special heat-treatable alloy coated on one side or both sides with the brazing alloy. The parts are formed by conventional methods and assembled for furnace brazing. The coating on the sheet provides the filler metal. If the heat-treatable alloy is employed, the properties of the part can be improved by a solution treatment followed by precipitation, after the practice specified for 6061 alloy. Special fluxes are provided for aluminum brazing when required, as in torch brazing.

A satisfactory bond is possible in brazing only with properly cleaned and fitted joints. Chemical and mechanical cleaning methods may be required to remove heavy scale and grease deposits. Fluxes serve to remove light oxide films, dissolve oxides formed in the brazing operation, and assist in the free-flowing qualities of the brazing alloy. Fused borax and boric acid, alone or in combination, are used as conventional brazing fluxes. Additions of alkali bifluorides may be required where refractory oxides are involved in the brazing of aluminum bronze, silicon bronze, beryllium-copper, and the stainless steels. Since most of these fluxes are strongly corrosive in the presence of moisture at room temperature, they should be thoroughly removed immediately after brazing.

TABLE 21-IV. SILVER BRAZING ALLOYS.
(*Metals Handbook*)†

COMPOSITION (%)						MELTING POINT	FLOW POINT
Ag	Cu	Zn	Cd	P	Sn	(°F)	(°F)
10	52	38	—	—	—	1450	1565
20	45	35	—	—	—	1430	1500
20	45	30	5	—	—	1140	1500
45	30	25	nil	—	—	1250	1370
50	34	16	nil	—	—	1240	1425
65	20	15	nil	—	—	1240	1305
70	20	10	nil	—	—	1335	1390
80	16	4	nil	—	—	1330	1490
50	15.5	16.5	18*	—	—	1160	1175
15	80	—	—	5**	—	1190	1300
30	38	32	—	—	—	1370	1410
40	36	24	—	—	—	1330	1445
60	25	15	—	—	—	1260	1325
72	28	—	—	—	—	1435	1435
56	22	17	—	—	5***	1165	1200

* Proprietary alloy, "Easy Flo."
** Proprietary alloy, "Sil Fos."
*** Proprietary alloy, "Ready Flo."
† Detail specifications are contained in ASTM Designation B260-56T.

In some brazing operations, a flux is not required. The phosphorus-copper brazing of copper and the hydrogen brazing of steel with commercially pure copper are examples of such operations.

In the brazing operation, bonding is obtained by the distribution of the nonferrous filler metal between closely fitted surfaces either by means of capillary attraction or by gravity. The most satisfactory brazing alloys are those that flow and easily "wet" the surface at a temperature at least 100 F (40 C) below the melting range of the parent metal and that do not react unfavorably to cause intercrystalline embrittlement. Some alloying may take place where liquid-state intersolubility exists between the brazing alloy and the base metal. Since brazing temperatures are higher than those employed in soldering, a greater degree of diffusion and alloying occurs.

The strength of a brazed joint may be higher than that of the brazing alloy, provided the thickness of filler metal is not greater than 0.001 or 0.003 in. Copper-brazed joints between steel parts are normally held to a maximum of 0.0005 in. in thickness. Greater thicknesses of filler metal tend to cause the joint strength to approach that of the brazing alloy. It may be necessary to provide greater clearances between mating parts in instances where the filler metal alloys readily with the base metal rather than being drawn into the joint by capillary attraction. In general, brazed joints possess greater ductility and higher strength than is possible by soldering methods. This is particularly true at elevated temperatures where the soft solders tend to soften with a resultant loss in strength.

Heating for brazing may be accomplished by torch, by dipping in molten alloy or salt bath, by furnace, or by electrical induction or resistance. The brazing alloy may be applied in various forms, including wire, rod, sheet, powder, lump, etc.

Torch brazing is used extensively as a low-cost method of fabrication and maintenance for relatively large parts, such as bicycles, propane cylinders, lawn mowers, furniture, electrical equipment, etc. *Dip brazing* permits rapid heat transfer and at the same time protects the surface of the part from oxidation. Metal baths have limited use in the brazing of electrical conductors and for assembly of copper tubes in heat exchangers. Molten salt baths are limited to applications where the filler metal can be positioned in the assembly prior to dipping, and the component parts may be held in place by suitable fixtures or other mechanical devices. *Furnace brazing* is very popular as a mass-production method of joining small parts where the filler metal can be added as sheet, wire, or powder, and the parts may be crimped, staked, pressed, etc., to prevent movement of the individual components in the furnace. A neutral or reducing atmosphere is usually provided to prevent oxidation at high brazing temperatures. Copper brazing of ferrous alloys in a hydrogen atmosphere is probably the most extensive application of the process. Many joints in subassemblies are made by furnace brazing with copper. An extensive study of the strength of joints of this type has been made.[3] The shear strength of the bond in 1-in. diameter tubing

[3] C. Rhyne, Jr. *Factors Controlling the Strength of Brazed Joints, from Final Report NRC 560, Welding Journal* **25,** July 1946, pp. 599-611.

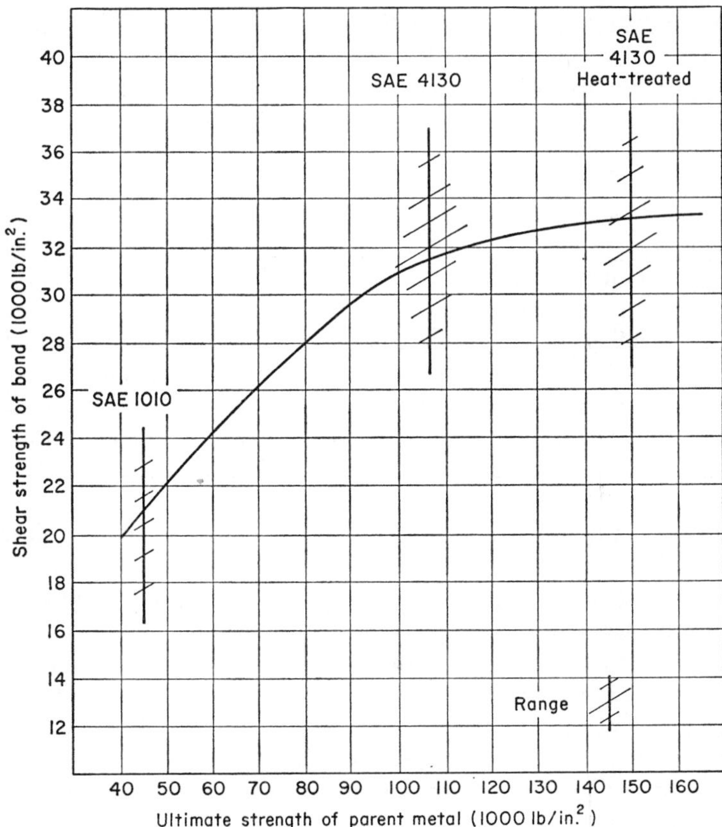

FIG. 21.1 Shear strength of copper brazed joints. (After Rhyne)

with a 0.120-in. wall thickness in relation to the strength of the parent metal is shown in Fig. 21.1. In these investigations, it was found that:

1. The strength of the joint is independent of the clearance between the two mating parts within the range of zero clearance and a press fit to 0.004 in. However, a press fit is preferred in order to maintain proper alignment of parts during the brazing process.
2. The amount of overlap in the joint is of importance in securing the proper stress distribution at the joint.
3. The strength of bond is slightly increased by an increase in the brazing time, probably resulting from an increase of alloying with the parent metal.

The solubility of copper in iron and of iron in copper is small, but even this slight solubility is of value in the copper brazing of steel parts. If the clearance between the two parts is too great, the strength will be approximately that of the copper filling the joint. *Electrical induction* or *resistance brazing* methods provide a rapid means of localized heat application with ease of control for the mass production of small parts.

Most of the common commercial metals and alloys may be brazed, provided their melting ranges are at least 100 F above that of the selected brazing metal, and provided proper cleaning and fluxing are employed. Close temperature control may be required with certain metals and their alloys, such as aluminum, to prevent fusion or softening of the parent metal or low melting-point alloy phases.

21.4 Welding. Welding differs from soldering and brazing in that the parent metal is heated to above its recrystallization temperature with the application of pressure or to above its fusion temperature with or without pressure. Welding methods may be broadly classed as (1) plastic or (2) fusion welding. A more detailed classification based on specific processes is shown in Fig. 21.2.

Plastic welding includes those processes in which metals are joined by the application of pressure at temperatures above the recrystallization temperature but below the melting range. Bonding takes place by diffusion and grain growth across the interface of the joint without intentional fusion. Plastic methods include forge, resistance, pressure, and cold welding.

Forge welding was the first welding method known to produce a union between metal parts, The parts are hammered or pressed together while at a proper temperature. In *resistance welding,* the relatively high electrical resistance of the joint to the passage of a high current causes localized heating at the contact surfaces with welding taking place by application of pressure through electrodes. Spot, seam, and projection resistance welding methods are of the plastic type. Flash, upset, and percussion methods are modifications in which an arc is struck between the pieces to be joined, causing incipient fusion at the interface, followed by pressure. *Pressure welding* employs the use of an oxyacetylene flame or a thermit reaction to furnish the heat necessary to raise the temperature of the surface to the plastic range without actual fusion.

Cold welding involves the cold-joining of properly cleaned metal surfaces by the application of high unit pressure or by ultrasonic vibrations applied to components clamped under low pressure. Although a temperature rise may actually take place at the interface, actual fusion does not occur and the temperature effect can be controlled to prevent undesired alloying or the formation of intermetallic compounds in the joining of dissimilar metals. In ultrasonic welding,[4] the vibratory energy is applied by a trans-

[4] Weare, Antonerich, and Morse, *Fundamental Studies of Ultrasonic Welding, Supplement to the Welding Journal* **39,** August 1960, p. 331-s.

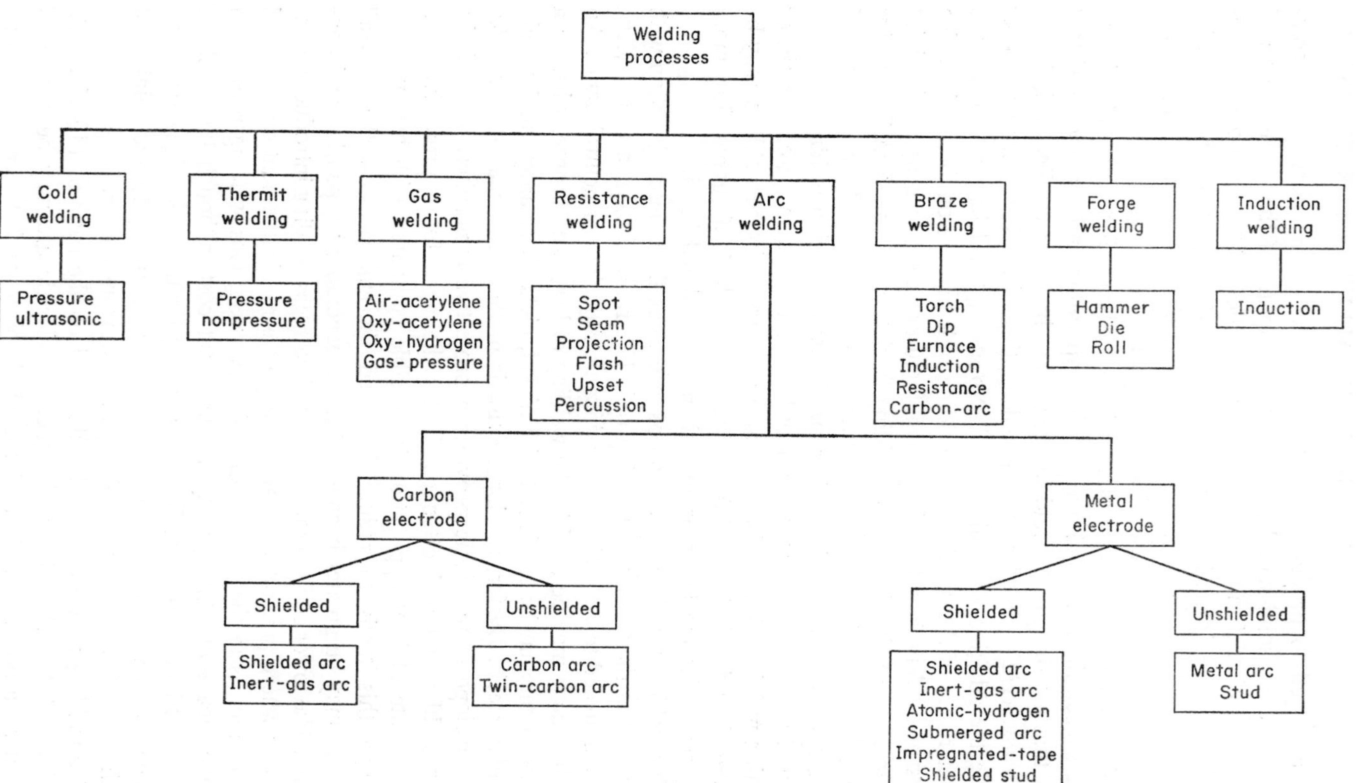

FIG. 21.2 Classification of welding processes.

ducer arranged to produce shear vibrations at the interface of the component parts. A solid-state bond occurs by rupture of the surface film and localized plastic deformation with interpenetration of the component members. The process is particularly adapted to the joining of thin foil and wire beyond the thickness limitations found in resistance welding. There are many applications in electronics, jewelry, heat exchangers, and airframe and missile components. This process is well adapted for joining aluminum, stainless steel, titanium, and many high-temperature and corrosion-resistant alloys that are difficult to weld to each other by other welding techniques.

Fusion welding involves the actual melting of the base metal and may include the addition of a filler metal. No attempt will be made to list the characteristics of the multitude of welding rods that are on the market. The interested student is referred to manufacturer's literature or to the joint specifications of the American Welding Society and American Society for Testing Materials for details of composition, mechanical properties, and coding system for the various filler metals used in fusion welding.

The fusion-welding processes may be classified as gas, arc, thermit, or induction welding. A variation of brazing, known as braze welding, may be considered under fusion welding, since the method of preparation more nearly resembles the latter process than conventional brazing.

Gas welding usually employs an oxyacetylene flame in conjunction with filler metal in the form of a welding rod. The ratio of oxygen to acetylene may be adjusted to yield either an oxidizing, reducing, or neutral flame. An oxidizing flame may be used in the welding of brass and bronze, whereas a reducing flame may be employed with nickel, nickel-base alloys, and certain high-alloy steels. By far, the largest use is made of the neutral flame. An outer envelope in the flame provides a protective atmosphere around the molten weld metal and often eliminates the need for a protective flux. A lower rate of heat application is possible compared with arc welding. Although this tends to reduce localized thermal stresses, the over-all distortion may be greater because of more generalized expansion and contraction. The lower cooling rate associated with gas welding introduces complications where precipitation effects may cause embrittlement or may lead to corrosion, as in the welding of stainless steels. Gas welding provides a portable means of welding at low equipment cost, although the need for highly skilled operators usually makes the operating cost higher than for arc welding. The process is particularly well adapted to the joining of thin sheet or plate and to the welding of nonferrous metals.

Arc-welding processes may be classified on the basis of (1) electrode material and (2) shielding technique. An electric arc may be generated between a carbon electrode and the work piece or between two carbon electrodes, with or without the addition of filler metal. A more common method is to strike an arc between the welding rod and the work.

Shielding of the arc is essential for producing sound welds having optimum strength and ductility. Shielding techniques include the use of fluxes in the form of coated rod, powder, paste, impregnated paper, or the direct use of neutral or reducing gases surrounding the arc. Fluxes serve to prevent contact of the molten metal with the atmosphere and, hence, the solution of oxygen and nitrogen; to remove impurities from the molten weld deposit; to provide a slag blanket to decrease the cooling rate; and sometimes to stabilize the arc. Atomic hydrogen, helium, and argon gas-shielding methods provide freedom from slag inclusions and protect light gauge alloy steels, aluminum, magnesium, stainless steel, nickel-base alloys, etc., from oxidation during the welding process.

Arc welding provides a versatile means of joining both thin and heavy sections and is adapted to the automatic welding required in mass production. In general, arc welding permits faster welding with more localized heating and greater depth of penetration than is possible in gas welding.

Thermit welding is actually a metallurgical reduction process for producing molten iron by the exothermic reaction brought about by igniting an intimate mixture of iron oxide and aluminum powders with the occasional addition of alloying elements to produce alloy steels. The thermit process has a specialized use in fabrication and maintenance where a large quality of molten weld metal is required for joining large parts, such as in railroad rails, stern frames of ships, rolling mill pinions, etc.

Induction welding provides a means of bringing metal surfaces of small steel parts up to welding temperature by means of induction heating. The process offers a rapid, highly efficient method of joining thin steel parts without excessive oxidation or discoloration, although the high initial cost of the equipment can usually be absorbed only on a production basis.

Braze welding differs from brazing in that filler metal in the form of conventional brazing alloys is applied to fill "V" joints or to produce fillets instead of being drawn into the joint by capillary attraction. The source of heat may be either the oxyacetylene torch or the electric arc. The lower temperatures employed in braze welding allow increased welding speed with a ductile filler metal capable of reducing residual welding stresses in such applications as the joining and repairing of iron castings. Oxyacteylene heating does not intentionally fuse the parent metal, and the bonding is governed largely by alloying through atomic diffusion and intergranular penetration. Some fusion of the base metal may result from electric arc heating.

21.5 Structure of Fusion Welds. The inherent characteristics of the casting process are present in the formation of a fusion weld. Nucleation and growth tendencies are influenced by the chilling action of the parent metal to produce a variable grain size within the weld. The structure of a ferrous weld is further influenced by the principles of heat treatment associated with the allotropic transformations taking place in both the filler and base metals.

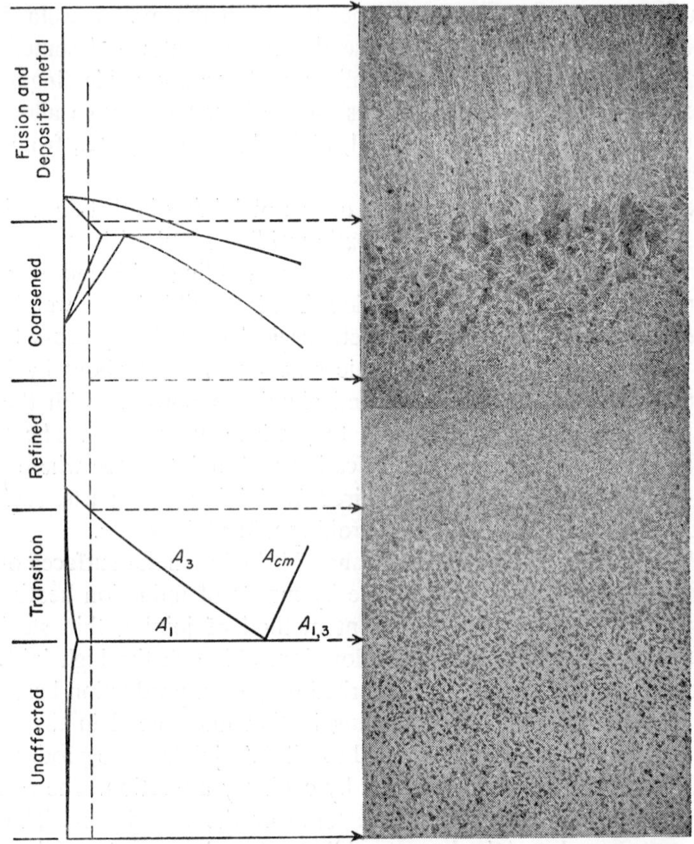

Fig. 21.3 Zones in fusion-welded, low-carbon steel joint (50×).

A polished and etched cross section of a fusion-welded, low-carbon steel joint is compared in Fig. 21.3 with a section of the iron-iron carbide diagram. Several distinct structural regions may be observed as representative of the unaffected, transition, refined, coarsened, fusion, and deposited metal zones. A study of the equilibrium diagram will explain the presence of each of these zones. In general, the several zones of a fusion weld are not sharply defined but blend together without a distinct line of demarcation.

They tend to be narrower when the heat application is highly concentrated, as in arc welding.

The *unaffected zone* represents the typical grain structure of the parent low-carbon steel which has not been heated to a high enough temperature to reach the critical range; hence, its structure is unchanged. In the *transition zone,* a temperature range exists between the A_1 and A_3 transformation temperatures where partial allotropic recrystallization takes place. The *refined zone* indicates a region heated to just above the A_3 temperature where grain refinement is completed and the finest grain structure exists. At higher temperatures above A_3, coarsening of the grain takes place as shown in the *coarsened zone.* At temperatures above the solidus, actual melting of the parent metal takes place as illustrated by the *fusion zone.* Whenever a filler metal is added to a fusion weld, a region will exist with a typical cast structure of coarse columnar grains known as the *deposited metal zone.* Each of these zones is shown at a higher magnification in Figs. 21.4 to 21.9 inclusive.

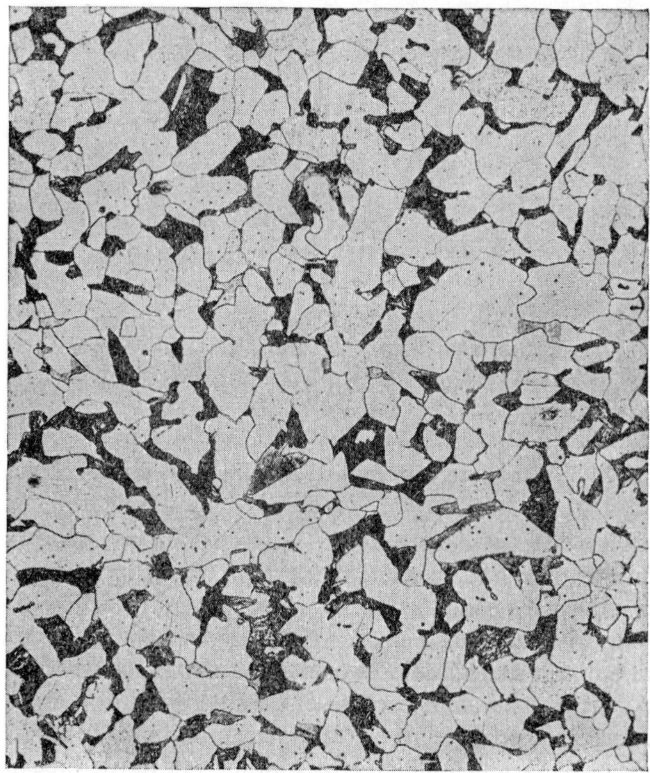

FIG. 21.4 Structure of unaffected zone in welded steel (250×).

Fig. 21.4 shows the structure of the unaffected zone consisting of a characteristic assembly of ferrite and pearlite. The transition zone is shown in Fig. 21.5 in which the ferrite grains have not been altered but the pearlite regions have been made much finer. This change was produced by heating into the critical range which transformed the pearlite into austenite, and by subsequent cooling reformed the pearlite. Complete recrystallization is

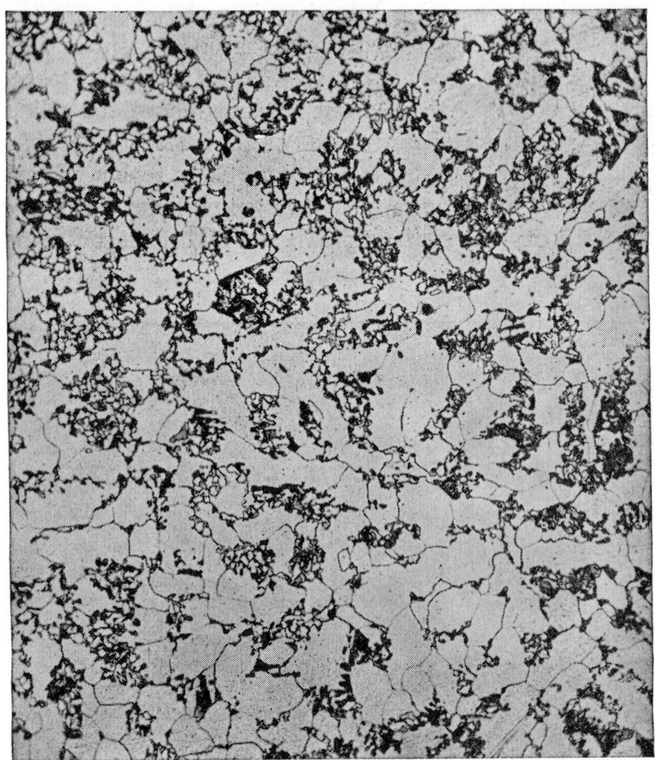

FIG. 21.5 Structure of transition zone in welded steel (250×).

shown in Fig. 21.6 in which the ferrite and pearlite areas are both much finer in forming from the austenite that existed at a temperature just above the upper critical temperature. The structure illustrated in Fig. 21.7 shows large regions of pearlite and smaller grains of ferrite. Examination of the pearlite at a higher magnification would show a finer structure than existed in the original pearlite areas because of the rate of cooling that prevailed. The very coarse structure near the fusion zone is shown in Fig. 21.8. The structure is of the Widmanstätten type with lines of ferrite breaking up the pearlite areas. This structure is common in a low-carbon steel in which transformation occurs from large austenite grains during a medium fast rate of cooling. The deposited metal, consisting of a columnar structure of ferrite and pearlite, is shown in Fig. 21.9.

These structures are characteristic of low-carbon steel and can be modified by the rate of cooling from the temperatures attained in the austenitic region. The presence of alloy elements in the steel and the carbon

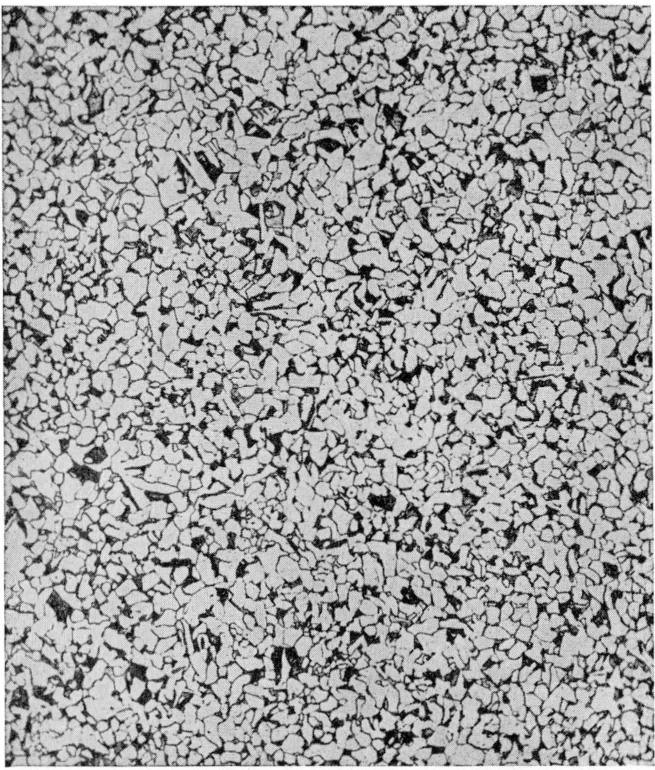

FIG. 21.6 Structure of refined zone in welded steel (250×).

content have a marked effect upon the structures produced in a weld for a given set of cooling conditions. This can be illustrated by referring to the continuous cooling transformation diagrams. Consider two steels, one a low-carbon steel and the other a low-alloy steel, for which the continuous cooling transformation diagrams are represented in Fig. 21.10 at *S*1 and *S*2. With cooling rate *A*, steel *S*1 will transform to pearlite on cooling while steel *S*2 will transform to martensite. In a weld, this latter condition will be unsatisfactory because of high internal stresses and the very brittle structure resulting from the low temperature transformation. Such a condition will usually produce cracks in the weld. In order to obtain satisfactory results, steel *S*2 must be cooled at a rate such as that indicated by curve *B*.

This brings up the problem of the composition of a steel in relation to welding. Plain carbon steel containing less than 0.30 per cent carbon and low-alloy steels with less than 0.15 per cent carbon are readily welded without the danger of forming brittle martensite. With larger proportions of carbon, difficulties are encountered which may be reduced by certain techniques. Any procedure that decreases the cooling rate tends to remove this difficulty. Probably the simplest method of reducing the cooling rate consists of putting more heat into the weld; however, there are limitations to this procedure. The joint can be preheated prior to welding; this will reduce the cooling rate considerably. These conditions are presented diagrammatically in Fig. 21.11.

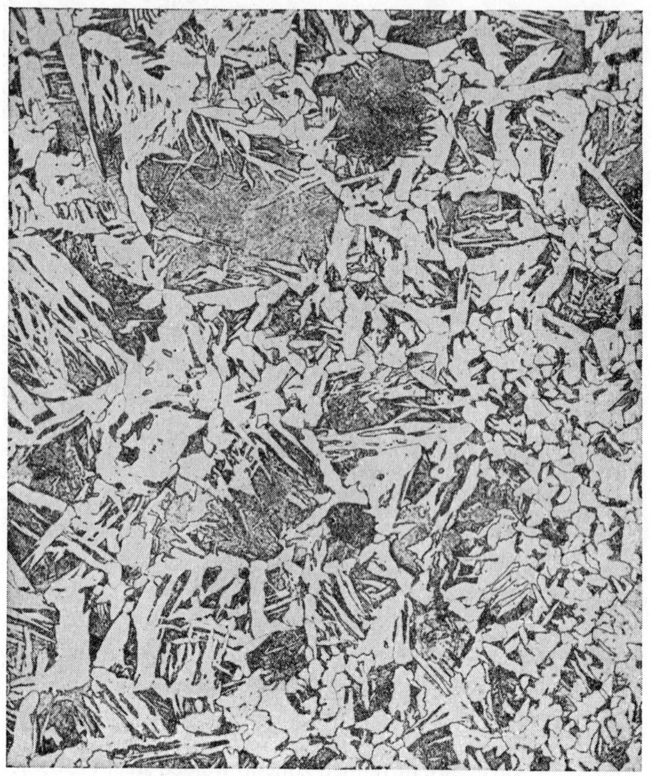

FIG. 21.7 Structure of coarsened zone in welded steel (250×).

The variable grain size and nonuniform structure associated with a fusion weld are not conducive to maximum strength and toughness. Some form of thermal treatment is generally employed to improve this condition in ferrous welds.

21.6 Thermal Treatment of Welds. The principal inherent weaknesses of the welding process are (1) localized heating and cooling which produces internal stresses or distortion and (2) coarse and variable grain size in the deposited and affected metal. Both of these factors lower the effective strength of the over-all structure and may cause immediate failure upon

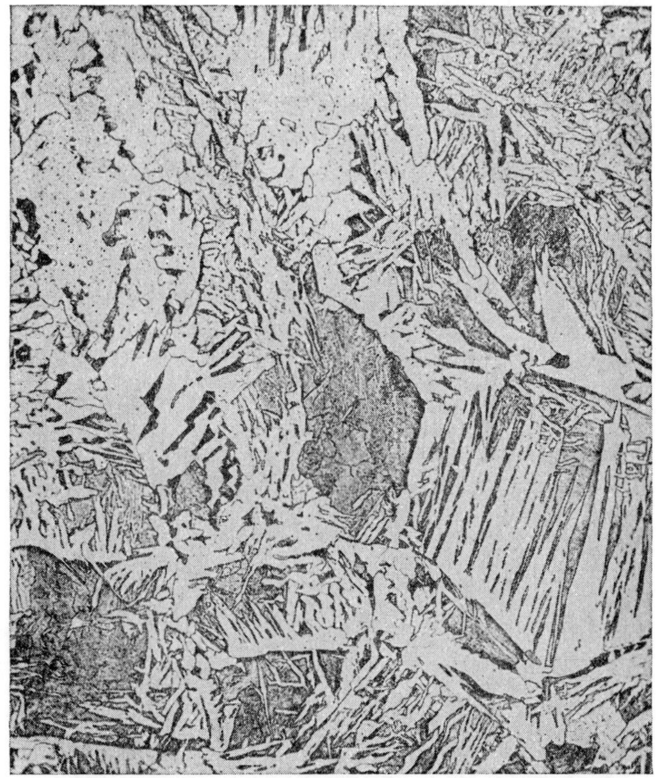

FIG. 21.8 Structure of coarsened and fusion zones in welded steel (250×).

cooling to room temperature or may exert a delayed reaction which becomes evident by application of service stresses. These conditions may be improved by certain thermal treatments, such as stress-relief annealing and full annealing.

Stress-relief annealing of ferrous alloys at temperatures below about 1000 F (540 C) followed by very slow cooling will relieve welding stresses which would otherwise cause possible cracking or distortion. No recrystallization of the grain structure occurs in this treatment. Mechanical peening of the weld deposit to distribute stress concentrations by plastic flow of the metal is often satisfactory in large structures for which furnaces are not available, such as buildings, bridges, ships, etc.

Full annealing of ferrous metals by heating to above the critical temperature range not only will effect complete stress relief but also will cause, at the same time, allotropic grain refinement of the weld. If the composition of the filler alloy is essentially the same as that of the base metal, then the

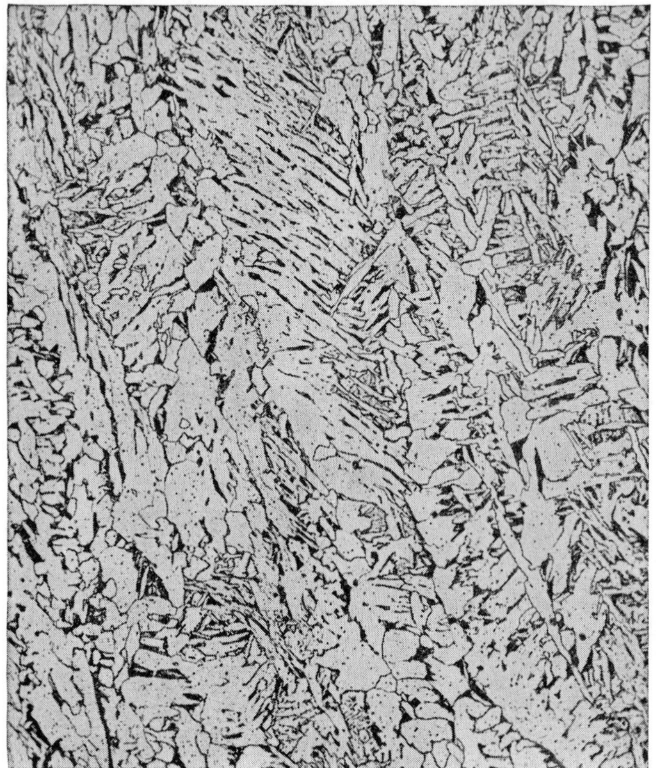

FIG. 21.9 Structure of deposited metal zone in welded steel (250×).

resultant grain size and structure should be homogeneous across the weld as a result of full annealing.

It should be recognized that it is not possible to refine the coarse as-cast grain structure of a nonferrous weld by any heat treatment. Furthermore, in the case of welding a cold-worked metal, a zone on either side of the weld will be recrystallized and subject to grain growth which cannot be subsequently corrected in nonferrous metals and alloys.

21.7 Weldability of Metals and Alloys. Weldability is a term which is used broadly to describe the capacity of a metal to be joined by welding into a structure that can perform in a satisfactory manner for an intended service. Most metallic materials can be joined by one or more of the welding processes. Although composition plays a dominant part in determining the weldability of a metal, other factors, including design, section size, service conditions, etc., influence the economical joining of metals by welding.

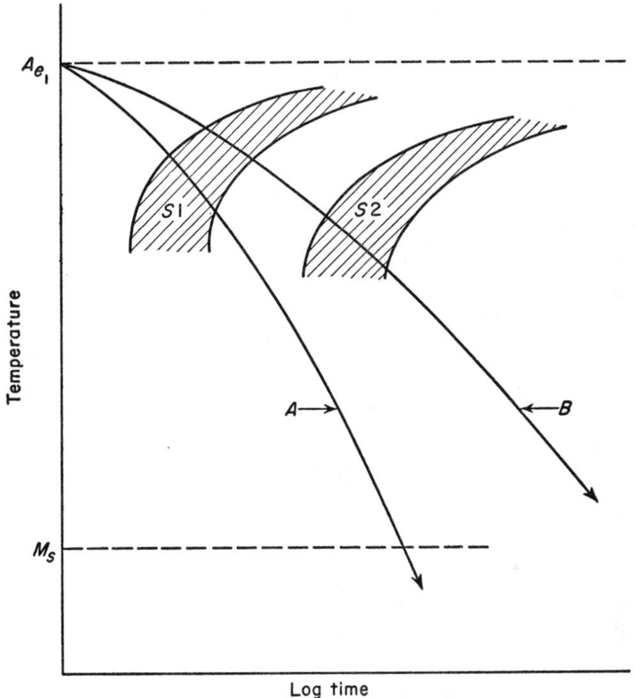

FIG. 21.10 Continuous cooling transformation diagrams for two steels (schematic).

The plain carbon steels containing less than 0.30 per cent carbon have excellent weldability and are the most important source of material for welded construction. Higher carbon steels require special techniques, such as preheating and postheating, to prevent cracking and the formation of martensite upon air cooling.

The influence of alloying elements on the weldability of hardenable alloy steels will be apparent from the earlier discussion of hardenability. The relatively soft, ductile welds produced in the low-carbon, low-alloy steels

Steel	Composition	Structures produced by cooling rate		
		A	B	C
I	Plain carbon - <0.30% C Low alloy - <0.15% C	Ferrite and Pearlite	Ferrite and Pearlite	Ferrite and Pearlite
II	Plain carbon - 0.30-0.50% C Low alloy - 0.15-0.30% C	Ferrite and Pearlite	Bainite	Martensite
III	Plain carbon - >0.50% C Low alloy - >0.30% C or alloy - >3.00%	Bainite	Martensite	Martensite

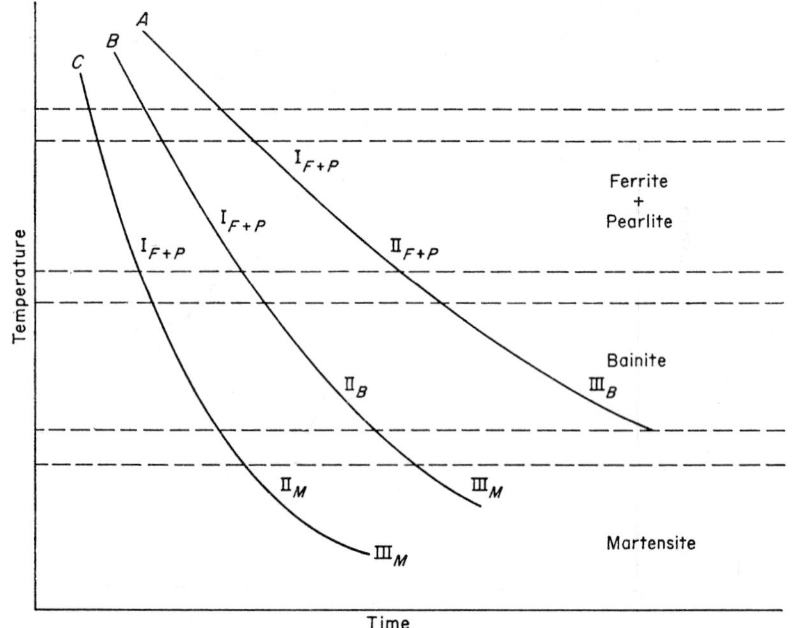

Cooling curve	Conditions producing cooling rates		
	Heat input	Section	Preheat
A	--	--	High
B	Large	Thin	Low
C	Small	Thick	None

FIG. 21.11 Effect of composition and cooling conditions on structures produced in welding.

permit relief of localized stresses by plastic flow of the weld metal. As the carbon content increases, difficulty is encountered in the welding of alloy steels.

Cast irons may be welded if precautions are taken to avoid localized overheating. Preheating is generally required, followed by very slow cooling to prevent chilling and nonuniform cooling which might lead to cracking. In view of the complex metallurgy of gray cast iron and its poor tensile and impact properties, the lower temperature braze-welding method is often employed to reduce the possibility of residual stresses.

The problems of sensitization and resulting intergranular corrosion in the austenitic stainless steels have been discussed in a previous chapter. The carbon content of these steels must be less than 0.08 per cent, or additions of stabilizing elements, such as columbium, are required to prevent sensitization adjacent to the weld. When carbides are precipitated, they may be redissolved by heating the part to a temperature of about 1850 F (1010 C), followed by quenching. With such a treatment, consideration must be given to possible warping.

The high-chromium ferritic stainless steels and irons suffer from possible precipitation of the sigma phase during welding and usually require a subsequent heat treatment to restore ductility.

Nonferrous metals may require special welding techniques to produce satisfactory joints, depending on the particular characteristics of the base metal, such as ease of oxidation, thermal conductivity, precipitation-hardening effects, affinity toward gases, etc.

21.8 Welding Defects. The defects usually associated with gas- or arc-welded construction may be divided into (1) dimensional, (2) structural, and (3) property deficiencies.

Dimensional defects include warpage, incorrect weld size and profile, etc. *Warpage* results from distortion caused by thermal stresses during the welding operation and may be controlled by the use of welding fixtures and suitable thermal treatment. Incorrect weld size and profile result from nonconformity with specifications and improper welding technique. In the case of fillet welds, *weld size* is specified as the length of the leg of the largest isosceles triangle that can be inscribed in the cross section of the weld, as illustrated by the ideal profile in Fig. 21.12. In groove welds, size usually refers to the depth of the groove below the surface of the members being joined. Undersized welds fail to satisfy the structural requirements of the joint and may be corrected by additional beads of filler metal. Oversized welds, on the other hand, tend to create a notch effect by undue localized stiffening of the members and introduce the additional problems of increased shrinkage and a tendency toward structural defects. *Weld profile* defects associated with fillet welds include excessive concavity, insufficient leg, undercut, overlap, excessive convexity, etc., as shown in Fig. 21.12. Although the simple fillet weld has been employed for purposes of illustration, many of these profile defects may occur in other types of welded joints. Excessive concavity and insufficient leg result in deficient strength in the fillet. Undercut and overlap produce a notch effect which tends to lower the overall joint strength by stress concentration. Undercutting is the localized

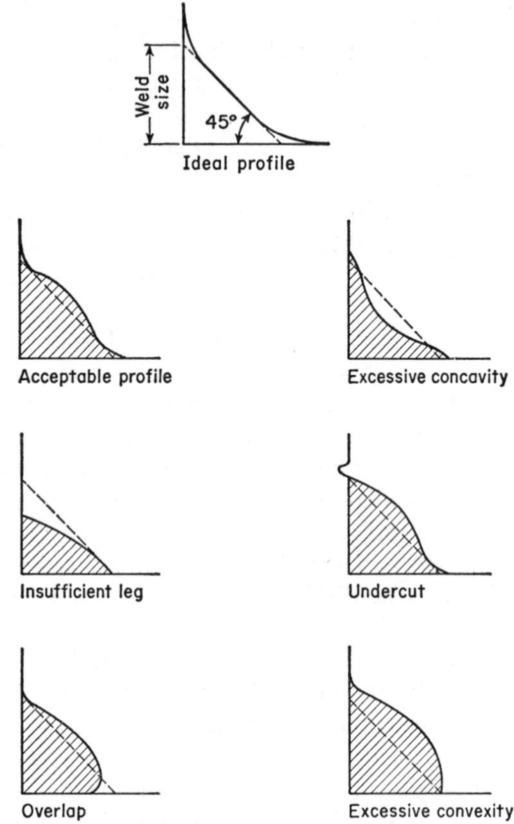

FIG. 21.12 Profile defects in fillet weld.

melting or oxidizing of the parent metal adjacent to the weld. Excessive convexity is more conducive to the entrapment of slag inclusions, the formation of voids, and shrinkage in the weld metal.

Structural defects include porosity, nonmetallic inclusions, incomplete fusion of the parent metal, lack of penetration, cracking, surface defects, etc. Fig. 21.13 shows a schematic representation of some of the more prominent structural defects found in fillet welds. *Porosity* may be caused by gaseous products of chemical reactions taking place in the molten weld metal or by the release of dissolved gases. *Nonmetallic inclusions* may exist in the form of oxides formed by reaction of the molten metal with the atmosphere or as actual slag particles produced by fluxes or rod coatings which are entrapped in large weld deposits. *Incomplete fusion* of the parent metal results from inadequate removal of oxide films, insufficient tem-

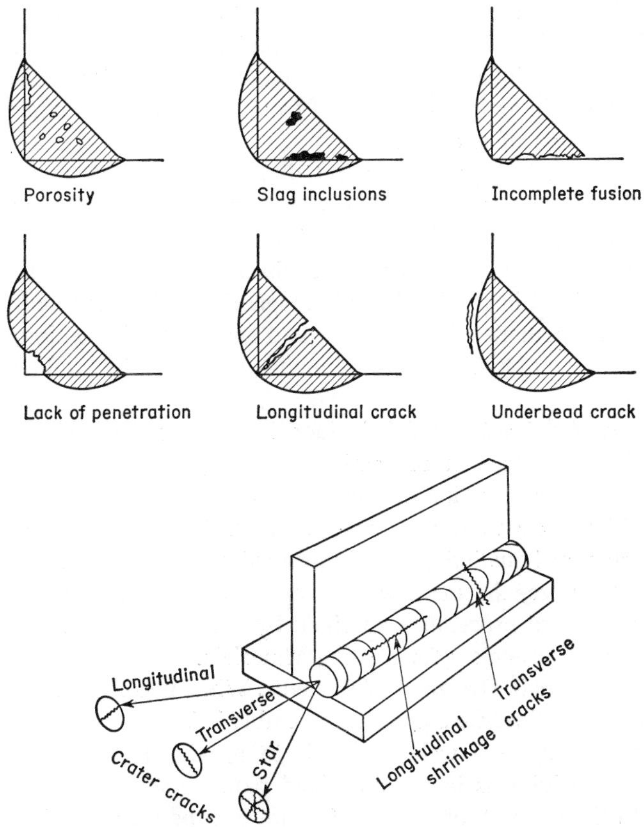

FIG. 21.13 Structural defects in fillet weld.

perature, or improper manipulation of the heat source. *Lack of penetration* occurs when the deposited metal and the parent metal are not completely fused together at the base or root of the joint. This condition is sometimes caused by the filler metal "bridging" between the members to be joined without attaining fusion in the root area. *Shrinkage cracks* may occur in the weld metal in the form of (1) transverse, (2) longitudinal, or (3) crater cracks. *Transverse cracks* usually form in rigid joints, whereas *longi-*

tudinal cracks originate by shrinkage of the weld metal or by propagation of crater cracks. *Crater cracks* are associated with the high heat concentration in arc welding. Cracks in the base metal resulting from the welding operation are usually longitudinal and are frequently found in connection with the welding of hardenable ferrous alloys, unless precautions are taken to preheat and postheat the metal. *Underbead cracking* is believed to be closely related to the presence of hydrogen in shielded arc welding; it may be reduced by the use of low-hydrogen electrodes. *Surface defects* include surface holes and surface irregularities. Surface holes are generally caused by improper technique in metal arc welding. Surface irregularities show poor workmanship and usually detract from the appearance of a weld without causing a loss of strength.

Visual inspection serves to detect dimensional defects and provides a guide in the preliminary evaluation of weld quality when coordinated with the past experience gained by other inspection methods. Nondestructive inspection techniques that are used to discover surface discontinuities not apparent on visual inspection include magnetic particle and fluorescent- or dye-penetrant methods. Radiographic and supersonic inspection are methods that may be employed to determine the presence of internal macroscopic defects.

Property deficiencies, which include inadequate mechanical properties or corrosion resistance in comparison with the parent metal, may result from the improper selection of filler metal or processing technique.

The ability of a welder to make a satisfactory weld is determined by means of standard qualifying tests prescribed by the American Welding Society. Furthermore, the mechanical properties of welded joints are determined by standard procedures specified by the American Welding Society and the American Society for Testing Materials.

21.9 Factors in Design of Weldments. When two or more components are joined by welding, the composite assembly is known as a *weldment.* The aim of every successful weld design is to provide at least an equivalent strength in the joint as in the unwelded parent metal. The ratio of weld strength to the strength of unwelded metal is expressed as *joint efficiency.*

A properly designed joint takes into account such factors as economical production, minimum residual stress, controlled distortion, freedom from corrosive attack, direction and nature of service stresses, etc. The actual choice of the best method of joint construction is influenced by cost of preparation, accessibility, penetration, stress distribution, strength requirements, etc. Detailed recommendations for joint construction and standard welding symbols have been prepared by the American Welding Society and are given in the *Welding Handbook,* together with codes and standards applicable to welding practice.

Welded construction should be of simple design with gradual changes in section size to provide for smooth stress flow through the weldment. Fillet welds should be small for maximum economy and strength. Butt welds are generally preferred where fatigue and impact resistance is important. Lap joints are usually eliminated in highly stressed members. Weldments should be designed to permit the economical use of welding fixtures for positioning component parts, and every effort should be made to permit welding in a horizontal position.

Weldments have gained an economical advantage over other methods of fabrication in many instances through the use of manual or machine oxygen cutting. In principle, *oxygen cutting* preheats the surface of the steel to its ignition temperature at about 1600 F (870 C) and then subjects the preheated surface to a concentrated jet of pure oxygen, which produces a cutting action by the rapid and localized oxidation of the steel. Close dimensional tolerances may be held by means of machine oxygen cutting, and this process competes with machining in the contour cutting of low-carbon steel plate. Steels containing more than 0.30 per cent carbon are flame-hardened in the cutting operation and require preheating and postheating torches to follow the path of the cutting blowtorch. Nonferrous metals and alloys are more difficult to cut by oxygen-cutting methods since they do not oxidize and produce fusible oxides as readily as steel, and the cutting action results largely from inefficient melting of the metal.

21.10 Weldments vs. Castings and Forgings. Welding offers certain advantages over casting and forging, provided the given part has been designed specifically for welding. The welding process is particularly adapted to:

1. Large and irregular shapes.
2. Lightweight construction.
3. Weight distribution.
4. Uninterrupted stress-flow design.
5. Experimental design at lower cost.
6. Possible multimetal construction.
7. Elimination of patterns or forging dies.
8. Reduction in machining cost.
9. Possible increased rate of production.

In contrast to the above advantages, certain disadvantages may detract from the economical effectiveness of weldments, such as:

1. Human element.
2. Metal preparation.

3. Welding fixtures.
4. Weld defects.
5. Special heat treatments.
6. Inspection.

REFERENCES

Henry and Claussen, *Welding Metallurgy,* revised by Linnert, American Welding Society, New York, 1949.

Procedure Handbook of Arc Welding Design and Practice, Lincoln Electric Co., Cleveland, Ohio, 1950.

The Oxyacetylene Handbook, Linde Air Products Co., New York, 1943.

Welding Handbook, 4th ed. Section 1, 1957; Section 2, 1958; Section 3, 1960; Section 4, 1961; American Welding Society, New York.

QUESTIONS

1. Define soldering.
2. How does brazing differ from soldering?
3. What is welding?
4. What are the two general types of solders?
5. What alloys are generally used for soft solders?
6. What is the function of a flux in soldering operations?
7. What materials are commonly used for soldering fluxes?
8. What are the three general types of brazing alloys?
9. What are the characteristics of the most satisfactory brazing alloys?
10. How is the thickness of the joint in brazing related to strength?
11. What are the three principal methods of brazing?
12. What are the two general classes of welding?
13. How are materials joined by plastic welding?
14. Indicate briefly the processes that are classed as plastic welding.
15. What processes are included under fusion welding?
16. Describe the structure of a weld in a low-carbon steel in relation to position in the weld.
17. What consideration must be given to the composition of steel that is to be fusion-welded?
18. What are the principal inherent weaknesses of the welding process?
19. What is the purpose of the stress-relief annealing of welds?
20. What factors must be considered in determining the weldability of a metal?
21. What are the three general types of defects found in fusion-welded joints?
22. What factors must be considered in the design of weldments?
23. Indicate some of the advantages and disadvantages of welding in comparison with casting and forging.

XXII

METALLURGY OF NUCLEAR ENGINEERING

22.1 Nuclear Engineering. The great developments in physics that have ultimately made possible the production of power through nuclear reactions has placed a tremendous demand upon the general field of materials, and in particular on the field of metallurgy. An entire field of engineering known as nuclear engineering has grown up which is very largely concerned with the design, construction, and operation of reactors. The aim of most of these reactors is for the production of power, either in stationary units or in a vehicle.

There are some very fundamental problems associated with materials for reactors. The limitations of these materials serve as the limitations for the entire design of the reactor, and the engineer is forced to make certain compromises. In making these compromises, he is faced with the need to give attention to health hazards that would be associated with a failure if it occurred.

In this chapter an attempt will be made to present some of the fundamental aspects of the utilization of metals in reactors. No attempt will be made to produce a catalogue of metals utilized for various parts of reactors, since they will change from year to year.

22.2 The Nucleus. A discussion of the atom was presented in Chapter II. The particles of which the nucleus is composed were discussed briefly. The *nucleus* contains two kinds of fundamental particles, protons and neutrons. The number of *protons* in the nucleus is always equal to the number of electrons outside the nucleus. Any change in the number of electrons calls for a change in the number of protons, and such a change effects an alteration in the chemical and physical properties of the element. Any change that may occur in the number of neutrons only changes the nuclear properties and not the chemical and physical properties.

The *neutron* is a particle without charge having a mass approximately the same as that of a proton. The mass of a proton is 1.00759 amu (atomic mass unit), while the mass of the neutron is 1.00898 amu. The proton has a single positive charge which balances the single negative charge of an electron.

The *atomic weight* of an element is determined by the sum of the number of protons and the number of neutrons. For example, fluorine has nine protons, nine electrons, and ten neutrons. The atomic number of fluorine is nine and the atomic weight is 19.

When two atoms have the same number of protons but a different number of neutrons, one is said to be an *isotope* of the other. Such atoms would have the same atomic number but different atomic weights; hence they would have the same chemical properties. Hydrogen, for example, may have three isotopes with masses one, two, and three, known as hydrogen, deuterium, and tritium, respectively.

22.3 Radiations. Radioactive elements or elements that are reacted upon by neutrons may emit three types of radiation: alpha, beta, and gamma rays.

Alpha rays are positively charged helium nuclei with two protons and two neutrons. The alpha rays are easily absorbed and travel a maximum distance of only some seven centimeters through air. Alpha rays can be stopped by a sheet of paper or an aluminum foil 0.06 millimeters thick.

The *beta rays* are electrons that move at a velocity near that of the velocity of light. They are singularly charged negative, the same as that of an electron, and they also have the mass of an electron. In general, beta particles have high speed and, therefore, have greater penetrating power than alpha rays. They will travel about 100 times as far in air before being slowed down sufficiently to ionize.

Gamma rays are electromagnetic radiation similar to X-rays, but they differ in their origin in contrast to X-rays. When a neutron is absorbed in a nucleus, or when an alpha particle is emitted, there is an energy imbalance and the nucleus is in an excited state. The excess energy that is released appears as gamma rays. In contrast to this, X-rays originate outside the nucleus. They are derived from the fact that an electron jumps from one shell to another. In the case of producing *X-rays,* an electron strikes an atom, dislodging an electron in a shell close to the nucleus. Another electron jumps from another shell to fill the vacancy and in so doing an X-ray is emitted. Gamma rays are emitted in discrete energies and the spectra consists of a series of sharply defined wavelengths. Gamma rays have the capacity of penetrating several inches of lead; thus adequate shielding against these rays is necessary.

22.4 Fission. The process of the *fission* of an atom involves the division of the nucleus into two approximately equal parts. For example, uranium is split into two fragments when it is bombarded by neutrons. When this split occurs, either two or three neutrons are emitted. Uranium-235 is fissionable by low kinetic energy neutrons, while uranium-238 is fissionable only by high kinetic energy neutrons. The fission of uranium-235 by a neutron converting mass to energy is indicated as follows:

Mass of U^{235}	235.124 amu
Mass of neutron added	1.009
Total	236.133
Mass after fission	
Mo^{95}	94.945 amu
La^{139}	138.955
2-neutrons	2.018
Total	235.918
Mass converted to energy	
Initial mass	236.133
Final mass	235.918
Net change	0.215 amu

Since 1 amu equals 931 Mev (1 million electron volts), therefore during the fission operation an energy of $0.215 \times 931 = 200$ Mev per atom is released. If properly controlled, this energy is available for utilization as a power source. The energy of most fission neutrons is approximately 2 Mev.

The *thermal neutrons,* which have an energy of about 0.025 ev, are the most effective in causing fission. Therefore it is desirable to slow down or to moderate the *fast neutrons* which normally have an energy of approximately 2 Mev. The moderation can be attained by mixing the U^{235} with graphite or water, which have low atomic weights. The moderator will absorb some of the neutrons.

The process, of course, is useless if only one reaction occurs. In order to produce power continuously it is necessary to have a chain reaction which requires a proper balance of the emission and absorption of neutrons. If a moderator absorbs too many neutrons or too many neutrons escape from the system, the reaction is quite likely to stop. Therefore, in a reactor, the self-sustaining chain reaction must be maintained under control in order to keep the reactor in proper operation. This is accomplished by *control rods* of boron or cadmium which absorb neutrons in large amounts. These control rods can be put into the reactor or withdrawn to produce the necessary control. When a reactor reaches stable control it is said to be *critical*. If more than one neutron is left over per fission cycle and provision has not been made to reduce the number of neutrons to one per fission cycle, then a supercritical condition will exist. An explosion may then result which will cause the core of the reactor to experience serious melting and damage.

When a neutron produces fission in U^{235}, either two or three neutrons will be produced by the fission process, and these neutrons can behave in one of three ways. The first alternative is that the neutron will pass through the moderator where it will be slowed down to be produced as a thermal neutron. It can then produce a second fission when it contacts the nucleus of a U^{235} atom. The second alternative is that the neutron will be absorbed by the control rod. This means that that neutron is lost. The third alternative is that the neutron may escape and be stopped by the shielding of the reactor. The neutron that takes the first alternate will produce a second fission process and therefore produce two or three more neutrons; again one may be absorbed in the structure by the poisons, that is, by those elements that absorb neutrons. It may not be moderated and, therefore, as a fast neutron striking a U^{238} atom it will be captured, but will produce neptunium which will decay in a very short time to plutonium. It may produce normal fission.

To minimize the loss of neutrons, the number lost will depend upon the surface area of the core. The lowest possible ratio of surface area to volume is secured by means of a sphere, but from a practical point of view this is not always the easiest type of vessel to produce and, therefore, in most instances a right cylinder is employed. It is frequently customary to surround the core with a *reflector,* which is a material that can reflect back a part of the neutrons that try to escape. This is not a process of very high efficiency. Actually it is not a reflector in the sense of optics, but produces a reflection through the process of diffusion.

22.5 Reactors. The reactor for the production of power serves as a heat source. Heat is derived from the energy that is liberated in the fission reaction. Most of this heat is produced by the process of slowing down the *fission fragments*. When fission occurs, the fragments are forced apart with tremendous energy and, in passing through the surrounding material, the kinetic energy of these fragments is transferred into heat. The heat is removed from the mass by an adequate coolant, which takes the heat from the core of the reactor to either another heat transfer point or directly to the unit that will use the heat.

Reactors may be classified first according to the energy of the neutrons employed, and second according to the physical configuration of the moderator and fuel. Three general neutron energies may be employed:

Neutron Energy	
Thermal	0.025-1.0 ev
Intermediate	1.0 ev-0.1 Mev
Fast	0.1 Mev or more

The thermal reactor may have fuel arranged in a heterogeneous or in a homogeneous manner. In the *heterogeneous reactor* the fuel is distributed in a geometrical lattice within the moderator, whereas in a *homogeneous reactor* the fuel is evenly dispersed throughout the moderator.

Reactors may also be classified in accordance with the coolant that is used, i.e., the method by which the heat is removed from the core. These may be gas, liquid metal, or water (pressurized or boiling).

22.6 Reactor Elements. The elements required in a reactor can be more simply visualized by reference to Fig. 22.1. While this is somewhat simplified, it gives an indication of the essential features of the reactor for which materials must be supplied. The key part of the reactor is the fuel,

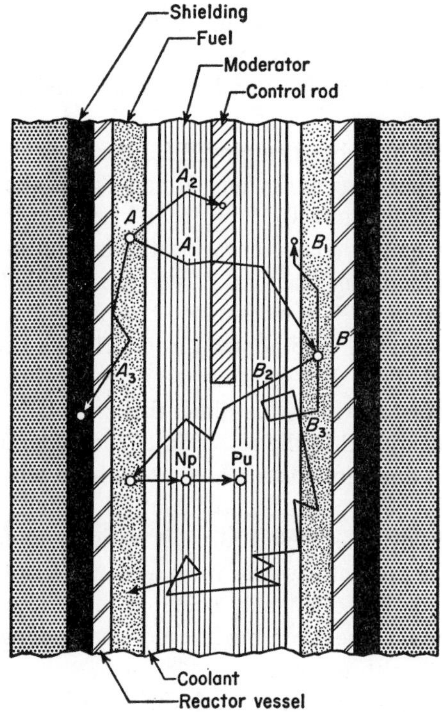

FIG. 22.1 Fissioning of U-235 atoms in a simplified reactor. (*Metals for Nuclear Reactors,* Am. Soc. for Metals, Metals Park, Ohio, 1959)

which in this instance may be U^{235} contained within some metallic substance called a *cladding*. There must be a coolant immediately adjacent to and surrounding the fuel elements. The character of the coolant will have an important bearing upon the cladding employed for the fuel.

A *moderator* must be provided for the purpose of slowing down the neutrons so that they can do their proper job. These moderators must have a minimum of neutron absorption capacity; therefore attention must be given to the material of which it is constructed. The control of the whole operation rests with the control rods, which must also be of the proper material. The entire unit consisting of the fuel, the moderator, and the control rod must be contained within a vessel that will hold it together with adequate safety. This assemblage may be designated as the reactor vessel. However, the reactor vessel will not absorb gamma radiation, and hence it must be surrounded by adequate shielding.

22.7 Cladding. The *cladding* that separates the fuel and the coolant is subjected to strenuous requirements. It must resist corrosion under rather unusual circumstances. It must withstand intense radiation. It must be a good thermal conductor in order to convey heat as efficiently and effectively as possible from the fuel to the coolant. It must not have the capability of capturing many neutrons. It must have sufficient strength within the temperature range encountered. It must be weldable and formable to permit fabrication.

Several materials have been employed for the cladding of fuel elements. Aluminum of 99.5 per cent purity and a magnesium alloy (magnox) containing 1 per cent aluminum, 0.05 per cent beryllium, and 0.05 per cent calcium have been employed for encasing the natural uranium slugs used in plutonium-manufacturing reactors. Power reactors that use uranium enriched with fissionable U^{235} have employed zircaloy-2 which is zirconium containing 1.5 per cent tin, 0.15 per cent iron, 0.10 per cent chromium, and 0.05 per cent nickel. Type 304 stainless steel has also been used in installations that utilize a fast flux, i.e., the direct fission of U^{235} or Pu^{239}.

Care must be exercised in the selection of the material for the cladding to make sure that it does not contain any atoms from which unstable radioactive isotopes may be produced upon neutron capture. If stainless steel, for example, is employed as the cladding, it must be completely free of cobalt. If any cobalt is present it would form a long-life isotope Co^{60}. In view of the fact that some corrosion may occur at the surface of the cladding, even a minute amount, the Co^{60} might enter the cooling system and thus eventually cause a health hazard.

22.8 Coolant Problems. In general, the size of the reactor core is limited by the heat transfer and not by nuclear consideration. A reactor in which natural uranium with 1 per cent U^{235} is employed is not limited by the cooling channels, because there is such a large mass of fuel. With U^{235} a large volume of fuel is required to make the reactor critical. The heat-transfer problem becomes quite serious. One method of securing cooling is by utilizing water. When the water boils, the fission process is automatically regulated, since the steam bubbles decrease the moderating effect of the liquid water, the reaction tends to stop. The steam produced in this case may be used directly in any power process.

A liquid metal offers a distinct advantage as the heat-transfer agent. However, this system usually requires a heat exchanger to convey the heat from the liquid metal to the working fluid. The liquid metal does not operate under high pressure, though, as does water.

Liquid metal is likely to lead to some corrosion problems. Sodium is usually employed because it has a low cross section and a low melting point. However, sodium becomes violently radioactive; therefore it is necessary to use an intermediate heat-transfer circuit. Liquid metals may attack the cladding material at the hottest point, transporting and depositing it at the position of the lowest temperature in the circuit. This action not only causes deterioration of the cladding but also plugging of the system. These difficulties can be eliminated if a cladding material is selected that will have low solubility in the coolant. Plugging can be reduced by utilizing cold traps by which the liquid metal is bled off through the trap at a lower temperature than any other part of the circuit. In this way the deposition occurs in the trap.

All the usual considerations of corrosion are present in these installations. However, even greater care must be exercised because of the radioactive nature of the coolant and the possible health hazard from any upset of the core.

22.9 Control Rods. The primary function of the *control rods* is to capture neutrons. These are moved into or out of the reactor core in order to control the reaction. The method of indicating neutron absorption capacity is to consider the comparative possibility of an atom absorbing incident neutrons.

The possibility of a neutron colliding with a nucleus of an atom is indicated by the nuclear *"cross section."* If a neutron enters a one-centimeter cube perpendicular to one face, each nucleus contained within the cubic centimeter provides a certain area designated by the symbol σ. The total area provided by the nuclei in the unit volume is $N\sigma$, where N is the number of atoms per unit volume. The chance of a collision is the ratio of area of the nuclei to the total area. Therefore the chance of collision is expressed as $N\sigma$. The cross section for thermal neutrons in U^{235} is $650 \times 10^{-24}\text{cm}^2$. For the sake of simplicity, the unit of cross section has been designated as a *"barn,"* which is equal to 10^{-24}cm^2.

Since the purpose of the control rods is to capture neutrons, the material utilized should have a very high cross section. For this purpose, boron or cadmium may be employed. Cadmium has a natural cross section of 3000 barns. B^{10} has a cross section of 3990 barns, while natural boron has a cross section of only 750 barns. Other materials, such as aluminum and stainless steel, have cross sections of only 0.22 barns and 2.75 barns, respectively. In some instances, steel containing about 2% boron has been employed for control rods. There are problems concerned with the materials for these control rods: for example, B^{10} transmutes to Li^7 plus an alpha particle when it captures a neutron. The alpha particle picks up an electron and becomes an atom of helium gas. After some time the helium collects in the porosities of the rod and tends to diffuse out, but this is likely to produce high internal pressures which may cause mechanical difficulties with the rod. There is a tendency to utilize cadmium for the capture of thermal neutrons.

22.10 Containing Vessel. To the casual observer, the material in which the reactor is encased may be inconsequential provided it possesses adequate strength. The vessel containing the reactor will be subjected to neutron bombardment, and possibly high temperature. Neutron bombardment may have adverse effects upon the properties of the material.

If a fast neutron hits an atom, it may knock it out of its proper position in its space lattice. When this occurs, the atom is called a *"knocked-on" atom*. This "knocked-on" atom may also knock a second atom, and every "knocked-on" atom creates a vacancy in the lattice. The "knocked-on" atom is forced into an interstitial position on the lattice, where it remains unless the temperature is high enough to permit its diffusion back into the vacancy that was created. If it remains in the interstitial position, causing distortion of the lattice, the material will be made harder and will have a lower thermal conductivity. This means that the material will have lower ductility, possibly somewhat higher yield strength and ultimate strength, but it will have a smaller percentage elongation. This is illustrated in Table 22-I.

TABLE 22-I. EFFECT OF RADIATION ON MECHANICAL PROPERTIES OF NICKEL, COPPER, AND ALUMINUM

(Neutron flux 10^{20} *nvt*, where *nv* is the average flux of neutrons per cm^2 per sec and *t* is the time in sec)

Property	Percentage Change in Properties		
	Nickel	Copper	Aluminum
Tensile strength	+56%	+11%	+3%
Elongation	−55%	−40%	−5%
Impact value	−23%	−59%	+62%
Hardness	+82%	+28%	−12%

The low-carbon steels, which exhibit the *ductile-brittle transition* characteristic, are markedly affected by radiation. This is illustrated in Fig. 22.2. The transition temperature has been increased from approximately +25 F to +125 F with an accompanying decrease in the impact values by radiation. Many of the effects of neutron bombardment can be corrected by heating the material so that vacancies can be eliminated by diffusion, although this is not always convenient to do when the material is a part of the reactor.

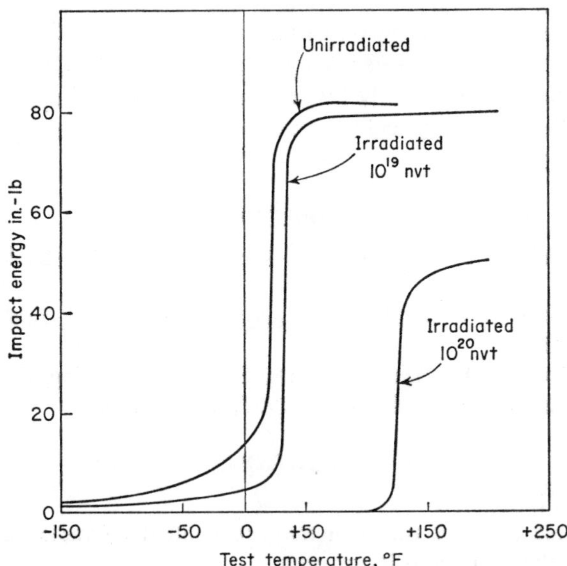

FIG. 22.2 Notch impact energy vs. temperature of pressure vessel steel before and after irradiation. (*Metals for Nuclear Reactors,* Am. Soc. for Metals, Metals Park, Ohio, 1959)

Neutron bombardment may also have an effect in bringing about transformation of metastable phases to stable phases. For example, if a material that is susceptible to precipitation hardening is used, it may be made to over-age by neutron bombardment. This means a loss in strength that could, of course, be serious. The neutron bombardment on stainless steels of certain compositions may effect a transition from austenite to ferrite or the sigma phase. This action can have adverse effects on the corrosion-resistant characteristics of the steel.

Another effect of radiation on some of the materials is to produce a local extremely high temperature which essentially causes localized melting for an extremely short time, followed by extremely rapid cooling. This thermally disturbed portion of the material is called a *"thermal spike."*

This region may consist of only a matter of a hundred atoms in diameter. The thermal spikes also lead to internal strain which can adversely affect the properties.

It is, therefore, obvious that one must be very careful in selecting the material that is to encase the core of the reactor. All of these factors must be taken into consideration.

22.11 Shielding. While the *shielding* of reactors is not primarily done with metals, it may be pointed out here that some means must be provided to absorb the gamma rays that are emitted from the fission fragments. Neutrons will escape through the vessel and must be absorbed. While neutrons are readily absorbed by water, they are most commonly absorbed by a concrete structure, particularly by stationary reactors. The shielding problem is one that is of particular concern in the utilization of reactors for mobile purposes.

REFERENCES

Metals for Nuclear Reactors, American Society for Metals, Metals Park, Ohio, 1959.

Murray, *Introduction to Nuclear Engineering,* Prentice-Hall, Englewood Cliffs, N. J., 1955.

Radiation Effects on Materials, American Society for Testing Materials, Philadelphia, Pa., Vol. 1, 1956; Vol. 2, 1957; Vol. 3, 1958.

QUESTIONS

1. Define an isotope.
2. What three types of radiation may be emitted by radioactive elements? Describe the characteristics of each.
3. Explain the difference in the mechanism of producing X-rays and gamma rays.
4. What is the process of fission?
5. How are fast neutrons moderated to produce thermal neutrons?
6. What is the condition for criticality in a reactor?
7. Explain the process of producing power by the nuclear reaction.
8. What is the function of control rods?
9. In what three ways may neutrons behave in the fission process?
10. How is the heat produced in a reactor?
11. How are reactors classified?
12. How may the fuel be arranged in a thermal reactor?
13. How are reactors classified according to the coolant employed?
14. What are the requirements of the fuel cladding utilized in a reactor?
15. What are the requirements for the cladding for fuel elements?
16. What materials are utilized for the cladding of fuel elements?
17. What is the objection to using any cobalt in the cladding material?

18. What is the advantage, and what are the disadvantages, of the use of liquid metal as a heat-transfer agent in reactors?
19. What is the function of control rods in a reactor?
20. What is meant by the "nuclear cross section"?
21. What is a barn?
22. What difficulties may be encountered if steel containing boron is employed for control rods?
23. What material is commonly used for the control rods?
24. What is meant by a "knocked-on" atom?
25. What is the effect of a "knocked-on" atom in a structure, insofar as properties are concerned?
26. What is the effect of radiation on the ductile-brittle transition characteristics of low-carbon steel?
27. What is a "thermal spike"?
28. What is the effect of the "thermal spike" on the properties of the material?
29. What considerations must be given to the material selected to encase the core of a reactor?
30. Why is it necessary to provide shielding for reactors, and what material is commonly used for this purpose?

APPENDIX A

AISI-SAE NONRESULFURIZED CARBON STEEL COMPOSITIONS—1957

AISI No.	SAE No.	Ladle Chemical Composition Limits, per cent			
		C	Mn	P max	S max
C 1008	1008	0.10 max.	0.25/0.50	0.040	0.050
C 1010	1010	0.08/0.13	0.30/0.60	0.040	0.050
C 1011	—	0.08/0.13	0.60/0.90	0.040	0.050
C 1012	1012	0.10/0.15	0.30/0.60	0.040	0.050
C 1015	1015	0.13/0.18	0.30/0.60	0.040	0.050
C 1016	1016	0.13/0.18	0.60/0.90	0.040	0.050
C 1017	1017	0.15/0.20	0.30/0.60	0.040	0.050
C 1018	1018	0.15/0.20	0.60/0.90	0.040	0.050
C 1019	1019	0.15/0.20	0.70/1.00	0.040	0.050
C 1020	1020	0.18/0.23	0.30/0.60	0.040	0.050
C 1021	1021	0.18/0.23	0.60/0.90	0.040	0.050
C 1022	1022	0.18/0.23	0.70/1.00	0.040	0.050
C 1023	1023	0.20/0.25	0.30/0.60	0.040	0.050
C 1024	1024	0.19/0.25	1.35/1.65	0.040	0.050
C 1025	1025	0.22/0.28	0.30/0.60	0.040	0.050
C 1026	1026	0.22/0.28	0.60/0.90	0.040	0.050
C 1027	1027	0.22/0.29	1.20/1.50	0.040	0.050
C 1029	—	0.25/0.31	0.60/0.90	0.040	0.050
C 1030	1030	0.28/0.34	0.60/0.90	0.040	0.050
C 1031	—	0.28/0.34	0.30/0.60	0.040	0.050
C 1033	1033	0.30/0.36	0.70/1.00	0.040	0.050
C 1035	1035	0.32/0.38	0.60/0.90	0.040	0.050
C 1036	1036	0.30/0.37	1.20/1.50	0.040	0.050
C 1037	1037	0.32/0.38	0.70/1.00	0.040	0.050
C 1038	1038	0.35/0.42	0.60/0.90	0.040	0.050
C 1039	1039	0.37/0.44	0.70/1.00	0.040	0.050
C 1040	1040	0.37/0.44	0.60/0.90	0.040	0.050
C 1041	1041	0.36/0.44	1.35/1.65	0.040	0.050
C 1042	1042	0.40/0.47	0.60/0.90	0.040	0.050
C 1043	1043	0.40/0.47	0.70/1.00	0.040	0.050
C 1045	1045	0.43/0.50	0.60/0.90	0.040	0.050
C 1046	1046	0.43/0.50	0.70/1.00	0.040	0.050
C 1049	1049	0.46/0.53	0.60/0.90	0.040	0.050
C 1050	1050	0.48/0.55	0.60/0.90	0.040	0.050
C 1051	—	0.45/0.56	0.85/1.15	0.040	0.050
C 1052	1052	0.47/0.55	1.20/1.50	0.040	0.050
C 1053	—	0.48/0.55	0.70/1.00	0.040	0.050
C 1055	1055	0.50/0.60	0.60/0.90	0.040	0.050
C 1060	1060	0.55/0.65	0.60/0.90	0.040	0.050
C 1070	1070	0.65/0.75	0.60/0.90	0.040	0.050
C 1078	1078	0.72/0.85	0.30/0.60	0.040	0.050
C 1080	1080	0.75/0.88	0.60/0.90	0.040	0.050
C 1084	1084	0.80/0.93	0.60/0.90	0.040	0.050
C 1085	1085	0.80/0.93	0.70/1.00	0.040	0.050
C 1086	1086	0.80/0.93	0.30/0.50	0.040	0.050
C 1090	1090	0.85/0.98	0.60/0.90	0.040	0.050
C 1095	1095	0.90/1.03	0.30/0.50	0.040	0.050

Silicon. When silicon is required, the following ranges and limits are commonly used:

Standard Steel Designations	Silicon Ranges or Limits
Up to C 1015 excl.	0.10 max
C 1015 to C 1025 incl.	0.10 max, 0.10/0.20, or 0.15/0.30
Over C 1025	0.10/0.20, or 0.15/0.30

Copper. When required, copper is specified as an added element to a standard steel.

Lead. When required, lead is specified as an added element to a standard steel.

APPENDIX B

AISI-SAE FREE-CUTTING STEEL COMPOSITIONS—1957

Acid Bessemer Resulfurized Carbon Steels

AISI No.	SAE No.	Ladle Chemical Composition Limits, per cent			
		C	Mn	P	S
B 1111	1111	0.13 max	0.60/0.90	0.07/0.12	0.08/0.15
B 1112	1112	0.13 max	0.70/1.00	0.07/0.12	0.16/0.23
B 1113	1113	0.13 max	0.70/1.00	0.07/0.12	0.24/0.33

Open-Hearth Resulfurized Carbon Steels

AISI No.	SAE No.	Ladle Chemical Composition Limits, per cent			
		C	Mn	P max	S
C 1108	1108	0.08/0.13	0.50/0.80	0.040	0.08/0.13
C 1109	1109	0.08/0.13	0.60/0.90	0.040	0.08/0.13
C 1110	—	0.08/0.13	0.30/0.60	0.040	0.08/0.13
C 1113	—	0.10/0.16	1.00/1.30	0.040	0.24/0.33
C 1115	1115	0.13/0.18	0.60/0.90	0.040	0.08/0.13
C 1116	1116	0.14/0.20	1.10/1.40	0.040	0.16/0.23
C 1117	1117	0.14/0.20	1.00/1.30	0.040	0.08/0.13
C 1118	1118	0.14/0.20	1.30/1.60	0.040	0.08/0.13
C 1119	1119	0.14/0.20	1.00/1.30	0.040	0.08/0.13
C 1120	1120	0.18/0.23	0.70/1.00	0.040	0.08/0.13
C 1125	—	0.22/0.28	0.60/0.90	0.040	0.08/0.13
C 1126	1126	0.23/0.29	0.70/1.00	0.040	0.08/0.13
C 1132	1132	0.27/0.34	1.35/1.65	0.040	0.08/0.13
C 1137	1137	0.32/0.39	1.35/1.65	0.040	0.08/0.13
C 1138	1138	0.34/0.40	0.70/1.00	0.040	0.08/0.13
C 1139	1139	0.35/0.43	1.35/1.65	0.040	0.12/0.20
C 1140	1140	0.37/0.44	0.70/1.00	0.040	0.08/0.13
C 1141	1141	0.37/0.45	1.35/1.65	0.040	0.08/0.13
C 1144	1144	0.40/0.48	1.35/1.65	0.040	0.24/0.33
C 1145	1145	0.42/0.49	0.70/1.00	0.040	0.04/0.07
C 1146	1146	0.42/0.49	0.70/1.00	0.040	0.08/0.13
C 1151	1151	0.48/0.55	0.70/1.00	0.040	0.08/0.13

Silicon. When silicon is required, the following ranges and limits are commonly used:

Standard Steel Designations	Silicon Ranges or Limits
Up to C 1113 excl.	0.10 max
C 1113 and over	0.10 max, 0.10/0.20 or 0.15/0.30

Lead. When required, lead is specified as an added element to a standard steel.

APPENDIX C

(See notes at end of Appendix C)

AISI-SAE ALLOY STEEL COMPOSITIONS—1958

AISI Number	Chemical Composition Ranges and Limits, per cent								Corresponding SAE Number
	C	Mn	P max	S max	Si	Ni	Cr	Mo	
				Manganese Steels					
1330	0.28/0.33	1.60/1.90	0.040	0.040	0.20/0.35	—	—	—	1330
1335	0.33/0.38	1.60/1.90	0.040	0.040	0.20/0.35	—	—	—	1335
1340	0.38/0.43	1.60/1.90	0.040	0.040	0.20/0.35	—	—	—	1340
1345	0.43/0.48	1.60/1.90	0.040	0.040	0.20/0.35	—	—	—	1345
				Nickel-Chromium Steels					
3140	0.38/0.43	0.70/0.90	0.040	0.040	0.20/0.35	1.10/1.40	0.55/0.75	—	3140
E3310	0.08/0.13	0.45/0.60	0.025	0.025	0.20/0.35	3.25/3.75	1.40/1.75	—	3310
				Molybdenum Steels					
4012	0.09/0.14	0.75/1.00	0.040	0.040	0.20/0.35	—	—	0.15/0.25	4012
4023	0.20/0.25	0.70/0.90	0.040	0.040	0.20/0.35	—	—	0.20/0.30	4023
4024	0.20/0.25	0.70/0.90	0.040	0.035/0.050	0.20/0.35	—	—	0.20/0.30	4024
4027	0.25/0.30	0.70/0.90	0.040	0.040	0.20/0.35	—	—	0.20/0.30	4027
4028	0.25/0.30	0.70/0.90	0.040	0.035/0.050	0.20/0.35	—	—	0.20/0.30	4028
4037	0.35/0.40	0.70/0.90	0.040	0.040	0.20/0.35	—	—	0.20/0.30	4037
4042	0.40/0.45	0.70/0.90	0.040	0.040	0.20/0.35	—	—	0.20/0.30	4042
4047	0.45/0.50	0.70/0.90	0.040	0.040	0.20/0.35	—	—	0.20/0.30	4047
4063	0.60/0.67	0.75/1.00	0.040	0.040	0.20/0.35	—	—	0.20/0.30	4063
				Chromium-Molybdenum Steels					
4118	0.18/0.23	0.70/0.90	0.040	0.040	0.20/0.35	—	0.40/0.60	0.08/0.15	4118
4130	0.28/0.33	0.40/0.60	0.040	0.040	0.20/0.35	—	0.80/1.10	0.15/0.25	4130
4135	0.33/0.38	0.70/0.90	0.040	0.040	0.20/0.35	—	0.80/1.10	0.15/0.25	4135
4137	0.35/0.40	0.70/0.90	0.040	0.040	0.20/0.35	—	0.80/1.10	0.15/0.25	4137
4140	0.38/0.43	0.75/1.00	0.040	0.040	0.20/0.35	—	0.80/1.10	0.15/0.25	4140
4142	0.40/0.45	0.75/1.00	0.040	0.040	0.20/0.35	—	0.80/1.10	0.15/0.25	4142
4145	0.43/0.48	0.75/1.00	0.040	0.040	0.20/0.35	—	0.80/1.10	0.15/0.25	4145
4147	0.45/0.50	0.75/1.00	0.040	0.040	0.20/0.35	—	0.80/1.10	0.15/0.25	4147
4150	0.48/0.53	0.75/1.00	0.040	0.040	0.20/0.35	—	0.80/1.10	0.15/0.25	4150
				Nickel-Chromium-Molybdenum Steels					
4320	0.17/0.22	0.45/0.65	0.040	0.040	0.20/0.35	1.65/2.00	0.40/0.60	0.20/0.30	4320
4337	0.35/0.40	0.60/0.80	0.040	0.040	0.20/0.35	1.65/2.00	0.70/0.90	0.20/0.30	4337
E4337	0.35/0.40	0.65/0.85	0.025	0.025	0.20/0.35	1.65/2.00	0.70/0.90	0.20/0.30	—
4340	0.38/0.43	0.60/0.80	0.040	0.040	0.20/0.35	1.65/2.00	0.70/0.90	0.20/0.30	4340
E4340	0.38/0.43	0.65/0.85	0.025	0.025	0.20/0.35	1.65/2.00	0.70/0.90	0.20/0.30	E4340
4422	0.20/0.25	0.70/0.90	0.040	0.040	0.20/0.35	—	—	0.35/0.45	4422
4427	0.24/0.29	0.70/0.90	0.040	0.040	0.20/0.35	—	—	0.35/0.45	4427
4520	0.18/0.23	0.45/0.65	0.040	0.040	0.20/0.35	—	—	0.45/0.60	4520
				Nickel-Molybdenum Steels					
4615	0.13/0.18	0.45/0.65	0.040	0.040	0.20/0.35	1.65/2.00	—	0.20/0.30	4615
4617	0.15/0.20	0.45/0.65	0.040	0.040	0.20/0.35	1.65/2.00	—	0.20/0.30	4617
4620	0.17/0.22	0.45/0.65	0.040	0.040	0.20/0.35	1.65/2.00	—	0.20/0.30	4620
4621	0.18/0.23	0.70/0.90	0.040	0.040	0.20/0.35	1.65/2.00	—	0.20/0.30	4621
4718	0.16/0.21	0.70/0.90	0.040	0.040	0.20/0.35	0.90/1.20	0.35/0.55	0.30/0.40	4718
4720	0.17/0.22	0.50/0.70	0.040	0.040	0.20/0.35	0.90/1.20	0.35/0.55	0.15/0.25	4720
4815	0.13/0.18	0.40/0.60	0.040	0.040	0.20/0.35	3.25/3.75	—	0.20/0.30	4815
4817	0.15/0.20	0.40/0.60	0.040	0.040	0.20/0.35	3.25/3.75	—	0.20/0.30	4817
4820	0.18/0.23	0.50/0.70	0.040	0.040	0.20/0.35	3.25/3.75	—	0.20/0.30	4820

APPENDIX C (*Continued*)

Chromium Steels

AISI Number	Chemical Composition Ranges and Limits, per cent								Corresponding SAE Number
	C	Mn	P max	S max	Si	Ni	Cr	Mo	
5015	0.12/0.17	0.30/0.50	0.040	0.040	0.20/0.35	—	0.30/0.50	—	5015
5046	0.43/0.50	0.75/1.00	0.040	0.040	0.20/0.35	—	0.20/0.35	—	5046
5115	0.13/0.18	0.70/0.90	0.040	0.040	0.20/0.35	—	0.70/0.90	—	5115
5120	0.17/0.22	0.70/0.90	0.040	0.040	0.20/0.35	—	0.70/0.90	—	5120
5130	0.28/0.33	0.70/0.90	0.040	0.040	0.20/0.35	—	0.80/1.10	—	5130
5132	0.30/0.35	0.60/0.80	0.040	0.040	0.20/0.35	—	0.75/1.00	—	5132
5135	0.33/0.38	0.60/0.80	0.040	0.040	0.20/0.35	—	0.80/1.05	—	5135
5140	0.38/0.43	0.70/0.90	0.040	0.040	0.20/0.35	—	0.70/0.90	—	5140
5145	0.43/0.48	0.70/0.90	0.040	0.040	0.20/0.35	—	0.70/0.90	—	5145
5147	0.45/0.52	0.70/0.95	0.040	0.040	0.20/0.35	—	0.85/1.15	—	5147
5150	0.48/0.53	0.70/0.90	0.040	0.040	0.20/0.35	—	0.70/0.90	—	5150
5155	0.50/0.60	0.70/0.90	0.040	0.040	0.20/0.35	—	0.70/0.90	—	5155
5160	0.55/0.65	0.75/1.00	0.040	0.040	0.20/0.35	—	0.70/0.90	—	5160
E50100	0.95/1.10	0.25/0.45	0.025	0.025	0.20/0.35	—	0.40/0.60	—	50100
E51100	0.95/1.10	0.25/0.45	0.025	0.025	0.20/0.35	—	0.90/1.15	—	51100
E52100	0.95/1.10	0.25/0.45	0.025	0.025	0.20/0.35	—	1.30/1.60	—	52100
Chromium-Vanadium Steel									
6118	0.16/0.21	0.50/0.70	0.040	0.040	0.20/0.35	—	0.50/0.70	0.10/0.15	6118
6120	0.17/0.22	0.70/0.90	0.040	0.040	0.20/0.35	—	0.70/0.90	0.10 Min	6120
6150	0.48/0.53	0.70/0.90	0.040	0.040	0.20/0.35	—	0.80/1.10	0.15 Min	6150
Nickel-Chromium-Molybdenum—Triple-Alloy Steels									
8115	0.13/0.18	0.70/0.90	0.040	0.040	0.20/0.35	0.20/0.40	0.30/0.50	0.08/0.15	8115
8615	0.13/0.18	0.70/0.90	0.040	0.040	0.20/0.35	0.40/0.70	0.40/0.60	0.15/0.25	8615
8617	0.15/0.20	0.70/0.90	0.040	0.040	0.20/0.35	0.40/0.70	0.40/0.60	0.15/0.25	8617
8620	0.18/0.23	0.70/0.90	0.040	0.040	0.20/0.35	0.40/0.70	0.40/0.60	0.15/0.25	8620
8622	0.20/0.25	0.70/0.90	0.040	0.040	0.20/0.35	0.40/0.70	0.40/0.60	0.15/0.25	8622
8625	0.23/0.28	0.70/0.90	0.040	0.040	0.20/0.35	0.40/0.70	0.40/0.60	0.15/0.25	8625
8627	0.25/0.30	0.70/0.90	0.040	0.040	0.20/0.35	0.40/0.70	0.40/0.60	0.15/0.25	8627
8630	0.28/0.33	0.70/0.90	0.040	0.040	0.20/0.35	0.40/0.70	0.40/0.60	0.15/0.25	8630
8637	0.35/0.40	0.75/1.00	0.040	0.040	0.20/0.35	0.40/0.70	0.40/0.60	0.15/0.25	8637
8640	0.38/0.43	0.75/1.00	0.040	0.040	0.20/0.35	0.40/0.70	0.40/0.60	0.15/0.25	8640
8642	0.40/0.45	0.75/1.00	0.040	0.040	0.20/0.35	0.40/0.70	0.40/0.60	0.15/0.25	8642
8645	0.43/0.48	0.75/1.00	0.040	0.040	0.20/0.35	0.40/0.70	0.40/0.60	0.15/0.25	8645
8650	0.48/0.53	0.75/1.00	0.040	0.040	0.20/0.35	0.40/0.70	0.40/0.60	0.15/0.25	8650
8655	0.50/0.60	0.75/1.00	0.040	0.040	0.20/0.35	0.40/0.70	0.40/0.60	0.15/0.25	8655
8660	0.55/0.65	0.75/1.00	0.040	0.040	0.20/0.35	0.40/0.70	0.40/0.60	0.15/0.25	8660
8720	0.18/0.23	0.70/0.90	0.040	0.040	0.20/0.35	0.40/0.70	0.40/0.60	0.20/0.30	8720
8735	0.33/0.38	0.75/1.00	0.040	0.040	0.20/0.35	0.40/0.70	0.40/0.60	0.20/0.30	—
8740	0.38/0.43	0.75/1.00	0.040	0.040	0.20/0.35	0.40/0.70	0.40/0.60	0.20/0.30	8740
8742	0.40/0.45	0.75/1.00	0.040	0.040	.20/0.35	0.40/0.70	0.40/0.60	0.20/0.30	8742
8822	0.20/0.25	0.75/1.00	0.040	0.040	0.20/0.35	0.40/0.70	0.40/0.60	0.30/0.40	8822
E9310	0.08/0.13	0.45/0.65	0.025	0.025	0.20/0.35	3.00/3.50	1.00/1.40	0.08/0.15	9310
9840	0.38/0.43	0.70/0.90	0.040	0.040	0.20/0.35	0.85/1.15	0.70/0.90	0.20/0.30	9840
9850	0.48/0.53	0.70/0.90	0.040	0.040	0.20/0.35	0.85/1.15	0.70/0.90	0.20/0.30	9850
Silicon-Manganese Steels									
9255	0.50/0.60	0.70/0.95	0.040	0.040	1.80/2.20	—	—	—	9255
9260	0.55/0.65	0.70/1.00	0.040	0.040	1.80/2.20	—	—	—	9260
9262	0.55/0.65	0.75/1.00	0.040	0.040	1.80/2.20	—	0.25/0.40	—	9262

APPENDIX C (*Continued*)

Boron Steels

These steels can be expected to have 0.0005 per cent minimum boron content.

AISI Number	Chemical Composition Ranges and Limits, per cent								Corresponding SAE Number
	C	Mn	P max	S max	Si	Ni	Cr	Mo	
50B40	0.38/0.43	0.75/1.00	0.040	0.040	0.20/0.35	—	0.40/0.60	—	50B40
50B44	0.43/0.48	0.75/1.00	0.040	0.040	0.20/0.35	—	0.40/0.60	—	50B44
50B46	0.43/0.50	0.75/1.00	0.040	0.040	0.20/0.35	—	0.20/0.35	—	50B46
50B50	0.48/0.53	0.75/1.00	0.040	0.040	0.20/0.35	—	0.40/0.60	—	50B50
50B60	0.55/0.65	0.75/1.00	0.040	0.040	0.20/0.35	—	0.40/0.60	—	50B60
51B60	0.55/0.65	0.75/1.00	0.040	0.040	0.20/0.35	—	0.70/0.90	—	51B60
81B45	0.43/0.48	0.75/1.00	0.040	0.040	0.20/0.35	0.20/0.40	0.35/0.55	0.08/0.15	81B45
86B45	0.43/0.48	0.75/1.00	0.040	0.040	0.20/0.35	0.40/0.70	0.40/0.60	0.15/0.25	86B45
94B15	0.13/0.18	0.75/1.00	0.040	0.040	0.20/0.35	0.30/0.60	0.30/0.50	0.08/0.15	94B15
94B17	0.15/0.20	0.75/1.00	0.040	0.040	0.20/0.35	0.30/0.60	0.30/0.50	0.08/0.15	94B17
94B30	0.28/0.33	0.75/1.00	0.040	0.040	0.20/0.35	0.30/0.60	0.30/0.50	0.08/0.15	94B30
94B40	0.38/0.43	0.75/1.00	0.040	0.040	0.20/0.35	0.30/0.60	0.30/0.50	0.08/0.15	94B40

NOTE 1. Grades shown in the above list with prefix letter E generally are manufactured by the basic electric furnace process. All others are normally manufactured by the basic open-hearth process but may be manufactured by the basic electric furnace process with adjustments in phosphorus and sulphur.

NOTE 2. The phosphorus and sulphur limitations for each process are as follows:

Basic electric furnace—0.025 maximum per cent   Acid electric furnace—0.050 maximum per cent
Basic open-hearth —0.040 maximum per cent   Acid open-hearth —0.050 maximum per cent

NOTE 3. Minimum silicon limit for acid open-hearth or acid electric furnace alloy steel is 0.15 per cent.

NOTE 4. Small quantities of certain elements are present in alloy steels which are not specified or required. These elements are considered as incidental and may be present to the following maximum amounts: copper, 0.35 per cent; nickel, 0.25 per cent; chromium, 0.20 per cent; and molybdenum, 0.06 per cent.

NOTE 5. Where minimum and maximum sulphur content is shown it is indicative of resulphurized steels.

APPENDIX D

TYPICAL HEAT TREATMENTS FOR AISI-SAE ALLOY STEELS

Steel No.	Normalizing Temperature (°F)	Annealing Temperature (°F)	Hardening Temperature (°F)	Quenching Medium
1330	1600-1700	1500-1600	1525-1575	Water or oil
1335, 1340	1600-1700	1500-1600	1525-1575	Oil
3130	1600-1700	—	1500-1550	Water or oil
3140	1600-1700	1450-1550	1500-1550	Oil
4037, 4042	1600-1700	1525-1575	1500-1575	Oil
4047	1600-1700	1450-1550	1500-1575	Oil
4063	1600-1700	1450-1550	1475-1550	Oil
4130	1600-1700	1450-1550	1600-1650	Water or oil
4137, 4140	1600-1700	1450-1550	1550-1600	Oil
4145, 4150	1600-1700	1450-1550	1500-1600	Oil
4340	1600-1700	1450-1550	1475-1525	Oil
5046	1600-1700	1450-1550	1475-1500	Oil
5130, 5132	1650-1750	1450-1550	1500-1550	Water, caustic solution, or oil
5135, 5140, 5145	1650-1750	1450-1550	1500-1550	Oil
5147, 5150	1650-1750	1450-1550	1475-1550	Oil
50100, 51100, 52100	—	1350-1450	1425-1475	Water
	—	1350-1450	1500-1600	Oil
6150	1650-1750	1550-1650	1600-1650	Oil
9255, 9260, 9262	1650-1750	1550-1650	1500-1650	Oil
8627, 8630	1600-1700	1450-1550	1550-1650	Water or oil
8637, 8640	1600-1700	1450-1550	1525-1575	Oil
8642, 8645, 8650	1600-1700	1450-1550	1500-1550	Oil
8655, 8660	1650-1750	1450-1550	1475-1550	Oil
8735, 8740	1600-1700	1450-1550	1525-1575	Oil
9840	1600-1700	1450-1550	1500-1550	Oil

APPENDIX E

HARDENABILITY BANDS FOR ALLOY STEELS

(*Courtesy of American Iron and Steel Institute*)

HARDENABILITY BAND 1330 H

C	Mn	Si				
.27 / .33	1.45 / 2.05	.20 / .35				

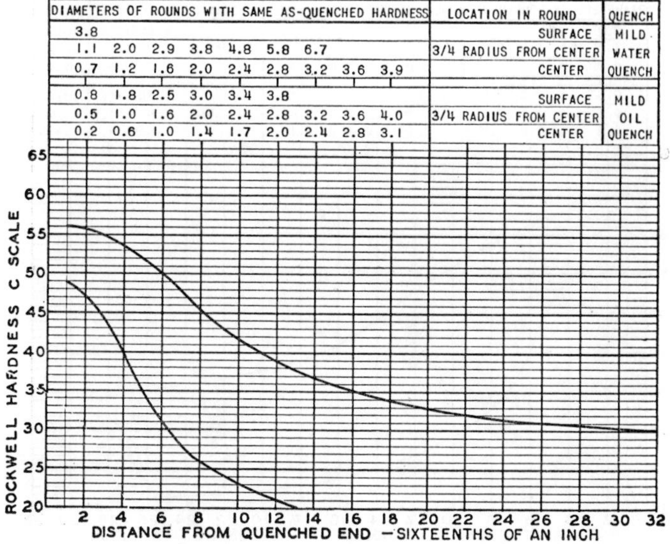

DIAMETERS OF ROUNDS WITH SAME AS-QUENCHED HARDNESS	LOCATION IN ROUND	QUENCH
3.8	SURFACE	MILD
1.1 2.0 2.9 3.8 4.8 5.8 6.7	3/4 RADIUS FROM CENTER	WATER
0.7 1.2 1.6 2.0 2.4 2.8 3.2 3.6 3.9	CENTER	QUENCH
0.8 1.8 2.5 3.0 3.4 3.8	SURFACE	MILD
0.5 1.0 1.6 2.0 2.4 2.8 3.2 3.6 4.0	3/4 RADIUS FROM CENTER	OIL
0.2 0.6 1.0 1.4 1.7 2.0 2.4 2.8 3.1	CENTER	QUENCH

HARDENABILITY BAND 1335 H

C	Mn	Si				
.32 / .38	1.45 / 2.05	.20 / .35				

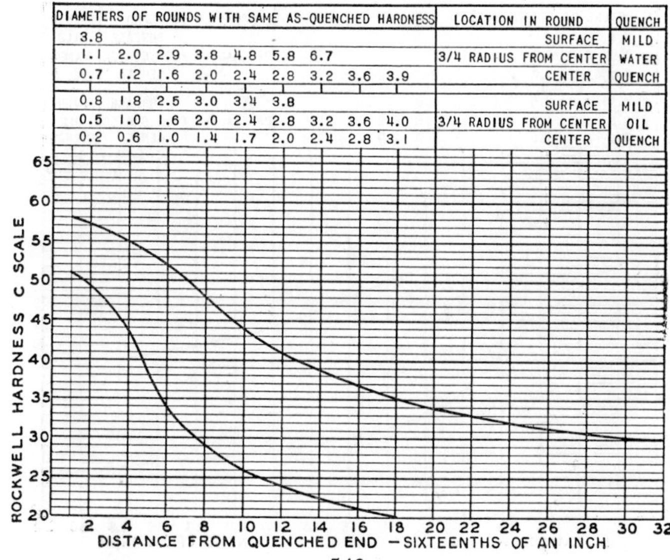

DIAMETERS OF ROUNDS WITH SAME AS-QUENCHED HARDNESS	LOCATION IN ROUND	QUENCH
3.8	SURFACE	MILD
1.1 2.0 2.9 3.8 4.8 5.8 6.7	3/4 RADIUS FROM CENTER	WATER
0.7 1.2 1.6 2.0 2.4 2.8 3.2 3.6 3.9	CENTER	QUENCH
0.8 1.8 2.5 3.0 3.4 3.8	SURFACE	MILD
0.5 1.0 1.6 2.0 2.4 2.8 3.2 3.6 4.0	3/4 RADIUS FROM CENTER	OIL
0.2 0.6 1.0 1.4 1.7 2.0 2.4 2.8 3.1	CENTER	QUENCH

HARDENABILITY BAND 1340 H

C	Mn	Si				
.37/.44	1.45/2.05	.20/.35				

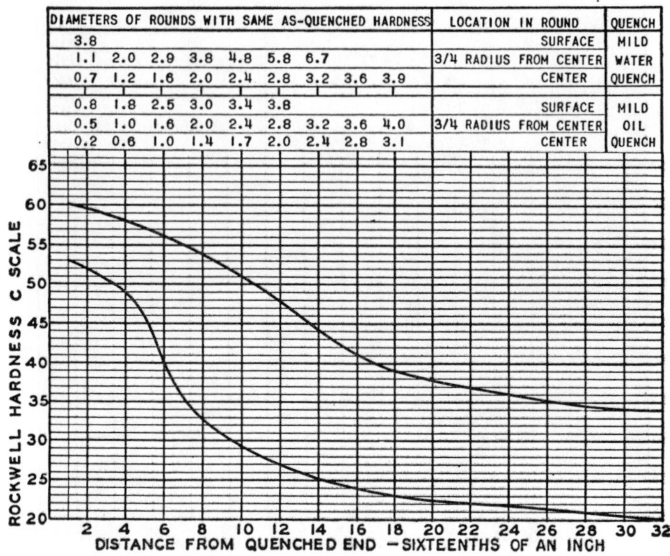

DIAMETERS OF ROUNDS WITH SAME AS-QUENCHED HARDNESS									LOCATION IN ROUND	QUENCH
3.8									SURFACE	MILD
1.1	2.0	2.9	3.8	4.8	5.8	6.7			3/4 RADIUS FROM CENTER	WATER
0.7	1.2	1.6	2.0	2.4	2.8	3.2	3.6	3.9	CENTER	QUENCH
0.8	1.8	2.5	3.0	3.4	3.8				SURFACE	MILD
0.5	1.0	1.6	2.0	2.4	2.8	3.2	3.6	4.0	3/4 RADIUS FROM CENTER	OIL
0.2	0.6	1.0	1.4	1.7	2.0	2.4	2.8	3.1	CENTER	QUENCH

HARDENABILITY BAND 3140 H

C	Mn	Si	Ni	Cr		
.37/.44	.60/1.00	.20/.35	1.00/1.45	.45/.85		

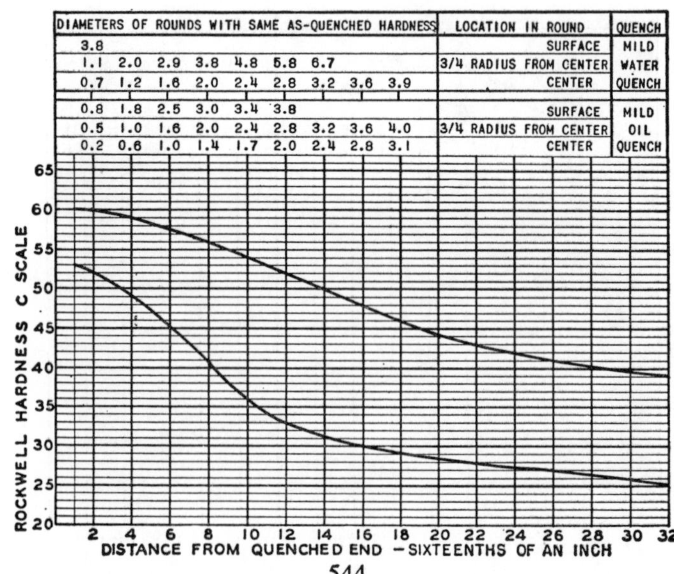

DIAMETERS OF ROUNDS WITH SAME AS-QUENCHED HARDNESS									LOCATION IN ROUND	QUENCH
3.8									SURFACE	MILD
1.1	2.0	2.9	3.8	4.8	5.8	6.7			3/4 RADIUS FROM CENTER	WATER
0.7	1.2	1.6	2.0	2.4	2.8	3.2	3.6	3.9	CENTER	QUENCH
0.8	1.8	2.5	3.0	3.4	3.8				SURFACE	MILD
0.5	1.0	1.6	2.0	2.4	2.8	3.2	3.6	4.0	3/4 RADIUS FROM CENTER	OIL
0.2	0.6	1.0	1.4	1.7	2.0	2.4	2.8	3.1	CENTER	QUENCH

HARDENABILITY BAND 3310 H

C	Mn	Si	Ni	Cr		
.07/.13	.30/.70	.20/.35	3.20/3.80	1.30/1.80		

DIAMETERS OF ROUNDS WITH SAME AS-QUENCHED HARDNESS	LOCATION IN ROUND	QUENCH
3.8	SURFACE	MILD
1.1 2.0 2.9 3.8 4.8 5.8 6.7	3/4 RADIUS FROM CENTER	WATER
0.7 1.2 1.6 2.0 2.4 2.8 3.2 3.6 3.9	CENTER	QUENCH
0.8 1.8 2.5 3.0 3.4 3.8	SURFACE	MILD
0.5 1.0 1.6 2.0 2.4 2.8 3.2 3.6 4.0	3/4 RADIUS FROM CENTER	OIL
0.2 0.6 1.0 1.4 1.7 2.0 2.4 2.8 3.1	CENTER	QUENCH

ROCKWELL HARDNESS C SCALE

65 60 55 50 45 40 35 30 25 20

2 4 6 8 10 12 14 16 18 20 22 24 26 28 30 32

DISTANCE FROM QUENCHED END — SIXTEENTHS OF AN INCH

HARDENABILITY BAND 4027 H & 4028 H •

C	Mn	Si			Mo	
.24/.30	.60/1.00	.20/.35			.20/.30	

•SULPHUR CONTENT 0.035/0.050

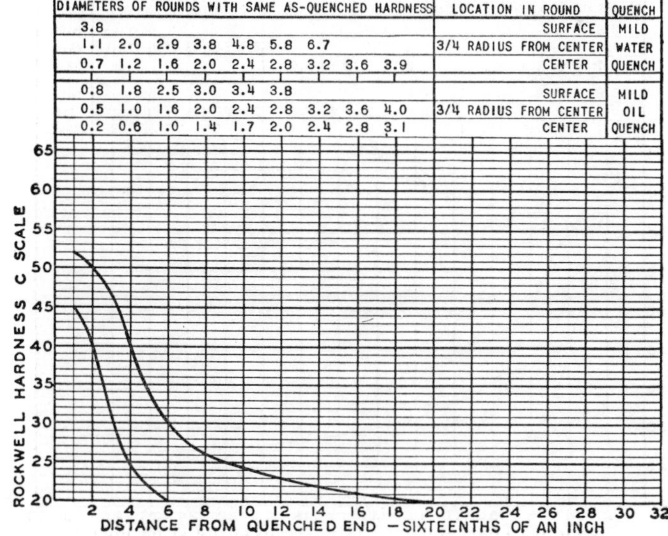

DIAMETERS OF ROUNDS WITH SAME AS-QUENCHED HARDNESS	LOCATION IN ROUND	QUENCH
3.8	SURFACE	MILD
1.1 2.0 2.9 3.8 4.8 5.8 6.7	3/4 RADIUS FROM CENTER	WATER
0.7 1.2 1.6 2.0 2.4 2.8 3.2 3.6 3.9	CENTER	QUENCH
0.8 1.8 2.5 3.0 3.4 3.8	SURFACE	MILD
0.5 1.0 1.6 2.0 2.4 2.8 3.2 3.6 4.0	3/4 RADIUS FROM CENTER	OIL
0.2 0.6 1.0 1.4 1.7 2.0 2.4 2.8 3.1	CENTER	QUENCH

HARDENABILITY BAND 4037 H

C	Mn	Si			Mo	
.34/.41	.60/1.00	.20/.35			.20/.30	

DIAMETERS OF ROUNDS WITH SAME AS-QUENCHED HARDNESS									LOCATION IN ROUND	QUENCH
3.8									SURFACE	MILD
1.1	2.0	2.9	3.8	4.8	5.8	6.7			3/4 RADIUS FROM CENTER	WATER
0.7	1.2	1.6	2.0	2.4	2.8	3.2	3.6	3.9	CENTER	QUENCH
0.8	1.8	2.5	3.0	3.4	3.8				SURFACE	MILD
0.5	1.0	1.6	2.0	2.4	2.8	3.2	3.6	4.0	3/4 RADIUS FROM CENTER	OIL
0.2	0.6	1.0	1.4	1.7	2.0	2.4	2.8	3.1	CENTER	QUENCH

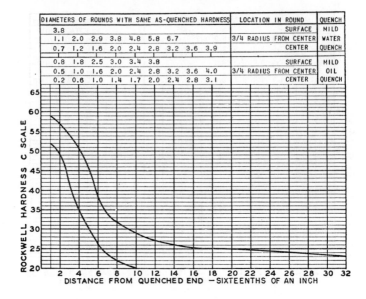

ROCKWELL HARDNESS C SCALE

65 60 55 50 45 40 35 30 25 20

2 4 6 8 10 12 14 16 18 20 22 24 26 28 30 32

DISTANCE FROM QUENCHED END — SIXTEENTHS OF AN INCH

HARDENABILITY BAND 4042 H

C	Mn	Si			Mo	
.39/.46	.60/1.00	.20/.35			.20/.30	

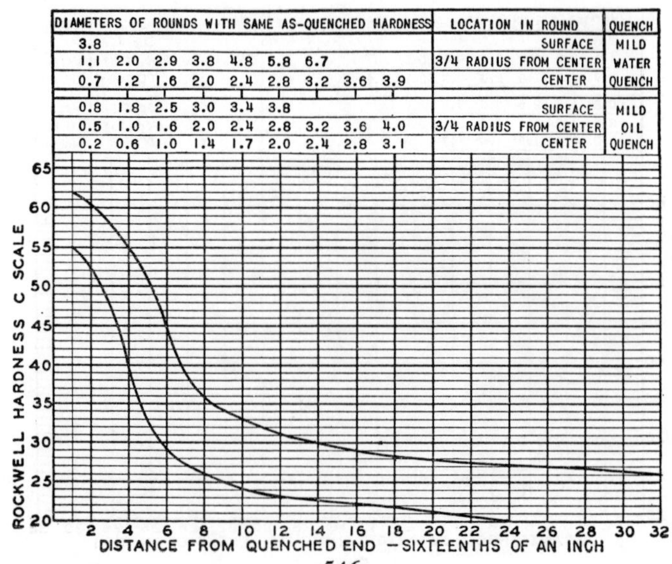

DIAMETERS OF ROUNDS WITH SAME AS-QUENCHED HARDNESS									LOCATION IN ROUND	QUENCH
3.8									SURFACE	MILD
1.1	2.0	2.9	3.8	4.8	5.8	6.7			3/4 RADIUS FROM CENTER	WATER
0.7	1.2	1.6	2.0	2.4	2.8	3.2	3.6	3.9	CENTER	QUENCH
0.8	1.8	2.5	3.0	3.4	3.8				SURFACE	MILD
0.5	1.0	1.6	2.0	2.4	2.8	3.2	3.6	4.0	3/4 RADIUS FROM CENTER	OIL
0.2	0.6	1.0	1.4	1.7	2.0	2.4	2.8	3.1	CENTER	QUENCH

HARDENABILITY BAND 4047 H

C	Mn	Si			Mo	
.44/.51	.60/1.00	.20/.35			.20/.30	

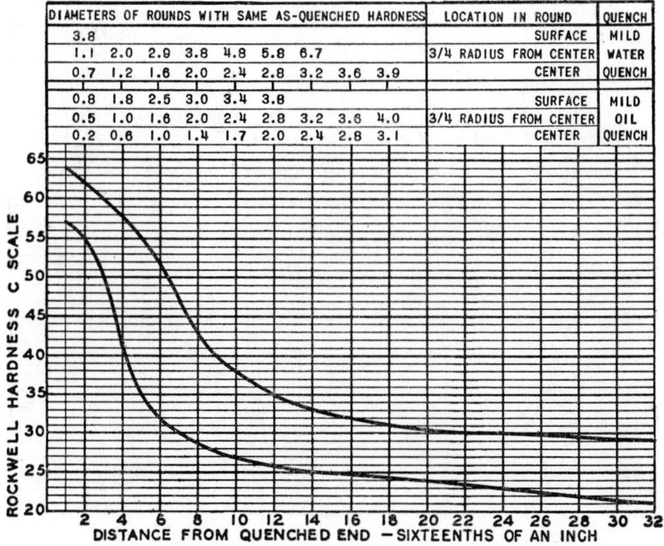

DIAMETERS OF ROUNDS WITH SAME AS-QUENCHED HARDNESS	LOCATION IN ROUND	QUENCH
3.8	SURFACE	MILD
1.1 2.0 2.9 3.8 4.8 5.8 6.7	3/4 RADIUS FROM CENTER	WATER
0.7 1.2 1.6 2.0 2.4 2.8 3.2 3.6 3.9	CENTER	QUENCH
0.8 1.8 2.5 3.0 3.4 3.8	SURFACE	MILD
0.5 1.0 1.6 2.0 2.4 2.8 3.2 3.6 4.0	3/4 RADIUS FROM CENTER	OIL
0.2 0.6 1.0 1.4 1.7 2.0 2.4 2.8 3.1	CENTER	QUENCH

HARDENABILITY BAND 4063 H

C	Mn	Si			Mo	
.59/.69	.65/1.10	.20/.35			.20/.30	

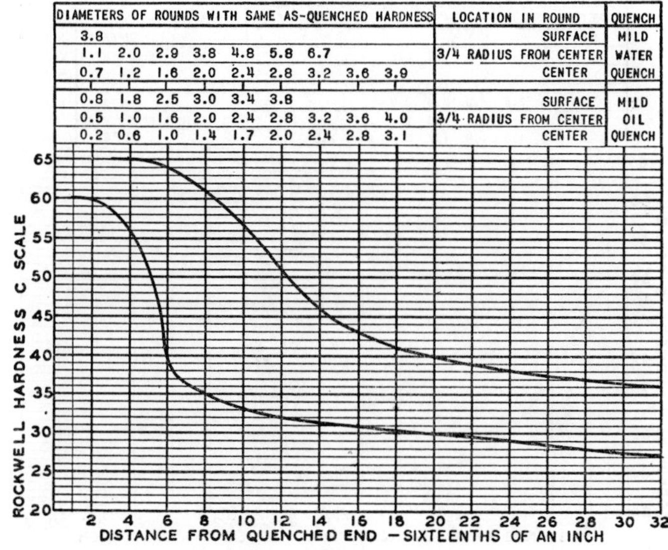

DIAMETERS OF ROUNDS WITH SAME AS-QUENCHED HARDNESS	LOCATION IN ROUND	QUENCH
3.8	SURFACE	MILD
1.1 2.0 2.9 3.8 4.8 5.8 6.7	3/4 RADIUS FROM CENTER	WATER
0.7 1.2 1.6 2.0 2.4 2.8 3.2 3.6 3.9	CENTER	QUENCH
0.8 1.8 2.5 3.0 3.4 3.8	SURFACE	MILD
0.5 1.0 1.6 2.0 2.4 2.8 3.2 3.6 4.0	3/4 RADIUS FROM CENTER	OIL
0.2 0.6 1.0 1.4 1.7 2.0 2.4 2.8 3.1	CENTER	QUENCH

HARDENABILITY BAND 4118 H

C	Mn	Si		Cr	Mo	
.17/.23	.60/1.00	.20/.35		.30/.70	.08/.15	

DIAMETERS OF ROUNDS WITH SAME AS-QUENCHED HARDNESS	LOCATION IN ROUND	QUENCH
3.8	SURFACE	MILD
1.1 2.0 2.9 3.8 4.8 5.8 6.7	3/4 RADIUS FROM CENTER	WATER
0.7 1.2 1.6 2.0 2.4 2.8 3.2 3.6 3.9	CENTER	QUENCH
0.8 1.8 2.5 3.0 3.4 3.8	SURFACE	MILD
0.5 1.0 1.6 2.0 2.4 2.8 3.2 3.6 4.0	3/4 RADIUS FROM CENTER	OIL
0.2 0.6 1.0 1.4 1.7 2.0 2.4 2.8 3.1	CENTER	QUENCH

ROCKWELL HARDNESS C SCALE: 20 25 30 35 40 45 50 55 60 65

2 4 6 8 10 12 14 16 18 20 22 24 26 28 30 32

DISTANCE FROM QUENCHED END —SIXTEENTHS OF AN INCH

HARDENABILITY BAND 4130 H

C	Mn	Si		Cr	Mo	
.27/.33	.30/.70	.20/.35		.75/1.20	.15/.25	

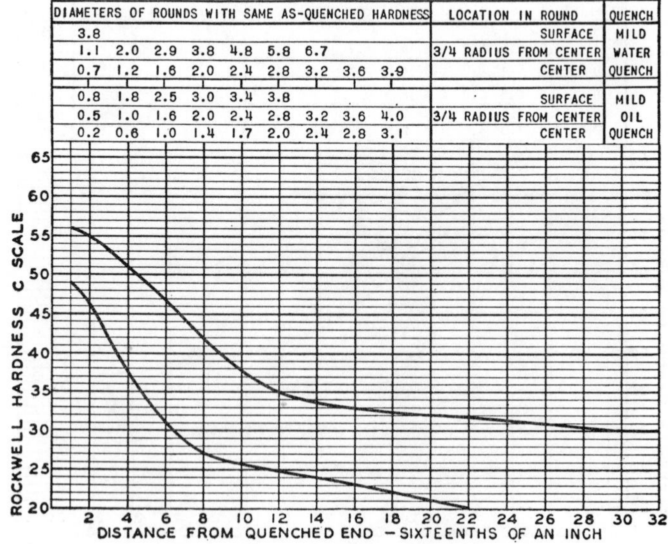

DIAMETERS OF ROUNDS WITH SAME AS-QUENCHED HARDNESS	LOCATION IN ROUND	QUENCH
3.8	SURFACE	MILD
1.1 2.0 2.9 3.8 4.8 5.8 6.7	3/4 RADIUS FROM CENTER	WATER
0.7 1.2 1.6 2.0 2.4 2.8 3.2 3.6 3.9	CENTER	QUENCH
0.8 1.8 2.5 3.0 3.4 3.8	SURFACE	MILD
0.5 1.0 1.6 2.0 2.4 2.8 3.2 3.6 4.0	3/4 RADIUS FROM CENTER	OIL
0.2 0.6 1.0 1.4 1.7 2.0 2.4 2.8 3.1	CENTER	QUENCH

HARDENABILITY BAND 4135 H

C	Mn	Si		Cr	Mo	
.32/.38	.60/1.00	.20/.35		.75/1.20	.15/.25	

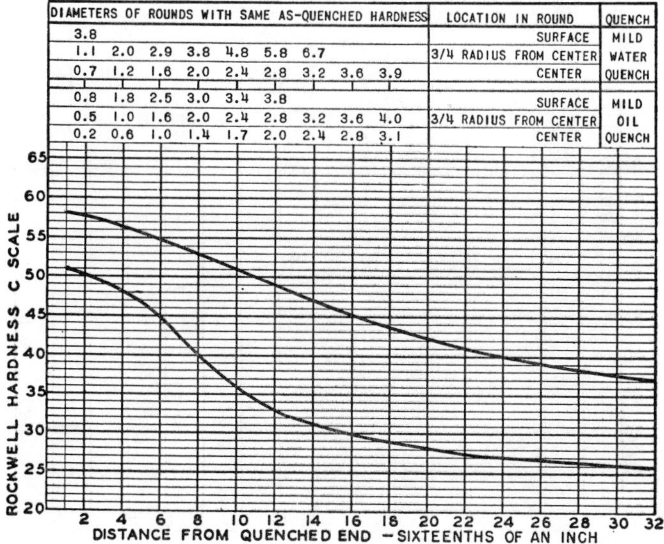

DIAMETERS OF ROUNDS WITH SAME AS-QUENCHED HARDNESS	LOCATION IN ROUND	QUENCH
3.8	SURFACE	MILD
1.1 2.0 2.9 3.8 4.8 5.8 6.7	3/4 RADIUS FROM CENTER	WATER
0.7 1.2 1.6 2.0 2.4 2.8 3.2 3.6 3.9	CENTER	QUENCH
0.8 1.8 2.5 3.0 3.4 3.8	SURFACE	MILD
0.5 1.0 1.6 2.0 2.4 2.8 3.2 3.6 4.0	3/4 RADIUS FROM CENTER	OIL
0.2 0.6 1.0 1.4 1.7 2.0 2.4 2.8 3.1	CENTER	QUENCH

HARDENABILITY BAND 4137 H

C	Mn	Si		Cr	Mo	
.34/.41	.60/1.00	.20/.35		.75/1.20	.15/.25	

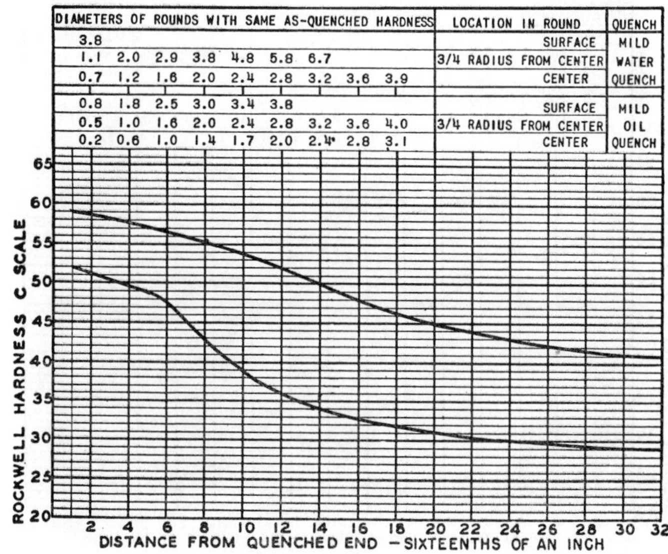

DIAMETERS OF ROUNDS WITH SAME AS-QUENCHED HARDNESS	LOCATION IN ROUND	QUENCH
3.8	SURFACE	MILD
1.1 2.0 2.9 3.8 4.8 5.8 6.7	3/4 RADIUS FROM CENTER	WATER
0.7 1.2 1.6 2.0 2.4 2.8 3.2 3.6 3.9	CENTER	QUENCH
0.8 1.8 2.5 3.0 3.4 3.8	SURFACE	MILD
0.5 1.0 1.6 2.0 2.4 2.8 3.2 3.6 4.0	3/4 RADIUS FROM CENTER	OIL
0.2 0.6 1.0 1.4 1.7 2.0 2.4 2.8 3.1	CENTER	QUENCH

HARDENABILITY BAND 4140 H

C	Mn	Si		Cr	Mo	
.37 / .44	.65 / 1.10	.20 / .35		.75 / 1.20	.15 / .25	

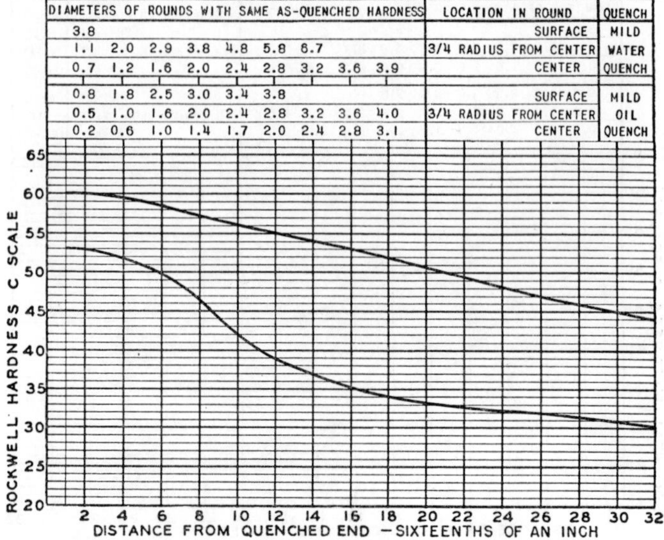

DIAMETERS OF ROUNDS WITH SAME AS-QUENCHED HARDNESS									LOCATION IN ROUND	QUENCH
3.8									SURFACE	MILD
1.1	2.0	2.9	3.8	4.8	5.8	6.7			3/4 RADIUS FROM CENTER	WATER
0.7	1.2	1.6	2.0	2.4	2.8	3.2	3.6	3.9	CENTER	QUENCH
0.8	1.8	2.5	3.0	3.4	3.8				SURFACE	MILD
0.5	1.0	1.6	2.0	2.4	2.8	3.2	3.6	4.0	3/4 RADIUS FROM CENTER	OIL
0.2	0.6	1.0	1.4	1.7	2.0	2.4	2.8	3.1	CENTER	QUENCH

HARDENABILITY BAND 4142 H

C	Mn	Si		Cr	Mo	
.39 / .46	.65 / 1.10	.20 / .35		.75 / 1.20	.15 / .25	

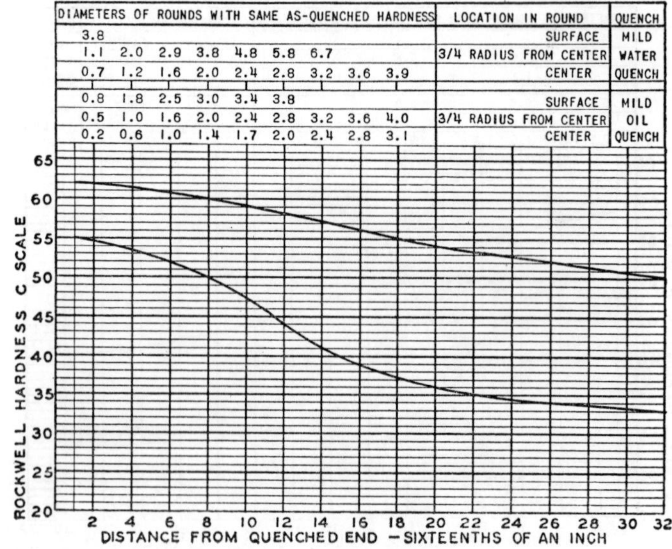

DIAMETERS OF ROUNDS WITH SAME AS-QUENCHED HARDNESS									LOCATION IN ROUND	QUENCH
3.8									SURFACE	MILD
1.1	2.0	2.9	3.8	4.8	5.8	6.7			3/4 RADIUS FROM CENTER	WATER
0.7	1.2	1.6	2.0	2.4	2.8	3.2	3.6	3.9	CENTER	QUENCH
0.8	1.8	2.5	3.0	3.4	3.8				SURFACE	MILD
0.5	1.0	1.6	2.0	2.4	2.8	3.2	3.6	4.0	3/4 RADIUS FROM CENTER	OIL
0.2	0.6	1.0	1.4	1.7	2.0	2.4	2.8	3.1	CENTER	QUENCH

HARDENABILITY BAND 4145 H

C	Mn	Si		Cr	Mo	
.42 / .49	.65 / 1.10	.20 / .35		.75 / 1.20	.15 / .25	

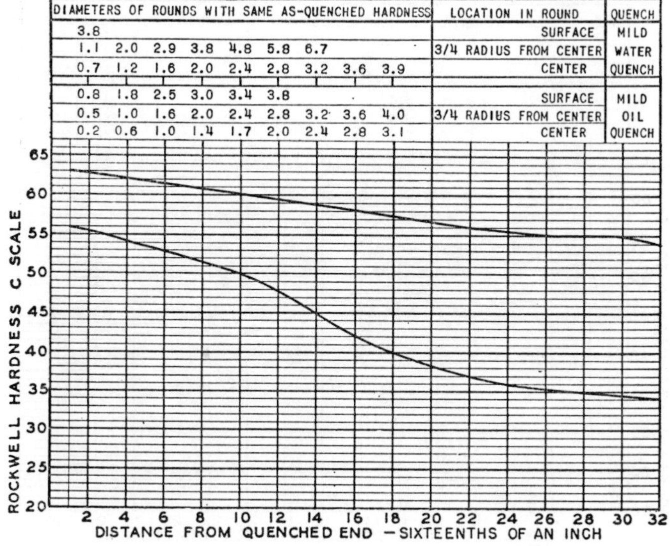

DIAMETERS OF ROUNDS WITH SAME AS-QUENCHED HARDNESS	LOCATION IN ROUND	QUENCH
3.8	SURFACE	MILD
1.1 2.0 2.9 3.8 4.8 5.8 6.7	3/4 RADIUS FROM CENTER	WATER
0.7 1.2 1.6 2.0 2.4 2.8 3.2 3.6 3.9	CENTER	QUENCH
0.8 1.8 2.5 3.0 3.4 3.8	SURFACE	MILD
0.5 1.0 1.6 2.0 2.4 2.8 3.2 3.6 4.0	3/4 RADIUS FROM CENTER	OIL
0.2 0.6 1.0 1.4 1.7 2.0 2.4 2.8 3.1	CENTER	QUENCH

HARDENABILITY BAND 4147 H

C	Mn	Si		Cr	Mo	
.44 / .51	.65 / 1.10	.20 / .35		.75 / 1.20	.15 / .25	

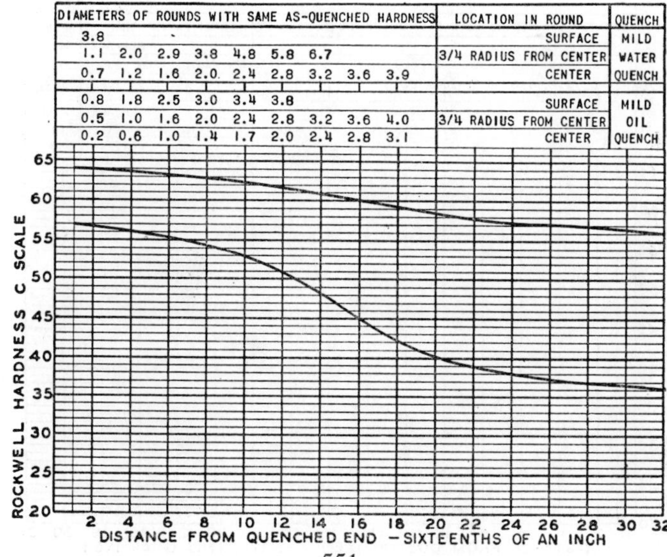

DIAMETERS OF ROUNDS WITH SAME AS-QUENCHED HARDNESS	LOCATION IN ROUND	QUENCH
3.8	SURFACE	MILD
1.1 2.0 2.9 3.8 4.8 5.8 6.7	3/4 RADIUS FROM CENTER	WATER
0.7 1.2 1.6 2.0 2.4 2.8 3.2 3.6 3.9	CENTER	QUENCH
0.8 1.8 2.5 3.0 3.4 3.8	SURFACE	MILD
0.5 1.0 1.6 2.0 2.4 2.8 3.2 3.6 4.0	3/4 RADIUS FROM CENTER	OIL
0.2 0.6 1.0 1.4 1.7 2.0 2.4 2.8 3.1	CENTER	QUENCH

HARDENABILITY BAND 4150 H

C	Mn	Si		Cr	Mo	
.47/.54	.65/1.10	.20/.35		.75/1.20	.15/.25	

DIAMETERS OF ROUNDS WITH SAME AS-QUENCHED HARDNESS									LOCATION IN ROUND	QUENCH
3.8									SURFACE	MILD
1.1	2.0	2.9	3.8	4.8	5.8	6.7			3/4 RADIUS FROM CENTER	WATER
0.7	1.2	1.6	2.0	2.4	2.8	3.2	3.6	3.9	CENTER	QUENCH
0.8	1.8	2.5	3.0	3.4	3.8				SURFACE	MILD
0.5	1.0	1.6	2.0	2.4	2.8	3.2	3.6	4.0	3/4 RADIUS FROM CENTER	OIL
0.2	0.6	1.0	1.4	1.7	2.0	2.4	2.8	3.1	CENTER	QUENCH

ROCKWELL HARDNESS C SCALE

65 60 55 50 45 40 35 30 25 20

2 4 6 8 10 12 14 16 18 20 22 24 26 28 30 32

DISTANCE FROM QUENCHED END — SIXTEENTHS OF AN INCH

HARDENABILITY BAND 4320 H

C	Mn	Si	Ni	Cr	Mo	
.17/.23	.40/.70	.20/.35	1.55/2.00	.35/.65	.20/.30	

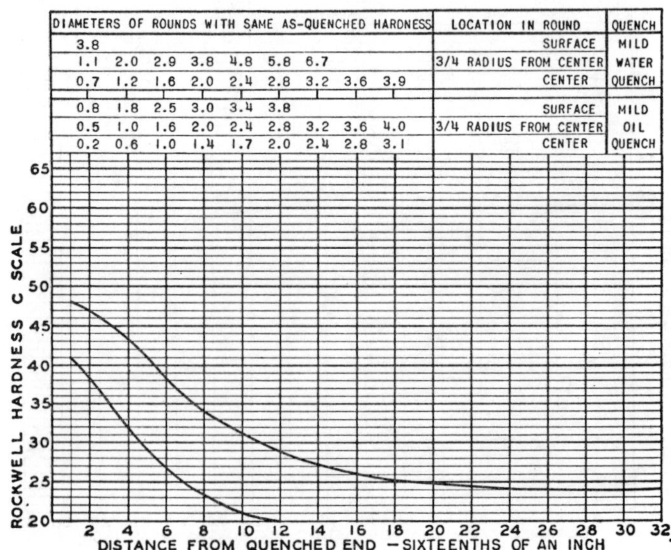

DIAMETERS OF ROUNDS WITH SAME AS-QUENCHED HARDNESS									LOCATION IN ROUND	QUENCH
3.8									SURFACE	MILD
1.1	2.0	2.9	3.8	4.8	5.8	6.7			3/4 RADIUS FROM CENTER	WATER
0.7	1.2	1.6	2.0	2.4	2.8	3.2	3.6	3.9	CENTER	QUENCH
0.8	1.8	2.5	3.0	3.4	3.8				SURFACE	MILD
0.5	1.0	1.6	2.0	2.4	2.8	3.2	3.6	4.0	3/4 RADIUS FROM CENTER	OIL
0.2	0.6	1.0	1.4	1.7	2.0	2.4	2.8	3.1	CENTER	QUENCH

HARDENABILITY BAND 4337 H

C	Mn	Si	Ni	Cr	Mo	
.34/.41	.55/.90	.20/.35	1.55/2.00	.65/.95	.20/.30	

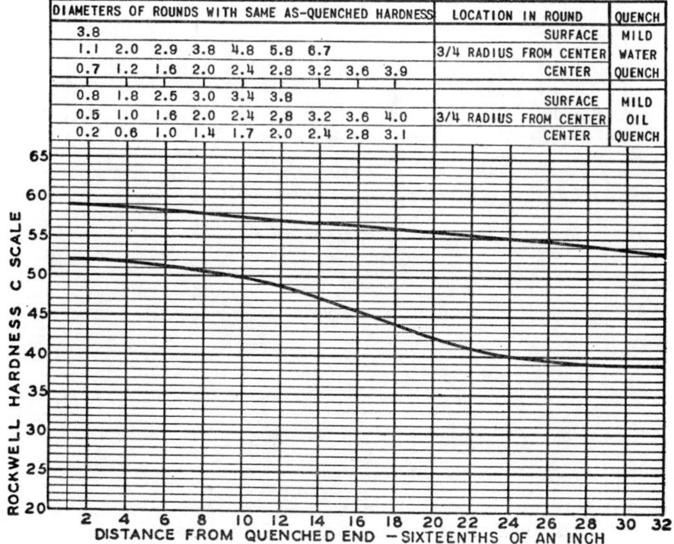

HARDENABILITY BAND 4340 H

C	Mn	Si	Ni	Cr	Mo	
.37/.44	.55/.90	.20/.35	1.55/2.00	.65/.95	.20/.30	

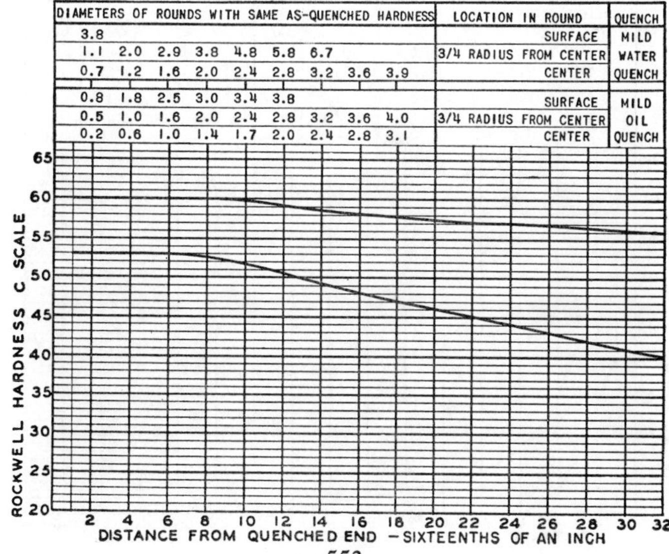

HARDENABILITY BAND E4340 H

C	Mn	Si	Ni	Cr	Mo	
.37/.44	.60/.95	.20/.35	1.55/2.00	.65/.95	.20/.30	

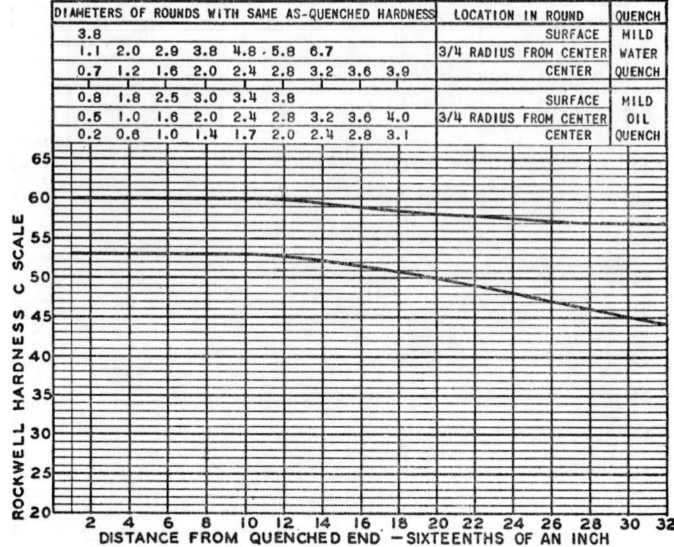

DIAMETERS OF ROUNDS WITH SAME AS-QUENCHED HARDNESS	LOCATION IN ROUND	QUENCH
3.8	SURFACE	MILD
1.1 2.0 2.9 3.8 4.8 5.8 6.7	3/4 RADIUS FROM CENTER	WATER
0.7 1.2 1.6 2.0 2.4 2.8 3.2 3.6 3.9	CENTER	QUENCH
0.8 1.8 2.5 3.0 3.4 3.8	SURFACE	MILD
0.5 1.0 1.6 2.0 2.4 2.8 3.2 3.6 4.0	3/4 RADIUS FROM CENTER	OIL
0.2 0.6 1.0 1.4 1.7 2.0 2.4 2.8 3.1	CENTER	QUENCH

HARDENABILITY BAND 4520H

C	Mn	Si			Mo	
.17/.23	.35/.75	.20/.35			.45/.60	

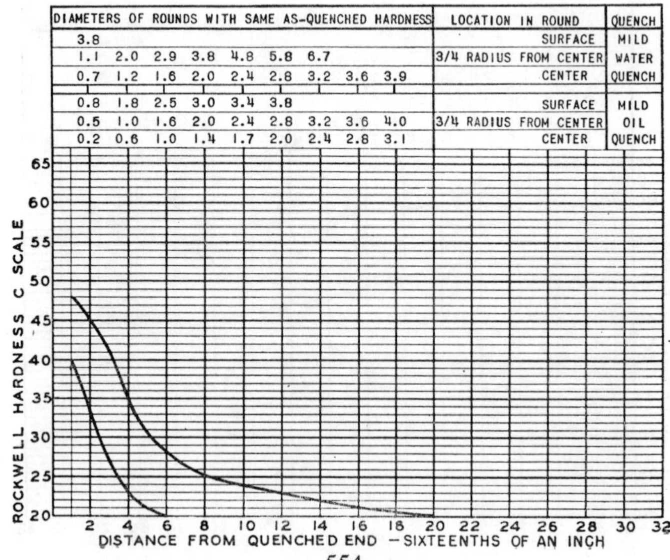

DIAMETERS OF ROUNDS WITH SAME AS-QUENCHED HARDNESS	LOCATION IN ROUND	QUENCH
3.8	SURFACE	MILD
1.1 2.0 2.9 3.8 4.8 5.8 6.7	3/4 RADIUS FROM CENTER	WATER
0.7 1.2 1.6 2.0 2.4 2.8 3.2 3.6 3.9	CENTER	QUENCH
0.8 1.8 2.5 3.0 3.4 3.8	SURFACE	MILD
0.5 1.0 1.6 2.0 2.4 2.8 3.2 3.6 4.0	3/4 RADIUS FROM CENTER	OIL
0.2 0.6 1.0 1.4 1.7 2.0 2.4 2.8 3.1	CENTER	QUENCH

HARDENABILITY BAND 4620 H

C	Mn	Si	Ni		Mo	
.17/.23	.35/.75	.20/.35	1.55/2.00		.20/.30	

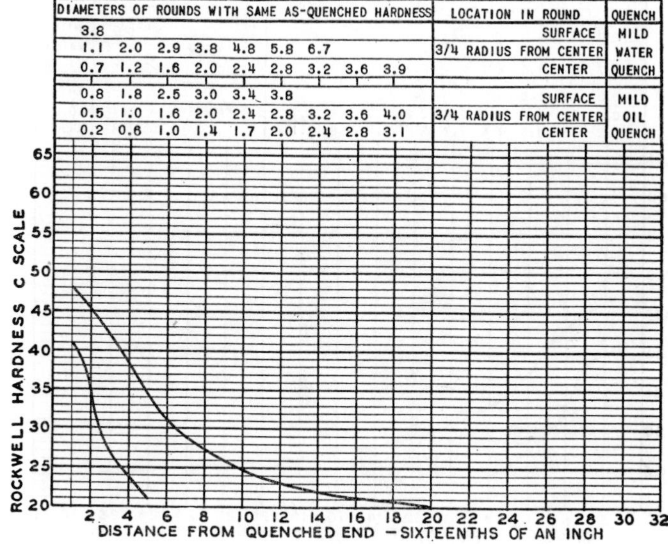

DIAMETERS OF ROUNDS WITH SAME AS-QUENCHED HARDNESS	LOCATION IN ROUND	QUENCH
3.8	SURFACE	MILD
1.1 2.0 2.9 3.8 4.8 5.8 6.7	3/4 RADIUS FROM CENTER	WATER
0.7 1.2 1.6 2.0 2.4 2.8 3.2 3.6 3.9	CENTER	QUENCH
0.8 1.8 2.5 3.0 3.4 3.8	SURFACE	MILD
0.5 1.0 1.6 2.0 2.4 2.8 3.2 3.6 4.0	3/4 RADIUS FROM CENTER	OIL
0.2 0.6 1.0 1.4 1.7 2.0 2.4 2.8 3.1	CENTER	QUENCH

HARDENABILITY BAND 4621 H

C	Mn	Si	Ni		Mo	
.17/.23	.60/1.00	.20/.35	1.55/2.00		.20/.30	

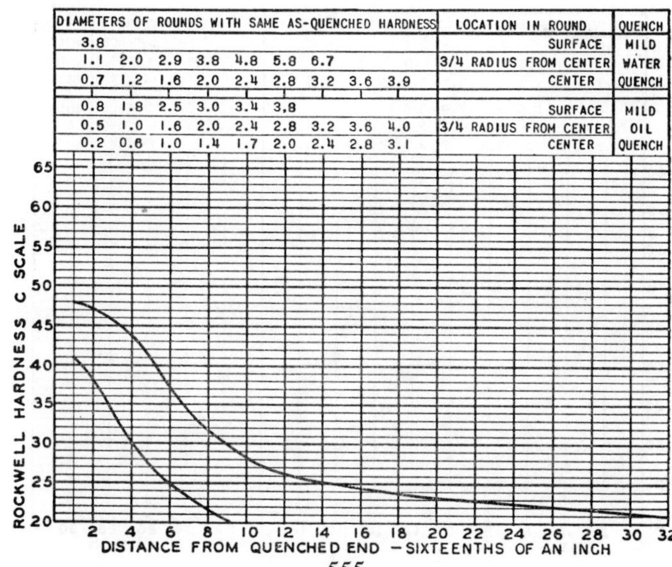

DIAMETERS OF ROUNDS WITH SAME AS-QUENCHED HARDNESS	LOCATION IN ROUND	QUENCH
3.8	SURFACE	MILD
1.1 2.0 2.9 3.8 4.8 5.8 6.7	3/4 RADIUS FROM CENTER	WATER
0.7 1.2 1.6 2.0 2.4 2.8 3.2 3.6 3.9	CENTER	QUENCH
0.8 1.8 2.5 3.0 3.4 3.8	SURFACE	MILD
0.5 1.0 1.6 2.0 2.4 2.8 3.2 3.6 4.0	3/4 RADIUS FROM CENTER	OIL
0.2 0.6 1.0 1.4 1.7 2.0 2.4 2.8 3.1	CENTER	QUENCH

HARDENABILITY BAND 4718H

C	Mn	Si	Ni	Cr	Mo	
.15 / .21	.60 / .95	.20 / .35	.85 / 1.25	.30 / .60	.30 / .40	

DIAMETERS OF ROUNDS WITH SAME AS-QUENCHED HARDNESS	LOCATION IN ROUND	QUENCH
3.8	SURFACE	MILD
1.1 2.0 2.9 3.8 4.8 5.8 6.7	3/4 RADIUS FROM CENTER	WATER
0.7 1.2 1.6 2.0 2.4 2.8 3.2 3.6 3.9	CENTER	QUENCH
0.8 1.8 2.5 3.0 3.4 3.8	SURFACE	MILD
0.5 1.0 1.6 2.0 2.4 2.8 3.2 3.6 4.0	3/4 RADIUS FROM CENTER	OIL
0.2 0.6 1.0 1.4 1.7 2.0 2.4 2.8 3.1	CENTER	QUENCH

ROCKWELL HARDNESS C SCALE

65 60 55 50 45 40 35 30 25 20

2 4 6 8 10 12 14 16 18 20 22 24 26 28 30 32

DISTANCE FROM QUENCHED END – SIXTEENTHS OF AN INCH

HARDENABILITY BAND 4720H

C	Mn	Si	Ni	Cr	Mo	
.17 / .23	.45 / .75	.20 / .35	.85 / 1.25	.30 / .60	.15 / .25	

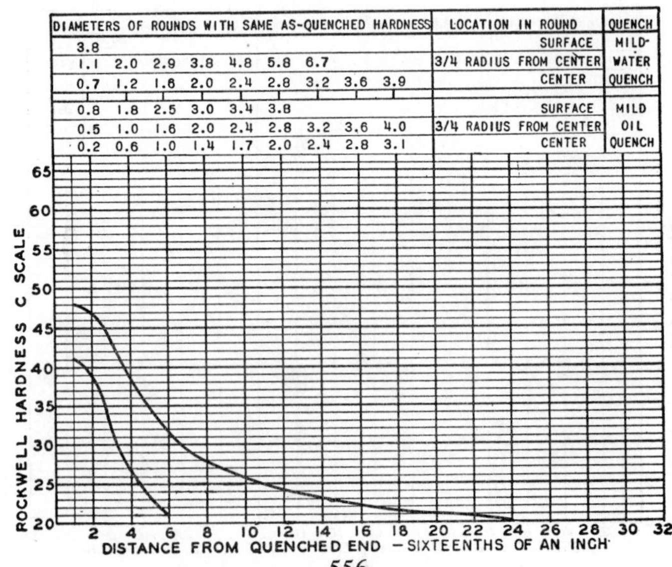

DIAMETERS OF ROUNDS WITH SAME AS-QUENCHED HARDNESS	LOCATION IN ROUND	QUENCH
3.8	SURFACE	MILD
1.1 2.0 2.9 3.8 4.8 5.8 6.7	3/4 RADIUS FROM CENTER	WATER
0.7 1.2 1.6 2.0 2.4 2.8 3.2 3.6 3.9	CENTER	QUENCH
0.8 1.8 2.5 3.0 3.4 3.8	SURFACE	MILD
0.5 1.0 1.6 2.0 2.4 2.8 3.2 3.6 4.0	3/4 RADIUS FROM CENTER	OIL
0.2 0.6 1.0 1.4 1.7 2.0 2.4 2.8 3.1	CENTER	QUENCH

HARDENABILITY BAND 4815 H

C	Mn	Si	Ni		Mo	
.12/.18	.30/.70	.20/.35	3.20/3.80		.20/.30	

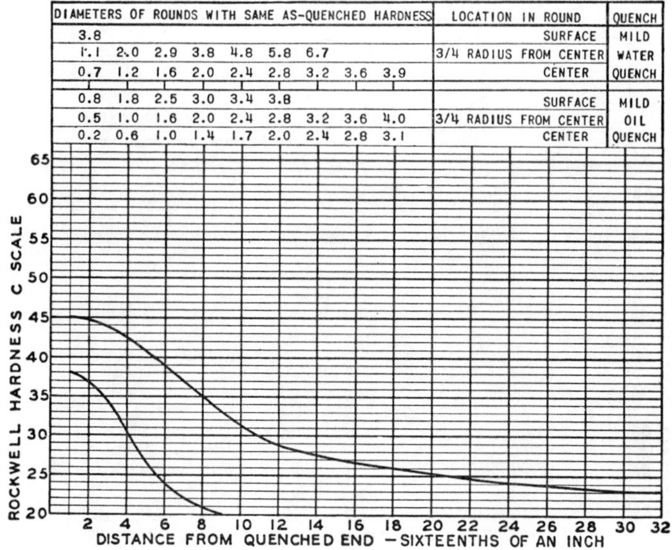

DIAMETERS OF ROUNDS WITH SAME AS-QUENCHED HARDNESS	LOCATION IN ROUND	QUENCH
3.8	SURFACE	MILD WATER QUENCH
1.1 2.0 2.9 3.8 4.8 5.8 6.7	3/4 RADIUS FROM CENTER	
0.7 1.2 1.6 2.0 2.4 2.8 3.2 3.6 3.9	CENTER	
0.8 1.8 2.5 3.0 3.4 3.8	SURFACE	MILD OIL QUENCH
0.5 1.0 1.6 2.0 2.4 2.8 3.2 3.6 4.0	3/4 RADIUS FROM CENTER	
0.2 0.6 1.0 1.4 1.7 2.0 2.4 2.8 3.1	CENTER	

HARDENABILITY BAND 4817 H

C	Mn	Si	Ni		Mo	
.14/.20	.30/.70	.20/.35	3.20/3.80		.20/.30	

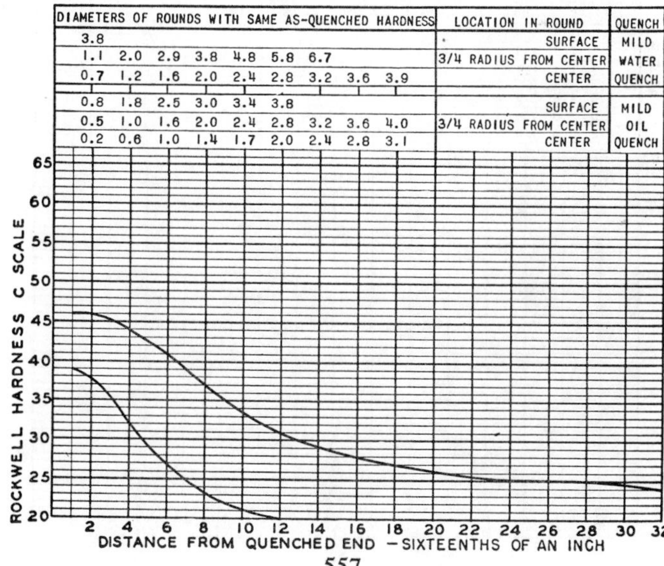

DIAMETERS OF ROUNDS WITH SAME AS-QUENCHED HARDNESS	LOCATION IN ROUND	QUENCH
3.8	SURFACE	MILD WATER QUENCH
1.1 2.0 2.9 3.8 4.8 5.8 6.7	3/4 RADIUS FROM CENTER	
0.7 1.2 1.6 2.0 2.4 2.8 3.2 3.6 3.9	CENTER	
0.8 1.8 2.5 3.0 3.4 3.8	SURFACE	MILD OIL QUENCH
0.5 1.0 1.6 2.0 2.4 2.8 3.2 3.6 4.0	3/4 RADIUS FROM CENTER	
0.2 0.6 1.0 1.4 1.7 2.0 2.4 2.8 3.1	CENTER	

HARDENABILITY BAND 4820 H

C	Mn	Si	Ni		Mo	
.17/.23	.40/.80	.20/.35	3.20/3.80		.20/.30	

DIAMETERS OF ROUNDS WITH SAME AS-QUENCHED HARDNESS									LOCATION IN ROUND	QUENCH
3.8									SURFACE	MILD
1.1	2.0	2.9	3.8	4.8	5.8	6.7			3/4 RADIUS FROM CENTER	WATER
0.7	1.2	1.6	2.0	2.4	2.8	3.2	3.6	3.9	CENTER	QUENCH
0.8	1.8	2.5	3.0	3.4	3.8				SURFACE	MILD
0.5	1.0	1.6	2.0	2.4	2.8	3.2	3.6	4.0	3/4 RADIUS FROM CENTER	OIL
0.2	0.6	1.0	1.4	1.7	2.0	2.4	2.8	3.1	CENTER	QUENCH

ROCKWELL HARDNESS C SCALE

65 60 55 50 45 40 35 30 25 20

2 4 6 8 10 12 14 16 18 20 22 24 26 28 30 32

DISTANCE FROM QUENCHED END — SIXTEENTHS OF AN INCH

HARDENABILITY BAND 5046H

C	Mn	Si		Cr		
.43/.50	.65/1.10	.20/.35		.13/.43		

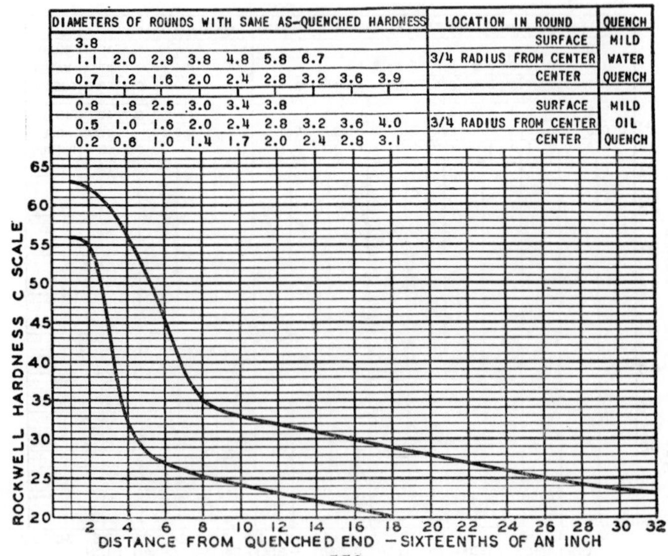

DIAMETERS OF ROUNDS WITH SAME AS-QUENCHED HARDNESS									LOCATION IN ROUND	QUENCH
3.8									SURFACE	MILD
1.1	2.0	2.9	3.8	4.8	5.8	6.7			3/4 RADIUS FROM CENTER	WATER
0.7	1.2	1.6	2.0	2.4	2.8	3.2	3.6	3.9	CENTER	QUENCH
0.8	1.8	2.5	3.0	3.4	3.8				SURFACE	MILD
0.5	1.0	1.6	2.0	2.4	2.8	3.2	3.6	4.0	3/4 RADIUS FROM CENTER	OIL
0.2	0.6	1.0	1.4	1.7	2.0	2.4	2.8	3.1	CENTER	QUENCH

HARDENABILITY BAND 5120 H

C	Mn	Si		Cr		
.17/.23	.60/1.00	.20/.35		.60/1.00		

DIAMETERS OF ROUNDS WITH SAME AS-QUENCHED HARDNESS									LOCATION IN ROUND	QUENCH
3.8									SURFACE	MILD
1.1	2.0	2.9	3.8	4.8	5.8	6.7			3/4 RADIUS FROM CENTER	WATER
0.7	1.2	1.6	2.0	2.4	2.8	3.2	3.6	3.9	CENTER	QUENCH
0.8	1.8	2.5	3.0	3.4	3.8				SURFACE	MILD
0.5	1.0	1.6	2.0	2.4	2.8	3.2	3.6	4.0	3/4 RADIUS FROM CENTER	OIL
0.2	0.6	1.0	1.4	1.7	2.0	2.4	2.8	3.1	CENTER	QUENCH

ROCKWELL HARDNESS C SCALE
65 60 55 50 45 40 35 30 25 20
2 4 6 8 10 12 14 16 18 20 22 24 26 28 30 32
DISTANCE FROM QUENCHED END — SIXTEENTHS OF AN INCH

HARDENABILITY BAND 5130 H

C	Mn	Si		Cr		
.27/.33	.60/1.00	.20/.35		.75/1.20		

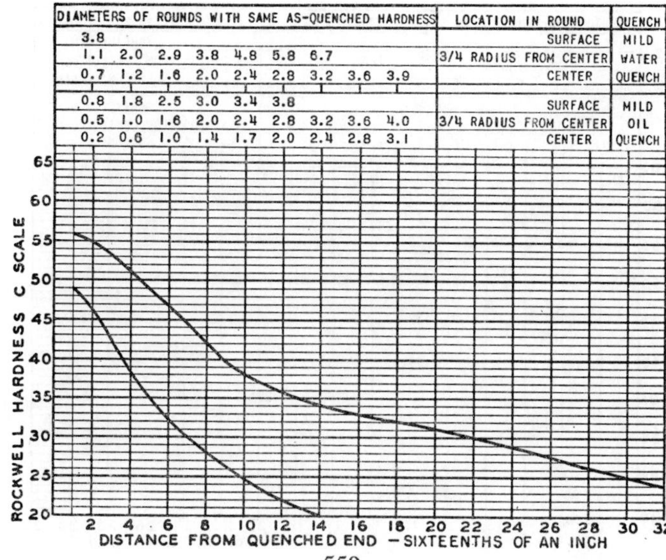

DIAMETERS OF ROUNDS WITH SAME AS-QUENCHED HARDNESS									LOCATION IN ROUND	QUENCH
3.8									SURFACE	MILD
1.1	2.0	2.9	3.8	4.8	5.8	6.7			3/4 RADIUS FROM CENTER	WATER
0.7	1.2	1.6	2.0	2.4	2.8	3.2	3.6	3.9	CENTER	QUENCH
0.8	1.8	2.5	3.0	3.4	3.8				SURFACE	MILD
0.5	1.0	1.6	2.0	2.4	2.8	3.2	3.6	4.0	3/4 RADIUS FROM CENTER	OIL
0.2	0.6	1.0	1.4	1.7	2.0	2.4	2.8	3.1	CENTER	QUENCH

HARDENABILITY BAND 5132 H

C	Mn	Si		Cr		
.29/.35	.50/.90	.20/.35		.65/1.10		

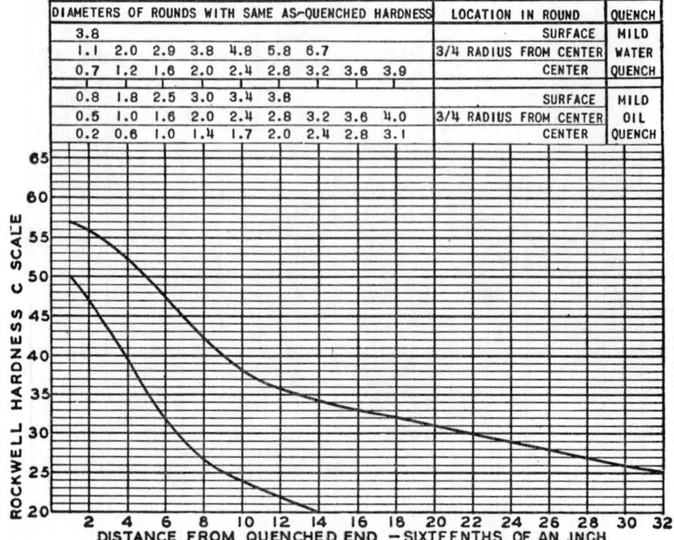

DIAMETERS OF ROUNDS WITH SAME AS-QUENCHED HARDNESS	LOCATION IN ROUND	QUENCH
3.8	SURFACE	MILD
1.1 2.0 2.9 3.8 4.8 5.8 6.7	3/4 RADIUS FROM CENTER	WATER
0.7 1.2 1.6 2.0 2.4 2.8 3.2 3.6 3.9	CENTER	QUENCH
0.8 1.8 2.5 3.0 3.4 3.8	SURFACE	MILD
0.5 1.0 1.6 2.0 2.4 2.8 3.2 3.6 4.0	3/4 RADIUS FROM CENTER	OIL
0.2 0.6 1.0 1.4 1.7 2.0 2.4 2.8 3.1	CENTER	QUENCH

HARDENABILITY BAND 5135 H

C	Mn	Si		Cr		
32/.38	.50/.90	.20/.35		.70/1.15		

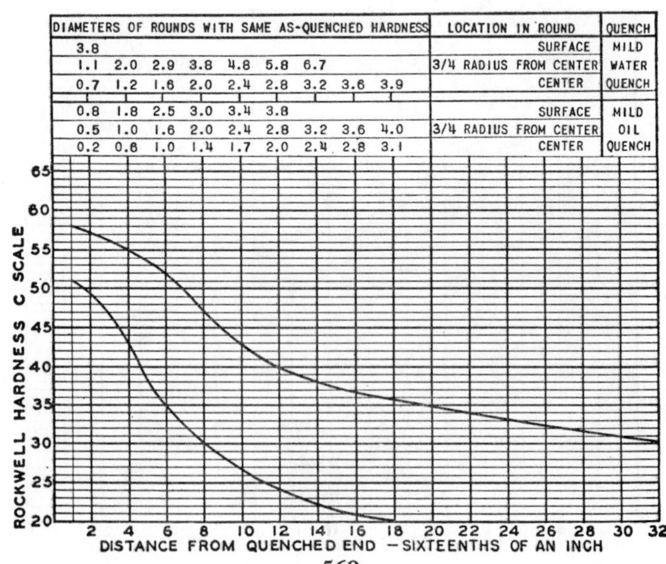

DIAMETERS OF ROUNDS WITH SAME AS-QUENCHED HARDNESS	LOCATION IN ROUND	QUENCH
3.8	SURFACE	MILD
1.1 2.0 2.9 3.8 4.8 5.8 6.7	3/4 RADIUS FROM CENTER	WATER
0.7 1.2 1.6 2.0 2.4 2.8 3.2 3.6 3.9	CENTER	QUENCH
0.8 1.8 2.5 3.0 3.4 3.8	SURFACE	MILD
0.5 1.0 1.6 2.0 2.4 2.8 3.2 3.6 4.0	3/4 RADIUS FROM CENTER	OIL
0.2 0.6 1.0 1.4 1.7 2.0 2.4 2.8 3.1	CENTER	QUENCH

HARDENABILITY BAND 5140 H

C	Mn	Si		Cr		
.37/.44	.60/1.00	.20/.35		.60/1.00		

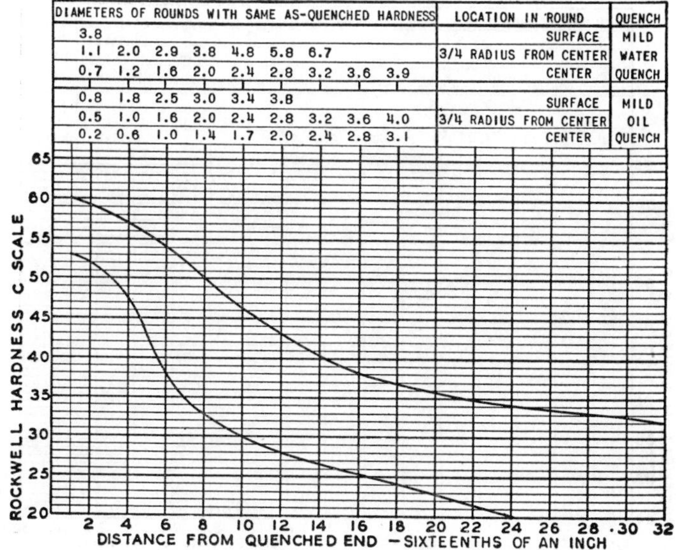

DIAMETERS OF ROUNDS WITH SAME AS-QUENCHED HARDNESS	LOCATION IN ROUND	QUENCH
3.8	SURFACE	MILD WATER QUENCH
1.1 2.0 2.9 3.8 4.8 5.8 6.7	3/4 RADIUS FROM CENTER	
0.7 1.2 1.6 2.0 2.4 2.8 3.2 3.6 3.9	CENTER	
0.8 1.8 2.5 3.0 3.4 3.8	SURFACE	MILD OIL QUENCH
0.5 1.0 1.6 2.0 2.4 2.8 3.2 3.6 4.0	3/4 RADIUS FROM CENTER	
0.2 0.6 1.0 1.4 1.7 2.0 2.4 2.8 3.1	CENTER	

HARDENABILITY BAND 5145 H

C	Mn	Si		Cr		
.42/.49	.60/1.00	.20/.35		.60/1.00		

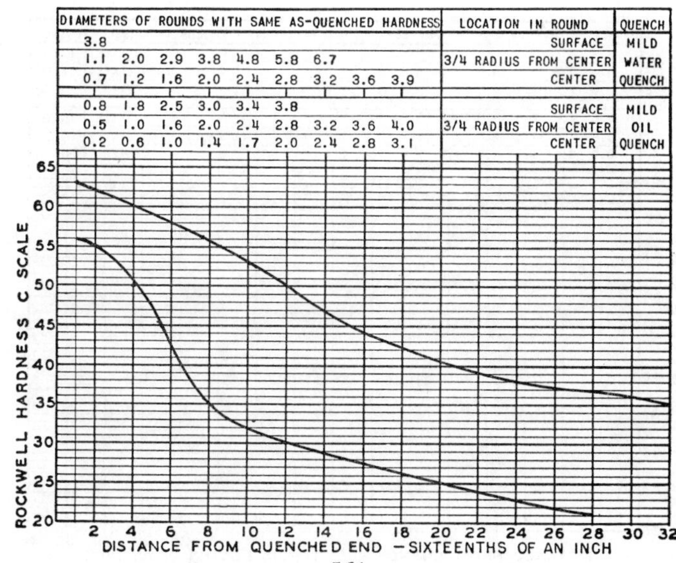

DIAMETERS OF ROUNDS WITH SAME AS-QUENCHED HARDNESS	LOCATION IN ROUND	QUENCH
3.8	SURFACE	MILD WATER QUENCH
1.1 2.0 2.9 3.8 4.8 5.8 6.7	3/4 RADIUS FROM CENTER	
0.7 1.2 1.6 2.0 2.4 2.8 3.2 3.6 3.9	CENTER	
0.8 1.8 2.5 3.0 3.4 3.8	SURFACE	MILD OIL QUENCH
0.5 1.0 1.6 2.0 2.4 2.8 3.2 3.6 4.0	3/4 RADIUS FROM CENTER	
0.2 0.6 1.0 1.4 1.7 2.0 2.4 2.8 3.1	CENTER	

HARDENABILITY BAND 5147 H

C	Mn	Si		Cr		
.45/.52	.60/1.05	.20/.35		.80/1.25		

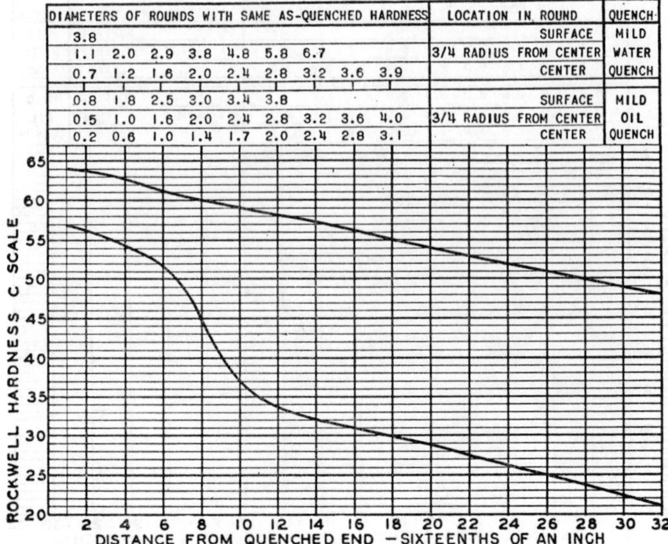

DIAMETERS OF ROUNDS WITH SAME AS-QUENCHED HARDNESS	LOCATION IN ROUND	QUENCH
3.8	SURFACE	MILD
1.1 2.0 2.9 3.8 4.8 5.8 6.7	3/4 RADIUS FROM CENTER	WATER
0.7 1.2 1.6 2.0 2.4 2.8 3.2 3.6 3.9	CENTER	QUENCH
0.8 1.8 2.5 3.0 3.4 3.8	SURFACE	MILD
0.5 1.0 1.6 2.0 2.4 2.8 3.2 3.6 4.0	3/4 RADIUS FROM CENTER	OIL
0.2 0.6 1.0 1.4 1.7 2.0 2.4 2.8 3.1	CENTER	QUENCH

HARDENABILITY BAND 5150 H

C	Mn	Si		Cr		
.47/.54	.60/1.00	.20/.35		.60/1.00		

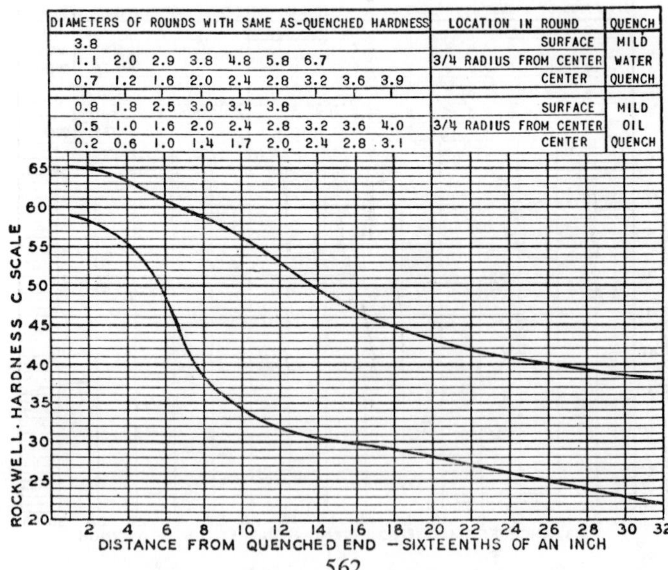

DIAMETERS OF ROUNDS WITH SAME AS-QUENCHED HARDNESS	LOCATION IN ROUND	QUENCH
3.8	SURFACE	MILD
1.1 2.0 2.9 3.8 4.8 5.8 6.7	3/4 RADIUS FROM CENTER	WATER
0.7 1.2 1.6 2.0 2.4 2.8 3.2 3.6 3.9	CENTER	QUENCH
0.8 1.8 2.5 3.0 3.4 3.8	SURFACE	MILD
0.5 1.0 1.6 2.0 2.4 2.8 3.2 3.6 4.0	3/4 RADIUS FROM CENTER	OIL
0.2 0.6 1.0 1.4 1.7 2.0 2.4 2.8 3.1	CENTER	QUENCH

HARDENABILITY BAND 5155 H

C	Mn	Si		Cr		
.50/.60	.60/1.00	.20/.35		.60/1.00		

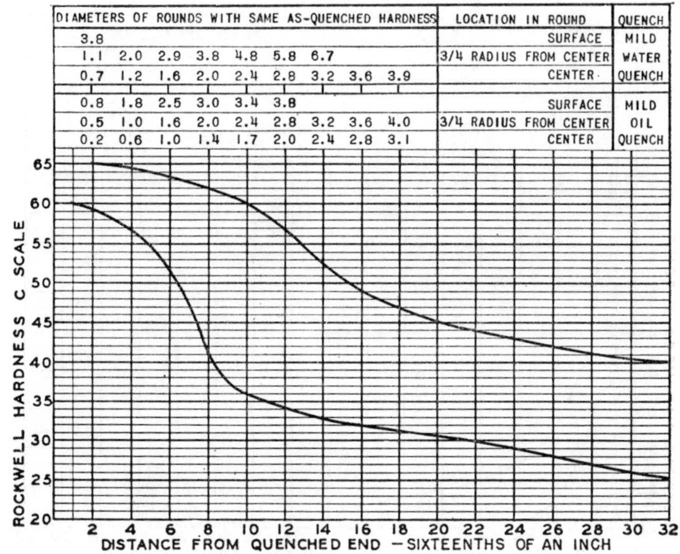

DIAMETERS OF ROUNDS WITH SAME AS-QUENCHED HARDNESS									LOCATION IN ROUND	QUENCH
3.8									SURFACE	MILD
1.1	2.0	2.9	3.8	4.8	5.8	6.7			3/4 RADIUS FROM CENTER	WATER
0.7	1.2	1.6	2.0	2.4	2.8	3.2	3.6	3.9	CENTER	QUENCH
0.8	1.8	2.5	3.0	3.4	3.8				SURFACE	MILD
0.5	1.0	1.6	2.0	2.4	2.8	3.2	3.6	4.0	3/4 RADIUS FROM CENTER	OIL
0.2	0.6	1.0	1.4	1.7	2.0	2.4	2.8	3.1	CENTER	QUENCH

HARDENABILITY BAND 5160 H

C	Mn	Si		Cr		
.55/.65	.65/1.10	.20/.35		.60/1.00		

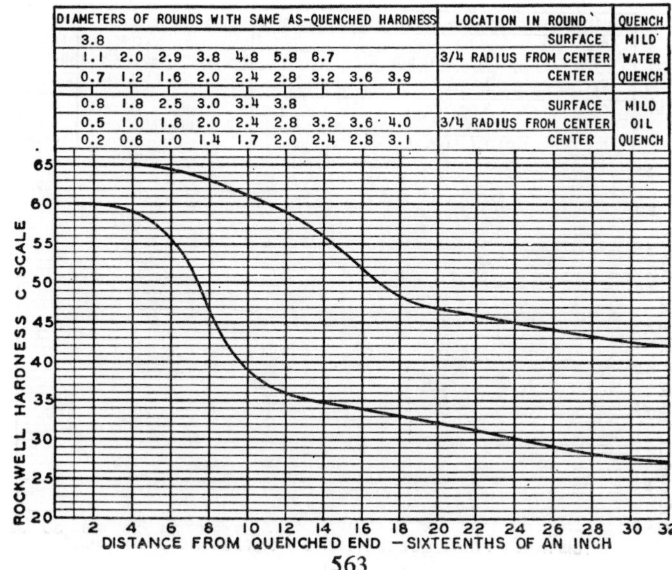

DIAMETERS OF ROUNDS WITH SAME AS-QUENCHED HARDNESS									LOCATION IN ROUND	QUENCH
3.8									SURFACE	MILD
1.1	2.0	2.9	3.8	4.8	5.8	6.7			3/4 RADIUS FROM CENTER	WATER
0.7	1.2	1.6	2.0	2.4	2.8	3.2	3.6	3.9	CENTER	QUENCH
0.8	1.8	2.5	3.0	3.4	3.8				SURFACE	MILD
0.5	1.0	1.6	2.0	2.4	2.8	3.2	3.6	4.0	3/4 RADIUS FROM CENTER	OIL
0.2	0.6	1.0	1.4	1.7	2.0	2.4	2.8	3.1	CENTER	QUENCH

HARDENABILITY BAND 6118H

C	Mn	Si		Cr		V
.15 / .21	.40 / .80	.20 / .35		.40 / .80		.10 / .15

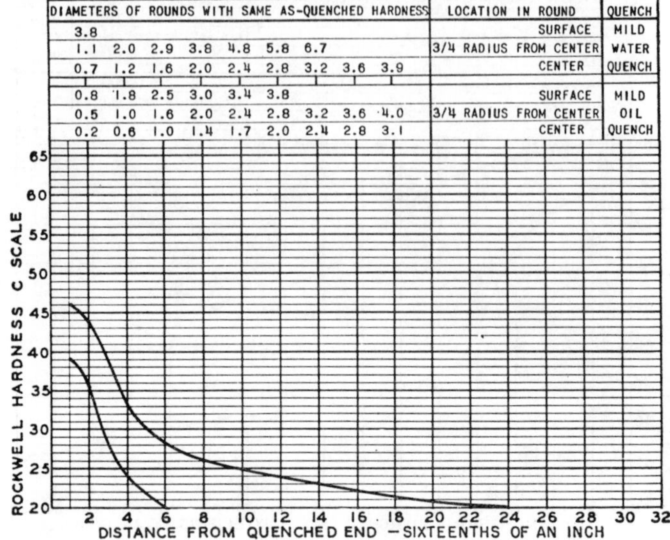

DIAMETERS OF ROUNDS WITH SAME AS-QUENCHED HARDNESS	LOCATION IN ROUND	QUENCH
3.8	SURFACE	MILD
1.1 2.0 2.9 3.8 4.8 5.8 6.7	3/4 RADIUS FROM CENTER	WATER
0.7 1.2 1.6 2.0 2.4 2.8 3.2 3.6 3.9	CENTER	QUENCH
0.8 1.8 2.5 3.0 3.4 3.8	SURFACE	MILD
0.5 1.0 1.6 2.0 2.4 2.8 3.2 3.6 4.0	3/4 RADIUS FROM CENTER	OIL
0.2 0.6 1.0 1.4 1.7 2.0 2.4 2.8 3.1	CENTER	QUENCH

HARDENABILITY BAND 6120 H

C	Mn	Si		Cr		V
.17 / .23	.60 / 1.00	.20 / .35		.60 / 1.00		.10 MIN.

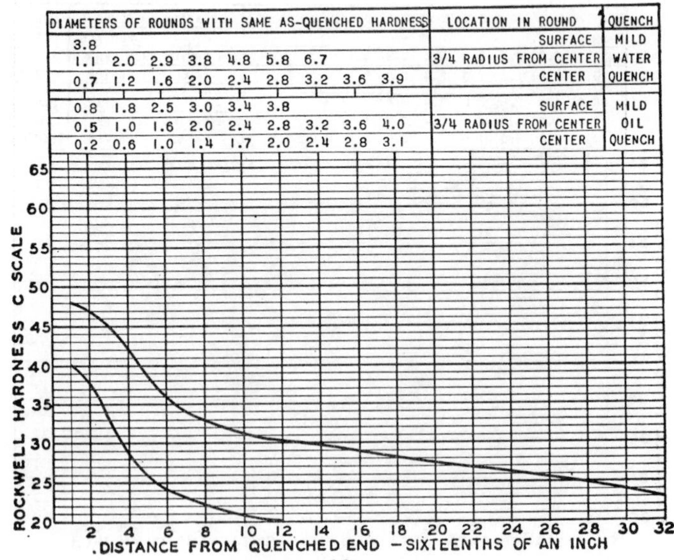

DIAMETERS OF ROUNDS WITH SAME AS-QUENCHED HARDNESS	LOCATION IN ROUND	QUENCH
3.8	SURFACE	MILD
1.1 2.0 2.9 3.8 4.8 5.8 6.7	3/4 RADIUS FROM CENTER	WATER
0.7 1.2 1.6 2.0 2.4 2.8 3.2 3.6 3.9	CENTER	QUENCH
0.8 1.8 2.5 3.0 3.4 3.8	SURFACE	MILD
0.5 1.0 1.6 2.0 2.4 2.8 3.2 3.6 4.0	3/4 RADIUS FROM CENTER	OIL
0.2 0.6 1.0 1.4 1.7 2.0 2.4 2.8 3.1	CENTER	QUENCH

HARDENABILITY BAND 6150 H

C	Mn	Si		Cr		V
.47/.54	.60/1.00	.20/.35		.75/1.20		.15 MIN.

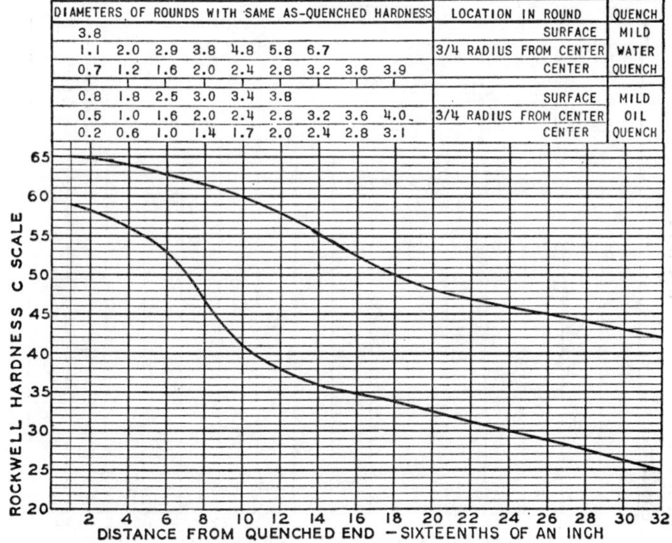

HARDENABILITY BAND 8617 H

C	Mn	Si	Ni	Cr	Mo	
.14/.20	.60/.95	.20/.35	.35/.75	.35/.65	.15/.25	

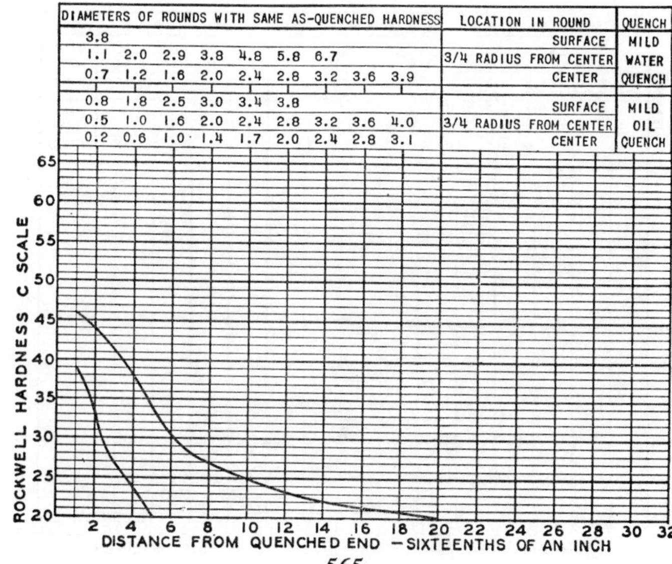

HARDENABILITY BAND 8620 H

C	Mn	Si	Ni	Cr	Mo	
.17/.23	.60/.95	.20/.35	.35/.75	.35/.65	.15/.25	

DIAMETERS OF ROUNDS WITH SAME AS-QUENCHED HARDNESS	LOCATION IN ROUND	QUENCH
3.8	SURFACE	MILD WATER QUENCH
1.1 2.0 2.9 3.8 4.8 5.8 6.7	3/4 RADIUS FROM CENTER	
0.7 1.2 1.6 2.0 2.4 2.8 3.2 3.6 3.9	CENTER	
0.8 1.8 2.5 3.0 3.4 3.8	SURFACE	MILD OIL QUENCH
0.5 1.0 1.6 2.0 2.4 2.8 3.2 3.6 4.0	3/4 RADIUS FROM CENTER	
0.2 0.6 1.0 1.4 1.7 2.0 2.4 2.8 3.1	CENTER	

ROCKWELL HARDNESS C SCALE

65 60 55 50 45 40 35 30 25 20

2 4 6 8 10 12 14 16 18 20 22 24 26 28 30 32

DISTANCE FROM QUENCHED END — SIXTEENTHS OF AN INCH

HARDENABILITY BAND 8622 H

C	Mn	Si	Ni	Cr	Mo	
.19/.25	.60/.95	.20/.35	.35/.75	.35/.65	.15/.25	

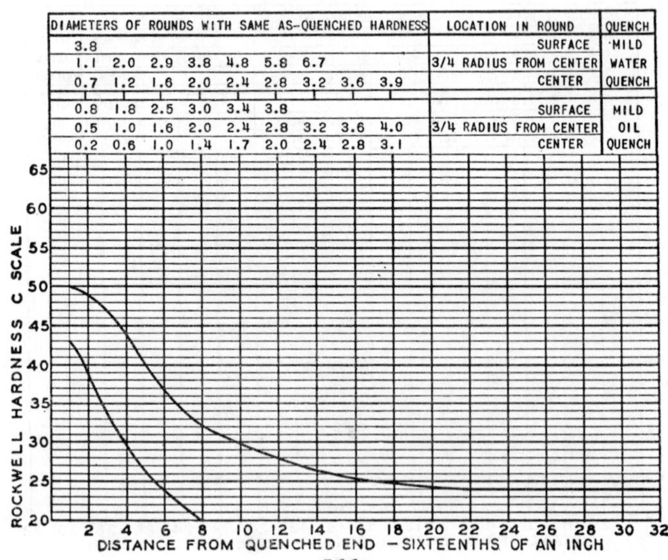

DIAMETERS OF ROUNDS WITH SAME AS-QUENCHED HARDNESS	LOCATION IN ROUND	QUENCH
3.8	SURFACE	MILD WATER QUENCH
1.1 2.0 2.9 3.8 4.8 5.8 6.7	3/4 RADIUS FROM CENTER	
0.7 1.2 1.6 2.0 2.4 2.8 3.2 3.6 3.9	CENTER	
0.8 1.8 2.5 3.0 3.4 3.8	SURFACE	MILD OIL QUENCH
0.5 1.0 1.6 2.0 2.4 2.8 3.2 3.6 4.0	3/4 RADIUS FROM CENTER	
0.2 0.6 1.0 1.4 1.7 2.0 2.4 2.8 3.1	CENTER	

HARDENABILITY BAND 8625 H

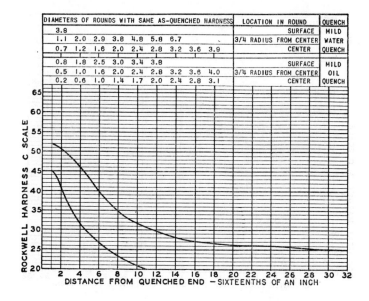

C	Mn	Si	Ni	Cr	Mo	
.22/.28	.60/.95	.20/.35	.35/.75	.35/.65	.15/.25	

DIAMETERS OF ROUNDS WITH SAME AS-QUENCHED HARDNESS	LOCATION IN ROUND	QUENCH
3.8	SURFACE	MILD WATER QUENCH
1.1 2.0 2.9 3.8 4.8 5.8 6.7	3/4 RADIUS FROM CENTER	
0.7 1.2 1.6 2.0 2.4 2.8 3.2 3.6 3.9	CENTER	
0.8 1.8 2.5 3.0 3.4 3.8	SURFACE	MILD OIL QUENCH
0.5 1.0 1.6 2.0 2.4 2.8 3.2 3.6 4.0	3/4 RADIUS FROM CENTER	
0.2 0.6 1.0 1.4 1.7 2.0 2.4 2.8 3.1	CENTER	

ROCKWELL HARDNESS C SCALE: 20 25 30 35 40 45 50 55 60 65

2 4 6 8 10 12 14 16 18 20 22 24 26 28 30 32

DISTANCE FROM QUENCHED END — SIXTEENTHS OF AN INCH

HARDENABILITY BAND 8627 H

C	Mn	Si	Ni	Cr	Mo	
.24/.30	.60/.95	.20/.35	.35/.75	.35/.65	.15/.25	

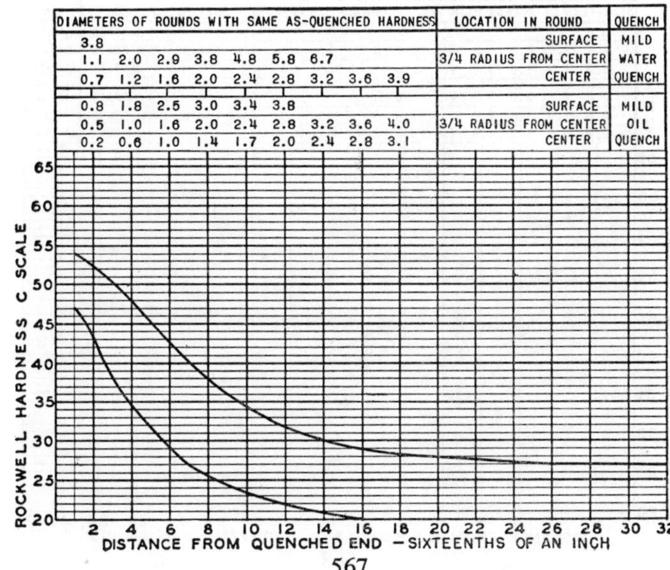

DIAMETERS OF ROUNDS WITH SAME AS-QUENCHED HARDNESS	LOCATION IN ROUND	QUENCH
3.8	SURFACE	MILD WATER QUENCH
1.1 2.0 2.9 3.8 4.8 5.8 6.7	3/4 RADIUS FROM CENTER	
0.7 1.2 1.6 2.0 2.4 2.8 3.2 3.6 3.9	CENTER	
0.8 1.8 2.5 3.0 3.4 3.8	SURFACE	MILD OIL QUENCH
0.5 1.0 1.6 2.0 2.4 2.8 3.2 3.6 4.0	3/4 RADIUS FROM CENTER	
0.2 0.6 1.0 1.4 1.7 2.0 2.4 2.8 3.1	CENTER	

HARDENABILITY BAND 8630 H

C	Mn	Si	Ni	Cr	Mo	
.27/.33	.60/.95	.20/.35	.35/.75	.35/.65	.15/.25	

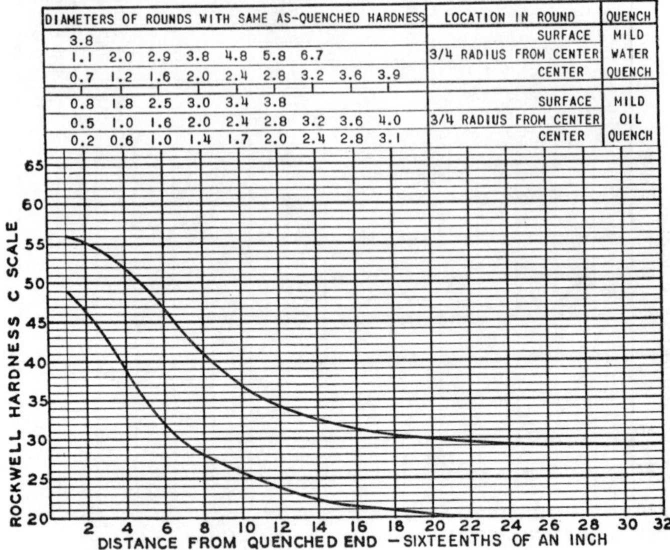

DIAMETERS OF ROUNDS WITH SAME AS-QUENCHED HARDNESS									LOCATION IN ROUND	QUENCH
3.8									SURFACE	MILD
1.1	2.0	2.9	3.8	4.8	5.8	6.7			3/4 RADIUS FROM CENTER	WATER
0.7	1.2	1.6	2.0	2.4	2.8	3.2	3.6	3.9	CENTER	QUENCH
0.8	1.8	2.5	3.0	3.4	3.8				SURFACE	MILD
0.5	1.0	1.6	2.0	2.4	2.8	3.2	3.6	4.0	3/4 RADIUS FROM CENTER	OIL
0.2	0.6	1.0	1.4	1.7	2.0	2.4	2.8	3.1	CENTER	QUENCH

HARDENABILITY BAND 8637 H

C	Mn	Si	Ni	Cr	Mo	
.34/.41	.70/1.05	.20/.35	.35/.75	.35/.65	.15/.25	

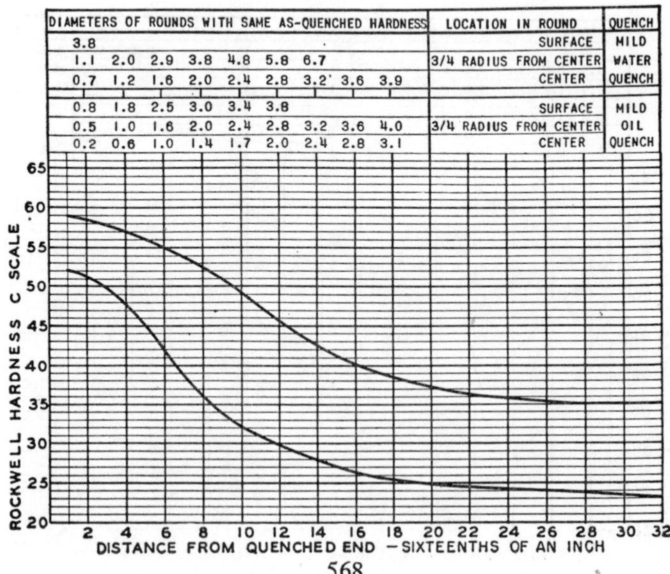

DIAMETERS OF ROUNDS WITH SAME AS-QUENCHED HARDNESS									LOCATION IN ROUND	QUENCH
3.8									SURFACE	MILD
1.1	2.0	2.9	3.8	4.8	5.8	6.7			3/4 RADIUS FROM CENTER	WATER
0.7	1.2	1.6	2.0	2.4	2.8	3.2	3.6	3.9	CENTER	QUENCH
0.8	1.8	2.5	3.0	3.4	3.8				SURFACE	MILD
0.5	1.0	1.6	2.0	2.4	2.8	3.2	3.6	4.0	3/4 RADIUS FROM CENTER	OIL
0.2	0.6	1.0	1.4	1.7	2.0	2.4	2.8	3.1	CENTER	QUENCH

HARDENABILITY BAND 8640 H

C	Mn	Si	Ni	Cr	Mo	
.37/.44	.70/1.05	.20/.35	.35/.75	.35/.65	.15/.25	

DIAMETERS OF ROUNDS WITH SAME AS-QUENCHED HARDNESS									LOCATION IN ROUND	QUENCH
3.8									SURFACE	MILD
1.1	2.0	2.9	3.8	4.8	5.8	6.7			3/4 RADIUS FROM CENTER	WATER
0.7	1.2	1.6	2.0	2.4	2.8	3.2	3.6	3.9	CENTER	QUENCH
0.8	1.8	2.5	3.0	3.4	3.8				SURFACE	MILD
0.5	1.0	1.6	2.0	2.4	2.8	3.2	3.6	4.0	3/4 RADIUS FROM CENTER	OIL
0.2	0.6	1.0	1.4	1.7	2.0	2.4	2.8	3.1	CENTER	QUENCH

ROCKWELL HARDNESS C SCALE

65 60 55 50 45 40 35 30 25 20

2 4 6 8 10 12 14 16 18 20 22 24 26 28 30 32

DISTANCE FROM QUENCHED END — SIXTEENTHS OF AN INCH

HARDENABILITY BAND 8642 H

C	Mn	Si	Ni	Cr	Mo	
.39/.46	.70/1.05	.20/.35	.35/.75	.35/.65	.15/.25	

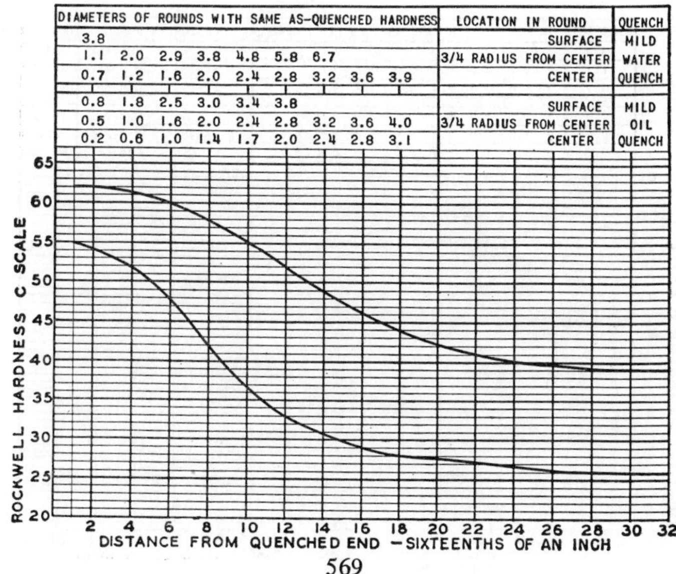

DIAMETERS OF ROUNDS WITH SAME AS-QUENCHED HARDNESS									LOCATION IN ROUND	QUENCH
3.8									SURFACE	MILD
1.1	2.0	2.9	3.8	4.8	5.8	6.7			3/4 RADIUS FROM CENTER	WATER
0.7	1.2	1.6	2.0	2.4	2.8	3.2	3.6	3.9	CENTER	QUENCH
0.8	1.8	2.5	3.0	3.4	3.8				SURFACE	MILD
0.5	1.0	1.6	2.0	2.4	2.8	3.2	3.6	4.0	3/4 RADIUS FROM CENTER	OIL
0.2	0.6	1.0	1.4	1.7	2.0	2.4	2.8	3.1	CENTER	QUENCH

HARDENABILITY BAND 8645 H

C	Mn	Si	Ni	Cr	Mo	
.42/.49	.70/1.05	.20/.35	.35/.75	.35/.65	.15/.25	

DIAMETERS OF ROUNDS WITH SAME AS-QUENCHED HARDNESS	LOCATION IN ROUND	QUENCH
3.8	SURFACE	MILD
1.1 2.0 2.9 3.8 4.8 5.8 6.7	3/4 RADIUS FROM CENTER	WATER
0.7 1.2 1.6 2.0 2.4 2.8 3.2 3.6 3.9	CENTER	QUENCH
0.8 1.8 2.5 3.0 3.4 3.8	SURFACE	MILD
0.5 1.0 1.6 2.0 2.4 2.8 3.2 3.6 4.0	3/4 RADIUS FROM CENTER	OIL
0.2 0.6 1.0 1.4 1.7 2.0 2.4 2.8 3.1	CENTER	QUENCH

ROCKWELL HARDNESS C SCALE

65 60 55 50 45 40 35 30 25 20

2 4 6 8 10 12 14 16 18 20 22 24 26 28 30 32

DISTANCE FROM QUENCHED END — SIXTEENTHS OF AN INCH

HARDENABILITY BAND 8650 H

C	Mn	Si	Ni	Cr	Mo	
.47/.54	.70/1.05	.20/.35	.35/.75	.35/.65	.15/.25	

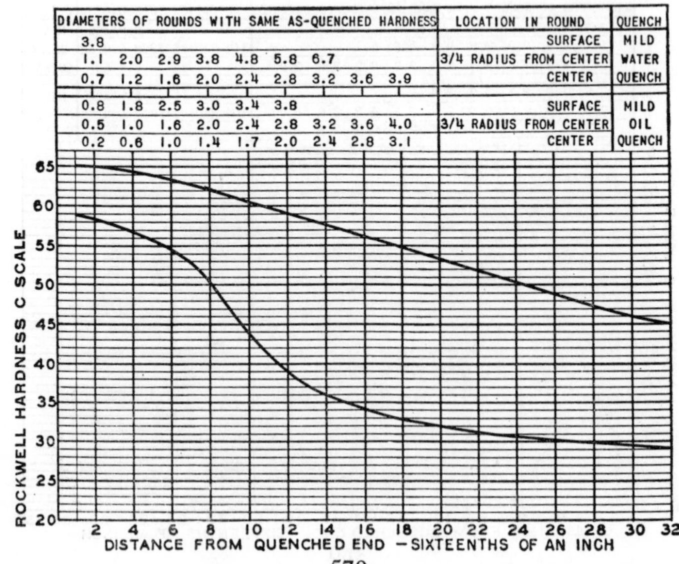

DIAMETERS OF ROUNDS WITH SAME AS-QUENCHED HARDNESS	LOCATION IN ROUND	QUENCH
3.8	SURFACE	MILD
1.1 2.0 2.9 3.8 4.8 5.8 6.7	3/4 RADIUS FROM CENTER	WATER
0.7 1.2 1.6 2.0 2.4 2.8 3.2 3.6 3.9	CENTER	QUENCH
0.8 1.8 2.5 3.0 3.4 3.8	SURFACE	MILD
0.5 1.0 1.6 2.0 2.4 2.8 3.2 3.6 4.0	3/4 RADIUS FROM CENTER	OIL
0.2 0.6 1.0 1.4 1.7 2.0 2.4 2.8 3.1	CENTER	QUENCH

HARDENABILITY BAND 8655 H

C	Mn	Si	Ni	Cr	Mo	
.50/.60	.70/1.05	.20/.35	.35/.75	.35/.65	.15/.25	

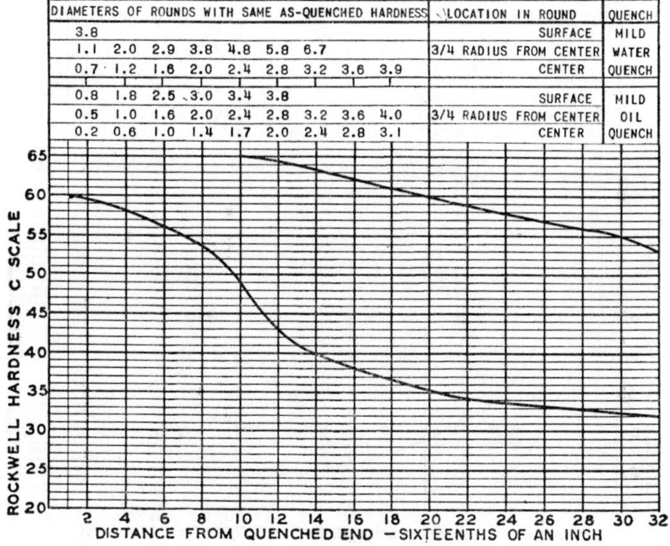

DIAMETERS OF ROUNDS WITH SAME AS-QUENCHED HARDNESS	LOCATION IN ROUND	QUENCH
3.8	SURFACE	MILD WATER QUENCH
1.1 2.0 2.9 3.8 4.8 5.8 6.7	3/4 RADIUS FROM CENTER	
0.7 1.2 1.6 2.0 2.4 2.8 3.2 3.6 3.9	CENTER	
0.8 1.8 2.5 3.0 3.4 3.8	SURFACE	MILD OIL QUENCH
0.5 1.0 1.6 2.0 2.4 2.8 3.2 3.6 4.0	3/4 RADIUS FROM CENTER	
0.2 0.6 1.0 1.4 1.7 2.0 2.4 2.8 3.1	CENTER	

HARDENABILITY BAND 8660 H

C	Mn	Si	Ni	Cr	Mo	
.55/.65	.70/1.05	.20/.35	.35/.75	.35/.65	.15/.25	

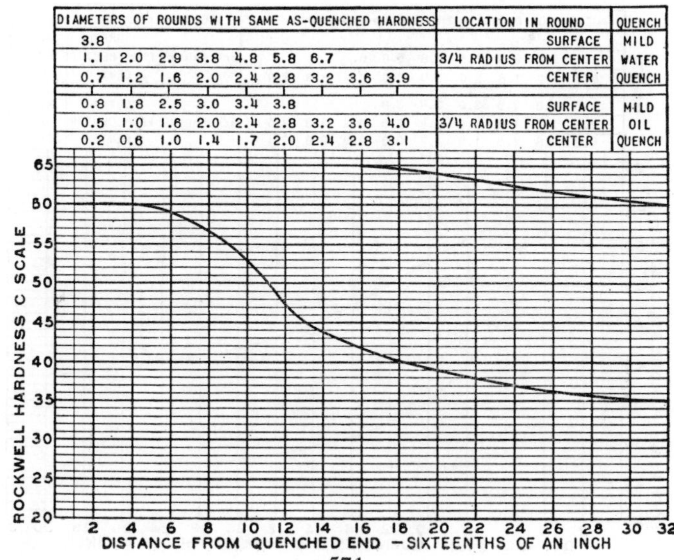

DIAMETERS OF ROUNDS WITH SAME AS-QUENCHED HARDNESS	LOCATION IN ROUND	QUENCH
3.8	SURFACE	MILD WATER QUENCH
1.1 2.0 2.9 3.8 4.8 5.8 6.7	3/4 RADIUS FROM CENTER	
0.7 1.2 1.6 2.0 2.4 2.8 3.2 3.6 3.9	CENTER	
0.8 1.8 2.5 3.0 3.4 3.8	SURFACE	MILD OIL QUENCH
0.5 1.0 1.6 2.0 2.4 2.8 3.2 3.6 4.0	3/4 RADIUS FROM CENTER	
0.2 0.6 1.0 1.4 1.7 2.0 2.4 2.8 3.1	CENTER	

HARDENABILITY BAND 8720 H

C	Mn	Si	Ni	Cr	Mo	
.17/.23	.60/.95	.20/.35	.35/.75	.35/.65	.20/.30	

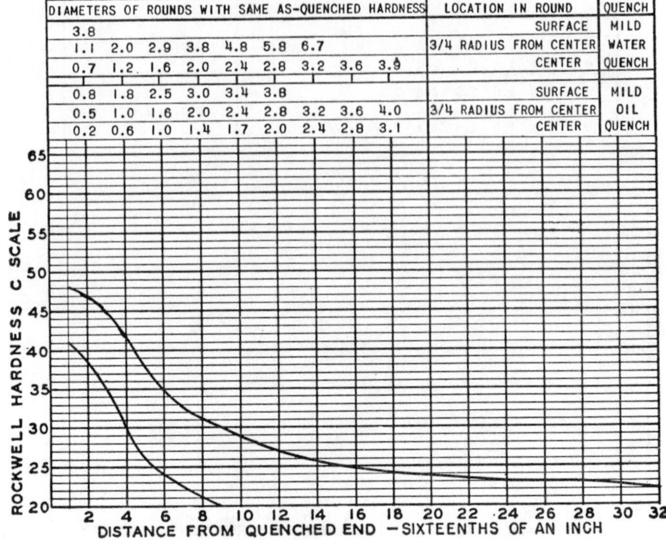

DIAMETERS OF ROUNDS WITH SAME AS-QUENCHED HARDNESS	LOCATION IN ROUND	QUENCH
3.8	SURFACE	MILD
1.1 2.0 2.9 3.8 4.8 5.8 6.7	3/4 RADIUS FROM CENTER	WATER
0.7 1.2 1.6 2.0 2.4 2.8 3.2 3.6 3.9	CENTER	QUENCH
0.8 1.8 2.5 3.0 3.4 3.8	SURFACE	MILD
0.5 1.0 1.6 2.0 2.4 2.8 3.2 3.6 4.0	3/4 RADIUS FROM CENTER	OIL
0.2 0.6 1.0 1.4 1.7 2.0 2.4 2.8 3.1	CENTER	QUENCH

HARDENABILITY BAND 8740 H

C	Mn	Si	Ni	Cr	Mo	
.37/.44	.70/1.05	.20/.35	.35/.75	.35/.65	.20/.30	

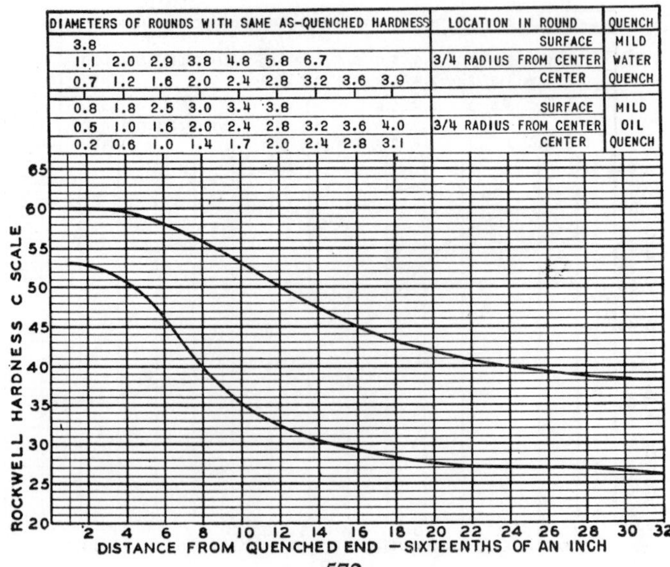

DIAMETERS OF ROUNDS WITH SAME AS-QUENCHED HARDNESS	LOCATION IN ROUND	QUENCH
3.8	SURFACE	MILD
1.1 2.0 2.9 3.8 4.8 5.8 6.7	3/4 RADIUS FROM CENTER	WATER
0.7 1.2 1.6 2.0 2.4 2.8 3.2 3.6 3.9	CENTER	QUENCH
0.8 1.8 2.5 3.0 3.4 3.8	SURFACE	MILD
0.5 1.0 1.6 2.0 2.4 2.8 3.2 3.6 4.0	3/4 RADIUS FROM CENTER	OIL
0.2 0.6 1.0 1.4 1.7 2.0 2.4 2.8 3.1	CENTER	QUENCH

HARDENABILITY BAND 8742 H

C	Mn	Si	Ni	Cr	Mo	
.39/.46	.70/1.05	.20/.35	.35/.75	.35/.65	.20/.30	

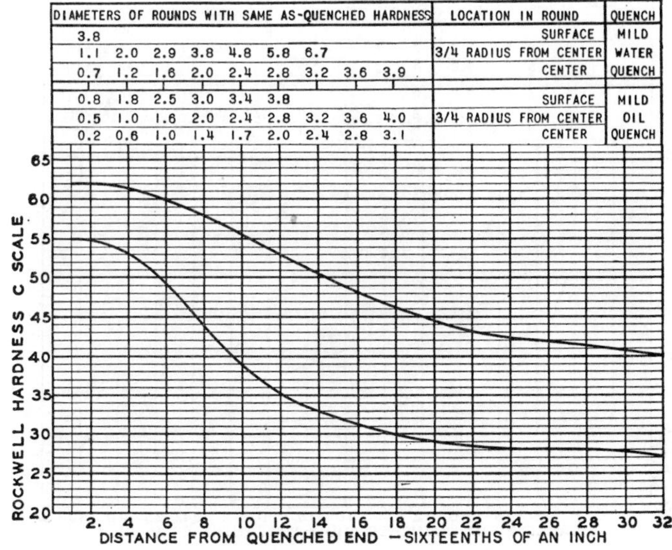

DIAMETERS OF ROUNDS WITH SAME AS-QUENCHED HARDNESS	LOCATION IN ROUND	QUENCH
3.8	SURFACE	MILD
1.1 2.0 2.9 3.8 4.8 5.8 6.7	3/4 RADIUS FROM CENTER	WATER
0.7 1.2 1.6 2.0 2.4 2.8 3.2 3.6 3.9	CENTER	QUENCH
0.8 1.8 2.5 3.0 3.4 3.8	SURFACE	MILD
0.5 1.0 1.6 2.0 2.4 2.8 3.2 3.6 4.0	3/4 RADIUS FROM CENTER	OIL
0.2 0.6 1.0 1.4 1.7 2.0 2.4 2.8 3.1	CENTER	QUENCH

HARDENABILITY BAND 8822H

C	Mn	Si	Ni	Cr	Mo	
.19/.25	.70/1.05	.20/.35	.35/.75	.35/.65	.30/.40	

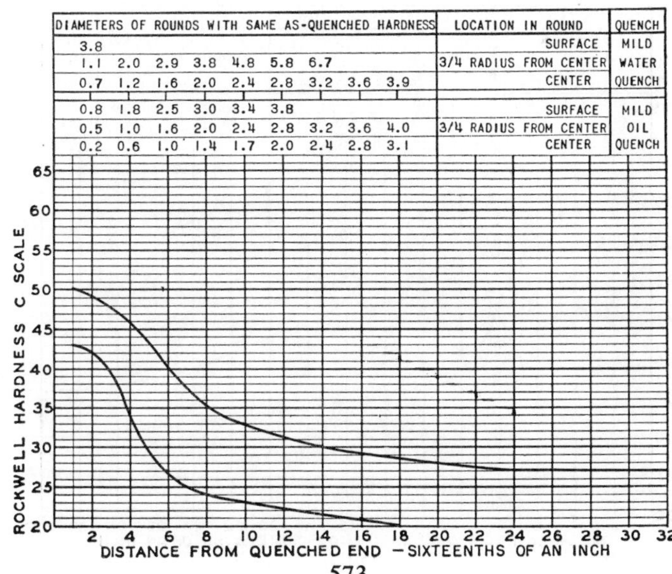

DIAMETERS OF ROUNDS WITH SAME AS-QUENCHED HARDNESS	LOCATION IN ROUND	QUENCH
3.8	SURFACE	MILD
1.1 2.0 2.9 3.8 4.8 5.8 6.7	3/4 RADIUS FROM CENTER	WATER
0.7 1.2 1.6 2.0 2.4 2.8 3.2 3.6 3.9	CENTER	QUENCH
0.8 1.8 2.5 3.0 3.4 3.8	SURFACE	MILD
0.5 1.0 1.6 2.0 2.4 2.8 3.2 3.6 4.0	3/4 RADIUS FROM CENTER	OIL
0.2 0.6 1.0 1.4 1.7 2.0 2.4 2.8 3.1	CENTER	QUENCH

HARDENABILITY BAND 9260 H

C	Mn	Si				
.55/.65	.65/1.10	1.70/2.20				

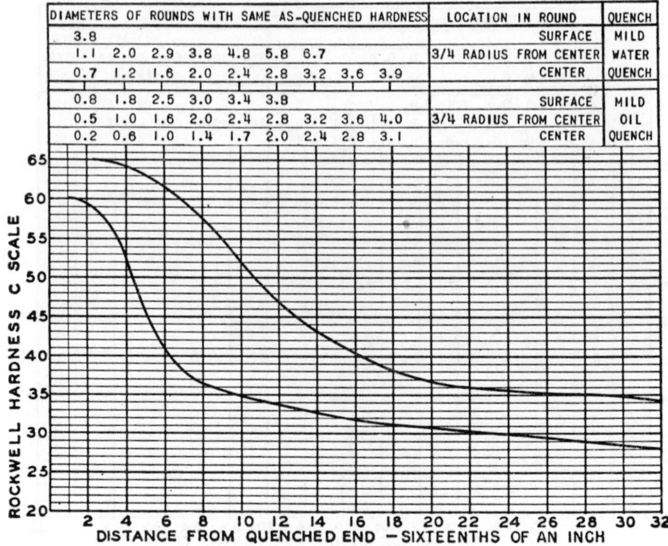

DIAMETERS OF ROUNDS WITH SAME AS-QUENCHED HARDNESS	LOCATION IN ROUND	QUENCH
3.8	SURFACE	MILD
1.1 2.0 2.9 3.8 4.8 5.8 6.7	3/4 RADIUS FROM CENTER	WATER
0.7 1.2 1.6 2.0 2.4 2.8 3.2 3.6 3.9	CENTER	QUENCH
0.8 1.8 2.5 3.0 3.4 3.8	SURFACE	MILD
0.5 1.0 1.6 2.0 2.4 2.8 3.2 3.6 4.0	3/4 RADIUS FROM CENTER	OIL
0.2 0.6 1.0 1.4 1.7 2.0 2.4 2.8 3.1	CENTER	QUENCH

HARDENABILITY BAND 9262 H

C	Mn	Si		Cr		
.55/.65	.65/1.10	1.70/2.20		.20/.50		

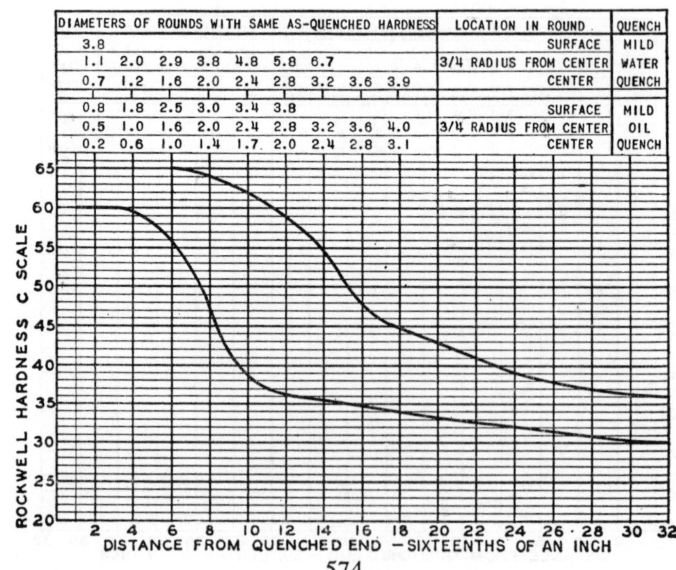

DIAMETERS OF ROUNDS WITH SAME AS-QUENCHED HARDNESS	LOCATION IN ROUND	QUENCH
3.8	SURFACE	MILD
1.1 2.0 2.9 3.8 4.8 5.8 6.7	3/4 RADIUS FROM CENTER	WATER
0.7 1.2 1.6 2.0 2.4 2.8 3.2 3.6 3.9	CENTER	QUENCH
0.8 1.8 2.5 3.0 3.4 3.8	SURFACE	MILD
0.5 1.0 1.6 2.0 2.4 2.8 3.2 3.6 4.0	3/4 RADIUS FROM CENTER	OIL
0.2 0.6 1.0 1.4 1.7 2.0 2.4 2.8 3.1	CENTER	QUENCH

HARDENABILITY BAND 9310 H

C	Mn	Si	Ni	Cr	Mo	
.07/.13	.40/.70	.20/.35	2.95/3.55	1.00/1.45	.08/.15	

DIAMETERS OF ROUNDS WITH SAME AS-QUENCHED HARDNESS									LOCATION IN ROUND	QUENCH
3.8									SURFACE	MILD
1.1	2.0	2.9	3.8	4.8	5.8	6.7			3/4 RADIUS FROM CENTER	WATER
0.7	1.2	1.6	2.0	2.4	2.8	3.2	3.6	3.9	CENTER	QUENCH
0.8	1.8	2.5	3.0	3.4	3.8				SURFACE	MILD
0.5	1.0	1.6	2.0	2.4	2.8	3.2	3.6	4.0	3/4 RADIUS FROM CENTER	OIL
0.2	0.6	1.0	1.4	1.7	2.0	2.4	2.8	3.1	CENTER	QUENCH

ROCKWELL HARDNESS C SCALE
65 60 55 50 45 40 35 30 25 20
2 4 6 8 10 12 14 16 18 20 22 24 26 28 30 32
DISTANCE FROM QUENCHED END – SIXTEENTHS OF AN INCH

HARDENABILITY BAND 9840 H

C	Mn	Si	Ni	Cr	Mo	
.37/.44	.60/.95	.20/.35	.80/1.20	.65/.95	.20/.30	

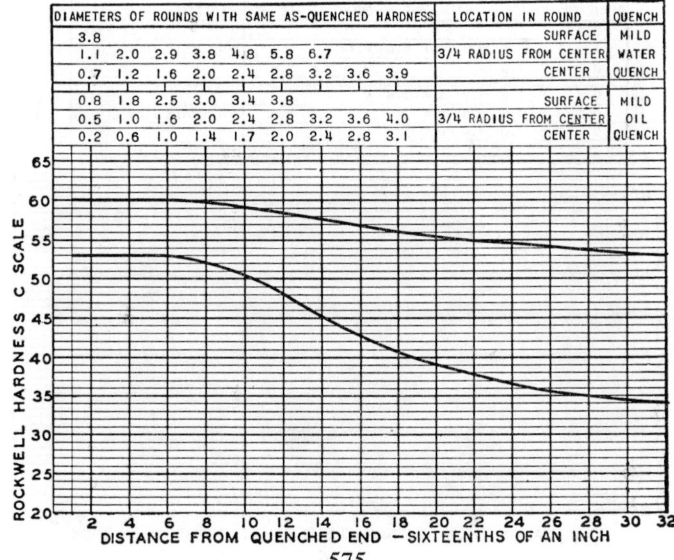

DIAMETERS OF ROUNDS WITH SAME AS-QUENCHED HARDNESS									LOCATION IN ROUND	QUENCH
3.8									SURFACE	MILD
1.1	2.0	2.9	3.8	4.8	5.8	6.7			3/4 RADIUS FROM CENTER	WATER
0.7	1.2	1.6	2.0	2.4	2.8	3.2	3.6	3.9	CENTER	QUENCH
0.8	1.8	2.5	3.0	3.4	3.8				SURFACE	MILD
0.5	1.0	1.6	2.0	2.4	2.8	3.2	3.6	4.0	3/4 RADIUS FROM CENTER	OIL
0.2	0.6	1.0	1.4	1.7	2.0	2.4	2.8	3.1	CENTER	QUENCH

HARDENABILITY BAND 9850 H

C	Mn	Si	Ni	Cr	Mo	
.47/.54	.60/.95	.20/.35	.80/1.20	.65/.95	.20/.30	

DIAMETERS OF ROUNDS WITH SAME AS-QUENCHED HARDNESS	LOCATION IN ROUND	QUENCH
3.8	SURFACE	MILD
1.1 2.0 2.9 3.8 4.8 5.8 6.7	3/4 RADIUS FROM CENTER	WATER
0.7 1.2 1.6 2.0 2.4 2.8 3.2 3.6 3.9	CENTER	QUENCH
0.8 1.8 2.5 3.0 3.4 3.8	SURFACE	MILD
0.5 1.0 1.6 2.0 2.4 2.8 3.2 3.6 4.0	3/4 RADIUS FROM CENTER	OIL
0.2 0.6 1.0 1.4 1.7 2.0 2.4 2.8 3.1	CENTER	QUENCH

ROCKWELL HARDNESS C SCALE: 20 25 30 35 40 45 50 55 60 65

2 4 6 8 10 12 14 16 18 20 22 24 26 28 30 32

DISTANCE FROM QUENCHED END — SIXTEENTHS OF AN INCH

HARDENABILITY BAND TS4140 H

C	Mn	Si		Cr	Mo	
.37/.44	.70/1.20	.20/.35		.85/1.30	.08/.15	

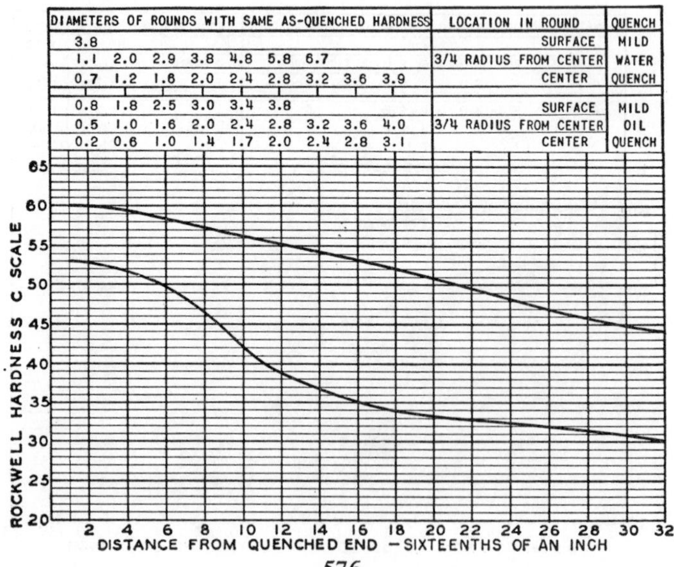

DIAMETERS OF ROUNDS WITH SAME AS-QUENCHED HARDNESS	LOCATION IN ROUND	QUENCH
3.8	SURFACE	MILD
1.1 2.0 2.9 3.8 4.8 5.8 6.7	3/4 RADIUS FROM CENTER	WATER
0.7 1.2 1.6 2.0 2.4 2.8 3.2 3.6 3.9	CENTER	QUENCH
0.8 1.8 2.5 3.0 3.4 3.8	SURFACE	MILD
0.5 1.0 1.6 2.0 2.4 2.8 3.2 3.6 4.0	3/4 RADIUS FROM CENTER	OIL
0.2 0.6 1.0 1.4 1.7 2.0 2.4 2.8 3.1	CENTER	QUENCH

HARDENABILITY BAND TS4150 H

C	Mn	Si		Cr	Mo	
.47/.54	.70/1.20	.20/.35		.85/1.30	.08/.15	

DIAMETERS OF ROUNDS WITH SAME AS-QUENCHED HARDNESS									LOCATION IN ROUND	QUENCH
3.8									SURFACE	MILD
1.1	2.0	2.9	3.8	4.8	5.8	6.7			3/4 RADIUS FROM CENTER	WATER
0.7	1.2	1.6	2.0	2.4	2.8	3.2	3.6	3.9	CENTER	QUENCH
0.8	1.8	2.5	3.0	3.4	3.8				SURFACE	MILD
0.5	1.0	1.6	2.0	2.4	2.8	3.2	3.6	4.0	3/4 RADIUS FROM CENTER	OIL
0.2	0.6	1.0	1.4	1.7	2.0	2.4	2.8	3.1	CENTER	QUENCH

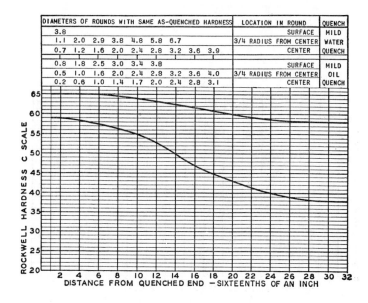

ROCKWELL HARDNESS C SCALE: 20, 25, 30, 35, 40, 45, 50, 55, 60, 65

2 4 6 8 10 12 14 16 18 20 22 24 26 28 30 32

DISTANCE FROM QUENCHED END — SIXTEENTHS OF AN INCH

HARDENABILITY BAND TS14B35 H

C	Mn	Si				B
.32/.38	.65/1.10	.20/.35				*

* Can be expected to have 0.0005 per cent minimum boron content.

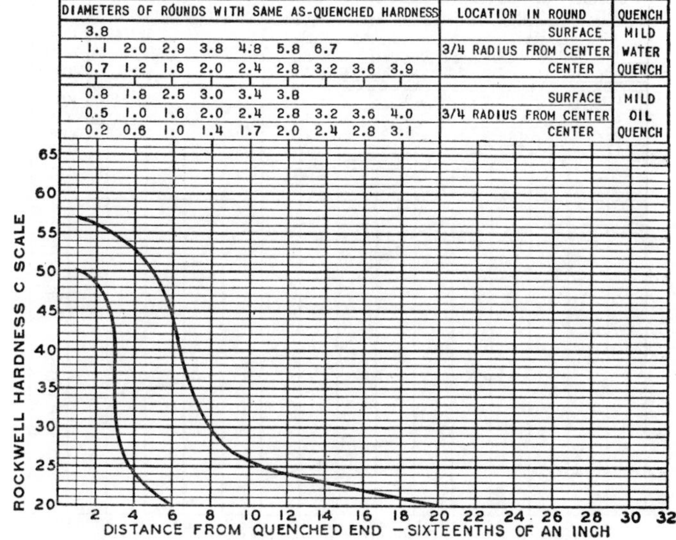

DIAMETERS OF ROUNDS WITH SAME AS-QUENCHED HARDNESS									LOCATION IN ROUND	QUENCH
3.8									SURFACE	MILD
1.1	2.0	2.9	3.8	4.8	5.8	6.7			3/4 RADIUS FROM CENTER	WATER
0.7	1.2	1.6	2.0	2.4	2.8	3.2	3.6	3.9	CENTER	QUENCH
0.8	1.8	2.5	3.0	3.4	3.8				SURFACE	MILD
0.5	1.0	1.6	2.0	2.4	2.8	3.2	3.6	4.0	3/4 RADIUS FROM CENTER	OIL
0.2	0.6	1.0	1.4	1.7	2.0	2.4	2.8	3.1	CENTER	QUENCH

HARDENABILITY BAND 50B40H

C	Mn	Si		Cr		B
.37/.44	.65/1.10	.20/.35		.30/.70		*

* Can be expected to have 0.0005 per cent minimum boron content.

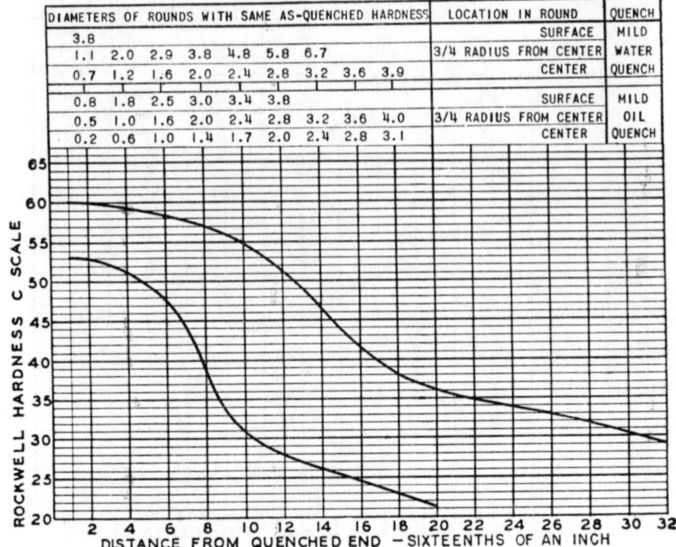

DIAMETERS OF ROUNDS WITH SAME AS-QUENCHED HARDNESS									LOCATION IN ROUND	QUENCH
3.8									SURFACE	MILD
1.1	2.0	2.9	3.8	4.8	5.8	6.7			3/4 RADIUS FROM CENTER	WATER
0.7	1.2	1.6	2.0	2.4	2.8	3.2	3.6	3.9	CENTER	QUENCH
0.8	1.8	2.5	3.0	3.4	3.8				SURFACE	MILD
0.5	1.0	1.6	2.0	2.4	2.8	3.2	3.6	4.0	3/4 RADIUS FROM CENTER	OIL
0.2	0.6	1.0	1.4	1.7	2.0	2.4	2.8	3.1	CENTER	QUENCH

HARDENABILITY BAND 50B44H

C	Mn	Si		Cr		B
.42/.49	.65/1.10	.20/.35		.30/.70		*

* Can be expected to have 0.0005 per cent minimum boron content.

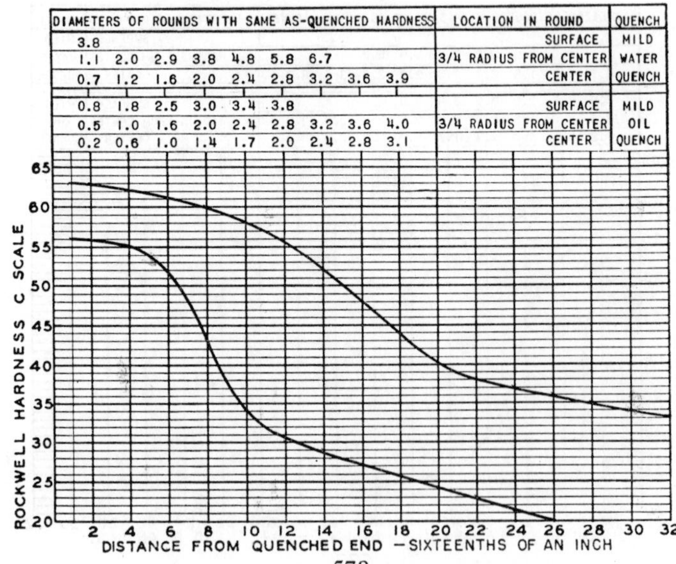

DIAMETERS OF ROUNDS WITH SAME AS-QUENCHED HARDNESS									LOCATION IN ROUND	QUENCH
3.8									SURFACE	MILD
1.1	2.0	2.9	3.8	4.8	5.8	6.7			3/4 RADIUS FROM CENTER	WATER
0.7	1.2	1.6	2.0	2.4	2.8	3.2	3.6	3.9	CENTER	QUENCH
0.8	1.8	2.5	3.0	3.4	3.8				SURFACE	MILD
0.5	1.0	1.6	2.0	2.4	2.8	3.2	3.6	4.0	3/4 RADIUS FROM CENTER	OIL
0.2	0.6	1.0	1.4	1.7	2.0	2.4	2.8	3.1	CENTER	QUENCH

HARDENABILITY BAND 50B46 H

C	Mn	Si		Cr		B
.43/.50	.65/1.10	.20/.35		.13/.43		*

* Can be expected to have 0.0005 per cent minimum boron content.

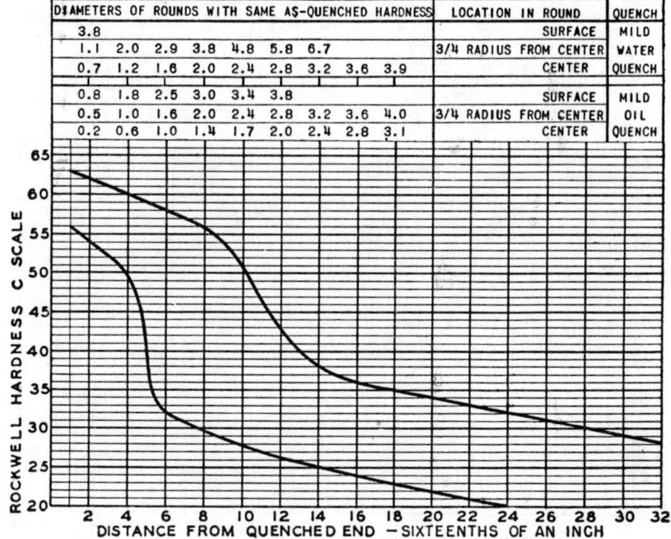

DIAMETERS OF ROUNDS WITH SAME AS-QUENCHED HARDNESS									LOCATION IN ROUND	QUENCH
3.8									SURFACE	MILD
1.1	2.0	2.9	3.8	4.8	5.8	6.7			3/4 RADIUS FROM CENTER	WATER
0.7	1.2	1.6	2.0	2.4	2.8	3.2	3.6	3.9	CENTER	QUENCH
0.8	1.8	2.5	3.0	3.4	3.8				SURFACE	MILD
0.5	1.0	1.6	2.0	2.4	2.8	3.2	3.6	4.0	3/4 RADIUS FROM CENTER	OIL
0.2	0.6	1.0	1.4	1.7	2.0	2.4	2.8	3.1	CENTER	QUENCH

HARDENABILITY BAND 50B50 H

C	Mn	Si		Cr		B
.47/.54	.65/1.10	.20/.35		.30/.70		*

* Can be expected to have 0.0005 per cent minimum boron content.

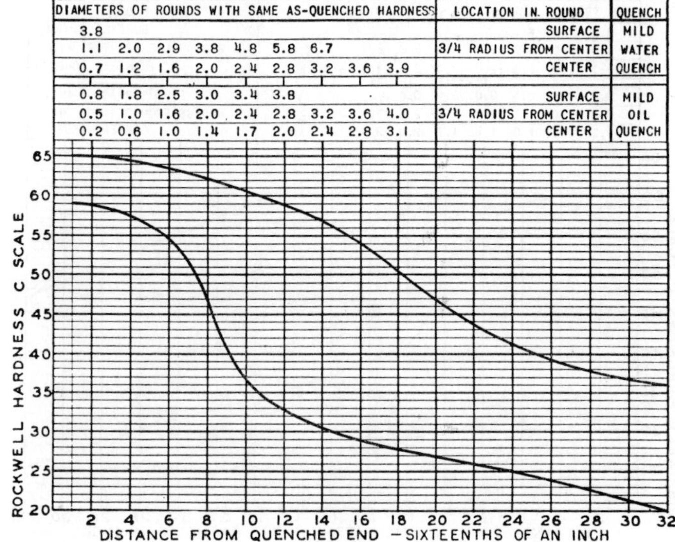

DIAMETERS OF ROUNDS WITH SAME AS-QUENCHED HARDNESS									LOCATION IN ROUND	QUENCH
3.8									SURFACE	MILD
1.1	2.0	2.9	3.8	4.8	5.8	6.7			3/4 RADIUS FROM CENTER	WATER
0.7	1.2	1.6	2.0	2.4	2.8	3.2	3.6	3.9	CENTER	QUENCH
0.8	1.8	2.5	3.0	3.4	3.8				SURFACE	MILD
0.5	1.0	1.6	2.0	2.4	2.8	3.2	3.6	4.0	3/4 RADIUS FROM CENTER	OIL
0.2	0.6	1.0	1.4	1.7	2.0	2.4	2.8	3.1	CENTER	QUENCH

HARDENABILITY BAND 50B60 H

C	Mn	Si		Cr		B
.55/.65	.65/1.10	.20/.35		.30/.70		*

* Can be expected to have 0.0005 per cent minimum boron content.

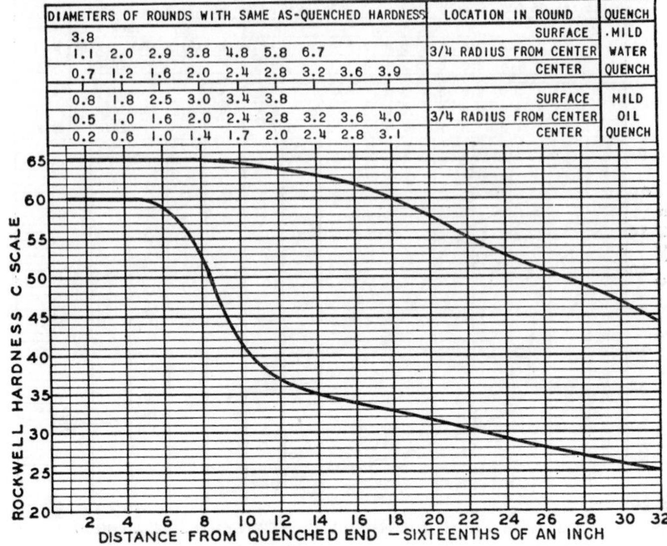

DIAMETERS OF ROUNDS WITH SAME AS-QUENCHED HARDNESS									LOCATION IN ROUND	QUENCH
3.8									SURFACE	MILD
1.1	2.0	2.9	3.8	4.8	5.8	6.7			3/4 RADIUS FROM CENTER	WATER
0.7	1.2	1.6	2.0	2.4	2.8	3.2	3.6	3.9	CENTER	QUENCH
0.8	1.8	2.5	3.0	3.4	3.8				SURFACE	MILD
0.5	1.0	1.6	2.0	2.4	2.8	3.2	3.6	4.0	3/4 RADIUS FROM CENTER	OIL
0.2	0.6	1.0	1.4	1.7	2.0	2.4	2.8	3.1	CENTER	QUENCH

HARDENABILITY BAND 51B60 H

C	Mn	Si		Cr		B
.55/.65	.65/1.10	.20/.35		.60/1.00		*

* Can be expected to have 0.0005 per cent minimum boron content.

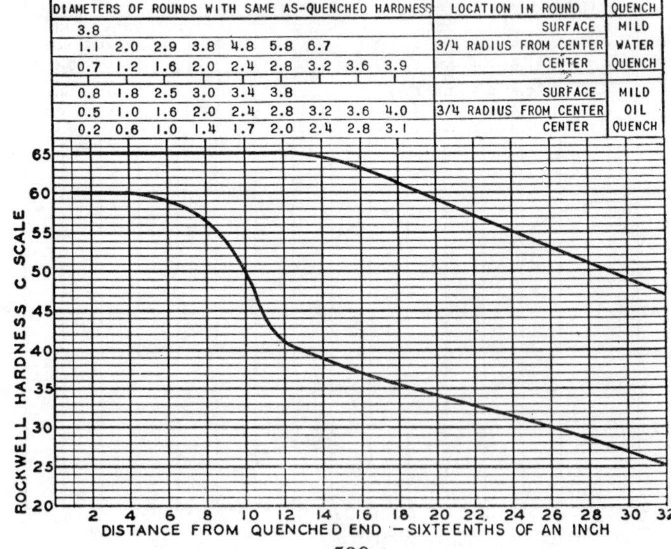

DIAMETERS OF ROUNDS WITH SAME AS-QUENCHED HARDNESS									LOCATION IN ROUND	QUENCH
3.8									SURFACE	MILD
1.1	2.0	2.9	3.8	4.8	5.8	6.7			3/4 RADIUS FROM CENTER	WATER
0.7	1.2	1.6	2.0	2.4	2.8	3.2	3.6	3.9	CENTER	QUENCH
0.8	1.8	2.5	3.0	3.4	3.8				SURFACE	MILD
0.5	1.0	1.6	2.0	2.4	2.8	3.2	3.6	4.0	3/4 RADIUS FROM CENTER	OIL
0.2	0.6	1.0	1.4	1.7	2.0	2.4	2.8	3.1	CENTER	QUENCH

HARDENABILITY BAND 81B45 H

C	Mn	Si	Ni	Cr	Mo	B
.42/.49	.70/1.05	.20/.35	.15/.45	.30/.60	.08/.15	*

* Can be expected to have 0.0005 per cent minimun boron content.

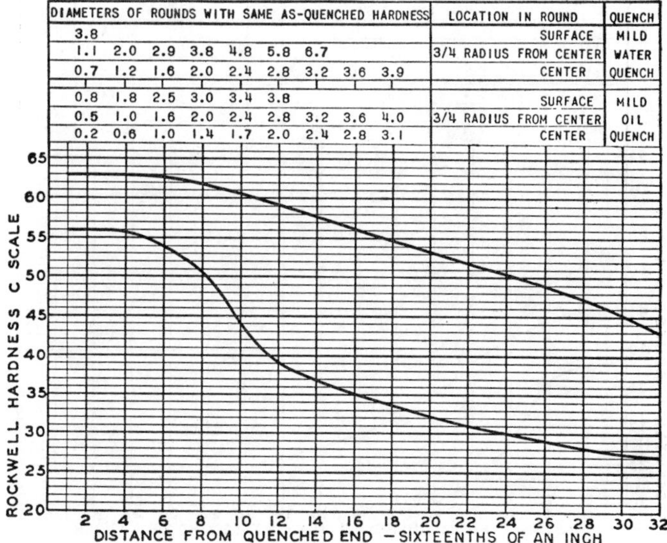

DIAMETERS OF ROUNDS WITH SAME AS-QUENCHED HARDNESS	LOCATION IN ROUND	QUENCH
3.8	SURFACE	MILD
1.1 2.0 2.9 3.8 4.8 5.8 6.7	3/4 RADIUS FROM CENTER	WATER
0.7 1.2 1.6 2.0 2.4 2.8 3.2 3.6 3.9	CENTER	QUENCH
0.8 1.8 2.5 3.0 3.4 3.8	SURFACE	MILD
0.5 1.0 1.6 2.0 2.4 2.8 3.2 3.6 4.0	3/4 RADIUS FROM CENTER	OIL
0.2 0.6 1.0 1.4 1.7 2.0 2.4 2.8 3.1	CENTER	QUENCH

HARDENABILITY BAND 86B45 H

C	Mn	Si	Ni	Cr	Mo	B
.42/.49	.70/1.05	.20/.35	.35/.75	.35/.65	.15/.25	*

* Can be expected to have 0.0005 per cent minimum boron content.

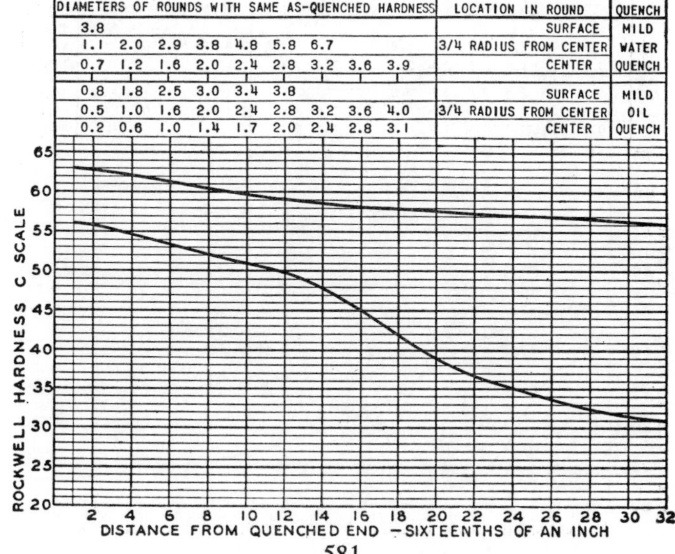

DIAMETERS OF ROUNDS WITH SAME AS-QUENCHED HARDNESS	LOCATION IN ROUND	QUENCH
3.8	SURFACE	MILD
1.1 2.0 2.9 3.8 4.8 5.8 6.7	3/4 RADIUS FROM CENTER	WATER
0.7 1.2 1.6 2.0 2.4 2.8 3.2 3.6 3.9	CENTER	QUENCH
0.8 1.8 2.5 3.0 3.4 3.8	SURFACE	MILD
0.5 1.0 1.6 2.0 2.4 2.8 3.2 3.6 4.0	3/4 RADIUS FROM CENTER	OIL
0.2 0.6 1.0 1.4 1.7 2.0 2.4 2.8 3.1	CENTER	QUENCH

HARDENABILITY BAND 94B15 H

C	Mn	Si	Ni	Cr	Mo	B
.12/.18	.70/1.05	.20/.35	.25/.65	.25/.55	.08/.15	*

* Can be expected to have 0.0005 per cent minimum boron content.

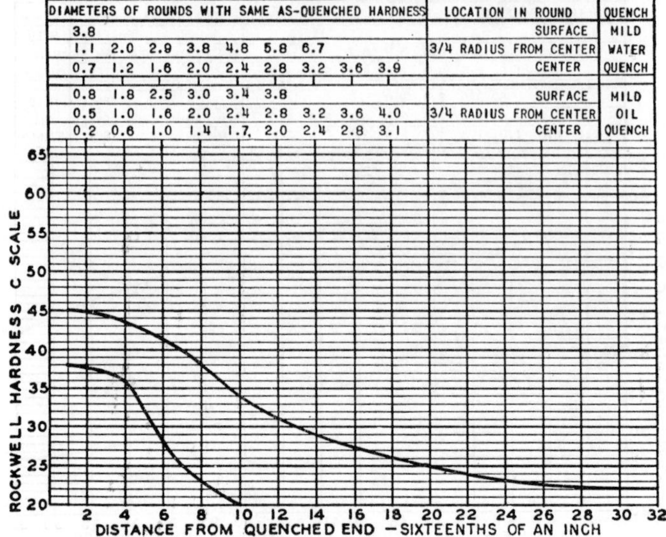

DIAMETERS OF ROUNDS WITH SAME AS-QUENCHED HARDNESS	LOCATION IN ROUND	QUENCH
3.8	SURFACE	MILD
1.1 2.0 2.9 3.8 4.8 5.8 6.7	3/4 RADIUS FROM CENTER	WATER
0.7 1.2 1.6 2.0 2.4 2.8 3.2 3.6 3.9	CENTER	QUENCH
0.8 1.8 2.5 3.0 3.4 3.8	SURFACE	MILD
0.5 1.0 1.6 2.0 2.4 2.8 3.2 3.6 4.0	3/4 RADIUS FROM CENTER	OIL
0.2 0.6 1.0 1.4 1.7 2.0 2.4 2.8 3.1	CENTER	QUENCH

HARDENABILITY BAND 94B17 H

C	Mn	Si	Ni	Cr	Mo	B
.14/.20	.70/1.05	.20/.35	.25/.65	.25/.55	.08/.15	*

Can be expected to have 0.0005 per cent minimum boron content.

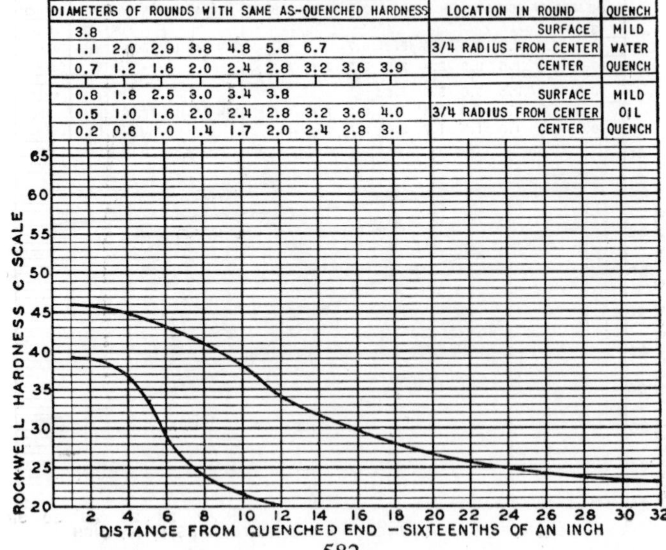

DIAMETERS OF ROUNDS WITH SAME AS-QUENCHED HARDNESS	LOCATION IN ROUND	QUENCH
3.8	SURFACE	MILD
1.1 2.0 2.9 3.8 4.8 5.8 6.7	3/4 RADIUS FROM CENTER	WATER
0.7 1.2 1.6 2.0 2.4 2.8 3.2 3.6 3.9	CENTER	QUENCH
0.8 1.8 2.5 3.0 3.4 3.8	SURFACE	MILD
0.5 1.0 1.6 2.0 2.4 2.8 3.2 3.6 4.0	3/4 RADIUS FROM CENTER	OIL
0.2 0.6 1.0 1.4 1.7 2.0 2.4 2.8 3.1	CENTER	QUENCH

HARDENABILITY BAND 94B30H

C	Mn	Si	Ni	Cr	Mo	B
.27/.33	.70/1.05	.20/.35	.25/.65	.25/.55	.08/.15	*

* Can be expected to have 0.0005 per cent minimum boron content.

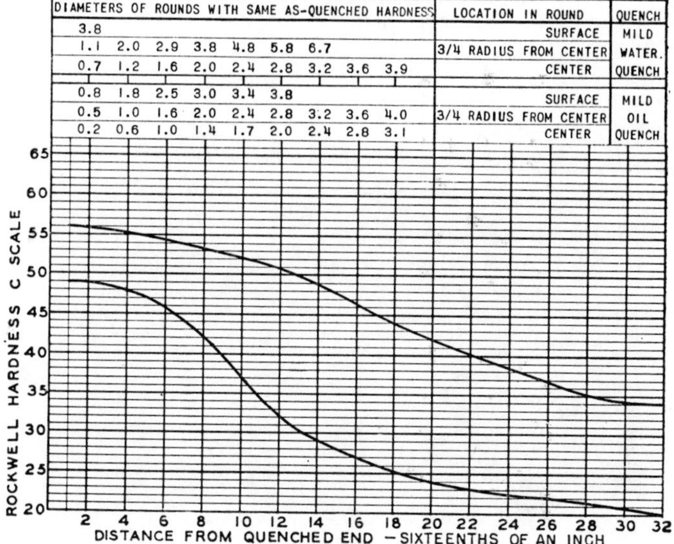

DIAMETERS OF ROUNDS WITH SAME AS-QUENCHED HARDNESS	LOCATION IN ROUND	QUENCH
3.8	SURFACE	MILD WATER QUENCH
1.1 2.0 2.9 3.8 4.8 5.8 6.7	3/4 RADIUS FROM CENTER	
0.7 1.2 1.6 2.0 2.4 2.8 3.2 3.6 3.9	CENTER	
0.8 1.8 2.5 3.0 3.4 3.8	SURFACE	MILD OIL QUENCH
0.5 1.0 1.6 2.0 2.4 2.8 3.2 3.6 4.0	3/4 RADIUS FROM CENTER	
0.2 0.6 1.0 1.4 1.7 2.0 2.4 2.8 3.1	CENTER	

HARDENABILITY BAND 94B40H

C	Mn	Si	Ni	Cr	Mo	B
.37/.44	.70/1.05	.20/.35	.25/.65	.25/.55	.08/.15	*

* Can be expected to have 0.0005 per cent minimum boron content.

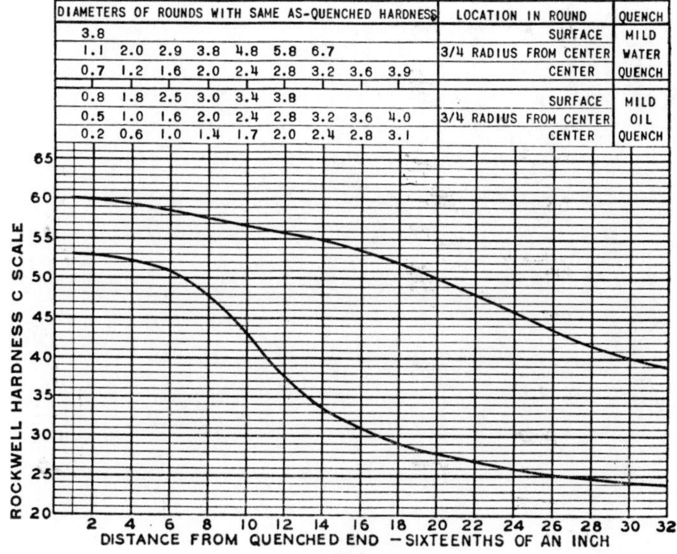

DIAMETERS OF ROUNDS WITH SAME AS-QUENCHED HARDNESS	LOCATION IN ROUND	QUENCH
3.8	SURFACE	MILD WATER QUENCH
1.1 2.0 2.9 3.8 4.8 5.8 6.7	3/4 RADIUS FROM CENTER	
0.7 1.2 1.6 2.0 2.4 2.8 3.2 3.6 3.9	CENTER	
0.8 1.8 2.5 3.0 3.4 3.8	SURFACE	MILD OIL QUENCH
0.5 1.0 1.6 2.0 2.4 2.8 3.2 3.6 4.0	3/4 RADIUS FROM CENTER	
0.2 0.6 1.0 1.4 1.7 2.0 2.4 2.8 3.1	CENTER	

APPENDIX F

TYPICAL END-QUENCH AND TEMPERED END-QUENCH CURVES

(Courtesy Joseph T. Ryerson and Son, Inc.)

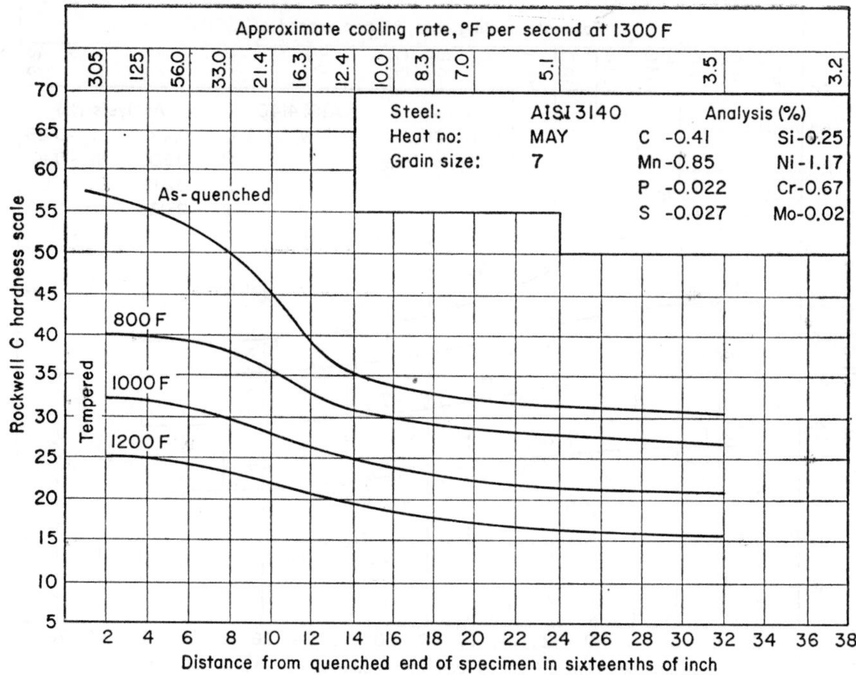
Approximate cooling rate, °F per second at 1300 F
305
125
56.0
33.0
21.4
16.3
12.4
10.0
8.3
7.0
5.1
3.5
3.2
Steel: AISI 3140
Heat no: MAY
Grain size: 7
Analysis (%)
C -0.41
Mn -0.85
P -0.022
S -0.027
Si-0.25
Ni-1.17
Cr-0.67
Mo-0.02
As-quenched
800 F
1000 F
1200 F
Tempered
Rockwell C hardness scale
Distance from quenched end of specimen in sixteenths of inch

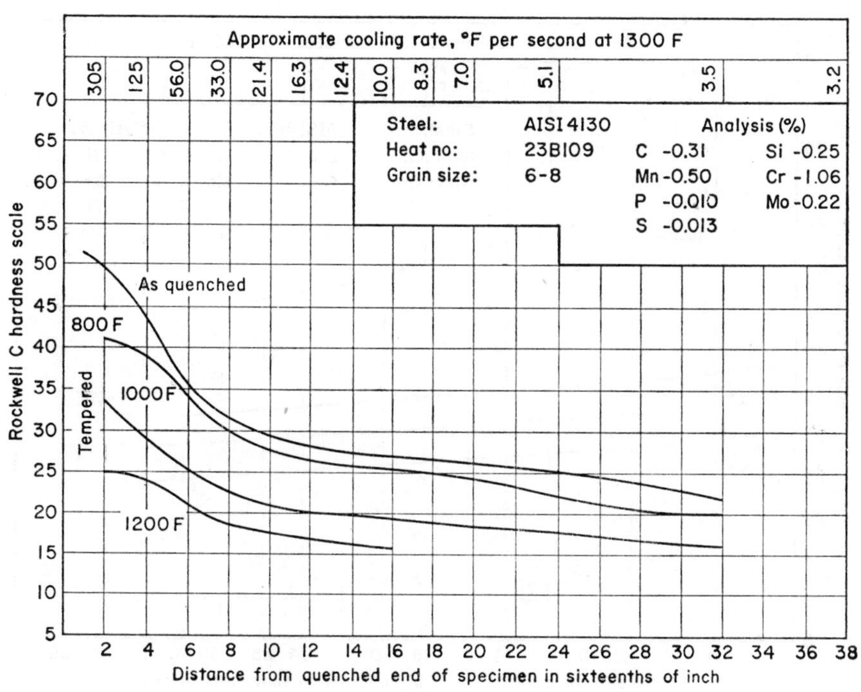
Approximate cooling rate, °F per second at 1300 F
305
125
56.0
33.0
21.4
16.3
12.4
10.0
8.3
7.0
5.1
3.5
3.2
Steel: AISI 4130
Heat no: 23B109
Grain size: 6-8
Analysis (%)
C -0.31
Mn -0.50
P -0.010
S -0.013
Si -0.25
Cr -1.06
Mo -0.22
As quenched
800 F
1000 F
1200 F
Tempered
Rockwell C hardness scale
Distance from quenched end of specimen in sixteenths of inch

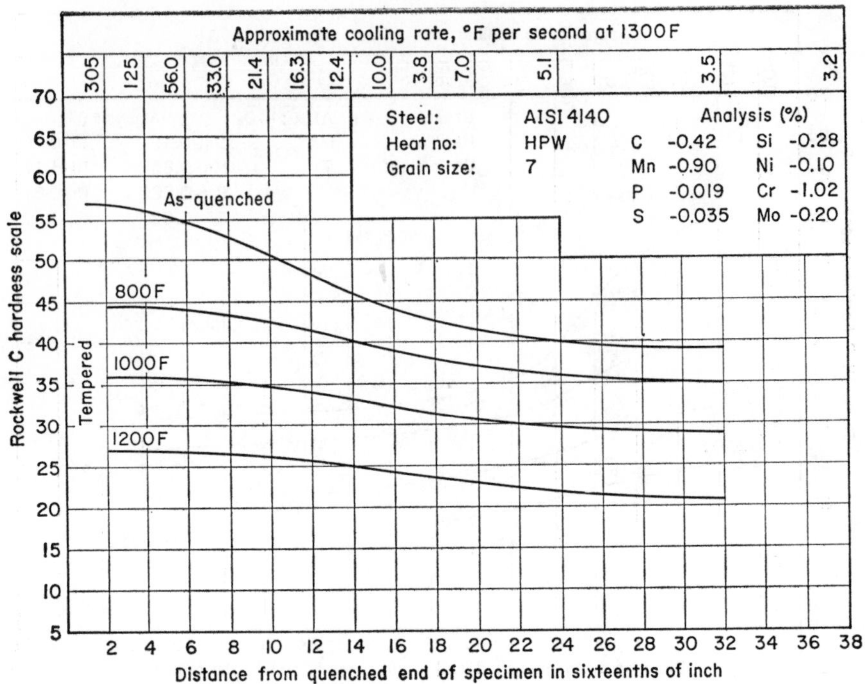

Approximate cooling rate, °F per second at 1300F
305
125
56.0
33.0
21.4
16.3
12.4
10.0
3.8
7.0
5.1
3.5
3.2
Steel: AISI 4140
Heat no: HPW
Grain size: 7
Analysis (%)
C -0.42 Si -0.28
Mn -0.90 Ni -0.10
P -0.019 Cr -1.02
S -0.035 Mo -0.20
As-quenched
800F
1000F
1200F
Tempered
Rockwell C hardness scale
Distance from quenched end of specimen in sixteenths of inch

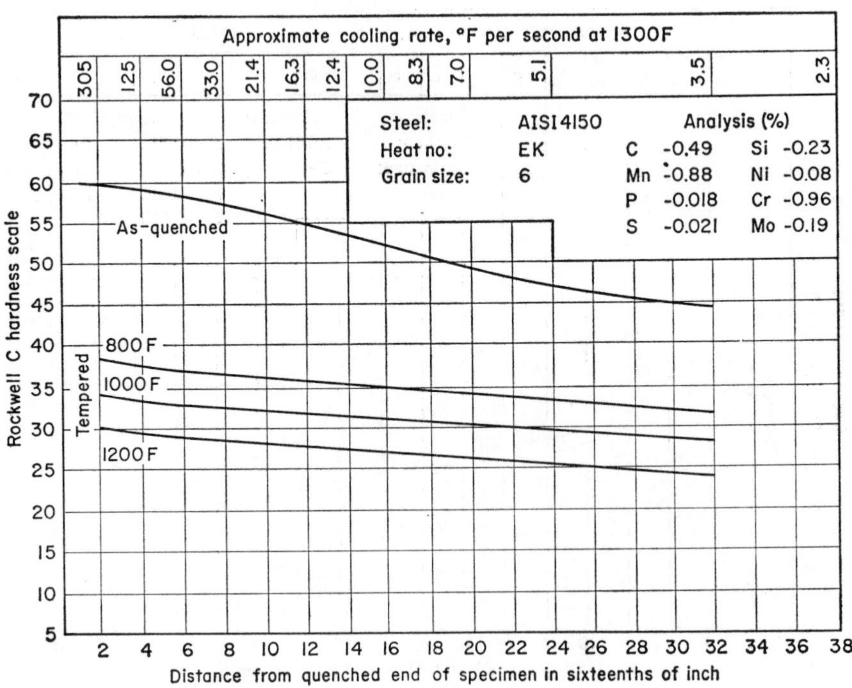

Approximate cooling rate, °F per second at 1300F
305
125
56.0
33.0
21.4
16.3
12.4
10.0
8.3
7.0
5.1
3.5
2.3
Steel: AISI 4150
Heat no: EK
Grain size: 6
Analysis (%)
C -0.49 Si -0.23
Mn -0.88 Ni -0.08
P -0.018 Cr -0.96
S -0.021 Mo -0.19
As-quenched
800F
1000F
1200F
Tempered
Rockwell C hardness scale
Distance from quenched end of specimen in sixteenths of inch

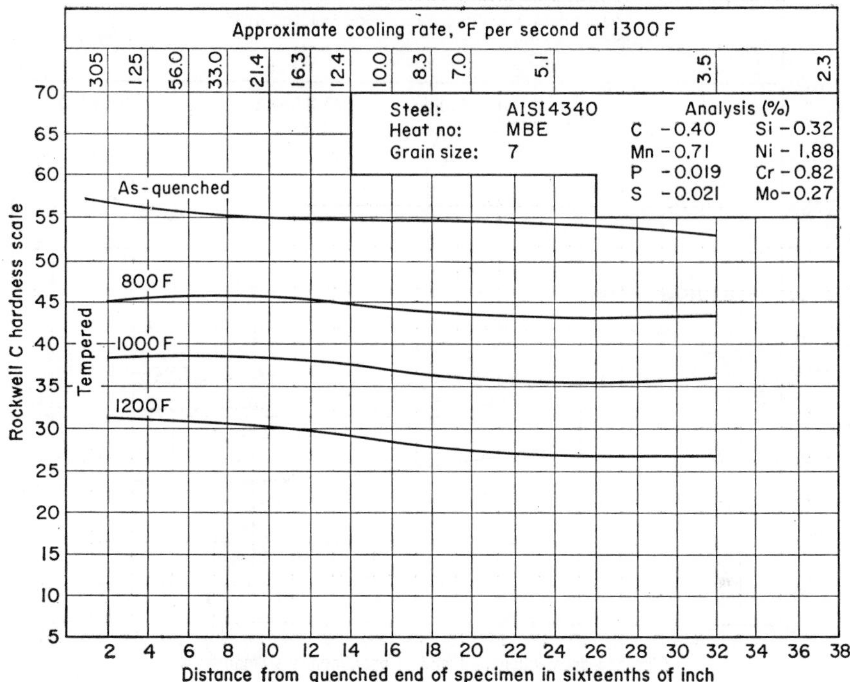
Approximate cooling rate, °F per second at 1300 F
305
125
56.0
33.0
21.4
16.3
12.4
10.0
8.3
7.0
5.1
3.5
2.3
Steel: AISI4340
Heat no: MBE
Grain size: 7
Analysis (%)
C -0.40
Mn -0.71
P -0.019
S -0.021
Si -0.32
Ni - 1.88
Cr -0.82
Mo-0.27
As-quenched
800 F
1000 F
1200 F
Tempered
Rockwell C hardness scale
Distance from quenched end of specimen in sixteenths of inch

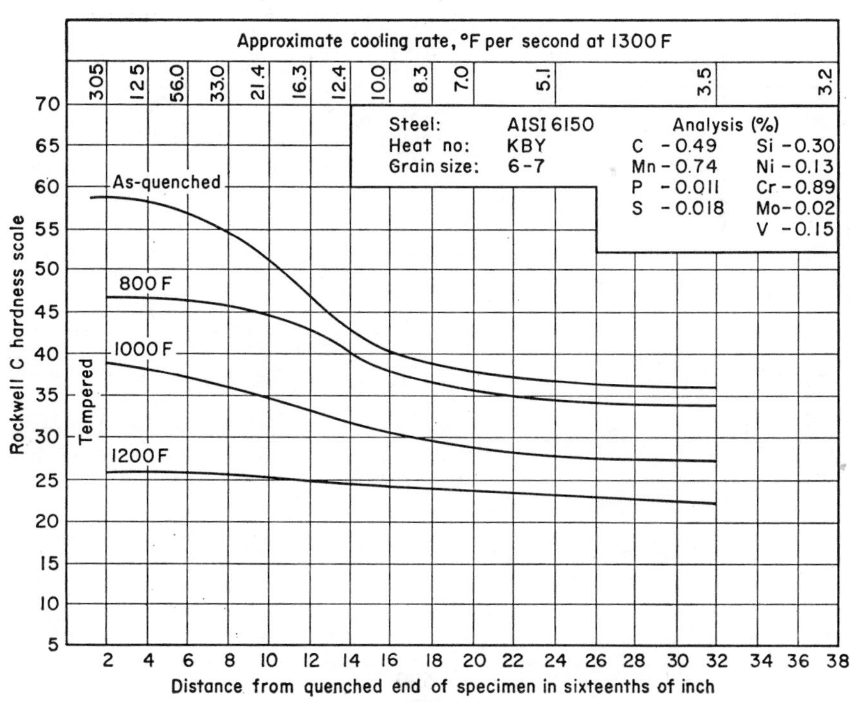
Approximate cooling rate, °F per second at 1300 F
305
125
56.0
33.0
21.4
16.3
12.4
10.0
8.3
7.0
5.1
3.5
3.2
Steel: AISI 6150
Heat no: KBY
Grain size: 6-7
Analysis (%)
C -0.49
Mn -0.74
P -0.011
S -0.018
Si -0.30
Ni -0.13
Cr -0.89
Mo-0.02
V -0.15
As-quenched
800 F
1000 F
1200 F
Tempered
Rockwell C hardness scale
Distance from quenched end of specimen in sixteenths of inch

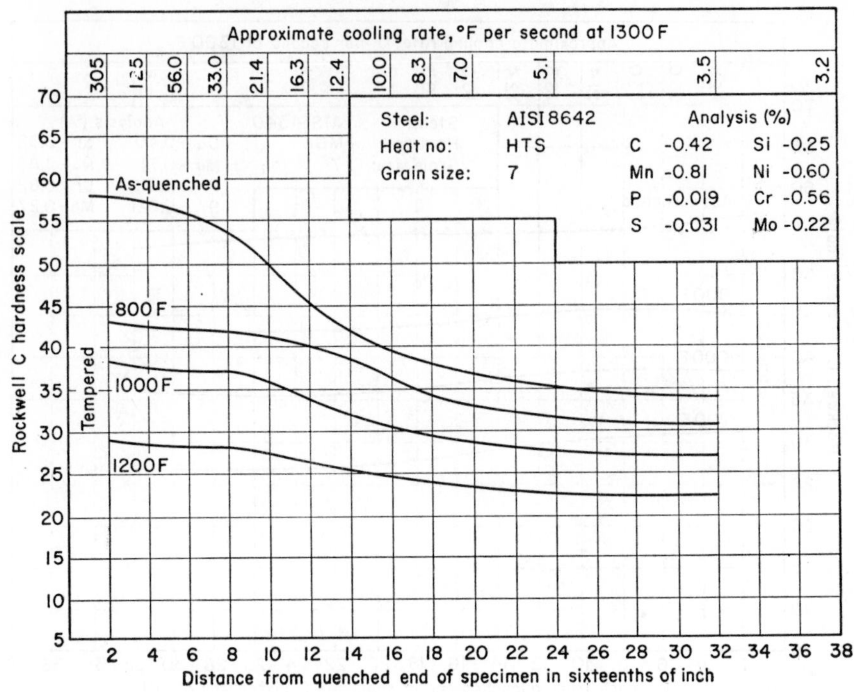

Approximate cooling rate, °F per second at 1300 F
305
125
56.0
33.0
21.4
16.3
12.4
10.0
8.3
7.0
5.1
3.5
3.2
Steel: AISI 8642
Heat no: HTS
Grain size: 7
Analysis (%)
C -0.42
Mn -0.81
P -0.019
S -0.031
Si -0.25
Ni -0.60
Cr -0.56
Mo -0.22
As-quenched
800 F
1000 F
1200 F
Tempered
Rockwell C hardness scale
Distance from quenched end of specimen in sixteenths of inch

APPENDIX G_1

APPROXIMATE HARDNESS CONVERSION TABLE FOR STEEL*

BRINELL HARDNESS NUMBER		STANDARD ROCKWELL		SUPERFICIAL ROCKWELL		Diamond Pyramid Hardness, Vickers (DPH)	Approximate Tensile Strength (Steel only) (1000 lb/in.²)
500 kg	3000 kg	B 1/16 in. Ball 100 kg	C Brale 150 kg	15-T 1/16 in. Ball 15 kg	15-N Brale 15 kg		
—	—	—	70.0	—	94.0	1076	—
—	—	—	69.0	—	93.5	1004	—
—	—	—	68.0	—	93.2	940	—
—	—	—	67.5	—	93.0	920	—
—	—	—	67.0	—	92.9	900	—
—	767*	—	66.4	—	92.7	880	—
—	757*	—	65.9	—	92.5	860	—
—	745*	—	65.3	—	92.3	840	—
—	733*	—	64.7	—	92.1	820	—
—	722*	—	64.0	—	91.8	800	—
—	710*	—	63.3	—	91.5	780	—
—	698*	—	62.5	—	91.2	760	—
—	682*	—	61.7	—	91.0	737	—
—	670*	—	61.0	—	90.7	720	—
—	653*	—	60.0	—	90.2	697	—
—	638*	—	59.2	—	89.8	680	329
—	627*	—	58.7	—	89.6	667	323
—	601*	—	57.3	—	89.0	640	309
—	578*	—	56.0	—	88.4	615	297
—	555*	—	54.7	—	87.8	591	285
—	534*	—	53.5	—	87.2	569	274
—	514*	—	52.1	—	86.5	547	263
—	495*	—	51.0	—	85.9	528	253
—	477	—	50.3	—	85.6	516	247
—	461	—	48.8	—	84.9	495	237
—	444	—	47.2	—	84.1	474	226
—	429	—	45.7	—	83.4	455	217
—	415	—	44.5	—	82.8	440	210
—	401	—	43.1	—	82.0	425	202
—	388	—	41.8	—	81.4	410	195
—	375	—	40.4	—	80.6	396	188
—	363	—	39.1	—	80.0	383	182
—	352	(110.0)	37.9	—	79.3	372	176
—	341	(109.0)	36.6	—	78.6	360	170
—	331	(108.5)	35.5	—	78.0	350	166
—	321	(108.0)	34.3	—	77.3	339	160
—	311	(107.5)	33.1	—	76.7	328	155
—	302	(107.0)	32.1	—	76.1	319	150
—	293	(106.0)	30.9	—	75.5	309	145
—	285	(105.5)	29.9	—	75.0	301	141
—	277	(104.5)	28.8	—	74.4	292	137
—	269	(104.0)	27.6	—	73.7	284	133
—	262	(103.0)	26.6	—	73.1	276	129
—	255	(102.0)	25.4	—	72.5	269	126
—	248	(101.0)	24.2	—	71.7	261	122
201	241	100.0	22.8	93.0	70.9	253	118
195	235	99.0	21.7	92.5	70.3	247	115

* For more details, see *Standard Hardness Conversion Tables for Metals* ASTM Designation E 140-58.

APPENDIX G_1 (*Continued*)

BRINELL HARDNESS NUMBER		STANDARD ROCKWELL		SUPERFICIAL ROCKWELL		Diamond Pyramid Hardness, Vickers (DPH)	Approximate Tensile Strength (Steel only) (1000 lb/in.²)
500 kg	3000 kg	B 1/16 in. Ball 100 kg	C Brale 150 kg	15-T 1/16 in. Ball 15 kg	15-N Brale 15 kg		
189	229	98.2	20.5	—	69.7	241	111
184	223	97.3	(18.8)	92.0	—	234	108
179	217	96.4	(17.5)	—	—	228	105
175	212	95.5	(16.0)	91.5	—	222	102
171	207	94.6	(15.2)	—	—	218	100
167	201	93.8	(13.8)	91.0	—	212	98
164	197	92.8	(12.7)	90.5	—	207	95
161	192	91.9	(11.5)	—	—	202	93
158	187	90.7	(10.0)	90.0	—	196	90
156	183	90.0	(9.0)	89.5	—	192	89
153	179	89.0	(8.0)	—	—	188	87
149	174	87.8	(6.4)	89.0	—	182	85
146	170	86.8	(5.4)	88.5	—	178	83
143	167	86.0	(4.4)	—	—	175	81
140	163	85.0	(3.3)	88.0	—	171	79
135	156	82.9	(0.9)	87.5	—	163	76
130	149	80.8	—	87.0	—	156	73
126	143	78.7	—	86.0	—	150	71
120	137	76.0	—	85.0	—	143	67
115	131	74.0	—	84.0	—	137	65
111	126	72.0	—	83.5	—	132	63
107	121	68.0	—	82.5	—	127	60
102	116	65.0	—	82.0	—	122	58
98	111	62.0	—	81.0	—	117	56
94	106	59.0	—	80.0	—	—	—
90	101	56.0	—	79.0	—	—	—
86	—	53.0	—	78.0	—	—	—
83	—	50.0	—	77.0	—	—	—
80	—	47.0	—	76.0	—	—	—
78	—	44.0	—	75.0	—	—	—
75	—	41.0	—	74.0	—	—	—
73	—	38.0	—	73.0	—	—	—
71	—	35.0	—	72.0	—	—	—
69	—	32.0	—	71.0	—	—	—
67	—	29.0	—	70.0	—	—	—
65	—	26.0	—	69.0	—	—	—
63	—	23.0	—	68.0	—	—	—
61	—	19.0	—	67.0	—	—	—
59	—	16.0	—	66.0	—	—	—
58	—	13.0	—	65.0	—	—	—
57	—	10.0	—	64.0	—	—	—
56	—	7.0	—	63.0	—	—	—
55	—	4.0	—	62.0	—	—	—
54	—	1.0	—	61.0	—	—	—

* Tungsten-carbide ball.
( ) Values beyond normal range for scale.

APPENDIX G_2

APPROXIMATE HARDNESS CONVERSION TABLE FOR NICKEL AND HIGH-NICKEL ALLOYS*

Brinell Hardness Number 3000 kg	STANDARD ROCKWELL B 1/16 in. Ball 100 kg	STANDARD ROCKWELL C Brale 150 kg	SUPERFICIAL ROCKWELL 15-T 1/16 in. Ball 15 kg	SUPERFICIAL ROCKWELL 15-N Brale 15 kg	Diamond Pyramid Hardness, Vickers (DPH)
450	—	48.0	—	84.5	481
425	—	46.0	—	83.5	452
403	—	44.0	—	82.5	427
382	—	42.0	—	81.5	404
346	—	38.0	—	79.5	362
313	—	34.0	—	77.5	326
275	(104)	28.5	94.0	75.0	285
241	100	22.5	92.5	72.0	248
215	96	(17.0)	91.0	69.0	220
194	92	(12.0)	89.5	66.5	198
176	88	(6.5)	88.0	64.0	179
161	84	(2.0)	87.0	61.5	164
149	80	—	85.5	—	151
139	76	—	84.0	—	140
129	72	—	82.5	—	130
121	68	—	81.0	—	122
111	62	—	79.0	—	112
103	56	—	77.0	—	103
93	48	—	74.0	—	93
85	40	—	71.0	—	85
79	34	—	69.0	—	79
77	30	—	67.5	—	77

* For more details see *Standard Hardness Conversion Tables for Metals*, ASTM Designation E140-58.

( ) Values beyond normal range for scale.

APPENDIX G_3

APPROXIMATE HARDNESS CONVERSION TABLE FOR CARTRIDGE BRASS (70% Cu-30% Zn)*

Brinell Hardness Number 500 kg	STANDARD ROCKWELL		SUPERFICIAL ROCKWELL	Diamond Pyramid Hardness (DPH)
	B 1/16 in. Ball 100 kg	F 1/16 in. Ball 60 kg	15-T 1/16 in. Ball 15 kg	
169	93.5	110.0	90.0	196
162	92.0	—	89.5	188
157	90.5	108.0	89.0	182
146	86.0	106.0	88.0	168
136	82.0	104.0	87.0	156
126	77.5	101.5	86.0	144
120	74.5	99.5	85.0	136
116	73.0	98.5	84.5	132
112	70.0	97.0	83.5	126
108	68.0	96.0	83.0	122
103	65.0	94.5	82.0	116
97	62.0	92.6	80.5	110
92	58.0	90.5	79.5	104
86	54.0	88.0	78.0	98
80	47.5	84.4	75.5	90
76	42.0	81.2	73.5	84
70	35.0	77.4	71.5	78
64	27.5	73.2	69.0	72
59	15.5	66.8	65.0	64
55	10.0	63.0	62.5	60
50	—	56.5	58.5	54
46	—	49.0	54.5	49
42	—	40.0	—	45

* For more details see *Standard Hardness Conversion Tables for Metals*, ASTM Designation E148-58.

INDEX